C0-AWS-291

Molecular Mechanisms of Neurodegenerative Diseases

Series Editors:
Ralph Lydic and
Helen A. Baghdoyan

Molecular Mechanisms of Neurodegenerative Diseases
edited by *Marie-Françoise Chesselet*, 2000

Contemporary Clinical Neuroscience

Molecular Mechanisms of Neurodegenerative Diseases

Marie-Françoise Chesselet, MD, PhD

Reed Neurological Research Center, School of Medicine, University of California, Los Angeles, Los Angeles, CA

Humana Press Totowa, New Jersey

This publication is printed on acid-free paper. ∞

ANSI Z39.48-1984 (American Standards Institute) Permanence of Paper for Printed Library Materials.

Cover design by Patricia F. Cleary.

Cover photos, clockwise from upper right: from Fig. 3, Chapter 12 (Opal and Paulson); from Fig. 3, Chapter 1 (Klein); from Fig. 1, Chapter 11 (Orr and Zoghbi); from Fig. 4, Chapter 1 (Klein).

For additional copies, pricing for bulk purchases, and/or information about other Humana titles, contact Humana at the above address or at any of the following numbers: Tel: 973-256-1699; Fax: 973-256-8341; E-mail: humana@humanapr.com, or visit our Website at www.humanapress.com

Printed in the United States of America. 10 9 8 7 6 5 4 3 2 1

Library of Congress Cataloging-in-Publication Data

Molecular mechanisms of neurodegenerative diseases / [edited by] Marie-Francoise Chesselet.
p. cm. -- (Contemporary clinical neuroscience)
Includes bibliographical references and index.
ISBN 0-89603-804-1 (alk. paper)
1. Nervous system--Degeneration--Molecular aspects. 2. Molecular neurobiology. I. Chesselet, Marie-Francoise. II. Series.
[DNLM: 1. Neurodegenerative Diseases--genetics 2. Neurodegenerative Diseases--physiopathology]
RC365 .M64 2000
616.8'047--dc21

00-032041

Preface

The field of neurodegenerative diseases is undergoing an unprecedented revolution. The past decade has seen the identification of new mutation mechanisms, such as triplet repeat expansions, and new genes causing familial forms of common neurodegenerative diseases, such as Parkinson's and Alzheimer's diseases. Cellular and animal models based on this genetic information are now available and, importantly, common mechanisms are rapidly emerging among diseases that were once considered unrelated. The field is poised for the development of new therapies based on high throughput screenings and a better understanding of the molecular and cellular mechanisms leading to neurodegeneration.

Molecular Mechanisms of Neurodegenerative Diseases reviews recent progress in this exploding field. By nature, such a book cannot be all inclusive. It focuses on Alzheimer's, Parkinson's, and CAG triplet repeat diseases. In the first chapter, Bill Klein reviews the role of Aβ toxicity in the pathophysiology of Alzheimer's disease. This controversial issue is further examined in the context of transgenic models of Alzheimer's disease by LaFerla and colleagues. Sue Griffin and Robert Mrak, and Caleb Finch and collaborators, then examine the role of glial cells and inflammation in Alzheimer's disease; a review of the role of proteolysis in the generation of abnormal protein fragments by Hook and Mende-Mueller follows. Therapeutic opportunities offered by a better understanding of Alzheimer's disease pathophysiology are examined by Perry Molinoff and his colleagues at Bristol-Myers Squibb.

The chapter on proteolysis by Hook and Mende-Mueller identifies one of the recurring themes that is appearing among neurodegenerative diseases: the formation of abnormal protein fragments, whose misfolding may lead to a cascade of cellular defects, ultimately leading to cell death. Similarities between pathological processes in Parkinson's, Alzheimer's, and related diseases is also the theme of the chapter by Virginia Lee, John Trojanowski, and collaborators, which discusses the role of Tau and synuclein. Despite the identification of mutations in synuclein, and the presence of synuclein in Lewy bodies, the pathophysiology of Parkinson's disease, however, remains poorly understood. Joel Perlmutter and his colleagues review the information we have recently gained on the progression of the disease from brain imaging studies.

BethAnn McLaughlin and Russell Swerdlow then examine the role of dopamine and of mitochondrial dysfunction, respectively, in neurodegeneration.

The last chapters of the book deal with different and complementary aspects of CAG repeat diseases, including SCA1 (Orr and Zoghbi), SCA3 (Opal and Paulson), SBMA (Merry), and Huntington's disease. Chesselet and Levine compare the different mouse models of Huntington's disease, MacDonald and colleagues review the role of proteins interacting with huntingtin, and George Jackson discusses the potential of fly genetics to identify the molecular mechanisms of neurodegenerative diseases.

Despite their differences in focus, many chapters of *Molecular Mechanisms of Neurodegenerative Diseases* overlap, presenting the variety of viewpoints that pervade this dynamic field. Evidently, since new data appear every day, the chapters in a book can only provide the basis for understanding ongoing research. It is hoped that the ideas and concepts presented here will lead, within a few short years, to therapies that prevent, delay the onset, slow the progression, or even cure these devastating neurodegenerative illnesses.

***Marie-Françoise Chesselet,* MD, PhD**

Dedication

This book is dedicated to the memory of Roger Chesselet, who believed that scientific discoveries happen at the junction of multidisciplinary fields,

and of John B. Penney, who dedicated his life to finding a cure for neurodegenerative diseases and whose untimely death prevented him from contributing to this book.

Contents

Contributors

MARIE-FRANÇOISE CHESSELET, MD, PhD • *Charles H. Markham Professor of Neurology, Reed Neurological Research Center, Department of Neurology, University of California, Los Angeles, School of Medicine, Los Angeles, CA*

KEVIN M. FELSENSTEIN, PhD • *Principal Scientist, Department of Neuroscience/ Genitourinary Drug Discovery, Bristol-Myers Squibb Pharmaceutical Research Institute, Wallingford, CT*

CALEB E. FINCH, PhD • *Andrus Gerontology Center and Department of Biological Sciences, University of Southern California, Los Angeles, CA*

BENOIT I. GIASSON, PhD • *Center for Neurodegenerative Disease Research, Department of Pathology and Laboratory Medicine, University of Pennsylvania, Philadelphia, PA*

MICHAEL GOLD, MD • *Director, Department of Clinical Pharmacology/ Experimental Medicine, Bristol-Myers Squibb Pharmaceutical Research Institute, Wallingford, CT*

W. SUE T. GRIFFEN, PhD • *Geriatrics and Mental Health Research Education and Clinical Center, Veterans Affairs Medical Center, and Department of Geriatrics, University of Arkansas for Medical Sciences, Little Rock, AR*

TAMARA HERSHEY, PhD • *Department of Psychiatry, Washington University School of Medicine, St. Louis, MO*

PAIGE HILDITCH-MAGUIRE, PhD • *Molecular Neurogenetics Unit, Massachusetts General Hospital, Charlestown, MA*

STEVEN F. HINTON • *Department of Neurobiology and Behavior, University of California, Irvine, Irvine, CA*

VIVIAN Y. H. HOOK, PhD • *Professor, Departments of Medicine and Neurosciences, University of California, San Diego, San Diego, CA*

GEORGE R. JACKSON, MD, PhD • *Assistant Professor, Department of Neurology, University of California, Los Angeles, School of Medicine, Los Angeles, CA*

WILLIAM L. KLEIN, PhD • *Department of Neurobiology and Physiology and Cognitive Neurology and Alzheimer's Disease Center, Northwestern University, Evanston, IL*

FRANK M. LAFERLA, PhD • *Department of Neurobiology and Behavior, University of California, Irvine, Irvine, CA*

VIRGINIA M. Y. LEE, PhD • *Director, Center for Neurodegenerative Disease Research, Department of Pathology and Laboratory Medicine, University of Pennsylvania, Philadelphia, PA*

MICHAEL S. LEVINE, PHD • *Neuropsychiatric Institute and Mental Retardation Research Center, University of California, Los Angeles, School of Medicine, Los Angeles, CA*
VALTER LONGO, PHD • *Andrus Gerontology Center and Department of Biological Sciences, University of Southern California, Los Angeles, CA*
MARCY E. MACDONALD, PHD • *Associate Professor of Neurology, Molecular Neurogenetics Unit, Massachusetts General Hospital, Charlestown, MA*
BETHANN MCLAUGHLIN, PHD • *Department of Neurobiology, University of Pittsburgh Medical School, Pittsburgh, PA*
LIANE MENDE-MUELLER, PHD • *Departments of Medicine and Neurosciences, University of California, San Diego, San Diego, CA*
DIANE E. MERRY, PHD • *Department of Biochemistry and Molecular Pharmacology, Thomas Jefferson University, Philadelphia, PA*
AYA MIYAO • *Andrus Gerontology Center and Department of Biological Sciences, University of Southern California, Los Angeles, CA*
STEPHEN M. MOERLEIN, PHD • *Department of Biochemistry and Molecular Biophysics, and Mallinckrodt Institute of Radiology, Washington University School of Medicine, St. Louis, MO*
PERRY MOLINOFF, MD • *Department of Neuroscience Drug Discovery, Bristol-Myers Squibb Pharmaceutical Research Institute, Wallingford, CT*
TODD E. MORGAN, PHD • *Andrus Gerontology Center and Department of Biological Sciences, University of Southern California, Los Angeles, CA*
ROBERT E. MRAK, MD, PHD • *Pathology and Laboratory Medicine Service, Veterans Affairs Medical Center, and Department of Pathology, University of Arkansas for Medical Sciences, Little Rock, AR*
PUNEET OPAL, MD, PHD • *Department of Neurology, Baylor College of Medicine, Houston, TX*
HARRY T. ORR, PHD • *Institute of Human Genetics, Department of Laboratory Medicine and Pathology, University of Minnesota, Minneapolis, MN*
LUCIUS PASSANI, PHD • *Molecular Neurogenetics Unit, Massachusetts General Hospital, Charlestown, MA*
HENRY PAULSON, MD, PHD • *Assistant Professor, Department of Neurology, University of Iowa College of Medicine, Iowa City, IA*
JOEL S. PERLMUTTER, MD • *Departments of Neurology and Neurological Surgery, and Anatomy and Neurobiology, and Mallinckrodt Institute of Radiology, Washington University School of Medicine, St. Louis, MO*
IRENA ROZOVSKY, PHD • *Andrus Gerontology Center and Department of Biological Sciences, University of Southern California, Los Angeles, CA*
YUBEI SOONG, PHD • *Andrus Gerontology Center and Department of Biological Sciences, University of Southern California, Los Angeles, CA*

MICHAEL C. SUGARMAN • *Department of Neurobiology and Behavior, University of California, Irvine, Irvine, CA*
RUSSELL H. SWERDLOW, MD • *Center for the Study of Neurodegenerative Diseases and Department of Neurology, University of Virginia Health System, Charlottesville, VA*
JOHN Q. TROJANOWSKI, MD, PhD • *Director, Center for Neurodegenerative Disease Research, Department of Pathology and Laboratory Medicine, University of Pennsylvania, Philadelphia, PA*
MIN WEI • *Andrus Gerontology Center and Department of Biological Sciences, University of Southern California, Los Angeles, CA*
CHRISTINA A. WILSON • *Center for Neurodegenerative Disease Research, Department of Pathology and Laboratory Medicine, University of Pennsylvania, Philadelphia, PA*
ZHONG XIE • *Andrus Gerontology Center and Department of Biological Sciences, University of Southern California, Los Angeles, CA*
HADI ZANJANI, PhD • *Andrus Gerontology Center and Department of Biological Sciences, University of Southern California, Los Angeles, CA*
HUDA Y. ZOGHBI, MD • *Department of Molecular and Human Genetics, Howard Hughes Medical Institute, Baylor College of Medicine, Houston, TX*

1
Aβ Toxicity in Alzheimer's Disease

William L. Klein

1.1. INTRODUCTION

The first diagnosed Alzheimer's disease patient was Auguste D (Fig. 1), a middle-aged woman cared for by Alsatian clinician-pathologist Alois Alzheimer from 1901 to 1906. Auguste D died in her middle fifties, and Alzheimer, aided by new histochemical methods, found her brain tissue corrupted by an abundance of extracellular and intracellular lesions, the now-familiar plaques and tangles *(1)*. Alzheimer's research provided the first direct evidence that dementia is the consequence of neurodegenerative mechanisms, not a simple fact of aging. A less diagnostic but equally apt description of the disease also came from Alzheimer's care of Auguste D. As he tracked his patient's progressively severe dementia, Alzheimer once asked Auguste D to write her name. She found the task impossible and replied, "I have lost myself" *(2)*. We now know that Alzheimer's disease (AD) is the most common cause of dementia in older individuals. The number of Alzheimer's disease patients has grown from the first diagnosed case in 1906 to an estimated 25 million worldwide *(3)*. Complications resulting from AD constitute the fourth leading cause of death in the United States, and the annual cost to the economy exceeds $140 billion *(4,5)*. As the elderly population continues to grow rapidly, AD represents an imminent social as well as medical problem.

The pathology of AD has been reviewed extensively and comprises multiple and varied factors *(6,7)*. As discovered by Alzheimer, two of the major hallmarks are plaques and tangles. The tangles are paired helical filaments made of hyperphosphorylated tau *(8)*. Tangles occur in living neurons but also are found as extracellular remains following nerve cell death. Plaques exhibit varied morphologies *(9–11)*. The most salient are senile plaques that show degenerating neurites in proximity to large extracellular amyloid deposits. Amyloid cores are made of polymerized amyloid beta (Aβ) peptide and contain a small variety of inflammatory

From: *Contemporary Clinical Neuroscience: Molecular Mechanisms of Neurodegenerative Diseases*
Edited by: M.-F. Chesselet

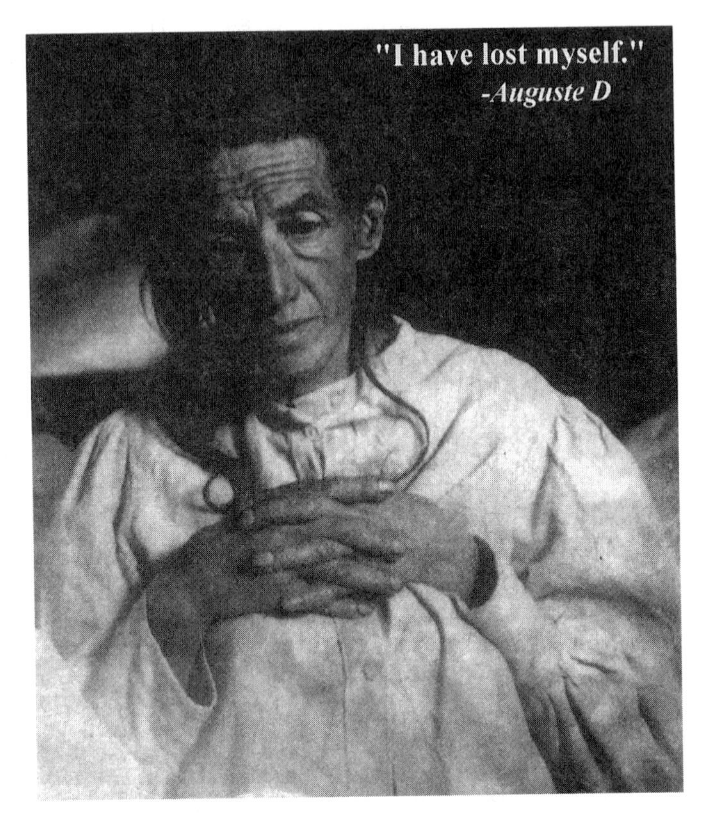

Fig. 1. Auguste D — the first Alzheimer's patient.

proteins *(12,13)*. Other major hallmarks of AD-afflicted brain are inflammatory gliosis *(14)* and selective nerve cell degeneration and death *(15–18)*, especially in limbic and cognitive centers.

Current thinking is that molecules associated with the hallmark pathologies play an active role in AD pathogenesis. Constituent molecules can act as toxins, toxin inducers, and toxin mediators. As yet, however, no consensus exists regarding the primary pathogenic molecules. In this chapter, evidence is reviewed that strongly implicates a role for neurotoxins derived from Aβ peptides (for earlier reviews, *see* refs. *19–22*). Nonetheless, central roles also can be argued for inflammatory processes *(23)* and for cytoskeletal dysfunction linked to aberrant tau phosphorylation *(24–26)*. Whichever molecular abnormality proves primary, the pathogenic phenomena are closely interrelated. Toxins from Aβ, for example, promote inflammatory gliosis *(27)*, and they stimulate cellular tau phosphorylation of the sort that occurs in AD *(28,29)*. Reciprocally, tau phosphorylation and microtubule dysfunction alter Aβ metabolism *(30,31)*, and inflammatory glial proteins influence the nature of Aβ-derived toxic aggregates (refs. *32* and *33*, and Subheading 1.2.).

This chapter reviews the composition of neurotoxins derived from Aβ peptides and considers how their toxic mechanisms may account for AD's early memory loss and progressively catastrophic dementia. Emerging data show that toxic Aβ-derived fibrils and oligomers exert a selective impact on signal transduction molecules that are coupled physiologically to mechanisms of apoptosis and synaptic plasticity. If these mechanisms ultimately are shown responsible for AD, they will provide molecular targets for drugs that block disease progression and perhaps even reverse early-stage memory loss.

1.2. ENDOGENOUS TOXINS DERIVED FROM Aβ

Aβ Cascade Hypothesis

Although lacking final consensus, the most prominent hypothesis for AD pathogenesis is the Aβ cascade, now nearly a decade old *(34)*. This hypothesis argues that neuron dysfunction and death in AD are caused by a cascade of reactions initiated by abnormally active Aβ peptides. The Aβ cascade is supported strongly by human pathology, transgenic modeling, and experimental nerve cell biology (*see* Milestones in Fig. 2; for earlier reviews, *see* refs. *35* and *36*).

Core Component of Alzheimer's Amyloid Is Aβ

The first molecular milestone en route to the Aβ cascade hypothesis was reached in 1984 by Glenner and Wong *(37)*. Their purification and analysis of the core component of Alzheimer's amyloid showed it to be a 4-kDa peptide, designated now as amyloid beta (Aβ). This seminal discovery, confirmed and extended by others *(12,38–40)*, enabled several groups to obtain the gene sequence for the Aβ precursor protein (APP). There is only a single APP gene, but it has five splice variants *(21,41–43)*, and related sequences show the existence of an amyloid precursor protein (APP) gene family *(44,45)*. APP itself is an integral membrane protein with one transmembrane domain. Within APP, the Aβ sequence comprises 28 amino acids of a hydrophilic juxtamembrane domain followed by 11–15 amino acids of a hydrophobic transmembrane α-helix. Aβ1–43 comprises amino acids 672–714 of human APP770, the longest splice variant. Although Aβ1–40 and Aβ1–42 are the predominant monomeric forms, peptides isolated from Alzheimer-afflicted brain tissue show length variations at both ends *(46)*. APP proteolysis involves complex trafficking that is highly regulated with multiple possible outcomes, including and excluding Aβ excision *(47,48)*.

- 1907-- Alzheimer reports *plaques and tangles*
- 1984-- Major chemical of plaques: Aβ
- 1987-- Gene found for Aβ precursor
- 1990-- Synthetic Aβ kills neurons
- 1991-- Mutated APP gene causes AD
- 1993-- Mutations increase Aβ
- 1994-- Aβ stimulates tau-P
- 1995-- New AD genes found, also increase Aβ
- 1996-- APP Transgene impairs learning
- 1998-- ADDLs-toxic Aβ oligomers

Fig. 2. Milestones in the development of the Aβ cascade hypothesis.

The peptide is secreted and also found within intracellular compartments *(49)*. The unusual tandem hydrophobic/hydrophilic domains of Aβ strongly influence its structure in solution and its biological properties (*see* Fibril Hypothesis, below).

Aβ Is Elevated by AD Genetic Risk Factors

Enormous interest in Aβ and APP sprang from the landmark discovery that APP point mutations account for particular subsets of familial AD (FAD). The first mutation identified was a highly conserved substitution in the APP transmembrane domain (Val – Ile) *(50)*. The locus at amino acid 717 is slightly C-terminal to the Aβ sequence. This single, simple change in the APP gene evokes full AD pathology, including formation of tangles as well as plaques. Several mutations in and around the Aβ sequence now are known to be pathogenic *(50–56)*. Discovery of APP mutations gave the first insight into a primary event in AD. The consequence has been a great impetus to understand the relationship between AD pathogenesis and Aβ/APP abnormalities.

Correlations between APP mutations and Alzheimer's phenotypes have been substantiated in transgenic animal models *(57–62)*. Although no single animal mimics all of AD pathology, current strains exhibit multiple aspects of neuronal dysfunction and degeneration and a variety of behavioral deficiencies.

Other genetically linked factors besides mutant APP have a role in AD and, in fact, are significantly more prevalent. Consistent with the Aβ cascade hypothesis, however, the established factors all contribute to anomalous accumulation of Aβ peptide. Aβ is elevated in familial AD, whether caused by presenilin 1, 2, or APP mutations *(63,64)* and is elevated also in sporadic AD associated with ApoE4 *(65)*. Increases occur most consistently with Aβ comprising 42 amino acids. This is a salient feature because the more hydrophobic Aβ1–42 has a particularly strong tendency to self-associate into neurotoxic multimers (*see* Fibril Hypothesis, below). The relationship of Aβ to more recently discovered genetic components of AD has not been established *(66)*.

Elevated Aβ accumulation is recapitulated in cell culture models that carry disease-associated transgenes *(67–70)*. Mechanisms responsible for this elevation are unclear, although the mutant presenilin effects likely are linked to the role of presenilin as APP protease *(71,72)*. Evidence that multiple pathogenetic factors lead to exaggerated Aβ production supports the hypothesis that Aβ is responsible for neuron dysfunction and degeneration in AD.

Aβ Is Toxic to Neurons

If Aβ were biologically inactive, its accumulation in AD brain could perhaps be dismissed as an epiphenomenon. This, however, is not the case, as shown in the capstone discovery by Yankner et al. *(73)* that an Aβ-containing fragment of APP is a potent neurotoxin. Extensive corroborating research with synthetic peptide has established Aβ-evoked neuronal cell death occurs in dissociated *(73–75)* and organotypic central nervous system (CNS) cultures *(33)*, human *(29)* and other mammalian neuronal cell lines *(76,77)*, and injected brain tissue *(78,79)*. Prior to death, affected neurons show increases in phosphotau antigens associated with neurofibrillary tangles *(28,29)*. In vivo, gliosis is induced *(79,80)*. Thus, the impact of Aβ in experimental models is consistent with at least a partial AD phenotype. Nerve cell biology experiments with Aβ provide a rational basis for neurodegeneration seen in AD and also provide powerful models for exploring mechanisms.

Fibril Hypothesis

The Aβ cascade hypothesis appears compelling: mutations and other agents that cause Alzheimer's disease do so by producing molecules that kill neurons. Nonetheless, debate over the Aβ cascade hypothesis has been contentious *(81,82)* and remains ongoing *(83)*. A key problem is the experimental difficulty in working with Aβ. Although Aβ can be neurotoxic, it also can be neurotrophic, and at other times, have no obvious impact at all *(84,85)*.

One factor in this problem is that Aβ is not a general cytotoxin. Some cells, such as glia, naturally resist the degenerative action of Aβ *(86)*, and subclones of neuronal cell lines can be selected for resistance *(87–89)*. The state of neuronal differentiation further determines the outcome of exposure to Aβ *(29,74)*. Factors that influence sensitivity may include cell cycle withdrawal, increased demand for trophic support, switchover from anaerobic to aerobic metabolism, or altered expression of signal transduction molecules.

A second factor is the inherent difficulty in controlling the conformation of monomeric Aβ, a molecule with tandem hydrophilic and hydrophobic domains. Many investigators have simplified experiments through use of Aβ25–35, a hydrophobic 11-mer that is a reliable toxin at high doses *(90,91)*. The 11-mer is not found naturally, however, and may only partially mimic the cellular impact of full-length Aβ. Synthetic Aβ1–40 and Aβ1–42 are available commercially now at reasonable cost, and means to obtain recombinant peptide have been described *(92)*. Current nerve cell biology paradigms favor use of Aβ42. Earlier purity problems with commercial Aβ42 have been largely overcome, although lot-to-lot variations in toxic efficacy continue to exist. Even with consistent purity, solution effects resulting from particular ions, oxidation, temperature, pH, Aβ concentration — and even mechanical manipulation — will influence structural and biological outcomes *(93)*. The potential of Aβ for alternative states is reminiscent of prions, which can be innocuous or deadly, depending on protein conformation *(94)*.

The third factor governing experimental Aβ toxicity is the most germane to AD pathogenesis. At issue is the relationship between neurotoxicity and Aβ self-association. Discovery of this relationship has played a significant role in overcoming objections to the Aβ cascade hypothesis.

Self-Association of Aβ Is Critical for Neurotoxicity

Major findings of Cotman, Yankner, and their colleagues established that synthetic Aβ is neurotoxic only after it self-associates into larger assemblies *(84,95)*. Solutions initially comprise monomeric peptide and are essentially innocuous. With time, the solutions become toxic, and the Aβ is prominently self-aggregated. Toxic solutions examined by electron microscopy or atomic force microscopy show abundant fibrils *(84,96,97)*. These appear analogous to fibrils that constitute amyloid in AD-afflicted brain tissue *(98,99)*. Solutions containing large amorphous aggregates appear nontoxic. It has been proposed, therefore, that fibrillar forms of Aβ are required for neurotoxicity.

Extrapolating to AD, the cascade responsible for dementia theoretically would be initiated by fibrils of neuritic amyloid plaques. Normal APP proteolysis would yield an innocuous Aβ monomer, but when conditions favored localized self-association, toxic fibrils would accumulate. One such condition, for example, would be anomalous abundance of monomer due to genetic factors.

Fibrillogenesis has been the subject of intense investigation (for reviews, *see* refs. *93* and *100–102*). For fibrils to form in vitro, Aβ40 must be above a critical concentration (50 μ*M* or more; *see* reviews in refs. *93*, *103*, and *104*). A lag phase also occurs (kinetic solubility), during which peptide slowly undergoes prerequisite associations. The more hydrophobic Aβ42 converts to fibrils at lower critical concentrations (5-μ*M* doses) and shows little or no kinetic solubility. Differences with respect to self-association are in parallel with findings that AD dementia correlates better with Aβ42 than Aβ40 *(56,105)*. It has been pointed out that critical concentrations for fibrillogenesis are orders of magnitude higher than concentrations found in cerebrospinal fluid (CSF) *(93,106,107)*. However, analogous to neurotransmitter levels, there is no obvious relationship between concentrations in CSF and concentrations that develop in a local extracellular milieu or inside cells. The abundance of amyloid fibril deposits in Alzheimer's brain and in transgenic animals gives *prima facie* evidence that local Aβ concentrations in vivo exceed critical concentration.

Some reports suggest that fibrils can kill neurons at nanomolar doses of Aβ *(74,91)*, but concentrations used in typical nerve cell biology experiments exceed 20 μ*M* (in total molarity of Aβ). Although this dose may seem high, molarity has little meaning with respect to insoluble assemblies such as fibrils. Moreover, even at 20 μ*M* Aβ, electron microscopy (EM) data show that neuronal cell surfaces are largely free of attached fibrils (Fig. 3). Reciprocally, only a small fraction of the fibrillar Aβ is attached. A majority of the fibrillar Aβ thus is bundled in nonproductive material. It also is clear from EM data that fibrils can interact strongly with neuronal cell surfaces. Such fibril–neuron contact in AD might be expected to produce local degeneration, as seen in neuritic plaques. Because immature or diffuse plaques, thought to comprise amorphous Aβ supramolecular assemblies, do not trigger local neuronal degeneration *(108–110)*, degenerative effects may depend on particular configurations of aggregated Aβ *(84)*.

The persuasiveness of the Aβ fibril hypothesis has motivated an intense search for compounds that inhibit fibril toxicity. Selkoe has identified five strategic drug classes based on this goal *(21)*. The most appealing would act furthest upstream, blocking fibril formation. Several promising fibril-blocker neuroprotectants *(111–113)* have been found, including certain dyes *(84,114,115)* and small peptides that act as β-sheet breakers *(116)*. However, as discussed in the following, compounds that only block fibril toxicity may prove insufficient as AD therapeutic drugs.

Fig. 3. Aβ fibrils attach to nerve cell surfaces. Electron microscopy shows fibrils from aggregated Aβ (arrows) extend to the plasma membrane of a neuron-like human cell line (arrowheads).

Fibril Hypothesis Is Powerful but Imperfect

An extensive literature focuses on fibril neurotoxicity. There are, however, well-known discrepancies and newly emerging results that make it difficult to accept fibrils as the sole basis for Aβ-evoked pathogenesis in AD.

A particular problem with the fibril hypothesis is the imperfect correlation between amyloid abundance and dementia. This issue has been discussed extensively *(25,117–120)*. Although postmortem analyses are not optimum for answering questions of cause-and-effect, attempts to correlate pathological markers with dementia have challenged as well as supported the Aβ fibril hypothesis. Some studies have concluded that decreased synaptic density and the abundance of tangles are more germane than plaques to the progression of dementia. Various explanations have been offered to account for the imperfect correlation *(121,122)*, including the argument that better data analysis and selection of plaque subtype show improved correlation *(123)*. It appears, however, that amyloid plaques can be abundant in individuals without dementia *(124–126)*. Moreover, examined closely in the hippocampus, the majority of neuron loss occurs in the absence of any proximal amyloid *(127,128)*.

Lack of correlation between amyloid levels and neurological deficits has been mimicked in several strains of APP transgenic animals. Most recently, animals carrying either the FAD (717v–f) or (670k–n)(671m–l) mutations *(129)* were found to exhibit large decreases in synapse abundance, functional synaptic communication, and neuron number before the emergence of any detectable amyloid. The mutations caused elevated Aβ in a manner uncoupled from total APP transgene expression. The authors conclude that neural loss is plaque independent. As an alternative, they hypothesize the presence of small diffusible toxins formed from Aβ, which might act either intracellularly or extracellularly. Other recent studies as well as earlier works also have reported amyloid-free transgenic mice that exhibit multiple aspects of pathology and behavioral anomalies *(60,130–138)*. Transgenic animals with neural deficits in the absence of amyloid once were considered poor models of AD pathogenesis. In fact, they may reflect a different aspect of Aβ-evoked pathogenesis, namely, one that involves nonfibrillar Aβ oligomers.

Small Oligomers as Molecular Alternatives to the Aβ Fibrillar State

Nonfibrillar forms of multimeric Aβ have not yet been detected in transgenic animals. Such diffusible oligomers, however, have been observed in various biochemical, cell culture, and human pathology studies and may even be present to varying degrees in classic fibril preparations *(97)*. Until recently, these small oligomers were considered transient intermediates, en route to fibrils, but new evidence indicates they exist as independent toxic entities.

Stable Oligomers

Self-association of Aβ into subfibrillar structures has been established for both Aβ42 and Aβ40. In cell culture, oligomers of Aβ42 accumulate spontaneously in conditioned medium of cells transfected with mutant APP transgenes *(139,140)*. The oligomers are upregulated in cells co-transfected with mutant presenilin genes *(70)*, as expected from AD pathology. The small oligomers exist in a fibril-free conditioned medium, consistent with biochemical stability. Thus, coupling of oligomers to fibrillogenesis is not an obligatory reaction. Solutions of synthetic Aβ40 also form oligomers, but they have been detected only after chemical crosslinking *(141)*. Crosslinking is not required to detect oligomers of Aβ42, which resist even relatively harsh sodium dodecyl sulfate (SDS) treatment *(33,140,142,143)*. Possible involvement of oligomers in pathogenesis is in harmony with the consistent upregulation of Aβ42 in AD *(105)*.

Several studies have now established the presence of small Aβ42 oligomers in human brain tissue. Oligomers first were detected in smooth muscle of CNS vessel walls as part of a study of amyloid angiopathy *(144)*. Their presence was thought to be a potential indicator of amyloidogenesis. In an analysis of brain parenchyma by Roher and colleagues *(145)*, comparison of normal versus AD-afflicted brains showed a 12-fold elevation of water-soluble Aβ42 oligomers in the disease state. The authors concluded that upregulation of oligomers most likely reflected ongoing amyloidogenesis, but speculated the oligomers might be bioactive. Extending these important findings, Masters, Beyreuther, and their colleagues recently have presented results showing that dementia correlates better with small oligomeric Aβ than with amyloid *(83)*. They conclude fibrils are unlikely to be the primary form of pathogenic Aβ in AD. The existence of small Aβ oligomers in human brain and their correlation with AD has given a new molecular dimension to the Aβ hypothesis. Because, as shown next, small oligomers can be neurotoxic, their occurrence in AD brain would explain the imperfect correlation between dementia and plaques and provide a new focus for therapeutic drug development.

Small Oligomers Are Potent CNS Neurotoxins

The fibril hypothesis predicts that Aβ will lose its neurotoxicity if fibrillogenesis is blocked. The prediction has potential practical value because fibril blockers would be prototypes for rationally designed therapeutic drugs. For a number of compounds, such as the peptide β-breakers, this prediction holds.

However, in some circumstances, the prediction fails. The implication, which is of rapidly emerging significance, is that fibrils are not the sole toxic Aβ entity. Smaller oligomeric forms of Aβ also are neurotoxins and could have a major part in AD pathogenesis.

The seminal discovery was made by Finch and colleagues. Their experiments used solutions of synthetic Aβ42 that contained clusterin as an additional component *(32,146)*. Even at a 5% molar ratio (1 mol clusterin to 20 mol Aβ42), clusterin caused a major reduction in fibril formation. Neurotoxicity, however, as assayed by a standard paradigm using reduction of 3-[4, 5 -dimethylthiazol-2-yl]-2, 5-diphenyltetrazolium bromide (MTT) in neuron-like PC12 cells, was not blocked. In fact, the slow-sedimenting molecules formed in Aβ–clusterin solutions were even more toxic than typical fibrils.

As a potential modifier of fibril formation in these experiments, clusterin was an apt choice. Clusterin is a multi-functional inflammatory protein released by glia and upregulated in AD brain (*see* ref. *147*; *see also* Chapter 2). Clusterin is found in senile plaques *(148)*, it binds well with Aβ *(149)*, and abundant Aβ–clusterin complexes occur in CSF *(150)*. Clusterin and Aβ42 thus encounter each other in Alzheimer-afflicted brain parenchyma. Somewhat surprisingly, in contrast to its impact on Aβ42, clusterin blocks the toxicity of Aβ40, even at substoichiometric doses *(151)*. The reasons for the differences have yet to be resolved.

ADDLs– Aβ-Derived Diffusible Ligands

The slow-sedimenting toxins fostered by clusterin more recently have been investigated in organotypic slice cultures prepared from mature mouse brain *(33)*. Such three-dimensional cultures provide physiologically relevant models for CNS and are frequently used in electrophysiological paradigms for learning and memory. Toxicity in slices is quantified by image analysis of dye uptake into living or dead cells (Fig. 4). In this paradigm, clusterin-induced Aβ toxins are extremely potent, with hippocampal nerve cell death significant even at nanomolar levels of Aβ.

Structural properties of the slow-sedimenting Aβ toxins have led to the suggestion that they be called ADDLs, an acronym for Aβ-derived diffusible ligands that induce dementia *(33)*. ADDLs, unlike fibrils, readily diffuse through filters that support brain slices in culture (as in Fig. 4). Inspected by atomic force microscopy (AFM) and electron microscopy, ADDL preparations are fibril-free (Fig. 4) and roughly comparable in size to globular proteins under 40 kDa. Molecules of similar size are evident following resolution by sodium dodecyl sulfate–polyacrylamide gel electrophoresis (SDS-PAGE). Predominant species comprise Aβ trimer through pentamer, although molecules as large as 24-mers are detectable *(152)*. Retention of oligomeric structure in SDS suggests considerable stability of the hydrophobic interactions between monomeric subunits. Whether the conformations are identical to oligomers detected in vivo is unknown, although certainly plausible. ADDL oligomers are too small to contain clusterin, which is an 80-kDa dimer. Because clusterin-free conditions have been found that also generate ADDL species, toxicity does not depend on a transient ADDL–clusterin complex. In nondenaturing gels, ADDL species migrate under 30 kDa, and preparations show no fibrils, confirming that oligomeric species seen in SDS-PAGE are not detergent-induced artifacts. In harmony with their molecular structure, ADDLs show selective ligandlike binding to cell surfaces, which may play a role in their toxicity (*see* following subsection). Overall, centrifugation, filtration, electrophoresis, and electron and atomic force microscopy show that toxic ADDLs are small fibril-free oligomers of Aβ.

Toxicity assays using extracts from AD brain have provided further evidence for nonfibrillar toxins derived from Aβ. The toxic entities are dimers *(153)*, which reportedly have no direct effect on neurons. Dimers apparently trigger glial responses that kill neurons indirectly, likely through NO-dependent reactions. The larger Aβ oligomers, as formed in vitro in the presence of clusterin, also activate glial cells *(86)*. Relationships between dimers and larger oligomers remain to be investigated.

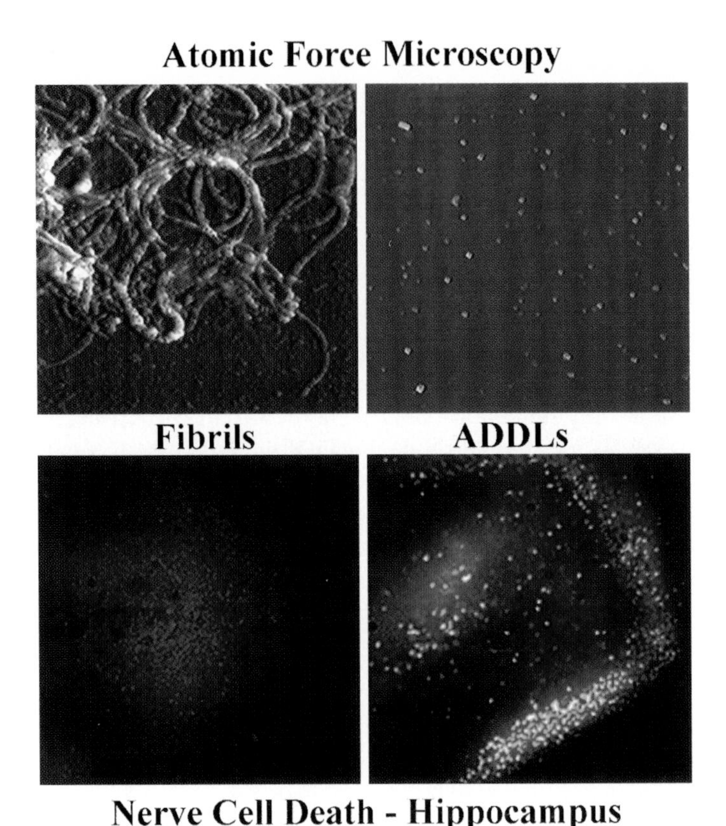

Fig. 4. Fibril-free ADDLs are toxic to neurons in hippocampal slice cultures. Atomic force microscopy (top) shows synthetic Aβ can self-associate into alternative structures such as large fibrils or ADDLs. In hippocampal slice cultures (bottom), ADDLs readily diffuse through a barrier filter and kill neurons, while fibrils do not. (Dead cells imaged by fluorescent dye.)

Critical Influence of Proteins

An important principle underscored by the clusterin experiments is that Aβ fibrillogenesis can be influenced drastically by heterogeneous proteins. Several other proteins besides clusterin exert a characterized impact on fibril formation, including glutamine synthetase, serum amyloid P, ApoE3, alpha chymotrypsin (ACT), and acetylcholinesterase (AChEase). Some of these proteins retard fibrillogenesis (clusterin, GS, serum amyloid P, ACT and ApoE3) *(32,146,154–157)*, whereas others stimulate it (AChEase) *(158)*. In vivo, scavenger effects associated with high-abundance Aβ-binding proteins may retard fibril formation. ApoE3, for example, binds Aβ tightly and may help clear Aβ from the intercellular milieu; ApoE4, on the other hand, has less affinity for Aβ, which might help explain why the ApoE4 allele is a risk factor for AD *(159–162)*. Scavenger function, however, cannot explain how proteins such as clusterin block fibril formation at extremely low molar doses, nor explain why some proteins stimulate fibril formation. Differing outcomes with respect to fibrillogenesis suggest that local protein milieu, by influencing Aβ self-association, could alter the particular course of the disease.

Biophysical models of Aβ fibrillogenesis incorporate the concept of critical concentration *(93,163)*. Below the critical concentration, *A*β is nonfibrillar. Above it, fibrillogenesis proceeds until the monomer again falls to a low level. The apparent critical concentration is in some way affected by the specific impact of proteins. At least two possibilities are appealing, with different mechanisms potentially associated with different proteins. First, proteins could act like Aβ chaperones. Different chaperones could favor Aβ conformations that, after release, could foster oligomers or foster fibrils. Precedent for stable, oligomer-favoring conformations is found in the effects of low temperature. In certain ice-cold solutions, ADDLs form from pure Aβ peptide, in the absence of any chaperone (ref. *33*; *see also* ref. *164*). The role of peptide conformation in determining subsequent aggregation state is a phenomenon well established for the pathogenic action of prions *(165,166)*. It is not clear whether Aβ oligomers or fibrils are thermodynamically more stable, although at low concentrations, oligomers exist in the absence of fibrils. In a second, somewhat related mechanism, an Aβ-binding protein could act like an antigen-presenting protein, holding the monomer within a surface pocket to facilitate interactions that produce oligomers or fibrils. The prion literature provides extensive precedent for mechanisms that require inducible protein conformation states to produce neurotoxic entities *(94)*. These mechanisms may provide principles pertinent to Aβ and Alzheimer's disease.

Protofibrils

In addition to ADDLs and large amyloid fibrils, there also exist intermediate-sized entities known as protofibrils, which can be isolated in structures up to 200 nm in length *(167,168)*. Protofibrils exhibit toxicity in assays for neuronal viability and electrophysiological activity *(169)*. Protofibrils are of variable length in high-resolution microscopy, but in SDS-PAGE, they appear to break down into ADDL-sized moieties. Factors that favor accumulation of linear protofibrils versus small globular ADDLs are not known. Given the SDS-PAGE findings, the possibility exists that ADDLs might be fundamental units of the more complex protofibril. The extent to which ADDLs, protofibrils, and amyloid coexist in a single synthetic preparation has not been examined carefully. Small ADDL-sized molecules can be seen, however, in images of some fibril preparations *(97)*.

Clearly, Aβ can assume various multimeric states in solution. A related question is the structural fate of ADDLs or other Aβ derivatives after attaching to cells. Do ADDLs, for example, remain stable after binding to cell surfaces? It is conceivable that surface binding might favor reorganization of ADDLs into larger protofibrils, a possibility consistent with recent fluorescent microscopy data *(170)*. The data also are consistent, however, with patch formation by putative ADDL receptors. Current data do not suggest that ADDLs on cell surfaces convert into large amyloid fibrils. Another question is whether ADDLs remain localized at the cell surface. Like many protein hormones *(171,172)*, ADDLs could trigger receptor-mediated internalization. Similar triggering might be envisioned for fibrils, perhaps with membrane complications leading to cytotoxicity.

Existence of Multiple Toxins Strengthens the Aβ Cascade Hypothesis

In summary, amyloid is not the only aggregated form of Aβ that accumulates in AD. In addition, there are Aβ oligomers, which cannot be detected by traditional methods used to image fibrils. Biochemical analyses of these physiological oligomers have suggested that their abundance correlates with AD dementia. In cell biology experiments, synthetic Aβ oligomers are highly neurotoxic. Small toxic entities derived from Aβ, whether ADDLs or protofibrils, could account for imperfect correlations between amyloid plaques and dementia in humans and also explain pathologies in APP-transgenic mice that lack fibril deposits. Quite likely, the presence of oligomers is fostered not only by elevated Aβ42 but also by upregulation of chaperone-like proteins that promote toxic oligomerization. Because ADDLs are diffusible as well as potent, their occurrence in the brain would be especially pernicious. To test rigorously the correlation between toxic oligomers and the progression of Alzheimer's dementia, it will be of value to develop antibodies that discriminate oligomers from other Aβ forms, freeing analysis from difficult biochemical procedures. Although prototype neuroprotective drugs have been developed for their ability to block fibril formation, advancements in therapeutic efficacy likely will require a parallel ability to block ADDLS or protofibrils. Overall, the Aβ cascade hypothesis, which only partially accounts for AD in its fibril-specific version, is significantly more robust with the inclusion of neurotoxic Aβ oligomers.

1.3. MECHANISMS OF Aβ TOXICITY

Signaling Mechanisms That Couple Aβ Toxicity with Tau Phosphorylation

Insights into the molecular basis for Aβ neurotoxicity have come largely from experiments with cell culture systems. For relevance to AD, an ideal experimental system would exhibit not only neuronal dysfunction and death but also neurofibrillary tangles. Tangles, however, have yet to be observed.

Nonetheless, a putative connection does exist between toxic Aβ and tangle formation. In various nerve cell models exposed to Aβ neurotoxins, evidence has been obtained for significant upregulation of Alzheimer-type tau phosphorylation *(28,29)*. Similar upregulation occurs in transgenic models *(60,62,173,174)*. Ectopic occurrence of phosphotau species once was considered to reflect an overall collapse of cell integrity. However, dying cells show reduced levels of AD-type phosphotau *(175,176)*. Evoked tau phosphorylation currently is thought to derive from a selective impact of Aβ on neuronal signal transduction *(19)*.

Signaling Mechanisms That Lead to the Phosphorylated State of Tau (Tau-P)

One approach to understanding the signal transduction impact of Aβ has been to seek clues from cell states that in vivo generate the same unusual forms of phosphotau as Aβ toxins. For example, immunoreactivity of the monoclonal PHF-1, which is diagnostic for an anomalous phosphotau epitope found in AD-afflicted neurons *(177)*, is upregulated in mature neuronal cultures exposed to Aβ toxins *(28,29)*. In developmental studies, the same PHF-1 epitopes are abundant in subpopulations of normal neurons, prior to their maturation *(178)*. Although transient developmental expression might theoretically reflect apoptotic pathways common to development and AD, PHF-1 immunoreactivity does not correlate with the timing of developmental programmed cell death *(178)*. Instead, PHF-1 epitopes are linked to mitosis in certain actively proliferating cells *(179)* and to nascent axon growth in postmitotic neurons *(178)*. Consistent with this linkage, laminin and fibronectin, which promote both mitosis and postmitotic axon growth, also upregulate PHF-1-type phosphorylation *(180)*. The effects of laminin and fibronectin are mediated by integrin receptors, which, intriguingly, have been found to bind Aβ *(181,182)*. The interaction with integrins may be of particular importance because intracellular signaling molecules controlled physiologically by integrins are modified by Aβ neurotoxins.

Focal Contact Signaling and Fyn

An early step in integrin signal transduction is the tyrosine phosphorylation and resultant activation of focal adhesion kinase (FAK), a protein tyrosine kinase that structurally and functionally is at the center of focal contacts *(183–185)*. Focal contacts are transient signal transduction organelles comprising a dynamic array of protein components. Focal contacts can be accessed by multiple receptors, not just integrins, and they influence an enormous range of cell physiology *(186)*.

Tyrosine phosphorylation of FAK is markedly increased in neuronal cells exposed to toxic Aβ *(77)*. The increase is evident in cells sensitive to Aβ toxicity but not in resistant ones. The impact of toxic Aβ requires G-protein activity and intact f-actin *(187)*, consistent with focal-contact-mediated signaling *(188)*. Also consistent with focal contact signaling, Aβ-treated neurons show increased levels of intermolecular complexes between FAK and Fyn *(189)*, a protein tyrosine kinase of the Src family *(190)*. Neuronal cells exposed to toxic Aβ also show a large increase in focal contact size *(191)*. The increase derives from the selective, Aβ-evoked translocation of focal contact constituents. Thus, Aβ can tap into pathways associated with focal contact signaling, perhaps through integrins or via alternative means. These findings complement early evidence that altered protein tyrosine phosphorylation is germane to Alzheimer's pathology *(192)*.

The coupling of Aβ toxins to the protein tyrosine kinase Fyn is noteworthy. Shirazi and Wood, in an immunocytochemical analysis of AD brain tissue *(193)*, have found that Fyn levels in tangle-expressing neurons are elevated nearly fivefold over unafflicted cells. An established link thus exists between AD pathology and Fyn. The ectopic abundance of Fyn in tangle-expressing neurons suggests possible involvement in pathogenic signal transduction. The relevance of Fyn to the cellular status of tau has been established in a recent study by Lee et al. who showed that Fyn and tau can be isolated from cells as an intermolecular complex *(194)*. A consequence of this interaction is the accumulation of tau at the plasma membrane. Lee's evidence suggests that tau–Fyn coupling could modulate distribution of microtubules in response to extracellular signals and thus play a role in axon outgrowth or regeneration. Disruption of tau distribution by amyloid peptides hypothetically could lead to the unusual neurite structure seen in AD. It is known, for example, that germline Fyn knockout causes aberrant axon structure *(195,196)*.

Coupling to Glycogen Synthase Kinase 3β

It recently has been found that serine phosphorylation of tau is upregulated by a signal transduction pathway dependent on Fyn activity. In addition to its physiological relevance, this pathway could be a mechanism that directly underlies formation of AD's tangles (*see* Chapter 7). A key relationship is between Fyn and glycogen synthase kinase 3β (GSK 3β). Imahori, Ishiguro, Takashima, and colleagues, in an elegant series of cell biological experiments and in vitro biochemical analyses, have shown that GSK 3β is essential for tau serine phosphorylation evoked by toxic Aβ *(197–200)*; most notably, fibrillar Aβ loses its impact when GSK 3β is eliminated by antisense oligodeoxynucleotides *(201)*. For GSK 3β to be active, it first must become tyrosine phosphorylated *(202)*, and recent studies of neuronal insulin receptors have shown that this activation step can be fulfilled by Fyn *(203)*. Insulin receptors initially stimulate Fyn, which tyrosine-phosphorylates and activates GSK 3β, which then serine-phosphorylates tau. Insulin activation of this pathway is transient. With chronic insulin, the effect is opposite — GSK 3β activity and tau-P levels decrease. Preliminary data have shown that cellular Fyn kinase activity also can be upregulated by ADDLs *(204)*. Unlike the insulin response, however, ADDL stimulation of Fyn activity is maintained for hours without desensitization. Thus, the chronic impact of toxic Aβ is opposite to that of trophic insulin. Because typical physiological kinase responses are transient, the nondesensitizing impact of ADDLs may be germane to its pathogenic effects. The Fyn–GSK 3β pathway provides an appealing hypothetical mechanism to unify the presence of toxic Aβ and neurofibrillary tangles in AD.

Signal Transduction Mechanisms for Aβ-Evoked Neuron Death

The signal transduction just discussed, which can lead to tangle-related tau phosphorylation, also is closely coupled to Aβ-evoked neuron death, and it is in harmony with an extensive literature that links AD and Aβ neurotoxicity to reactive oxygen species (ROS). Steps in the pathway, moreover, may underlie dysfunctional synaptic plasticity (shown in the next section). The emerging picture suggests an economical molecular-level hypothesis for Alzheimer's memory loss and dementia.

FYN and GSK 3β in Cell Death

Involvement of Fyn–GSK 3β signal transduction in Aβ toxicity is in harmony with multiple studies that place Fyn upstream in apoptotic mechanisms. Fyn knockout mice, for example, have an overabundance of hippocampal granule cells in the dentate gyrus and pyramidal cells in the CA3 region, indicating a defect in developmental apoptosis *(205)*. Too much as well as too little Fyn is deleterious, as knockout of Csk, a kinase that negatively regulates Fyn and other Src-family kinases, causes neural tube defects and death at mid-gestation *(206)*. In the lethal Csk knockouts, Fyn and Src kinase activities are 10-fold higher than normal. Immune system cells physiologically couple Fyn to the various T-cell receptor apoptotic cascades, which are used to rid the system of potentially harmful T-cells *(207–209)*. The death receptor Fas, moreover, stimulates tyrosine phosphorylation of multiple proteins in a manner that correlates with activation of Fyn (and Lck, another Src-family kinase) *(210)*. Fas binds Fyn protein, and Fas-mediated thymocyte death is significantly decreased in *fyn* knockout mice *(211)*.

Consistent with known Fyn participation in apoptotic pathways, it has been demonstrated recently that Fyn is essential for Aβ neurotoxicity. Hippocampal neurons in brain slices exposed to ADDLs undergo massive cell death, but this death is completely blocked in slices from *fyn* knockout mice *(33)*. The *fyn* knockout data complement an earlier study in which neurons become resistant to Aβ toxicity following antisense reduction in GSK 3β *(201,212)*. Thus, elimination of either Fyn or GSK 3β protects neurons against toxic Aβ, establishing that components of a signaling pathway linked to tau phosphorylation also figure prominently in neuronal loss.

Excellent candidates exist for steps that lead from Fyn to neurodegeneration. PI3 kinase, for example, is a Fyn-associated molecule whose activity is required to inhibit apoptosis through several trophic receptor pathways *(213–215)*. Fyn and PI3K bind to the same domain of FAK *(216,217)*, suggesting that competition might play a role in lowering the anti-apoptotic effectiveness of PI3K. Another candidate is the IP3 receptor. When tyrosine-phosphorylated by Fyn in some cell types, the IP3 receptor releases more Ca^{2+} from the endoplasmic reticulum *(218)*. This could contribute to the slow buildup in cytoplasmic Ca^{2+} evoked by Aβ *(219,220)*. Focal adhesion kinase, moreover, is a Fyn substrate associated with programmed cell death *(221,222)* and responsive to Aβ *(77,187,189)*. Pathways involving Fyn and FAK are reported to upregulate cyclin D *(223,224)*, and cyclin D and other cell cycle proteins have been found to be anomalously high in AD *(225)*. Because Fyn can mediate growth factor signals and also act as a transforming kinase *(224,226,227)*, ectopic Fyn in neurons hypothetically could trigger a mitotic catastrophe leading to apoptosis *(228,229)*. Finally, Fyn and other closely related Src-family kinases participate in signal transduction that generates reactive oxygen species, molecules that are closely tied to Aβ toxicity.

Reactive Oxygen Species

A dominant theme in the molecular analysis of AD and other neurodegenerative disease is the role played by ROS *(230,231)*. The brain appears especially susceptive to oxidative stress *(232)*, and, compounding the problem, aging brains undergo a decline in antioxidant capacity. The possible relevance of oxidative stress to AD was first inferred from metabolic data *(233)*, but recent data have been more direct. Alzheimer-afflicted brain contains significantly elevated levels of oxidative end products *(234–236)*, including oxidative changes in proteins, DNA, and membrane lipids.

Several potential sources of oxygen free radicals in AD brain have been identified. First, inflammatory reactions associated with microglia have been suggested to play a prominent role, with nitric oxide radicals of microglia potentially involved in degeneration of proximal neurons *(237)*. Microglia show a chemotactic response to Aβ *(238)*. Second, Aβ fibrils, through their intrinsic chemistry, initiate local radicalization comprising a cascade from a superoxide radical through hydrogen peroxide to a hydroxyl radical *(232)*. Extracellular formation of these radicals would be expected to cause localized membrane damage *(231)*. Third, Aβ generates oxidative damage within neurons *(232,239–241)*. Fluorescent probes reveal Aβ-evoked increases in intracellular ROS *(87)*; lipophilic antioxidants such as vitamin E and estrogen are neuroprotective *(91,242,243)*; and resistance to Aβ toxicity in certain PC12 subclones correlates with elevated levels of the antioxidant enzymes catalase and glutathione peroxidase and the redox-sensitive transcription factor NF-κB *(244,245)*. A basis for the ability of Aβ to generate intracellular ROS may be via receptors that act as oxidative toxicity transducers *(246)*, although this idea remains unconfirmed.

ROS and Signal Transduction

Two theoretical alternatives *a priori* would be consistent with the dual impact of Aβ on ROS and signal transduction. First, dysfunction in signal transduction could be downstream from an impact on ROS. One could imagine, for example, that ROS damage to nerve cell surfaces exerts multiple cytotoxic consequences, including disrupted signaling. Signaling dysfunction thus could be a peripheral effect, perhaps not germane to neurodegeneration. Alternatively, signaling dysfunctions caused by Aβ might have a prerequisite role in cellular degeneration, even acting upstream from ROS.

This second alternative is supported by results from knockout experiments involving GSK 3β and Fyn. When GSK 3β is reduced by antisense oligodeoxynucleotide *(201,212)* or when Fyn is eliminated by germline knockout *(33)*, Aβ neurotoxicity is strikingly reduced. Thus, Aβ-evoked changes in signal transduction cascades are essential for evoked neuronal death, not epiphenomena.

For a number of physiological, receptor-mediated responses, upregulation of ROS is an intrinsic signal transduction event *(247)*. It is plausible, therefore, that intracellular ROS could be elevated by Aβ as a consequence of prior signaling steps. A potentially relevant pathway involves the small guanosine triphosphate (GTP)-binding proteins Ras and Rac1. Multiple studies with non-neuronal cells have implicated these transduction molecules in signaling-evoked generation of intracellular ROS *(248–251)*. Activity in the Ras and Rac 1 pathway, moreover, mediates apoptosis induced by Fas in the Jurkat T-cell line *(252,253)*, in a cascade putatively downstream from Fyn *(210,211,254)*. Very recent experiments have obtained analogous results in PC12 neuronlike cells, wherein dominant negative mutants of Rac 1 block oxidative damage and cell death resulting from Aβ fibrils and ADDLs *(255)*. In PC12 cells, Ras and Rac 1 act downstream from Src-family kinases *(256–258)*. Moreover, excessive Rac 1 induces apoptosis in neuronal cells *(259)*, even in the presence of nerve growth factor (NGF), whereas dominant negative Ras mutants decrease apoptosis after NGF or serum withdrawal *(260)*. Current data thus support the hypothesis that Aβ toxins generate harmful amounts of ROS downstream from Src-family protein tyrosine kinases in a pathway involving Rac 1. Consistent with a possible pathogenic role, Rac 1 is highly expressed in hippocampal neurons at risk for AD *(261)*.

Certainly, signal transduction pathways are so interconnected that one could draw a circuitous path to explain nearly anything. Thus, there exists somewhat of a Gordian knot of hypothetical possibilities to connect Aβ to apoptosis. It may be possible, however, to cut the knot by focusing on key alternatives such as the link between Fyn and GSK 3β. Both are associated with tau phosphorylation, both are associated with apoptosis pathways, both are associated with Aβ toxicity, and both, moreover, are associated with each other. The high degree of consilience suggests the hypothetical pathway from Aβ through Fyn and GSK 3β is plausible and could promote both tangle formation and nerve cell death. Fyn may be the linchpin. In addition to upregulating GSK 3β activity, it also can activate Rac 1 and hence generate intracellular ROS. At present, the mechanism is based on a patchwork of conclusions from multiple experimental systems and has not been subject to rigorous tests within a single AD paradigm. The toxic Fyn-initiated sequence does, however, have the additional merit of being consistent with the rapid impact of Aβ on synaptic plasticity, discussed in the following section.

Dysfunction Before Nerve Cell Death: A Potential for Memory-Restoring Drugs?

Alzheimer's dementia is characterized by its progressive nature. End-stage dementia is severe, and afflicted individuals can lack the most primitive cognitive abilities, even the capacity to eat, bathe, and dress. Early-stage dementia, however, is much less drastic, usually presenting as specific deficits in new memory formation. Both early- and end-stage dementia typically have been attributed to nerve cell death, with worsening dementia thought to be the consequence of increasingly widespread cell loss.

An intriguing alternative is that the progressive stages of AD might be separable in terms of cellular mechanism. Early-stage memory loss, for example, could be caused by neuronal dysfunction rather than death. This hypothesis is supported by recent evidence that some APP transgenic mice can exhibit behavioral or neurological anomalies in the absence of obvious cell degeneration *(262)*. Cellular etiology that separates neuron dysfunction and neuron death would imply, theoretically at least, that drug therapy might one day actually reverse AD's early-stage deficits. Targets for such drugs again may turn out to be bioactive molecular forms of Aβ. Fibrils, protofibrils, and ADDLs have rapid electrophysiological and cellular actions that might undermine memory mechanisms well in advance of degenerative cell loss *(33,169,263,264)*.

Rapid Impact on Cholinergic Signaling and on LTP

Results from two different nerve cell biology paradigms indicate that putative memory mechanisms can specifically be disrupted by Aβ. In the first case, Aβ decreases the ability of cultured neurons to make and release acetylcholine (ACh) *(265)*. Although the particular form of Aβ involved is not certain, the mechanism involves protein tyrosine kinase activity. In a related observation, nonaggregated Aβ42 lowers ACh by an impact of GSK 3β and pyruvate dehydrogenase *(199)*. Deleterious impact of Aβ on cholinergic transmission is especially relevant because cholinergic neurons are at high risk in AD and are known to be essential for learning and memory *(266,267)*. In the second case, ADDLs inhibit hippocampal long-term potentiation (LTP), a classic electrophysiological paradigm for synaptic plasticity and a model for learning and memory *(268,269)*. Loss of LTP resulting from ADDLs occurs in organotypic hippocampal slices *(33)* and in living mice (Fig. 5). The pattern of inhibition attributed to ADDLs is not unlike that caused by inhibitors of signal transduction through PKC *(270)* or integrin function *(271)*. Inhibition essentially is instantaneous, indicating loss of neuron function rather than viability. As seen in Fig. 5, ADDL-exposed neurons retain capacity for evoked action potentials and even a capacity for short-term potentiation.

Disruption of signal transduction potentially is reversible. The rapid, nonlethal impact of small Aβ toxins on memory mechanisms such as cholinergic transmission and synaptic plasticity suggests it may be possible one day to treat early-stage AD with drugs that restore normal capacities. It is interesting that normal elderly, who often have memory difficulties, show ADDL-sized oligomers of Aβ42, at levels about one-twelfth of those found in AD-afflicted individuals *(142)*. Whether memory dysfunction seen in individuals aging "normally" also could be the result of low levels of ADDLs, and thus potentially responsive to drug reversal, is speculative but an intriguing possibility.

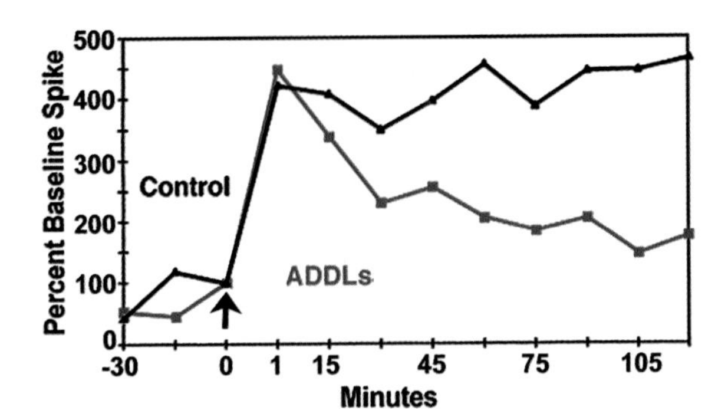

Fig. 5. Memory dysfunction precedes neuron death. Long-term potentiation in living mice is rapidly blocked by stereotaxic injection of ADDLs (ADDLs at t = –60 min; tetanic stimulation at t = 0). Block of LTP occurs without neurodegeneration.

Disrupted Signal Transduction May Account for Rapid Memory Dysfunction

Fyn and Memory Mechanisms

Fyn, in addition to its relevance to nerve cell death, is closely associated with memory and synaptic plasticity. Two different paradigms establish this conclusion. First, germline knockout of *fyn*, which disrupts developmental neuronal apoptosis and blocks ADDL toxicity, also produces animals with impaired hippocampal LTP *(205,272)*. These animals show normal short-term synaptic plasticity, so the impairment is not the result of globally disrupted neurotransmission. Because LTP is rescued by re-expressing the kinase in Fyn-deficient postnatal mice *(196)*, the compromised LTP also is not the result of dysfunctional synaptogenesis. Knockout and recovery data are in harmony with a linkage between Fyn signal transduction and synaptic plasticity mechanisms.

The second paradigm involves murine experimental autoimmune-deficiency syndrome (AIDS), induced by replication-defective leukemia virus *(273)*. Mice with AIDS exhibit impaired spatial learning and memory. At the molecular level, deficits are evident in Fyn signal transduction. Normally, Fyn is stimulated by glutamate receptor activity, a key component of hippocampal memory mechanisms. In tissue from infected mice, Fyn no longer responds to glutamate *(274)*. In this paradigm, Fyn protein levels are normal, but distribution is ectopic. In uninfected cells, Fyn accumulates in postsynaptic densities *(275,276)* and caveolae-like complexes enriched in signal transduction molecules, including plasticity-associated GAP-43 *(277–279)*. The virus blocks this normal trafficking.

Although the mechanism is unknown, Fyn trafficking to the plasma membrane is dependent on a rapidly reversible fatty acylation *(280)*, which could present a site for pathogenic activity.

Viral disruption of memory and Fyn signaling could provide a clue into the basis of rapid disruption of LTP by ADDLs. Like the virus, ADDLs could disrupt local control of Fyn distribution and render the kinase useless for mechanisms of plasticity. Larger disturbances in Fyn signaling, either in subcellular location, total magnitude, or duration, could lead to apoptosis and account for the ultimate neuronal cell loss in AD.

How Fyn is coupled to memory mechanisms remains to be established. The known cellular and molecular activities of Fyn, however, are in harmony with memory functions. In development, for example, Fyn plays a role in synaptic sculpting, acting downstream from Src in receptor clustering and postsynaptic membrane assembly *(281)*. Fyn also is coupled to a recently discovered family of synaptic cadherins *(282,283)*. Before synapses form, Fyn is associated with growth-promoting adhesion molecules *(284)*. In *fyn* mutants, abnormalities are evident in neural cell adhesion molecule (NCAM)-dependent neurite outgrowth and guidance *(195)*. In detergent extracts of embryonic brain *(285)*, NCAM and Fyn coimmunoprecipitate. Fyn also coimmunoprecipitates with FAK. Although they are developmentally downregulated, NCAM and FAK, like Fyn itself, are retained by mature neurons capable of plasticity, and all three molecules have been implicated in the mechanism of LTP *(286–288)*. Recent evidence also has implicated adhesion molecules of the integrin family in LTP *(271,289,290)*, a significant finding because of integrin linkage to FAK, Fyn, and Aβ *(77,187,189,291)*.

With respect to neurotransmission, LTP is glutamatergic *(292)*, glutamate stimulates Fyn kinase *(274)*, and NMDA-type glutamate receptors alter net protein tyrosine phosphorylation *(293,294)*. NMDA receptors, moreover, are phosphorylated by Fyn, leading to enhanced channel activity *(275,276,287,295,296)*. Depolarization *per se* also stimulates Fyn kinase *(297)*. These multiple interactions suggest a positive feedback loop that could be important in potentiation. Hypoactive glutamate receptor responses seen in APP transgenic mice *(298)*, which lead to disturbed behavior, conceivably may be the result of disrupted Fyn activity. On the other hand, uncontrolled positive feedback could lead to neurodegeneration, and FAK and NMDA-type glutamate receptors have been linked to apoptosis as well as LTP *(221,299–301)*. Involvement of Fyn transduction components in both LTP and apoptosis may account for the particular vulnerability of certain neurons to degeneration in AD.

The Question of Receptors

Receptor-Dependent and Receptor-Independent Mechanisms

Early concepts concerning Aβ toxicity were not consistent with receptor-dependent mechanisms. If, for example, physical nerve cell damage could be attributed to enormous, ROS-producing Aβ fibrils, there would be little reason to invoke specific toxin receptors. Receptor mechanisms become more appealing when selective impact on signal transduction becomes relevant. The toxicity of small oligomers, moreover, with their ligand-like properties, could derive from adventitious binding to particular proteins. Even fibril attachment (as shown in Fig. 3) appears restricted to a small portion of surface membrane molecules. The first cell surface proteins found to associate with Aβ were the integrin extracellular matrix receptors *(181,182,302)*. As discussed earlier, integrins have been linked to pathways mediating LTP, apoptosis, and Aβ toxicity *(77,187,189,290,303,304)*. Recently, a number of other candidate molecules have been identified as potential cell surface mediators of Aβ toxicity. These include p75-NTR, RAGE, and scavenger receptors, the latter found on glia *(246,305–308)*. The possibility that cytotoxicity is generated by intracellular forms of Aβ also is under investigation, and an intracellular Aβ binding protein has been identified *(309,310)*.

If Aβ toxins were to trigger degenerative cascades by binding to specific cell surfaces or intracellular molecules, then therapeutic antagonists would be foreseeable. At present, however, it remains possible that Aβ is neurotoxic to cells via alternative triggering mechanisms. Aβ toxins, for example, appear competent under some circumstances to form membrane channels leaky to various ions, including Ca^{2+} *(311,312)*. Cascades initiated by localized oxidation could have negative impact on vital ATPases *(20)* or transmembrane signal transduction molecules such as Fyn. Even simple binding by pathogenic proteins may disrupt normal functions. A middle T-antigen of murine polyomavirus, for example, ectopically stimulates Src kinases, leading to cell transformation *(313,314)*. Hamster polyomavirus shows specificity for Fyn *(315,316)*. Thus, even without receptor involvement, it is possible to trigger specific intracellular cascades. As for now, it is uncertain whether Aβ toxins evoke neuron dysfunction and death nonselectively or through particular proteins that act as toxin receptors. Overall, precedents exist for a wide range of different mechanisms, including binding to specific receptors, channel creation, lipid and protein peroxidation, and local perturbations of membrane enzymes.

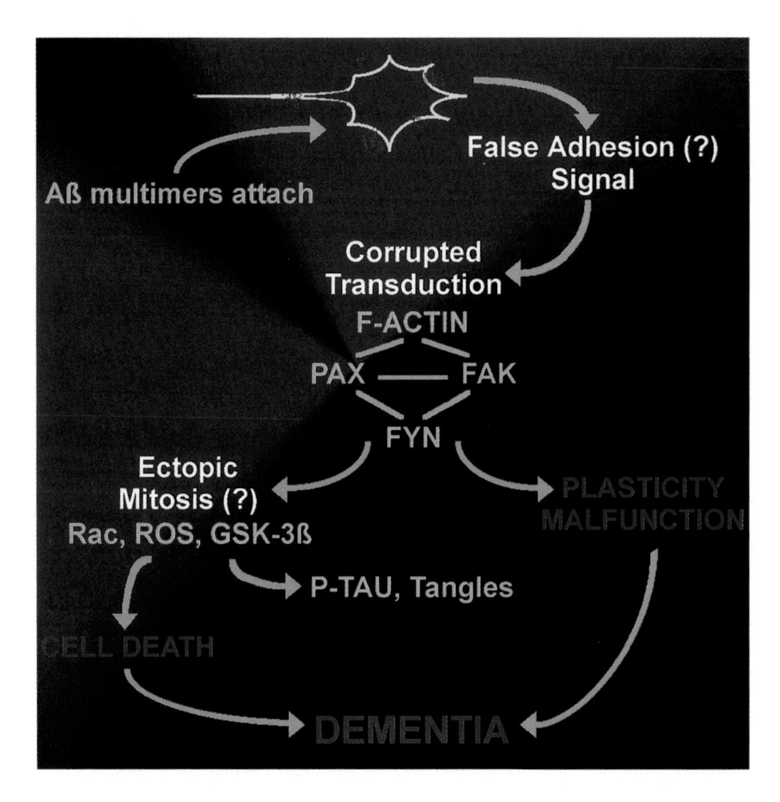

Fig. 6. Hypothetical mechanism for neuronal dysfunction and death caused by Aβ neurotoxins.

1.4. CONCLUSION

Approximately 3500 articles pertinent to Aβ have appeared since Selkoe's comprehensive review in 1994 *(21)*. Consistent with this intense level of research, the Aβ cascade remains a strong hypothesis for neuron dysfunction and death in AD. There is an emerging recognition that fibrillar amyloid is not the only toxic form of Aβ, perhaps not even the most relevant form. Considerable interest recently has focused on small oligomeric Aβ toxins. These small toxins may be the missing link that accounts for the imperfect correlation between amyloid and disease progression. Relative stabilities and structural interrelationships among fibrils, protofibrils, and ADDLs require further investigation. Conceivably, the same toxin unit occurs in each. How toxic multimers of Aβ cause neuron dysfunction and death is still a Gordian knot of possibilities. Nonetheless, what appeared as three alternative mechanisms as diagrammed by Yankner in 1996 *(19)* now can be posed as hypothetical attributes of an integrated cascade (Fig. 6).

Plausible lines can be drawn from Aβ toxins through Fyn protein tyrosine kinase to tangles, ROS, Ca^{2+}, dysfunctional synaptic plasticity, and nerve cell death. A high degree of consilience in this model suggests it merits rigorous scrutiny. The five levels of therapeutic intervention based on the Aβ cascade, as laid out by Selkoe in 1994, remain unchanged, but approaches must now take into consideration the several forms of toxic Aβ. Fibril blockers may not function as ADDL blockers, for example, as shown dramatically with clusterin. Whether fibrils, protofibrils, and ADDLs attach to cells in the same way, or even to the same cells, and whether their downstream actions are identical or idiosyncratic all remain to be determined.

New findings continue to substantiate the relevance of Aβ toxicity to Alzheimer's pathogenesis. *(i.) Aβ and cytoskeletal pathology*. Experimental stimulation of AD-like tau phosphorylation by toxic Aβ has been linked to a MAP kinase-dependent signaling pathway *(317)*. Consistent with the idea that Aβ toxins may contribute to tangle formation in AD, it has been established that Aβ levels in cortical regions increase before onset of cytoskeletal pathology; Aβ levels themselves correlate with Alzheimer's cognitive decline *(318)*. *(ii.) Loss of neural function before neurodegeneration*. Apoptosis appears to account for the slow neurodegeneration induced experimentally by Aβ, and new pathways involving particular caspases have been identified *(319)*. Sub-neurotoxic doses of Aβ, however, continue to be implicated in the rapid inhibition of synaptic plasticity *(320)*. *(iii.) Involvement of non-fibrillar Aβ toxins*. Further evidence has shown that soluble forms of Aβ are elevated in AD and correlate better than fibrillar amyloid with dementia *(321)*. These forms include prominent oligomers of 13 and 47 kDa *(322)*. APP transgenic animals continue to show CNS pathology in the absence of amyloid deposits, implicating non-fibrillar toxins *(323,324)*. Candidate toxins besides ADDLs now include protofibrils, which have been found to exert a rapid electrophysiological impact and a slow induction of nerve cell death *(325,326)*. *(iv.) Prospective new therapeutics*. The apparent value of cigaret smoking as an antidote to Alzheimer's disease could derive in part from the association of Aβ toxins with brain nicotine receptors *(327)*. Vaccination against Aβ has proven surprisingly effective at reducing CNS pathology in animal model experiments *(328)*, while enzymes that release Aβ from its precursor protein have been identified as promising new drug targets *(329)*. An alternative means for reducing Aβ accumulation is suggested by intriguing experiments showing that Aβ secretion is inhibited by testosterone *(330)*.

REFERENCES

1. Alzheimer, A. (1907) Über eine eigenartige Erkankung der Hirnrinde. Allg. Zeitschr. Psychiatrie Phych.-Gerichtl. Med., (Berlin) **64,** 146–148.
2. Leroux, C. (22 July 1997) The case of Auguste D. *Chicago Tribune* Sect. Tempo, pp. 1–4.
3. Geldmacher, D. S. and Whitehouse, P. J., Jr. (1997) Differential diagnosis of Alzheimer's disease. *Neurology* **48(5 Suppl 6),** S2–S9.
4. Ernst, R. L. and Hay, J. W. (1994) The US economic and social costs of Alzheimer's disease revisited. *Am. J. Public Health* **84(8),** 1261–1264.
5. Rice, D. P., Fox, P. J., Max, W., Webber, P. A., Lindeman, D. A., Hauck, W. W., et al. (1993) The economic burden of Alzheimer's disease care. *Health Aff (Millwood)* **12(2),** 164–176.
6. Jellinger, K. A. and Bancher, C. (1998) Neuropathology of Alzheimer's disease: a critical update. *J. Neural Transm.* **54(Suppl.),** 77–95.
7. Smith, M. A. and Perry, G. (1998) What are the facts and artifacts of the pathogenesis and etiology of Alzheimer disease? *J. Chem. Neuroanat.* **16(1),** 35–41.
8. Kosik, K. S., Joachim, C. L., and Selkoe, D. J. (1986) Microtubule-associated protein tau (tau) is a major antigenic component of paired helical filaments in Alzheimer disease. *Proc. Natl. Acad. Sci. USA* **83(11),** 4044–4048.
9. Yamaguchi, H., Hirai, S., Morimatsu, M., Shoji, M., and Ihara, Y. (1988) A variety of cerebral amyloid deposits in the brains of the Alzheimer- type dementia demonstrated by beta protein immunostaining. *Acta. Neuropathol. (Berl.)* **76(6),** 541–549.
10. Yamaguchi, H., Hirai, S., Shoji, M., Harigaya, Y., Okamoto, Y., and Nakazato, Y. (1989) Alzheimer type dementia: diffuse type of senile plaques demonstrated by beta protein immunostaining. *Prog. Clin. Biol. Res.* **317,** 467–474.
11. Probst, A., Brunnschweiler, H., Lautenschlager, C., and Ulrich, J. (1987) A special type of senile plaque, possibly an initial stage. *Acta Neuropathol. (Berl.)* **74(2),** 133–141.
12. Masters, C. L., Simms, G., Weinman, N. A., Multhaup, G., McDonald, B. L., and Beyreuther, K. (1985) Amyloid plaque core protein in Alzheimer disease and Down syndrome. *Proc. Natl. Acad. Sci. USA* **82(12),** 4245–4249.
13. McGeer, P. L., Klegeris, A., Walker, D. G., Yasuhara, O., and McGeer, E. G. (1994) Pathological proteins in senile plaques. *Tohoku. J. Exp. Med.* **174(3),** 269–277.
14. McGeer, P. L., Akiyama, H., Itagaki, S., and McGeer, E. G. (1989) Immune system response in Alzheimer's disease. *Can. J. Neurol. Sci.* **16(4 Suppl),** 516–527.
15. Pearce, B. R., Palmer, A. M., Bowen, D. M., Wilcock, G. K., Esiri, M. M., and Davison, A. N. (1984) Neurotransmitter dysfunction and atrophy of the caudate nucleus in Alzheimer's disease. *Neurochem. Pathol.* **2(4),** 221–232.
16. Whitehouse, P. J., Price, D. L., Struble, R. G., Clark, A. W., Coyle, J. T., and Delon, M. R. (1982) Alzheimer's disease and senile dementia, loss of neurons in the basal forebrain. *Science* **215(4537),** 1237–1239.

17. Davies, P. and Maloney, A. J. (1976) Selective loss of central cholinergic neurons in Alzheimer's disease. *Lancet* **2(8000),** 1403.
18. Bondareff, W., Mountjoy, C. Q., and Roth, M. (1981) Selective loss of neurones of origin of adrenergic projection to cerebral cortex (nucleus locus coeruleus) in senile dementia. *Lancet* **1(8223),** 783–784.
19. Yankner, B. A. 1996) Mechanisms of neuronal degeneration in Alzheimer's disease. *Neuron* **16(5),** 921–932.
20. Mattson, M. P. (1997) Cellular actions of beta–amyloid precursor protein and its soluble and fibrillogenic derivatives. *Physiol. Rev.* **77(4),** 1081–1132.
21. Selkoe, D. J. (1994) Normal and abnormal biology of the beta–amyloid precursor protein. *Annu. Rev. Neurosci.* **17,** 489–517.
22. Cotman, C. W., Pike, C. J., and Copani, A. (1992) beta–Amyloid neurotoxicity, a discussion of in vitro findings. *Neurobiol. Aging* **13(5),** 587–590.
23. McGeer, P. L. and McGeer, E. G. (1999) Inflammation of the brain in Alzheimer's disease, implications for therapy. *J. Leukocyte Biol.* **65(4),** 409–415.
24. Mandelkow, E. M. and Mandelkow, E. (1998) Tau in Alzheimer's disease. *Trends Cell Biol.* **8(11),** 425–427.
25. Arriagada, P. V., Growdon, J. H., Hedley-Whyte, E. T., and Hyman, B. T. (1992) Neurofibrillary tangles but not senile plaques parallel duration and severity of Alzheimer's disease. *Neurology* **42(3 Pt 1),** 631–639.
26. Spillantini, M. G. and Goedert, M. (1998) Tau protein pathology in neurodegenerative diseases. *Trends Neurosci.* **21(10),** 428–433.
27. Pike, C. J., Cummings, B. J., Monzavi, R., and Cotman, C. W. (1994) Beta-amyloid-induced changes in cultured astrocytes parallel reactive astrocytosis associated with senile plaques in Alzheimer's disease. *Neuroscience* **63(2),** 517–531.
28. Busciglio, J., Lorenzo, A., Yeh, J., and Yankner, B. A. (1995) beta-amyloid fibrils induce tau phosphorylation and loss of microtubule binding. *Neuron* **14(4),** 879–888.
29. Lambert, M. P., Stevens, G., Sabo, S., Barber, K., Wang, G., Wade, W., et al. (1994) Beta/A4–evoked degeneration of differentiated SH–SY5Y human neuroblastoma cells. *J. Neurosci. Res.* **39(4),** 377–385.
30. LeBlanc, A. C. and Goodyer, C. G. (1999) Role of endoplasmic reticulum, endosomal–lysosomal compartments, and microtubules in amyloid precursor protein metabolism of human neurons. *J. Neurochem.* **72(5),** 1832–1842.
31. Sheetz, M. P., Pfister, K. K., Bulinski, J. C., and Cotman, C. W. (1998) Mechanisms of trafficking in axons and dendrites, implications for development and neurodegeneration. *Prog. Neurobiol.* **55(6),** 577–594.
32. Oda, T., Wals, P., Osterburg, H. H., Johnson, S. A., Pasinetti, G. M., Morgan, T. E., et al. (1995) Clusterin (apoJ) alters the aggregation of amyloid beta-peptide (A beta 1–42) and forms slowly sedimenting A beta complexes that cause oxidative stress. *Exp. Neurol.* **136(1),** 22–31.
33. Lambert, M. P., Barlow, A. K., Chromy, B. A., Edwards, C., Freed, R., Liosatos, M., et al. (1998) Diffusible, nonfibrillar ligands derived from Abeta1-42 are potent central nervous system neurotoxins. *Proc. Natl. Acad. Sci. USA* **95(11),** 6448–6453.

34. Hardy, J. A. and Higgins, G. A. (1992) Alzheimer's disease, the amyloid cascade hypothesis. *Science* **256(5054),** 184–185.
35. Selkoe, D. J. (1999) Translating cell biology into therapeutic advances in Alzheimer's disease. *Nature* **399(6738 Suppl),** A23–A31.
36. Hardy, J. (1997) Amyloid, the presenilins and Alzheimer's disease. *Trends Neurosci.* **20(4),** 154–159.
37. Glenner, G. G. and Wong, C. W. (1984) Alzheimer's disease, initial report of the purification and characterization of a novel cerebrovascular amyloid protein. *Biochem. Biophys. Res. Commun.* **120(3),** 885–890.
38. Wong, C. W., Quaranta, V., and Glenner, G. G. (1985) Neuritic plaques and cerebrovascular amyloid in Alzheimer disease are antigenically related. *Proc. Natl. Acad. Sci. USA* **82(24),** 8729–8732.
39. Joachim, C. L., Duffy, L. K., Morris, J. H., and Selkoe, D. J. (1988) Protein chemical and immunocytochemical studies of meningovascular beta- amyloid protein in Alzheimer's disease and normal aging. *Brain Res.* **474(1),** 100–111.
40. Roher, A., Wolfe, D., Palutke, M., and KuKuruga, D. (1986) Purification, ultrastructure, and chemical analysis of Alzheimer disease amyloid plaque core protein. *Proc. Natl. Acad. Sci. USA* **83(8),** 2662–2666.
41. Goldgaber, D., Lerman, M. I., McBride, O. W., Saffiotti, U., and Gajdusek, D. C. (1987) Characterization and chromosomal localization of a cDNA encoding brain amyloid of Alzheimer's disease. *Science* **235(4791),** 877–880.
42. Tanzi, R. E., Gusella, J. F., Watkins, P. C., Bruns, G. A., George-Hyslop, P., Van Keuren, M. L., et al. (1987) Amyloid beta protein gene, cDNA, mRNA distribution, and genetic linkage near the Alzheimer locus. *Science* **235(4791),** 880–884.
43. Kang, J., Lemaire, H. G., Unterbeck, A., Salbaum, J. M., Masters, C. L., Grzeschik, K. H., et al. (1987) The precursor of Alzheimer's disease amyloid A4 protein resembles a cell–surface receptor. *Nature* **325(6106),** 733–736.
44. McNamara, M. J., Ruff, C. T., Wasco, W., Tanzi, R. E., Thinakaran, G., and Hyman, B. T. (1998) Immunohistochemical and in situ analysis of amyloid precursor-like protein-1 and amyloid precursor-like protein-2 expression in Alzheimer disease and aged control brains. *Brain Res.* **804(1),** 45–51.
45. Wasco, W., Bupp, K., Magendantz, M., Gusella, J. F., Tanzi, R. E., and Solomon, F. (1992) Identification of a mouse brain cDNA that encodes a protein related to the Alzheimer disease-associated amyloid beta protein precursor. *Proc. Natl. Acad. Sci. USA* **89(22),** 10,758–10,762.
46. Miller, D. L., Papayannopoulos, I. A., Styles, J., Bobin, S. A., Lin, Y. Y., Biemann, K., et al. (1993) Peptide compositions of the cerebrovascular and senile plaque core amyloid deposits of Alzheimer's disease. *Arch. Biochem. Biophys.* **301(1),** 41–52.
47. Selkoe, D. J. (1998) The cell biology of beta–amyloid precursor protein and presenilin in Alzheimer's disease. *Trends Cell Biol.* **8(11),** 447–453.
48. Mills, J. and Reiner, P. B. (1999) Regulation of amyloid precursor protein cleavage. *J. Neurochem.* **72(2),** 443–460.
49. De Strooper, B., Simons, M., Multhaup, G., Van Leuven, F., Beyreuther, K., and Dotti, C. G. (1995) Production of intracellular amyloid-containing frag-

ments in hippocampal neurons expressing human amyloid precursor protein and protection against amyloidogenesis by subtle amino acid substitutions in the rodent sequence. *EMBO J* **14(20),** 4932–4938.
50. Goate, A., Chartier-Harlin, M. C., Mullan, M., Brown, J., Crawford, F., Fidani, L., et al. (1991) Segregation of a missense mutation in the amyloid precursor protein gene with familial Alzheimer's disease. *Nature* **349(6311),** 704–706.
51. Chartier-Harlin, M. C., Crawford, F., Houlden, H., Warren, A., Hughes, D., Fidani, L., et al. (1991) Early-onset Alzheimer's disease caused by mutations at codon 717 of the beta-amyloid precursor protein gene. *Nature* **353(6347),** 844–846.
52. Murrell, J., Farlow, M., Ghetti, B., and Benson, M. D. (1991) A mutation in the amyloid precursor protein associated with hereditary Alzheimer's disease. *Science* **254(5028),** 97–99.
53. Kamino, K., Orr, H. T., Payami, H., Wijsman, E. M., Alonso, M. E., Pulst, S. M., et al. (1992) Linkage and mutational analysis of familial Alzheimer disease kindreds for the APP gene region. *Am. J. Hum. Genet.* **51(5),** 998–1014.
54. Hendriks, L., van Duijn, C. M., Cras, P., Cruts, M., van Hul, W., van Harskamp, F., et al. (1992) Presenile dementia and cerebral haemorrhage linked to a mutation at codon 692 of the beta–amyloid precursor protein gene. *Nat. Genet.* **1(3),** 218–221.
55. Mullan, M., Crawford, F., Axelman, K., Houlden, H., Lilius, L., Winblad, B., et al. (1992) A pathogenic mutation for probable Alzheimer's disease in the APP gene at the N-terminus of beta-amyloid. *Nat. Genet.* **1(5),** 345–347.
56. Ancolio, K., Dumanchin, C., Barelli, H., Warter, J. M., Brice, A., Campion, D., et al. (1999) Unusual phenotypic alteration of beta amyloid precursor protein (betaAPP) maturation by a new Val-715 —> Met betaAPP-770 mutation responsible for probable early-onset Alzheimer's disease. *Proc. Natl. Acad. Sci. USA* **96(7),** 4119–4124.
57. Hsiao, K., Chapman, P., Nilsen, S., Eckman, C., Harigaya, Y., Younkin, S., et al. (1996) Correlative memory deficits, Abeta elevation, and amyloid plaques in transgenic mice. *Science* **274(5284),** 99–102.
58. Games, D., Adams, D., Alessandrini, R., Barbour, R., Berthelette, P., Blackwell, C., et al. (1995) Alzheimer-type neuropathology in transgenic mice overexpressing V717F beta-amyloid precursor protein. *Nature* **373(6514),** 523–527.
59. Higgins, L. S., Holtzman, D. M., Rabin, J., Mobley, W. C., and Cordell, B. (1994) Transgenic mouse brain histopathology resembles early Alzheimer's disease. *Ann. Neurol.* **35(5),** 598–607.
60. Moechars, D., Dewachter, I., Lorent, K., Reverse, D., Baekelandt, V., Naidu, A., et al. (1999) Early phenotypic changes in transgenic mice that overexpress different mutants of amyloid precursor protein in brain. *J. Biol. Chem.* **274(10),** 6483–6492.
61. Quon, D., Wang, Y., Catalano, R., Scardina, J. M., Murakami, K., and Cordell, B. (1991) Formation of beta-amyloid protein deposits in brains of transgenic mice. *Nature* **352(6332),** 239–241.

62. Sturchler-Pierrat, C., Abramowski, D., Duke, M., Wiederhold, K. H., Mistl, C., Rothacher, S., et al. (1997) Two amyloid precursor protein transgenic mouse models with Alzheimer disease-like pathology. *Proc. Natl. Acad. Sci. USA* **94(24),** 13,287–13,292.
63. Scheuner, D., Eckman, C., Jensen, M., Song, X., Citron, M., Suzuki, N., et al. Secreted amyloid beta-protein similar to that in the senile plaques of Alzheimer's disease is increased in vivo by the presenilin 1 and 2 and APP mutations linked to familial Alzheimer's disease. *Nat. Med.* **2(8),** 864–870.
64. Selkoe, D. J. (1997) Alzheimer's disease, genotypes, phenotypes, and treatments. *Science* **275(5300),** 630–631.
65. Schmechel, D. E., Saunders, A. M., Strittmatter, W. J., Crain, B. J., HuLett.e, C. M., Joo, S. H., et al. (1993) Increased amyloid beta-peptide deposition in cerebral cortex as a consequence of apolipoprotein E genotype in late–onset Alzheimer disease. *Proc. Natl. Acad. Sci. USA* **90(20),** 9649–9653.
66. Blacker, D., Wilcox, M. A., Laird, N. M., Rodes, L., Horvath, S. M., Go, R. C., et al. (1998) Alpha–2 macroglobulin is genetically associated with Alzheimer disease. *Nat. Genet.* **19(4),** 357–360.
67. Citron, M., Oltersdorf, T., Haass, C., McConlogue, L., Hung, A. Y., Seubert, P., et al. (1992) Mutation of the beta-amyloid precursor protein in familial Alzheimer's disease increases beta-protein production. *Nature* **360(6405),** 672–674.
68. Mehta, N. D., Refolo, L. M., Eckman, C., Sanders, S., Yager, D., Perez-Tur, J., Younkin, S., et al. (1998) Increased Abeta42(43) from cell lines expressing presenilin 1 mutations. *Ann. Neurol.* **43(2),** 256–258.
69. Borchelt, D. R., Thinakaran, G., Eckman, C. B., Lee, M. K., Davenport, F., Ratovitsky, T., et al. (1996) Familial Alzheimer's diseaselinked presenilin 1 variants elevate Abeta1-42/1–40 ratio in vitro and in vivo. *Neuron* **17(5),** 1005–1013.
70. Xia, W., Zhang, J., Kholodenko, D., Citron, M., Podlisny, M. B., Teplow, D. B., et al. (1997) Enhanced production and oligomerization of the 42-residue amyloid beta– protein by Chinese hamster ovary cells stably expressing mutant presenilins. *J. Biol. Chem.* **272(12),** 7977–7982.
71. De Strooper, B., Saftig, P., Craessaerts, K., Vanderstichele, H., Guhde, G., Annaert, W., et al. (1998) Deficiency of presenilin-1 inhibits the normal cleavage of amyloid precursor protein. *Nature* **391(6665),** 387–390.
72. Wolfe, M. S., Xia, W., Ostaszewski, B. L., Diehl, T. S., Kimberly, W. T., and Selkoe, D. J. (1999) Two transmembrane aspartates in presenilin-1 required for presenilin endoproteolysis and gamma-secretase activity. *Nature* **398(6727),** 513–517.
73. Yankner, B. A., Dawes, L. R., Fisher, S., Villa-Komaroff, L., Oster-Granite, M. L., and Neve, R. L. (1989) Neurotoxicity of a fragment of the amyloid precursor associated with Alzheimer's disease. *Science* **245(4916),** 417–420.
74. Yankner, B. A., Caceres, A., and Duffy, L. K. (1990) Nerve growth factor potentiates the neurotoxicity of beta amyloid. *Proc. Natl. Acad. Sci. USA* **87(22),** 9020–9023.

75. Loo, D. T., Copani, A., Pike, C. J., Whittemore, E. R., Walencewicz, A. J., and Cotman, C.W. (1993) Apoptosis is induced by beta-amyloid in cultured central nervous system neurons. *Proc. Natl. Acad. Sci. USA* **90(17),** 7951–7955.
76. Gschwind, M. and Huber, G. (1995) Apoptotic cell death induced by beta-amyloid 1-42 peptide is cell type dependent. *J. Neurochem.* **65(1),** 292–300.
77. Zhang, C., Lambert. M. P., Bunch. C., Barber. K., Wade. W. S., Krafft, G. A., et al. (1994) Focal adhesion kinase expressed by nerve cell lines shows increased tyrosine phosphorylation in response to Alzheimer's A beta peptide. *J. Biol. Chem.* **269(41),** 25247–25250.
78. Malouf, A. T. (1992) Effect of beta amyloid peptides on neurons in hippocampal slice cultures. *Neurobiol. Aging* **13(5),** 543–551.
79. Kowall, N. W., McKee, A. C., Yankner, B. A., and Beal, M. F. (1992) In vivo neurotoxicity of beta-amyloid [beta(1–40)] and the beta(25–35) fragment. *Neurobiol. Aging* **13(5),** 537–542.
80. Canning, D. R., McKeon, R. J., DeWitt, D. A., Perry, G., Wujek, J. R., Frederickson, R. C., et al. (1993) beta–Amyloid of Alzheimer's disease induces reactive gliosis that inhibits axonal outgrowth. *Exp. Neurol.* **124(2),** 289–298.
81. Fowler, C. J., Cowburn, R. F., and Joseph, J. A. (1997) Alzheimer's, ageing and amyloid, an absurd allegory? *Gerontology* **43(1–2),** 132–142.
82. Marx, J. (1992) Alzheimer's debate boils over. *Science* **257(5075),** 1336–1338.
83. Small, D. H. (1998) The Sixth International Conference on Alzheimer's disease, Amsterdam, The Netherlands, July 1998. The amyloid cascade hypothesis debate, emerging consensus on the role of A beta and amyloid in Alzheimer's disease. *Amyloid* **5(4),** 301–304.
84. Lorenzo, A. and Yankner, B. A. (1994) Beta-amyloid neurotoxicity requires fibril formation and is inhibited by congo red. *Proc. Natl. Acad. Sci. USA* **91(25),** 12,243–12,247.
85. Whitson, J. S., Glabe, C. G., Shintani, E., Abcar, A., and Cotman, C. W. (1990) Beta-amyloid protein promotes neuritic branching in hippocampal cultures. *Neurosci. Lett.* **110(3),** 319–324.
86. Hu, J., Akama, K. T., Krafft, G. A., Chromy, B. A., and Van Eldik, L. J. (1998) Amyloid-beta peptide activates cultured astrocytes, morphological alterations, cytokine induction and nitric oxide release. *Brain Res.* **785(2),** 195–206.
87. Behl, C., Davis, J. B., Lesley, R., and Schubert, D. (1994) Hydrogen peroxide mediates amyloid beta protein toxicity. *Cell* **77(6),** 817–827.
88. Boland, K., Behrens, M., Choi, D., Manias, K., and Perlmutter, D. H. (1996) The serpin-enzyme complex receptor recognizes soluble, nontoxic amyloid-beta peptide but not aggregated, cytotoxic amyloid-beta peptide. *J. Biol. Chem.* **271(30),** 18,032–18,044.
89. Mazziotti, M. and Perlmutter, D. H. (1998) Resistance to the apoptotic effect of aggregated amyloid-beta peptide in several different cell types including neur. *Biochem. J.* **332 (Pt 2),** 517–524.
90. Forloni, G., Chiesa, R., Smiroldo, S., Verga, L., Salmona, M., Tagliavini, F., et al. (1993) Apoptosis mediated neurotoxicity induced by chronic application of beta amyloid fragment 25-35. *Neuroreport* **4(5),** 523–526.

91. Behl, C., Davis, J., Cole, G. M., and Schubert, D. (1992) Vitamin E protects nerve cells from amyloid beta protein toxicity. *Biochem. Biophys. Res. Commun.* **186(2),** 944–950.
92. Dobeli, H., Draeger, N., Huber, G., Jakob, P., Schmidt, D., Seilheimer, B., et al. (1995) A biotechnological method provides access to aggregation competent monomeric Alzheimer's 1-42 residue amyloid peptide. *Biotechnology (N.Y.)* **13(9),** 988–993.
93. Harper, J. D. and Lansbury, P. T., Jr. (1997) Models of amyloid seeding in Alzheimer's disease and scrapie: mechanistic truths and physiological consequences of the time-dependent solubility of amyloid proteins. *Annu. Rev. Biochem.* **66,** 385–407.
94. Cohen, F. E. and Prusiner, S. B. (1998) Pathologic conformations of prion proteins. *Annu. Rev. Biochem.* **67,** 793–819.
95. Pike, C. J., Walencewicz, A. J., Glabe, C. G., and Cotman, C. W. (1991) In vitro aging of beta-amyloid protein causes peptide aggregation and neurotoxicity. *Brain Res.* **563(1-2),** 311–314.
96. Stine, W. B., Jr., Snyder, S. W., Ladror, U. S., Wade, W. S., Miller, M. F., Perun, T. J., et al. (1996) The nanometer-scale structure of amyloid-beta visualized by atomic force microscopy. *J. Protein Chem.* **15(2),** 193–203.
97. HowLett., D. R., Jennings, K. H., Lee, D. C., Clark, M. S., Brown, F., Wetzel, R., et al. (1995) Aggregation state and neurotoxic properties of Alzheimer beta-amyloid peptide. *Neurodegeneration* **4(1),** 23–32.
98. Kirschner, D. A., Inouye, H., Duffy, L. K., Sinclair, A., Lind, M., and Selkoe, D. J. (1987) Synthetic peptide homologous to beta protein from Alzheimer disease forms amyloid-like fibrils in vitro. *Proc. Natl. Acad. Sci. USA* **84(19),** 6953–6957.
99. Merz, P. A., Wisniewski, H. M., Somerville, R. A., Bobin, S. A., Masters, C. L., and Iqbal, K. (1983) Ultrastructural morphology of amyloid fibrils from neuritic and amyloid plaques. *Acta Neuropathol. (Berl.)* **60(1-2),** 113–124.
100. Allsop, D., HowLett., D., Christie, G., and Karran, E. (1998) Fibrillogenesis of beta-amyloid. *Biochem. Soc. Trans.* **26(3),** 459–463.
101. Kelly, J. W. (1998) The alternative conformations of amyloidogenic proteins and their multi– step assembly pathways. *Curr. Opin. Struct. Biol.* **8(1),** 101–106.
102. Lansbury, P. T., Jr. (1997) Structural neurology, are seeds at the root of neuronal degeneration? *Neuron* **19(6),** 1151–1154.
103. Jarrett, J. T., Berger, E. P., and Lansbury, P.T., Jr. (1993) The carboxy terminus of the beta amyloid protein is critical for the seeding of amyloid formation, implications for the pathogenesis of Alzheimer's disease. *Biochemistry* **32(18),** 4693–4697.
104. Soreghan, B., Kosmoski, J., and Glabe, C. (1994) Surfactant properties of Alzheimer's A beta peptides and the mechanism of amyloid aggregation. *J. Biol. Chem.* **269(46),** 28,551–28,554.
105. Younkin, S. G. (1998) The role of A beta 42 in Alzheimer's disease. *J. Physiol. Paris* **92(3-4),** 289–292.
106. Nitsch, R. M., Rebeck, G. W., Deng, M., Richardson, U. I., Tennis, M., Schenk, D. B., et al. (1995) Cerebrospinal fluid levels of amyloid beta-protein in

Alzheimer's disease, inverse correlation with severity of dementia and effect of apolipoprotein E genotype. *Ann. Neurol.* **37(4),** 512–518.
107. van Gool, W. A., Kuiper, M. A., Walstra, G. J., Wolters, E. C., and Bolhuis, P. A. (1995) Concentrations of amyloid beta protein in cerebrospinal fluid of patients with Alzheimer's disease. *Ann. Neurol.* **37(2),** 277–279.
108. Delaere, P., Duyckaerts, C., Masters, C., Beyreuther, K., Piette, F., and Hauw, J. J. (1990) Large amounts of neocortical beta A4 deposits without neuritic plaques nor tangles in a psychometrically assessed, non-demented person. *Neurosci Lett.* **116(1-2),** 87–93.
109. Joachim, C. L., Morris, J. H., and Selkoe, D. J. (1989) Diffuse senile plaques occur commonly in the cerebellum in Alzheimer's disease. *Am. J. Pathol.* **135(2),** 309–319.
110. Yamaguchi, H., Hirai, S., Morimatsu, M., Shoji, M., and Harigaya, Y. (1988) Diffuse type of senile plaques in the brains of Alzheimer-type dementia. *Acta Neuropathol. (Berl.)* **77(2),** 113–119.
111. HowLett., D. R., Perry, A. E., Godfrey, F., Swatton, J. E., Jennings, K. H., Spitzfaden, C., et al. (1999) Inhibition of fibril formation in beta-amyloid peptide by a novel series of benzofurans. *Biochem. J.* **340(Pt. 1),** 283–289.
112. Tomiyama, T., Shoji, A., Kataoka, K., Suwa, Y., Asano, S., Kaneko, H., and Endo, N. (1996) Inhibition of amyloid beta protein aggregation and neurotoxicity by rifampicin. Its possible function as a hydroxyl radical scavenger. *J. Biol. Chem.* **271(12),** 6839–6844.
113. Merlini, G., Ascari, E., Amboldi, N., Bellotti, V., Arbustini, E., Perfetti, V., et al. (1995) Interaction of the anthracycline 4'–iodo–4'–deoxydoxorubicin with amyloid fibrils, inhibition of amyloidogenesis. *Proc. Natl. Acad. Sci. USA* **92(7),** 2959–2963.
114. Klunk, W. E., Debnath, M. L., and Pettegrew, J. W. (1994) Development of small molecule probes for the beta-amyloid protein of Alzheimer's disease. *Neurobiol. Aging* **15(6),** 691–698.
115. Pollack, S. J., Sadler, I. I., Hawtin, S. R., Tailor, V. J., and Shearman, M. S. (1995) Sulfonated dyes attenuate the toxic effects of beta-amyloid in a structure-specific fashion. *Neurosci. Lett.* **197(3),** 211–214.
116. Soto, C., Sigurdsson, E. M., Morelli, L., Kumar, R. A., Castano, E. M., and Frangione, B. (1998) Beta-sheet breaker peptides inhibit fibrillogenesis in a rat brain model of amyloidosis, implications for Alzheimer's therapy. *Nat. Med.* **4(7),** 822–826.
117. Terry, R. D. (1998) The cytoskeleton in Alzheimer disease. *J. Neural. Transm.* **53(Suppl.),** 141–145.
118. Hyman, B. T., Marzloff, K., and Arriagada, P. V. (1993) The lack of accumulation of senile plaques or amyloid burden in Alzheimer's disease suggests a dynamic balance between amyloid deposition and resolution. *J. Neuropathol. Exp. Neurol.* **52(6),** 594–600.
119. Armstrong, R. A. (1994) beta-Amyloid (A beta) deposition in elderly non-demented patients and patients with Alzheimer's disease. *Neurosci. Lett.* **178(1),** 59–62.

120. Giannakopoulos, P., Hof, P. R., Michel, J. P., Guimon, J., and Bouras, C. (1997) Cerebral cortex pathology in aging and Alzheimer's disease, a quantitative survey of large hospital-based geriatric and psychiatric cohorts. *Brain Res. Brain Res. Rev.* **25(2),** 217–245.
121. Geddes, J. W., Tekirian, T. L., Soultanian, N. S., Ashford, J. W., Davis, D. G., and Markesbery, W. R. (1997) Comparison of neuropathologic criteria for the diagnosis of Alzheimer's disease. *Neurobiol. Aging* **18(4 Suppl),** S99–105.
122. Duyckaerts, C. and Hauw, J. J. (1997) Diagnosis and staging of Alzheimer disease. *Neurobiol. Aging* **18(4 Suppl),** S33–S42.
123. Guillozet, A. L., Smiley, J. F., Mash, D. C., and Mesulam, M. M. (1997) Butyrylcholinesterase in the life cycle of amyloid plaques. *Ann. Neurol.* **42(6),** 909–918.
124. Katzman, R., Terry, R., DeTeresa, R., Brown, T., Davies, P., Fuld, P., et al. (1988) Clinical, pathological, and neurochemical changes in dementia: a subgroup with preserved mental status and numerous neocortical plaques. *Ann. Neurol.* **23(2),** 138–144.
125. Crystal, H., Dickson, D., Fuld, P., Masur, D., Scott, R., Mehler, M., et al. (1988) Clinico-pathologic studies in dementia, nondemented subjects with pathologically confirmed Alzheimer's disease. *Neurology* **38(11),** 1682–1687.
126. McKee, A. C., Kosik, K. S., and Kowall, N. W. (1991) Neuritic pathology and dementia in Alzheimer's disease. *Ann. Neurol.* **30(2),** 156–165.
127. Salehi, A., Bakker, J. M., Mulder, M., and Swaab, D. F. (1998) Limited effect of neuritic plaques on neuronal density in the hippocampal CA1 area of Alzheimer patients. *Alzheimer Dis. Assoc. Disord.* **12(2),** 77–82.
128. Einstein, G., Buranosky, R., and Crain, B. J. (1994) Dendritic pathology of granule cells in Alzheimer's disease is unrelated to neuritic plaques. *J. Neurosci.* **14(8),** 5077–5088.
129. Hsia, A. Y., Masliah, E., McConlogue, L., Yu, G. Q., Tatsuno, G., Hu, K., Kholodenko, D., et al. (1999) Plaque-independent disruption of neural circuits in Alzheimer's disease mouse models. *Proc. Natl. Acad. Sci. USA* **96(6),** 3228–3233.
130. Hsiao, K. (1998) Strain dependent and invariant features of transgenic mice expressing Alzheimer amyloid precursor proteins. *Prog. Brain Res.* **117,** 335–341.
131. Yamaguchi, F., Richards, S. J., Beyreuther, K., Salbaum, M., Carlson, G. A., and Dunnett, S. B. (1991) Transgenic mice for the amyloid precursor protein 695 isoform have impaired spatial memory. *Neuroreport* **2(12),** 781–784.
132. Czech, C., Masters, C., and Beyreuther, K. (1994) Alzheimer's disease and transgenic mice. *J. Neural. Transm.* **44(Suppl),** 219–230.
133. Hsiao, K. K., Borchelt, D. R., Olson, K., Johannsdottir, R., Kitt, C., Yunis, W., et al. (1995) Age-related CNS disorder and early death in transgenic FVB/N mice overexpressing Alzheimer amyloid precursor proteins. *Neuron* **15(5),** 1203–1218.
134. D'Hooge, R., Nagels, G., Westland, C. E., Mucke, L., and De Deyn, P. P. (1996) Spatial learning deficit in mice expressing human 751-amino acid beta-amyloid precursor protein. *Neuroreport* **7(15-17),** 2807–2811.

135. Holcomb, L., Gordon, M. N., McGowan, E., Yu, X., Benkovic, S., Jantzen, P., et al. (1998) Accelerated Alzheimer-type phenotype in transgenic mice carrying both mutant amyloid precursor protein and presenilin 1 transgenes. *Nat. Med.* **4(1),** 97–100.
136. Moechars, D., Lorent, K., Dewachter, I., Baekelandt, V., De Strooper, B., and Van Leuven, F. (1998) Transgenic mice expressing an alpha-secretion mutant of the amyloid precursor protein in the brain develop a progressive CNS disorder. *Behav. Brain Res.* **95(1),** 55–64.
137. Moechars, D., Gilis, M., Kuiperi, C., Laenen, I., and Van Leuven, F. (1998) Aggressive behaviour in transgenic mice expressing APP is alleviated by serotonergic drugs. *Neuroreport* **9(16),** 3561–3564.
138. Chui, D. H., Tanahashi, H., Ozawa, K., Ikeda, S., Checler, F., Ueda, O., et al. (1999) Transgenic mice with Alzheimer presenilin 1 mutations show accelerated neurodegeneration without amyloid plaque formation. *Nat. Med.* **5(5),** 560–564.
139. Podlisny, M. B., Walsh, D. M., Amarante, P., Ostaszewski, B. L., Stimson, E. R., Maggio, J. E., et al. (1998) Oligomerization of endogenous and synthetic amyloid beta-protein at nanomolar levels in cell culture and stabilization of monomer by Congo red. *Biochemistry* **37(11),** 3602–3611.
140. Podlisny, M. B., Ostaszewski, B. L., Squazzo, S. L., Koo, E. H., Rydell, R. E., Teplow, D. B., et al. Aggregation of secreted amyloid beta-protein into sodium dodecyl sulfate-stable oligomers in cell culture. *J. Biol. Chem.* **270(16),** 9564–9570.
141. Levine, H., III. (1995) Soluble multimeric Alzheimer beta(1-40) pre-amyloid complexes in dilute solution. *Neurobiol. Aging* **16(5),** 755–764.
142. Burdick, D., Soreghan, B., Kwon, M., Kosmoski, J., Knauer, M., Henschen, A., et al. (1992) Assembly and aggregation properties of synthetic Alzheimer's A4/beta amyloid peptide analogs. *J. Biol. Chem.* **267(1),** 546–554.
143. Klein, W. L., Barlow, A., Chromy, B., Edwards, C., Freed, R., Lambert, M. P., et al. (1997) "ADDLs" — Soluble Aß oligomers that cause biphasic loss of hippocampal neuron function and survival. *Soc. Neurosci. Abstr.* **23,** 1662.
144. Frackowiak, J., Zoltowska, A., and Wisniewski, H. M. (1994) Non-fibrillar beta-amyloid protein is associated with smooth muscle cells of vessel walls in Alzheimer disease. *J. Neuropathol. Exp. Neurol.* **53(6),** 637–645.
145. Kuo, Y. M., Emmerling, M. R., Vigo-Pelfrey, C., Kasunic, T. C., Kirkpatrick, J. B., Murdoch, G. H., et al. (1996) Water-soluble Abeta (N-40, N-42) oligomers in normal and Alzheimer disease brains. *J. Biol. Chem.* **271(8),** 4077–4081.
146. Oda, T., Pasinetti, G. M., Osterburg, H. H., Anderson, C., Johnson, S. A., and Finch, C. E. (1994) Purification and characterization of brain clusterin. *Biochem. Biophys. Res. Commun.* **204(3),** 1131–1136.
147. Lidstrom, A. M., Bogdanovic, N., Hesse, C., Volkman, I., Davidsson, P., and Blennow, K. (1998) Clusterin (apolipoprotein J) protein levels are increased in hippocampus and in frontal cortex in Alzheimer's disease. *Exp. Neurol.* **154(2),** 511–521.
148. Kida, E., Choi-Miura, N. H., and Wisniewski, K. E. (1995) Deposition of apolipoproteins E and J in senile plaques is topographically determined in

both Alzheimer's disease and Down's syndrome brain. *Brain Res.* **685(1–2),** 211–216.
149. Matsubara, E., Soto, C., Governale, S., Frangione, B., and Ghiso, J. (1996) Apolipoprotein J and Alzheimer's amyloid beta solubility. *Biochem. J.* **316(Pt. 2),** 671–679.
150. Golabek, A., Marques, M. A., Lalowski, M., and Wisniewski, T. (1995) Amyloid beta binding proteins in vitro and in normal human cerebrospinal fluid. *Neurosci. Lett.* **191(1-2),** 79–82.
151. Boggs, L. N., Fuson, K. S., Baez, M., Churgay, L., McClure, D., Becker, G., et al. (1996) Clusterin (Apo J) protects against in vitro amyloid-beta (1-40) neurotoxicity. *J. Neurochem.* **67(3),** 1324–1327.
152. Chromy, B. A., Nowak, R. J., Finch, C. E., Krafft, G. A., and Klein, W. L. (1999) Stability of small oligomers of $A\beta_{1-42}$ (ADDLs). *Soc. Neurosci. Abstr.* **25,** 2129.
153. Roher, A. E., Chaney, M. O., Kuo, Y. M., Webster, S. D., Stine, W. B., Haverkamp, L. J., et al. (1996) Morphology and toxicity of Abeta-(1-42) dimer derived from neuritic and vascular amyloid deposits of Alzheimer's disease. *J. Biol. Chem.* **271(34),** 20,631–20,635.
154. Evans, K. C., Berger, E. P., Cho, C. G., Weisgraber, K. H., and Lansbury, P. T., Jr. (1995) Apolipoprotein E is a kinetic but not a thermodynamic inhibitor of amyloid formation, implications for the pathogenesis and treatment of Alzheimer disease. *Proc. Natl. Acad. Sci. USA* **92(3),** 763–767.
155. Aksenov, M. Y., Aksenova, M. V., Butterfield, D. A., Hensley, K., Vigo-Pelfrey, C., and Carney, J. M. (1996) Glutamine synthetase-induced enhancement of beta-amyloid peptide A beta (1-40) neurotoxicity accompanied by abrogation of fibril formation and A beta fragmentation. *J. Neurochem.* **66(5),** 2050–2056.
156. Janciauskiene, S., Garcia, D. F., Carlemalm, E., Dahlback, B., and Eriksson, S. (1995) Inhibition of Alzheimer beta-peptide fibril formation by serum amyloid P component. *J. Biol. Chem.* **270(44),** 26,041–26,044.
157. Eriksson, S., Janciauskiene, S., and Lannfelt, L. (1995) Alpha 1-antichymotrypsin regulates Alzheimer beta-amyloid peptide fibril formation. *Proc. Natl. Acad. Sci. USA* **92(6),** 2313–2317.
158. Alvarez, A., Bronfman, F., Perez, C. A., Vicente, M., Garrido, J., and Inestros, N. C. (1995) Acetylcholinesterase, a senile plaque component, affects the fibrillogenesis of amyloid-beta-peptides. *Neurosci. Lett.* **201(1),** 49–52.
159. Beffert, U., Aumont, N., Dea, D., Lussier-Cacan, S., Davignon, J., and Poirier, J. (1999) Apolipoprotein E isoform-specific reduction of extracellular amyloid in neuronal cultures. *Brain Res. Mol. Brain Res.* **68(1-2),** 181–185.
160. Russo, C., Angelini, G., Dapino, D., Piccini, A., Piombo, G., Schettini, G., et al. (1998) Opposite roles of apolipoprotein E in normal brains and in Alzheimer's disease. *Proc. Natl. Acad. Sci. USA* **95(26),** 15,598–15,602.
161. Yang, D. S., Smith, J. D., Zhou, Z., Gandy, S. E., and Martins, R. N. (1997) Characterization of the binding of amyloid-beta peptide to cell culture-derived native apolipoprotein E_2, E_3, and E_4 isoforms and to isoforms from human plasma. *J. Neurochem.* **68(2),** 721–725.

162. Strittmatter, W. J., Saunders, A. M., Schmechel, D., Pericak-Vance, M., Enghild, J., Salvesen, G. S., et al. (1993) Apolipoprotein E, high-avidity binding to beta-amyloid and increased frequency of type 4 allele in late-onset familial Alzheimer disease. *Proc. Natl. Acad. Sci. USA* **90(5),** 1977–1981.
163. Lansbury, P. T., Jr. (1999) Evolution of amyloid, what normal protein folding may tell us about fibrillogenesis and disease [comment]. *Proc. Natl. Acad. Sci. USA* **96(7),** 3342–3344.
164. Kusumoto, Y., Lomakin, A., Teplow, D. B., and Benedek, G. B. (1998) Temperature dependence of amyloid beta-protein fibrillization. *Proc. Natl. Acad. Sci. USA* **95(21),** 12,277–12,282.
165. Harrison, P. M., Bamborough, P., Daggett, V., Prusiner, S. B., and Cohen, F. E. (1997) The prion folding problem. *Curr. Opin. Struct. Biol.* **7(1),** 53–59.
166. Gasset, M., Baldwin, M. A., Lloyd, D. H., Gabriel, J. M., Holtzman, D. M., Cohen, F., et al. (1992) Predicted alpha-helical regions of the prion protein when synthesized as peptides form amyloid. *Proc. Natl. Acad. Sci. USA* **89(22),** 10,940–10,944.
167. Walsh, D. M., Lomakin, A., Benedek, G. B., Condron, M. M., and Teplow, D.B. (1997) Amyloid beta-protein fibrillogenesis. Detection of a protofibrillar intermediate. *J. Biol. Chem.* **272(35),** 22,364–22,372.
168. Harper, J. D., Wong, S. S., Lieber, C. M., and Lansbury, P. T. (1997) Observation of metastable Abeta amyloid protofibrils by atomic force microscopy. *Chem. Biol.* **4(2),** 119–125.
169. Hartley, D. M. (1999) Keystone Symposium on Alzheimer's Disease.
170. Viola, K. L., Lambert, M. P., Morgan, T. E., Finch, C. E., Krafft, G. A., Menco, B. Ph. M., et al. (1999) Immunolocalization of oligomeric Aβ42 binding to primary mouse hippocampal cells and B103 rat neuroblastoma cells. *Soc. Neurosci. Abstr.* **25,** 2130.
171. Perrot-Applanat, M., Gualillo, O., Buteau, H., Edery, M., and Kelly, P. A. (1997) Internalization of prolactin receptor and prolactin in transfected cells does not involve nuclear translocation. *J. Cell Sci.* **110(Pt. 9),** 1123–1132.
172. Klein, W. L. and Wolf, M. (1982) Regulation of cell surface receptors. In *Membrane Abnormalities and Disease.* (Tao, M., ed.), Boca Raton, FL: CRC Press, Inc., 97–144.
173. Chen, K. S., Masliah, E., Grajeda, H., Guido, T., Huang, J., Khan, K., et al. (1998) Neurodegenerative Alzheimer-like pathology in PDAPP 717V—>F transgenic mice. *Prog. Brain Res.* **117,** 327–334.
174. Moran, P. M., Higgins, L. S., Cordell, B., and Moser, P. C. (1995) Age-related learning deficits in transgenic mice expressing the 751-amino acid isoform of human beta–amyloid precursor protein. *Proc. Natl. Acad. Sci. USA* **92(12),** 5341–5345.
175. Pope, W. B. (1994) Developmental regulation of tau phosphorylation and expression. Ph. D. dissertation, Northwestern University.
176. Enam, S. A. (1991) Cell biology of Alzheimer's Disease. Ph. D. dissertation, Northwestern University.

177. Greenberg, S. G. and Davies, P. (1990) A preparation of Alzheimer paired helical filaments that displays distinct tau proteins by polyacrylamide gel electrophoresis. *Proc. Natl. Acad. Sci. USA* **87(15),** 5827–5831.
178. Pope, W., Enam, S. A., Bawa, N., Miller, B. E., Ghanbari, H. A., and Klein, W. L. (1993) Phosphorylated tau epitope of Alzheimer's disease is coupled to axon development in the avian central nervous system. *Exp. Neurol.* **120(1),** 106–113.
179. Pope, W. B., Lambert, M. P., Leypold, B., Seupaul, R., Sletten, L., Krafft, G., et al. Microtubule-associated protein tau is hyperphosphorylated during mitosis in the human neuroblastoma cell line SH-SY5Y. *Exp. Neurol.* **126(2),** 185–194.
180. Martin, H., Lambert, M. P., Barber, K., Hinton, S., and Klein, W. L. (1995) Alzheimer's-associated phospho-tau epitope in human neuroblastoma cell cultures, up-regulation by fibronectin and laminin. *Neuroscience* **66(4),** 769–779.
181. Ghiso, J., Rostagno, A., Gardella, J. E., Liem, L., Gorevic, P. D., and Frangione, B. (1992) A 109-amino-acid C-terminal fragment of Alzheimer's-disease amyloid precursor protein contains a sequence, -RHDS-, that promotes cell adhesion. *Biochem. J.* **288 (Pt. 3),** 1053–1059.
182. Sabo, S., Lambert, M. P., Kessey, K., Wade, W., Krafft, G., and Klein, W. L. (1995) Interaction of beta-amyloid peptides with integrins in a human nerve cell line. *Neurosci Lett.* **184(1),** 25–28.
183. Clark, E. A. and Brugge, J. S. (1995) Integrins and signal transduction pathways: the road taken. *Science* **268(5208),** 233–239.
184. Burridge, K. and Chrzanowska-Wodnicka, M. (1996) Focal adhesions, contractility, and signaling. *Annu. Rev. Cell Dev. Biol.* **12,** 463–518.
185. Otey, C. A. (1996) pp125FAK in the focal adhesion. *Int. Rev. Cytol.* **167,** 161–183.
186. Girault, J. A., Costa, A., Derkinderen, P., Studler, J. M., and Toutant, M. (1999) FAK and PYK2/CAKbeta in the nervous system: a link between neuronal activity, plasticity and survival? *Trends Neurosci.* **22(6),** 257–263.
187. Zhang, C., Qiu, H. E., Krafft, G. A., and Klein, W. L. (1996) Protein kinase C and F-actin are essential for stimulation of neuronal FAK tyrosine phosphorylation by G-proteins and amyloid beta protein. *FEBS Lett.* **386(2-3),** 185–188.
188. Seufferlein, T. and Rozengurt, E. (1995) Sphingosylphosphorylcholine rapidly induces tyrosine phosphorylation of p125FAK and paxillin, rearrangement of the actin cytoskeleton and focal contact assembly. Requirement of p21rho in the signaling pathway. *J. Biol. Chem.* **270(41),** 24,343–24,351.
189. Zhang, C., Qiu, H. E., Krafft, G. A., and Klein, W. L. (1996) A beta peptide enhances focal adhesion kinase/Fyn association in a rat CNS nerve cell line. *Neurosci. Lett.* **211(3),** 187–190.
190. Resh, M. D. (1998) Fyn, a Src family tyrosine kinase. *Int. J. Biochem. Cell. Biol.* **30(11),** 1159–1162.
191. Berg, M. M., Krafft, G. A., and Klein, W. L. (1997) Rapid impact of beta-amyloid on paxillin in a neural cell line. *J. Neurosci. Res.* **50(6),** 979–989.
192. Shapiro, I. P., Masliah, E., and Saitoh, T. (1991) Altered protein tyrosine phosphorylation in Alzheimer's disease. *J. Neurochem.* **56(4),** 1154–1162.
193. Shirazi, S. K. and Wood, J. G. (1993) The protein tyrosine kinase, fyn, in Alzheimer's disease pathology. *Neuroreport* **4(4),** 435–437.

194. Lee, G., Newman, S. T., Gard, D. L., Band, H., and Panchamoorthy, G. (1998) Tau interacts with src-family non-receptor tyrosine kinases. *J. Cell Sci.* **111(Pt. 21),** 3167–3177.
195. Beggs, H. E., Soriano, P., and Maness, P. F. (1994) NCAM-dependent neurite outgrowth is inhibited in neurons from Fyn-minus mice. *J. Cell Biol.* **127(3),** 825–833.
196. Kojima, N., Wang, J., Mansuy, I. M., Grant, S. G., Mayford, M., and Kandel, E. R. (1997) Rescuing impairment of long-term potentiation in *fyn*-deficient mice by introducing Fyn transgene. *Proc. Natl. Acad. Sci. USA* **94(9),** 4761–4765.
197. Michel, G., Mercken, M., Murayama, M., Noguchi, K., Ishiguro, K., Imahori, K., et al. (1998) Characterization of tau phosphorylation in glycogen synthase kinase- 3beta and cyclin dependent kinase-5 activator (p23) transfected cells. *Biochim. Biophys. Acta.* **1380(2),** 177–182.
198. Ishiguro, K., Shiratsuchi, A., Sato, S., Omori, A., Arioka, M., Kobayashi, S., Uchida, T., et al. Glycogen synthase kinase 3 beta is identical to tau protein kinase I generating several epitopes of paired helical filaments. *FEBS Lett.* **325(3),** 167–172.
199. Imahori, K., Hoshi, M., Ishiguro, K., Sato, K., Takahashi, M., Shiurba, R., Yet al. (1998) Possible role of tau protein kinases in pathogenesis of Alzheimer's disease. *Neurobiol. Aging* **19(1 Suppl.),** S93–S98.
200. Takashima, A., Honda, T., Yasutake, K., Michel, G., Murayama, O., Murayama, M., et al. (1998) Activation of tau protein kinase I/glycogen synthase kinase-3beta by amyloid beta peptide (25-35) enhances phosphorylation of tau in hippocampal neurons. *Neurosci. Res.* **31(4),** 317–323.
201. Takashima, A., Noguchi, K., Sato, K., Hoshino, T., and Imahori, K. (1993) Tau protein kinase I is essential for amyloid beta–protein–induced neurotoxicity. *Proc. Natl. Acad. Sci. USA* **90(16),** 7789–7793.
202. Wang, Q. M., Fiol, C. J., DePaoli-Roach, A. A., and Roach, P. J. (1994) Glycogen synthase kinase-3 beta is a dual specificity kinase differentially regulated by tyrosine and serine/threonine phosphorylation. *J. Biol. Chem.* **269(20),** 14,566–14,574.
203. Lesort, M., Jope, R. S., and Johnson, G. V. (1999) Insulin transiently increases tau phosphorylation, involvement of glycogen synthase kinase-3beta and Fyn tyrosine kinase. *J. Neurochem.* **72(2),** 576–584.
204. Lambert, M. P., Barlow, A., Lin, J., Priebe, G., Viola, K., Zhang, C., et al. (1999) Neuron dysfunction and death caused by small Aβ oligomers: role of signal transduction. *Soc. Neurosci. Abstr.* **25,** 2129.
205. Grant, S. G., O'Dell, T. J., Karl, K. A., Stein, P. L., Soriano, P., and Kandel, E. R. (1992) Impaired long-term potentiation, spatial learning, and hippocampal development in fyn mutant mice. *Science* **258(5090),** 1903–1910.
206. Imamoto, A. and Soriano, P. (1993) Disruption of the csk gene, encoding a negative regulator of Src family tyrosine kinases, leads to neural tube defects and embryonic lethality in mice. *Cell* **73(6),** 1117–1124.
207. Stein, P. L., Lee, H. M., Rich, S., and Soriano, P. (1992) pp59fyn mutant mice display differential signaling in thymocytes and peripheral T cells. *Cell* **70(5),** 741–750.

208. Migita, K., Eguchi, K., Kawabe, Y., and Nagataki, S. (1995) Tyrosine phosphorylation participates in peripheral T-cell activation and programmed cell death in vivo. *Immunology* **85(4),** 550–555.
209. Lancki, D. W., Fields, P., Qian, D., and Fitch, F. W. (1995) Induction of lytic pathways in T cell clones derived from wild-type or protein tyrosine kinase Fyn mutant mice. *Immunol. Rev.* **146,** 117–144.
210. Schlottmann, K. E., Gulbins, E., Lau, S. M., and Coggeshall, K. M. (1996) Activation of Src-family tyrosine kinases during Fas-induced apoptosis. *J. Leukoc. Biol.* **60(4),** 546–554.
211. Atkinson, E. A., Ostergaard, H., Kane, K., Pinkoski, M. J., Caputo, A., Olszowy, M. W., et al. (1996) A physical interaction between the cell death protein Fas and the tyrosine kinase p59fynT. *J. Biol. Chem.* **271(11),** 5968–5971.
212. Takashima, A., Yamaguchi, H., Noguchi, K., Michel, G., Ishiguro, K., Sato, K., et al. (1995) Amyloid beta peptide induces cytoplasmic accumulation of amyloid protein precursor via tau protein kinase I/glycogen synthase kinase-3 beta in rat hippocampal neurons. *Neurosci. Lett.* **198(2),** 83–86.
213. Takashima, A., Noguchi, K., Michel, G., Mercken, M., Hoshi, M., Ishiguro, K., et al. (1996) Exposure of rat hippocampal neurons to amyloid beta peptide (25-35) induces the inactivation of phosphatidyl inositol-3 kinase and the activation of tau protein kinase I/glycogen synthase kinase-3 beta. *Neurosci. Lett.* **203(1),** 33–36.
214. Pap, M. and Cooper, G.M. (1998) Role of glycogen synthase kinase-3 in the phosphatidylinositol 3- Kinase/Akt cell survival pathway. *J. Biol. Chem.* **273(32),** 19,929–19,932.
215. Ishiguro, K. (1998) Involvement of tau protein kinase in amyloid-beta-induced neurodegeneration. *Rinsho Byori* **46(10),** 1003–1007.
216. Reiske, H. R., Kao, S. C., Cary, L. A., Guan, J. L., Lai, J. F., and Chen, H. C. (1999) Requirement of phosphatidylinositol 3-kinase in focal adhesion kinase- promoted cell migration. *J. Biol. Chem.* **274(18),** 12,361–12,366.
217. Chen, H. C., Appeddu, P. A., Isoda, H., and Guan, J. L. (1996) Phosphorylation of tyrosine 397 in focal adhesion kinase is required for binding phosphatidylinositol 3-kinase. *J. Biol. Chem.* **271(42),** 26,329–26,334.
218. Jayaraman, T., Ondrias, K., Ondriasova, E., and Marks, A. R. (1996) Regulation of the inositol 1,4,5-trisphosphate receptor by tyrosine phosphorylation. *Science* **272(5267),** 1492–1494.
219. Mattson, M. P., Tomaselli, K. J., and Rydel, R. E. (1993) Calcium-destabilizing and neurodegenerative effects of aggregated beta-amyloid peptide are attenuated by basic FGF. *Brain Res.* **621(1),** 35–49.
220. Archuleta, M. M., Schieven, G. L., Ledbetter, J. A., Deanin, G. G., and Burchiel, S. W. (1993) 7,12-Dimethylbenz[a]anthracene activates protein-tyrosine kinases Fyn and Lck in the HPB-ALL human T-cell line and increases tyrosine phosphorylation of phospholipase C–gamma 1, formation of inositol 1,4,5-trisphosphate, and mobilization of intracellular calcium. *Proc. Natl. Acad. Sci. USA* **90(13),** 6105–6109.
221. Ilic, D., Almeida, E. A., Schlaepfer, D. D., Dazin, P., Aizawa, S., and Damsky, C. H. (1998) Extracellular matrix survival signals transduced by focal adhesion kinase suppress p53–mediated apoptosis. *J. Cell Biol.* **143(2),** 547–560.

222. Grant, S. G., Karl, K. A., Kiebler, M. A., and Kandel, E. R. (1995) Focal adhesion kinase in the brain, novel subcellular localization and specific regulation by Fyn tyrosine kinase in mutant mice. *Genes Dev.* **9(15),** 1909–1921.
223. Zhao, J. H., Reiske, H., and Guan, J. L. (1998) Regulation of the cell cycle by focal adhesion kinase. *J. Cell Biol.* **143(7),** 1997–2008.
224. Wary, K. K., Mariotti, A., Zurzolo, C., and Giancotti, F. G. (1998) A requirement for caveolin-1 and associated kinase Fyn in integrin signaling and anchorage-dependent cell growth. *Cell* **94(5),** 625–634.
225. Busser, J., Geldmacher, D. S., and Herrup, K. (1998) Ectopic cell cycle proteins predict the sites of neuronal cell death in Alzheimer's disease brain. *J. Neurosci.* **18(8),** 2801–2807.
226. Roche, S., Fumagalli, S., and Courtneidge, S. A. (1995) Requirement for Src family protein tyrosine kinases in G2 for fibroblast cell division. *Science* **269(5230),** 1567–1569.
227. Semba, K., Kawai, S., Matsuzawa, Y., Yamanashi, Y., Nishizawa, M., and Toyoshima, K. (1990) Transformation of chicken embryo fibroblast cells by avian retroviruses containing the human Fyn gene and its mutated genes. *Mol. Cell Biol.* **10(6),** 3095–3104.
228. Rubin, L. L., Gatchalian, C. L., Rimon, G., and Brooks, S. F. (1994) The molecular mechanisms of neuronal apoptosis. *Curr. Opin. Neurobiol.* **4(5),** 696–702.
229. Heintz, N. (1993) Cell death and the cell cycle: a relationship between transformation and neurodegeneration? *Trends Biochem. Sci.* **18(5),** 157–159.
230. Multhaup, G., Ruppert, T., Schlicksupp, A., Hesse, L., Beher, D., Masters, C. L., et al. (1997) Reactive oxygen species and Alzheimer's disease. *Biochem. Pharmacol.* **54(5),** 533–539.
231. Butterfield, D. A. (1997) beta-Amyloid-associated free radical oxidative stress and neurotoxicity, implications for Alzheimer's disease. *Chem. Res. Toxicol.* **10(5),** 495–506.
232. Behl, C. (1997) Amyloid beta-protein toxicity and oxidative stress in Alzheimer's disease. *Cell Tissue Res.* **290(3),** 471–480.
233. Martins, R. N., Harper, C. G., Stokes, G. B., and Masters, C. L. (1986) Increased cerebral glucose-6-phosphate dehydrogenase activity in Alzheimer's disease may reflect oxidative stress. *J. Neurochem.* **46(4),** 1042–1045.
234. Subbarao, K. V., Richardson, J. S., and Ang, L. C. (1990) Autopsy samples of Alzheimer's cortex show increased peroxidation in vitro. *J. Neurochem.* **55(1),** 342–345.
235. Smith, M. A., Rudnicka-Nawrot, M., Richey, P. L., Praprotnik, D., Mulvihill, P., Miller, C. A., et al. (1995) Carbonyl-related posttranslational modification of neurofilament protein in the neurofibrillary pathology of Alzheimer's disease. *J. Neurochem.* **64(6),** 2660–2666.
236. Mecocci, P., MacGarvey, U., and Beal, M. F. (1994) Oxidative damage to mitochondrial DNA is increased in Alzheimer's disease. *Ann. Neurol.* **36(5),** 747–751.
237. Weldon, D. T., Rogers, S. D., Ghilardi, J. R., Finke, M. P., Cleary, J. P., O'Hare, E., et al. (1998) Fibrillar beta-amyloid induces microglial phagocy-

tosis, expression of inducible nitric oxide synthase, and loss of a select population of neurons in the rat CNS in vivo. *J. Neurosci.* **18(6),** 2161–2173.
238. Davis, J. B., McMurray, H. F., and Schubert, D. (1992) The amyloid beta-protein of Alzheimer's disease is chemotactic for mononuclear phagocytes. *Biochem. Biophys. Res. Commun.* **189(2),** 1096–1100.
239. Kruman, I., Bruce-Keller, A. J., Bredesen, D., Waeg, G., and Mattson, M. P. (1997) Evidence that 4-hydroxynonenal mediates oxidative stress-induced neuronal apoptosis. *J. Neurosci.* **17(13),** 5089–5100.
240. Behl, C. and Sagara, Y. (1997) Mechanism of amyloid beta protein induced neuronal cell death, current concepts and future perspectives. *J. Neural Transm.* **49(Suppl),** 125–134.
241. Keller, J. N., Kindy, M. S., Holtsberg, F. W., St. Clair, D. K., Yen, H. C., Germeyer, A., et al. (1998) Mitochondrial manganese superoxide dismutase prevents neural apoptosis and reduces ischemic brain injury: suppression of peroxynitrite production, lipid peroxidation, and mitochondrial dysfunction. *J. Neurosci.* **18(2),** 687–697.
242. Mattson, M. P., Robinson, N., and Guo, Q. (1997) Estrogens stabilize mitochondrial function and protect neural cells against the pro-apoptotic action of mutant presenilin-1. *Neuroreport* **8(17),** 3817–3821.
243. Behl, C., Skutella, T., Lezoualc'h, F., Post, A., Widmann, M., Newton, C. J., et al. (1997) Neuroprotection against oxidative stress by estrogens, structure-activity relationship. *Mol. Pharmacol.* **51(4),** 535–541.
244. Lezoualc'h, F., Sagara, Y., Holsboer, F., and Behl, C. (1998) High constitutive NF-kappaB activity mediates resistance to oxidative stress in neuronal cells. *J. Neurosci.* **18(9),** 3224–3232.
245. Sagara, Y., Dargusch, R., Klier, F. G., Schubert, D., and Behl, C. (1996) Increased antioxidant enzyme activity in amyloid beta protein-resistant cells. *J. Neurosci.* **16(2),** 497–505.
246. Mattson, M. P. and Rydel, R. E. (1996) Alzheimer's disease. Amyloid ox-tox transducers [news; comment]. *Nature* **382(6593),** 674–675.
247. Burdon, R. H. (1995) Superoxide and hydrogen peroxide in relation to mammalian cell proliferation. *Free Radical Biol. Med.* **18(4),** 775–794.
248. Sundaresan, M., Yu, Z. X., Ferrans, V. J., Sulciner, D. J., Gutkind, J. S., Irani, K., et al. (1996) Regulation of reactive-oxygen-species generation in fibroblasts by Rac1. *Biochem J.* **318(Pt. 2),** 379–382.
249. Cool, R. H., Merten, E., Theiss, C., and Acker, H. (1998) Rac1, and not Rac2, is involved in the regulation of the intracellular hydrogen peroxide level in HepG2 cells. *Biochem. J.* **332(Pt. 1),** 5–8.
250. Lee, A. C., Fenster, B. E., Ito, H., Takeda, K., Bae, N. S., Hirai, T., et al. (1999) Ras proteins induce senescence by altering the intracellular levels of reactive oxygen species. *J. Biol. Chem.* **274(12),** 7936–7940.
251. Dorseuil, O., Vazquez, A., Lang, P., Bertoglio, J., Gacon, G., and Leca, G. (1992) Inhibition of superoxide production in B lymphocytes by rac antisense oligonucleotides. *J. Biol. Chem.* **267(29),** 20,540–20,542.
252. Gulbins, E., Coggeshall, K. M., Brenner, B., Schlottmann, K., Linderkamp, O., and Lang, F. (1996) Fas-induced apoptosis is mediated by activation of a

Ras and Rac protein-regulated signaling pathway. *J. Biol. Chem.* **271(42),** 26,389–26,394.
253. Brenner, B., Koppenhoefer, U., Weinstock, C., Linderkamp, O., Lang, F., and Gulbins, E. (1997) Fas- or ceramide-induced apoptosis is mediated by a Rac1-regulated activation of Jun N-terminal kinase/p38 kinases and GADD153. *J. Biol. Chem.* **272(35),** 22,173–22,181.
254. Hane, M., Lowin, B., Peitsch, M., Becker, K., and Tschopp, J. (1995) Interaction of peptides derived from the Fas ligand with the Fyn-SH3 domain. *FEBS Lett.* **373(3),** 265–268.
255. Longo, V. D., Klein, W. L., and Finch, C. E. Nonfibrillar Aβ1-42 (ADDL) causes aconitase inactivation and iron-dependent neurotoxicity. *J. Neurosci,* in press.
256. Altun-Gultekin, Z. F. and Wagner, J. A. (1996) Src, ras, and rac mediate the migratory response elicited by NGF and PMA in PC12 cells. *J. Neurosci. Res.* **44(4),** 308–327.
257. D'Arcangelo, G. and Halegoua, S. (1993) A branched signaling pathway for nerve growth factor is revealed by Src- , Ras-, and Raf-mediated gene inductions. *Mol. Cell Biol.* **13(6),** 3146–3155.
258. Rusanescu, G., Qi, H., Thomas, S. M., Brugge, J. S., and Halegoua, S. (1995) Calcium influx induces neurite growth through a Src-Ras signaling cassette. *Neuron* **15(6),** 1415–1425.
259. Bazenet, C. E., Mota, M. A., and Rubin, L. L. (1998) The small GTP-binding protein Cdc42 is required for nerve growth factor withdrawal-induced neuronal death. *Proc. Natl. Acad. Sci. USA* **95(7),** 3984–3989.
260. Ferrari, G. and Greene, L. A. (1994) Proliferative inhibition by dominant-negative Ras rescues naive and neuronally differentiated PC12 cells from apoptotic death. *EMBO J.* **13(24),** 5922–5928.
261. Olenik, C., Barth, H., Just, I., Aktories, K., and Meyer, D. K. (1997) Gene expression of the small GTP-binding proteins RhoA, RhoB, Rac1, and Cdc42 in adult rat brain. *Brain Res. Mol. Brain Res.* **52(2),** 263–269.
262. Chapman, P. F., White, G. L., Jones, M. W., Cooper-Blacketer, D., Marshall, V. J., Irizarry, M., et al. (1999) Impaired synaptic plasticity and learning in aged amyloid precursor protein transgenic mice. *Nat. Neurosci.* **2(3),** 271–276.
263. Sanderson, K. L., Butler, L., and Ingram, V. M. (1997) Aggregates of a beta-amyloid peptide are required to induce calcium currents in neuron-like human teratocarcinoma cells, relation to Alzheimer's disease. *Brain Res.* **744(1),** 7–14.
264. Good, T. A., Smith, D. O., and Murphy, R. M. (1996) Beta-amyloid peptide blocks the fast-inactivating K+ current in rat hippocampal neurons. *Biophys. J.* **70(1),** 296–304.
265. Pedersen, W. A., Kloczewiak, M. A., and Blusztajn, J. K. (1996) Amyloid beta-protein reduces acetylcholine synthesis in a cell line derived from cholinergic neurons of the basal forebrain. *Proc. Natl. Acad. Sci. USA* **93(15),** 8068–8071.
266. Mesulam, M. M. (1998) Some cholinergic themes related to Alzheimer's disease, synaptology of the nucleus basalis, location of m2 receptors, interactions with amyloid metabolism, and perturbations of cortical plasticity. *J. Physiol. (Paris)* **92(3-4),** 293–298.

267. Smith, C. M. and Swash, M. (1978) Possible biochemical basis of memory disorder in Alzheimer disease. *Ann. Neurol.* **3(6),** 471–473.
268. Eccles, J. C. (1986) Mechanisms of long-term memory. *J. Physiol. (Paris)* **81(4),** 312–317.
269. Jeffery, K. J. (1997) LTP and spatial learning—where to next? *Hippocampus* **7(1),** 95–110.
270. Lovinger, D. M., Wong, K. L., Murakami, K., and Routtenberg, A. (1987) Protein kinase C inhibitors eliminate hippocampal long-term potentiation. *Brain Res.* **436(1),** 177–183.
271. Staubli, U., Chun, D., and Lynch, G. (1998) Time-dependent reversal of long-term potentiation by an integrin antagonist. *J. Neurosci.* **18(9),** 3460–3469.
272. Grant, S. G. and Silva, A. J. (1994) Targeting learning. *Trends Neurosci.* **17(2),** 71–75.
273. Sei, Y., Kustova, Y., Li, Y., Morse, H. C., III, Skolnick, P., and Basile, A. S. (1998) The encephalopathy associated with murine acquired immunodeficiency syndrome. *Ann. NY Acad. Sci.* **840,** 822–834.
274. Sei, Y., Whitesell, L., Kustova, Y., Paul, I. A., Morse, H. C., III, Skolnick, P., et al. (1996) Altered brain fyn kinase in a murine acquired immunodeficiency syndrome. *FASEB J* **10(2),** 339–344.
275. Suzuki, T. and Okumura-Noji, K. (1995) NMDA receptor subunits epsilon 1 (NR2A) and epsilon 2 (NR2B) are substrates for Fyn in the postsynaptic density fraction isolated from the rat brain. *Biochem. Biophys. Res. Commun.* **216(2),** 582–588.
276. Tezuka, T., Umemori, H., Akiyama, T., Nakanishi, S., and Yamamoto, T. (1999) PSD-95 promotes Fyn-mediated tyrosine phosphorylation of the N-methyl-D- aspartate receptor subunit NR2A. *Proc. Natl. Acad. Sci. USA* **96(2),** 435–440.
277. Maekawa, S., Kumanogoh, H., Funatsu, N., Takei, N., Inoue, K., Endo, Y., et al. (1997) Identification of NAP-22 and GAP-43 (neuromodulin) as major protein components in a Triton insoluble low density fraction of rat brain. *Biochim. Biophys. Acta.* **1323(1),** 1–5.
278. Henke, R. C., Seeto, G. S., and Jeffrey, P. L. (1997) Thy-1 and AvGp50 signal transduction complex in the avian nervous system: c-Fyn and G alpha i protein association and activation of signalling pathways. *J. Neurosci. Res.* **49(6),** 655–670.
279. Gorodinsky, A. and Harris, D. A. (1995) Glycolipid-anchored proteins in neuroblastoma cells form detergent- resistant complexes without caveolin. *J. Cell Biol.* **129(3),** 619–627.
280. Wolven, A., Okamura, H., Rosenblatt, Y., and Resh, M. D. (1997) Palmitoylation of p59fyn is reversible and sufficient for plasma membrane association. *Mol. Biol. Cell* **8(6),** 1159–1173.
281. Fuhrer, C. and Hall, Z. W. (1996) Functional interaction of Src family kinases with the acetylcholine receptor in C2 myotubes. *J. Biol. Chem.* **271(50),** 32,474–32,481.
282. Kohmura, N., Senzaki, K., Hamada, S., Kai, N., Yasuda, R., Watanabe, M., et al. (1998) Diversity revealed by a novel family of cadherins expressed in neurons at a synaptic complex. *Neuron* **20(6),** 1137–1151.

283. Hagler, D. J., Jr. and Goda, Y. (1998) Synaptic adhesion: the building blocks of memory? *Neuron* **20(6),** 1059–1062.
284. Maness, P. F. (1992) Nonreceptor protein tyrosine kinases associated with neuronal development. *Dev. Neurosci.* **14(4),** 257–270.
285. Beggs, H. E., Baragona, S. C., Hemperly, J. J., and Maness, P. F. (1997) NCAM140 interacts with the focal adhesion kinase p125(fak) and the SRC-related tyrosine kinase p59(fyn). *J. Biol. Chem.* **272(13),** 8310–8319.
286. Schachner, M. (1997) Neural recognition molecules and synaptic plasticity. *Curr. Opin. Cell Biol.* **9(5),** 627–634.
287. Grant, S. G. (1996) Analysis of NMDA receptor mediated synaptic plasticity using gene targeting, roles of Fyn and FAK non-receptor tyrosine kinases. *J. Physiol. (Paris)* **90(5-6),** 337–338.
288. Rose, S.P. (1995) Cell-adhesion molecules, glucocorticoids and long-term-memory formation. *Trends Neurosci.* **18(11),** 502–506.
289. Bahr, B. A., Staubli, U., Xiao, P., Chun, D., Ji, Z. X., Esteban, E. T., et al. (1997) Arg-Gly-Asp-Ser-selective adhesion and the stabilization of long-term potentiation, pharmacological studies and the characterization of a candidate matrix receptor. *J. Neurosci.* **17(4),** 1320–1329.
290. Xiao, P., Bahr, B. A., Staubli, U., Vanderklish, P. W., and Lynch, G. (1991) Evidence that matrix recognition contributes to stabilization but not induction of LTP. *Neuroreport* **2(8),** 461–464.
291. Cotman, C. W., Hailer, N. P., Pfister, K. K., Soltesz, I., and Schachner, M. (1998) Cell adhesion molecules in neural plasticity and pathology, similar mechanisms, distinct organizations? *Prog. Neurobiol.* **55(6),** 659–669.
292. Muller, D., Joly, M., and Lynch, G. (1988) Contributions of quisqualate and NMDA receptors to the induction and expression of LTP. *Science* **242(4886),** 1694–1697.
293. Bading, H. and Greenberg, M. E. (1991) Stimulation of protein tyrosine phosphorylation by NMDA receptor activation. *Science* **253(5022),** 912–914.
294. Gurd, J. W. (1997) Protein tyrosine phosphorylation: implications for synaptic function. *Neurochem Int.* **31(5),** 635–649.
295. Kohr, G. and Seeburg, P. H. (1996) Subtype-specific regulation of recombinant NMDA receptor-channels by protein tyrosine kinases of the src family. *J. Physiol. (London)* **492(Pt. 2),** 445–452.
296. Sala, C. and Sheng, M. (1999) The fyn art of N-methyl-D-aspartate receptor phosphorylation. *Proc. Natl. Acad. Sci. USA* **96(2),** 335–337.
297. Kobayashi, S., Okumura, N., Okada, M., and Nagai, K. (1998) Depolarization-induced tyrosine phosphorylation of p130(cas). *J. Biochem. (Tokyo)* **123(4),** 624–629.
298. Moechars, D., Lorent, K., De Strooper, B., Dewachter, I., and Van Leuven, F. (1996) Expression in brain of amyloid precursor protein mutated in the alpha-secretase site causes disturbed behavior, neuronal degeneration and premature death in transgenic mice. *EMBO J* **15(6),** 1265–1274.
299. Levkau, B., Herren, B., Koyama, H., Ross, R., and Raines, E. W. (1998) Caspase-mediated cleavage of focal adhesion kinase pp125FAK and disas-

sembly of focal adhesions in human endothelial cell apoptosis. *J. Exp. Med.* **187(4),** 579–586.
300. Bonfoco, E., Krainc, D., Ankarcrona, M., Nicotera, P., and Lipton, S. A. (1995) Apoptosis and necrosis, two distinct events induced, respectively, by mild and intense insults with N-methyl-D-aspartate or nitric oxide/ superoxide in cortical cell cultures. *Proc. Natl. Acad. Sci. USA* **92(16),** 7162–7166.
301. Qin, Z. H., Wang, Y., and Chase, T. N. (1996) Stimulation of N-methyl-D-aspartate receptors induces apoptosis in rat brain. *Brain Res.* **725(2),** 166–176.
302. Matter, M. L., Zhang, Z., Nordstedt, C., and Ruoslahti, E. (1998) The alpha5beta1 integrin mediates elimination of amyloid-beta peptide and protects against apoptosis. *J. Cell Biol.* **141(4),** 1019–1030.
303. Staubli, U., Vanderklish, P., and Lynch, G. (1990) An inhibitor of integrin receptors blocks long–term potentiation. *Behav. Neural Biol.* **53(1),** 1–5.
304. Ruoslahti, E. and Reed, J. C. (1994) Anchorage dependence, integrins, and apoptosis. *Cell* **77(4),** 477–478.
305. Yaar, M., Zhai, S., Pilch, P. F., Doyle, S. M., Eisenhauer, P. B., Fine, R. E., et al. (1997) Binding of beta-amyloid to the p75 neurotrophin receptor induces apoptosis. A possible mechanism for Alzheimer's disease. *J. Clin. Invest.* **100(9),** 2333–2340.
306. Yaar, M. and Gilchrest, B. A. (1997) Human melanocytes as a model system for studies of Alzheimer disease. *Arch. Dermatol.* **133(10),** 1287–1291.
307. Yan, S. D., Chen, X., Fu, J., Chen, M., Zhu, H., Roher, A., et al. (1996) RAGE and amyloid-beta peptide neurotoxicity in Alzheimer's disease. *Nature* **382(6593),** 685–691.
308. El Khoury, J., Hickman, S. E., Thomas, C. A., Cao, L., Silverstein, S. C., and Loike, J. D. (1996) Scavenger receptor-mediated adhesion of microglia to beta-amyloid fibrils. *Nature* **382(6593),** 716–719.
309. Oppermann, U. C., Salim, S., Tjernberg, L. O., Terenius, L., and Jornvall, H. (1999) Binding of amyloid beta-peptide to mitochondrial hydroxyacyl-CoA dehydrogenase (ERAB): regulation of an SDR enzyme activity with implications for apoptosis in Alzheimer's disease. *FEBS Lett.* **451(3),** 238–242.
310. Yan, S. D., Fu, J., Soto, C., Chen, X., Zhu, H., Al Mohanna, F., et al. (1997) An intracellular protein that binds amyloid-beta peptide and mediates neurotoxicity in Alzheimer's disease. *Nature* **389(6652),** 689–695.
311. Fraser, S. P., Suh, Y. H., and Djamgoz, M. B. (1997) Ionic effects of the Alzheimer's disease beta–amyloid precursor protein and its metabolic fragments. *Trends Neurosci.* **20(2),** 67–72.
312. Arispe, N., Rojas, E., and Pollard, H. B. (1993) Alzheimer disease amyloid beta protein forms calcium channels in bilayer membranes: blockade by tromethamine and aluminum. *Proc. Natl. Acad. Sci. USA* **90(2),** 567–571.
313. Courtneidge, S. A. (1986) Transformation by polyoma virus middle T antigen. *Cancer Surveys* **5(2),** 173–182.
314. Dunant, N. M., Senften, M., and Ballmer-Hofer, K. (1996) Polyomavirus middle-T antigen associates with the kinase domain of Src-related tyrosine kinases. *J. Virol.* **70(3),** 1323–1330.

315. Goutebroze, L., Dunant, N. M., Ballmer-Hofer, K., and Feunteun, J. (1997) The N terminus of hamster polyomavirus middle T antigen carries a determinant for specific activation of p59c-Fyn. *J. Virol.* **71(2),** 1436–1442.
316. Dunant, N. M., Messerschmitt, A. S., and Ballmer-Hofer, K. (1997) Functional interaction between the SH2 domain of Fyn and tyrosine 324 of hamster polyomavirus middle-T antigen. *J. Virol.* **71(1),** 199–206.
317. Rapoport, M. and Ferreira, A. (2000) PD98059 prevents neurite degeneration induced by fibrillar beta-amyloid in mature hippocampal neurons. *J. Neurochem.* **74,** 125–133.
318. Naslund, J., Haroutunian, V., Mohs, R., Davis, K. L., Davies, P., Greengard, P., and Buxbaum, J. D. (2000) Correlation between elevated levels of amyloid beta-peptide in the brain and cognitive decline. *JAMA* **283,** 1571–1577.
319. Nakagawa, T., Zhu, H., Morishima, N., Li, E., Xu, J., Yankner, B. A., and Yuan, J. (2000) Caspase-12 mediates endoplasmic-reticulum-specific apoptosis and cytotoxicity by amyloid-beta. *Nature* **403,** 98–103.
320. Chen, Q. S., Kagan, B. L., Hirakura, Y., and Xie, C. W. (2000) Impairment of hippocampal long-term potentiation by Alzheimer's amyloid beta-peptides. *J. Neurosci. Res.* **60,** 65–72.
321. McLean, C. A, Cherny, R. A., Fraser, F. W., Fuller, S. J., Smith, M. J., Beyreuther, K., Bush A. I., and Masters, C. L. (1999) Soluble pool of Aβ amyloid as a determinant of severity of neurodegeneration in Alzheimer's disease. *Ann. Neurol.* **46,** 860.
322. Guerette, P. A., Legg, J. T., Cherny, R. A., McLean, C. A., Masters, C. L., Beyreuther, K., and Bush, A. I. (1999) Oligomeric Aβ in PBS-soluble extracts of human Alzheimer brain. *Soc. Neurosci. Abs.* **25,** 2129.
323. Dodart, J. C., Mathis, C., Bales, K. R., Paul, S. M., and Ungerer, A. (2000) Behavioral deficits in APP(V717F) transgenic mice deficient for the apolipoprotein E gene. *Neuroreport* **11,** 603–607.
324. Mucke, L., Masliah, E., Yuo, G. Q., Mallory, M., Rockenstein, E. M., Tatsuno, G., Hu, K., Kholodenko, D., Johnson-Wood K., and McConlogue, L. (2000) High-level neuronal expression of A beta(1-42) in wild-type human amyloid protein precursor transgenic mice: Synaptotoxicity without plaque formation. *J. Neurosci.* **20,** 4050–4058.
325. Walsh, D. M., Hartley, D. M., Kusumoto, Y., Fezoui, Y., Condron, M. M., Lamakin, A., Benedek, G. B., Selkoe, D. J., and Teplow, D. B. (1999) Amyloid beta-protein fibrillogenesis. Structure and biological activity of protofibrillar intermediates. *J. Biol. Chem.* **274,** 2945–2952.
326. Hartley, D. M., Walsh, D. M. Ye, C. P., Diehl, T., Vasquez, S., Vassilev, P. M. Teplow, D. B., and Selkoe, D. J. (1999) Protofibrillar intermediates of amyloid beta-protein induce acute electrophysiological changes and progressive neurotoxicity in cortical neurons. *J. Neurosci.* **19,** 8876–8884.
327. Wang, H. Y., Lee, D. H., D'Andrea, M. R., Peterson, P. A., Shank, R. P., and Reitz, A. B. (2000) beta-Amyloid(1-42) binds to alpha7 nicotinic acetylcholine receptor with high affinity. Implications for Alzheimer's disease pathology. *J Biol. Chem.* **275,** 5626–5632.

328. Schenk, D., Barbour, R., Dunn, W., Gordon, G., Grajeda, H., Guido, T., Hu, K., Huang, J., Johnson-Wood, K., Khan, K., Kholodenko, D., Lee, M., Liao, Z., Lieberburg, I., Motter, R., Mutter, L., Soriano, F., Shopp, G., Vasquez, N., Vandevert, C., Walker, S., Wogulis, M., Yednock, T., Games, D., and Seubert, P. (1999) Immunization with amyloid-beta attenuates Alzheimer-disease-like pathology in the PDAPP mouse. *Nature* **400,** 173–177.
329. Vassar, R., Bennett, B. D., Babu-Khan, S., Kahn, S., Mendiaz, E. A., Denis, P., Teplow, D. B., Ross, S., Amarante, P., Loeloff, R., Luo, Y., Fisher, S., Fuller, J., Edenson, S., Lile, J., Jarosinski, M. A., Biere, A. L., Curran, E., Burgess, T., Louis, J. C., Collins, F., Treanor, J., Rogers, G., and Citron, M. (1999) Beta-secretase cleavage of Alzheimer's amyloid precursor protein by the transmembrane aspartic protease BACE. *Science* **286,** 735–741.
330. Gouras, G. K., Xu, H., Gross, R. S., Greenfield, J. P., Hai, B., Wang, R., and Greengard, P. (2000) Testosterone reduces neuronal secretion of Alzheimer's beta-amyloid peptides. *Proc. Natl. Acad. Sci. USA* **97,** 1202–1205.

2

Transgenic Models of Alzheimer's Disease

Michael C. Sugarman, Steven F. Hinton, and Frank M. LaFerla

2.1. INTRODUCTION

The ability to introduce foreign genes into an animal's genome or to modify or delete existing genes provides a powerful means to study their impact in an intact organism. This technology has been extensively exploited to study the pathogenesis of Alzheimer's disease (AD), a progressive neurodegenerative disease characterized by memory loss and cognitive decline. At the neuropathological level, the AD brain is marked by three principle features: (1) diffuse and neuritic plaques composed primarily of the β-amyloid (Aβ) peptide, (2) intracellular neurofibrillary tangles (NFTs), which consist of hyperphosphorylated tau protein, and (3) neuronal and synaptic loss *(1)*. Thus, to faithfully model AD, an animal model must contain these three principle neuropathological features and the accompanying deficits in memory and cognition. To date, none of the existing models truly fulfills these criteria. Nevertheless, this should not imply that these existing models do not have value, because replicating one or more aspects of the disease provides a valuable experimental system to investigate the underlying pathogenic mechanisms and to evaluate potential therapeutic interventions.

As with virtually any transgenic strategy, approaches to model AD in mice have capitalized on the remarkable advances made in elucidating the genetics of this complex disorder. Therefore, we begin this chapter by presenting a brief overview of the genetics underlying familial AD (FAD). AD can be broadly divided into early-onset or late-onset classifications depending on whether the disease is acquired before or after 60 yr of age. Approximately half of all cases of early-onset autosomal dominant AD can be attributed to missense mutations in three genes encoding transmembrane proteins (Fig. 1): *amyloid precursor protein (APP)* gene on chromosome 21,

From: *Contemporary Clinical Neuroscience: Molecular Mechanisms of Neurodegenerative Diseases*
Edited by: M.-F. Chesselet © Humana Press Inc., Totowa, NJ

presenilin-1 (PS1) gene on chromosome 14, and *presenilin-2 (PS2)* gene on chromosome 1 *(3)*.

2.2. AUTOSOMAL-DOMINANT FAD GENES

The *APP* gene was the first gene conclusively linked to FAD and to date, seven pathological mutations have been described *(4)*. An important characteristic of all the *APP* mutations identified thus far is that they occur within or in close proximity to the Aβ coding region, consistent with the interpretation that the biological effect of these mutations is an alteration of APP processing, leading to increased Aβ production. The first mutation discovered, a glutamate-to-glutamine substitution at codon 693 (using 770 nomenclature, corresponding to residue 22 of the Aβ sequence), ironically was found not in an AD kindred, but in a family with autosomal-dominant hereditary cerebral hemorrhage with amyloidosis, Dutch type *(5,6)*. The first AD-associated mutation in *APP* was found at codon 717 near the γ-secretase site, and resulted in a valine-to-isoleucine substitution *(7,8)*; since this seminal finding, different missense mutations at this same codon have also been documented in other families, resulting in the substitution of valine with either phenylalanine *(9)* or glycine *(10)*. Another missense mutation at codon 715 near the γ-secretase site has recently been described that results in a valine-to-methionine substitution *(11)*. A double missense mutation near the β-secretase site at codons 670 and 671 was identified in two separate Swedish families and results in a lysine-methionine to asparagine-leucine substitution *(12)*. Another mutation within the Aβ sequence has also been described that results in a glycine-to-alanine substitution at residue 21 *(13)*.

Only 2-3% of early–onset FAD cases are attributable to mutations within *APP*. Genetic linkage analysis revealed another locus implicated in early-onset FAD, which culminated in the identification of the *PS1* gene by positional cloning *(14)*. Shortly thereafter, a homologous gene called *PS2* was identified and cloned *(15–17)*. Nearly half of all FAD cases have been associated with mutations in the presenilin genes, mainly in the *PS1* gene, which accounts for up to 50% of early-onset FAD cases *(3)*. All of the more than 75 mutations in the *PS1* gene described to date are missense mutations, including the "Δexon 9" mutation *(18)*. Mutations in *PS2* are more rare than *PS1* and, thus far, only three missense mutations have been identified: asparagine-to-isoleucine at codon 141 in Volga-Germans, methionine-to-valine at codon 239 in Italian families *(15–17)*, and arginine-to-histidine at codon 62, which is found in a sporadic AD case and may represent a polymorphism with or without biological implications *(19)*.

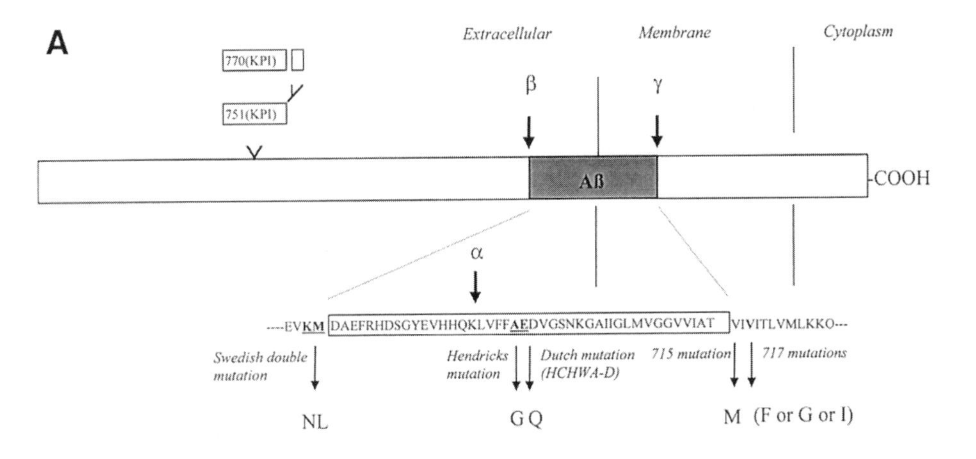

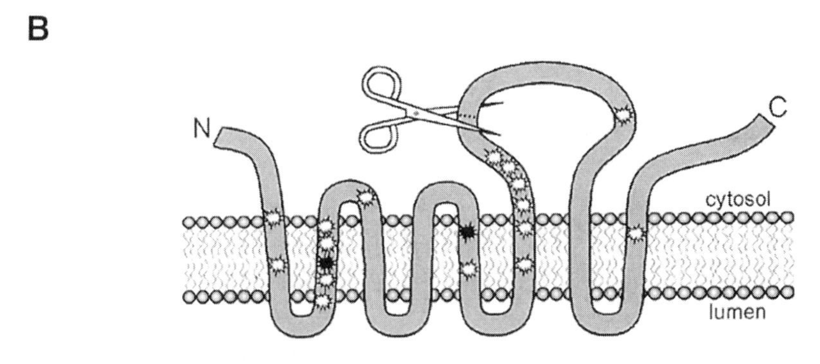

Fig. 1. Schematic diagram of the APP and PS molecules. **(A)** The position of the Aβ sequence within the APP molecule is indicated by the shaded box, with the 43-amino-acid version of the peptide indicated by single-letter amino acid code within the boxed area. The relative position of the α-, β-, γ- secretase cleavage sites are shown. The amino acid substitutions that have been identified in families with *APP* mutations are shown with the arrow pointing to the substituted amino acids. (Adapted from ref. *2*) **(B)** Schematic diagram of the presenilin molecule, showing multiple transmembrane domains. Some FAD *PS1* mutations are indicated in white and *PS2* mutations are indicated in black.

2.3. RISK FACTOR GENES

The characterization of a subset of FAD families that developed late-onset AD led to the linkage of a predisposing gene located on chromosome 19, which was identified as *apolipoprotein E* (*ApoE*) *(20–22)*. Three different allelic forms of the *ApoE* gene, referred to as ε2, ε3, and ε4, are present in the population. In 1993, Allen Roses and colleagues recognized that there was an increased frequency of the ε4 allele among AD patients 65 years of age and older *(21,23)*. In short, possession of the ApoEε4 allele increases the risk of developing AD in a dose-dependent manner, whereas the ε2 allele confers protection from AD *(24)*. Depending on the combination of ApoE alleles inherited, individuals are susceptible to AD to varying degrees. Thus, the presence of each additional ε4 allele leads to an earlier onset of symptoms. In sum, *ApoE* is a genetic risk factor, not a causal agent, because even individuals without an ε4 allele develop AD and vice versa (*see* ref. *25* for review).

The genetics of late-onset AD have not been described as well as the genetics of early-onset FAD. The reason for this may be twofold. First, a significant number of late-onset AD cases may be sporadic (i.e., nongenetic). The second problem is the inherent difficulty associated with studying an aged population. Consequently, if a sizable proportion of late-onset cases are familial, it may be difficult to identify individuals destined to develop late-onset FAD, because they or their kin may die before the disease is manifested. Nevertheless, some candidate genes are emerging, including the identification of a gene on chromosome 12 called α2-macroglobulin, which may be associated with some late-onset FAD cases *(26)*.

2.4. A CENTRAL PATHOLOGIC ROLE FOR Aβ

The Aβ peptide is a heterogenous molecule that can display variability at both the N- and C-termini *(27)*. The significance of the C-terminal heterogenity is particularly notable, as it imparts profound biological consequences of significance to AD pathogenesis. A marked biophysical difference between these two species of Aβ is that the longer form ($A\beta_{42(43)}$) tends to be more amyloidogenic, forming fibrils in vitro more readily than the shorter form ($A\beta_{40}$) *(28)*. In vivo analyses also confirm the more pathogenic nature of the longer form, as immunohistochemical studies of human AD and brains of individuals with Down's Syndrome indicate that the earliest form of Aβ deposits consist of $A\beta_{42(43)}$. Thus far, a consistent theme to emerge is that all mutations within the *APP*, *PS1* and *PS2* genes linked to early-onset FAD eventually lead to enhanced production of either total Aβ levels or the more insoluble $A\beta_{42/43}$ *(29–37)*. The effects of the FAD-linked genes on Aβ has been observed in a variety of experimental systems including cell culture (both transfected cells overexpressing mutant proteins and fibroblasts from FAD patients) and transgenic mice (to be discussed later) and has provided strong evidence for the critical role of Aβ in AD neurodegeneration.

2.5. Aβ TRANSGENIC MICE

Although in vitro experiments had indicated that Aβ was deleterious to a variety of cell types grown in culture *(38)*, the in vivo toxicity of Aβ was less certain and somewhat controversial because direct injection of the peptide into the CNS of various animals yielded conflicting results (*see*, for example, ref. *39*). Therefore, we felt that a transgenic approach might be the most appropriate way to test the neurotoxicity of Aβ in an in vivo context. Rather than express the entire *APP* sequence, we opted to express only the Aβ peptide *(40)*. Moreover, in designing our transgenic model, we also wanted to determine whether Aβ toxicity was mediated as a result of intracellular accumulation or as part of its accumulation in the extracellular milieu. We hypothesized that toxicity might be concentration dependent (i.e., for efficient nucleation of Aβ to occur), and thus we rationalized that the Aβ concentration might reach higher levels inside a contained environment, such as inside the cell. Consequently, two sets of transgenic mice were developed that expressed a cDNA sequence encoding only the mouse Aβ peptide; the only difference between the two sets was that the "extracellular" mice contained the neural cell adhesion molecule (NCAM) signal sequence incorporated into the transgene to allow Aβ to be targeted extracellularly.

The neurologic phenotype observed in these mice has been extensively described *(40,41)*. In short, a surprising observation emerged from the study of these Aβ transgenic mice: Only the mice expressing Aβ intracellularly developed pathology. The pathophenotype that developed in the intracellular Aβ transgenic mice consisted of seizures, astrogliosis, neuronal cell death, and extracellular amyloid deposition. Perhaps even more surprising was the lack of neuropathology in the transgenic mice in which the NCAM signal sequence was incorporated; this was true despite the fact that these mice expressed the NCAM–Aβ transgene. These findings indicated that accumulation of Aβ intracellularly can have deleterious neurotoxic effects and highlight a previously underappreciated role of intracellular Aβ in the pathogenesis of AD.

Expressing Aβ intraneuronally initiated a cascade of pathological events that occurred in an age-dependent and region-specific fashion in the "intracellular" Aβ transgenic mice. The earliest phenotypic changes that we observed were changes in neuronal morphology that included neuritic degeneration and cytoplasmic vacuolization. A large percentage (approx 75%) of the transgenic mice also developed profound astrogliosis by 6 mo of age, providing confirmatory evidence of underlying central nervous system (CNS) injury evoked as a result of intracellular Aβ expression. At later time-points, other evidence of neuronal damage were apparent, including fragmentation of nuclear DNA following TUNEL. Although TUNEL staining, by itself does not discriminate between cells dying by necrosis versus apoptosis *(42)*, we combined terminal deoxynucleotide transferase-mediated dUTP-biotin nick end labeling (TUNEL) with ultrastructural analysis which revealed neurons that were indeed morphologically altered in a manner consistent with cells undergoing apoptosis. Therefore, intracellular Aβ expression triggers neuronal apoptosis. This finding is consistent with apoptotic-inducing properties of Aβ administered in vitro *(43)* and with recent demonstrations that one of the pathologic mechanisms by which certain genes linked to FAD (e.g., presenilins) may induce disease is through enhancement of neuronal apoptosis (*see*, ref. *44* for review).

In addition to the primary neuronal injury, inflammatory or reactive processes were also apparent in the Aβ transgenic mice. Besides the prominent astrogliosis in the transgenic brains, reactive microglia were also present following histochemical staining with the lectin, RCA. Notably, as with the neuronal degeneration, both the astrogliosis and microgliosis occurred in brain regions such as the neocortex and hippocampus, which are major sites of AD pathology. Thus, expression of Aβ intraneuronally is able to trigger many reactive cellular processes within the brain. Finally, the occurrence of these reactive processes in the transgenic brains is relevant because they also represent a significant aspect of AD pathology *(45)*.

The profound extent of neuronal cell loss that occurred in the intracellular Aβ transgenic model is one feature that distinguishes it from other transgenic models. In addition, evidence from light and electron microscopy indicated that the cell death was consistent with an apoptotic pathway. More specifically, we found that neurons were dying by an apoptotic pathway that required p53 expression *(41)*. p53 is a pleiotropic molecule that plays an important role in inducing cellular apoptosis *(46)*. There are now several reports that indicate that p53 expression is elevated in AD brains *(47–49)*; thus, it is quite likely that, as was the case for the Aβ transgenic mice, p53 may play a comparable role in mediating cell death in the AD brain.

Although the occurrence of extracellular Aβ deposits was not as robust in our intracellular Aβ transgenic mice as in some of the mutant APP mice (*see* Subheading 2.6), an important hypothesis regarding the genesis of Aβ plaques has emerged nevertheless. Specifically, we proposed that extracellular deposition of Aβ occurs following neuronal cell death. In part, this hypothesis is supported by our observations that extracellular Aβ deposition occurred in areas in which there was prominent TUNEL labeling in the transgenic brains, indicating that cell death precedes extracellular amyloid deposition in this transgenic model *(41)*.

The finding that expression of intracellular Aβ leads to neuronal cell death via apoptosis in the transgenic brains prompted us to investigate human AD brain postmortem samples to determine if neuronal loss also occurred by apoptosis. We found that indeed there was evidence for neuronal cell death occurring by apoptosis *(50)*, which agrees with numerous other reports *(51,52)*. Curiously, the number of TUNEL-positive cells in the AD brain is substantially higher than one might have predicted *a priori*, given the rapid demise the apoptotic cell undergoes in vitro. The reasons for this are unclear, but many of the cells containing fragmented DNA also paradoxically express bcl-2 *(53)*, which is counterintuitive given that bcl-2 normally represses cell death. Thus, it is plausible that despite the extent of DNA damage, mechanisms exist to retard the loss of postmitotic cells to perhaps allow repair so that the cell can survive.

Histological analysis in combination with the TUNEL procedure allowed us to identify neurons with different degrees of DNA fragmentation, which we interpret to reflect different stages in the cell death pathway. More importantly, from this analysis, we observed that degenerating neurons exhibiting DNA damage also contained *intracellular* Aβ accumulation, despite the absence of any extracellular Aβ. Furthermore, these same cells also contained elevated expression of apoE, which likely stabilizes the hydrophobic Aβ. Therefore, we concluded that, as in the transgenic brains, intracellular accumulation represents a key feature in the pathogenesis of AD that precedes the extracellular accumulation of amyloid plaques *(50)*.

In sum, although expressing Aβ intracellularly in transgenic mice represented an unorthodox approach, these mice did develop some pathological changes that are consistent with those observed in the AD brain. Since then, the suggestion that intracellular Aβ may play an important role in the disease process has been growing. Aβ has traditionally been regarded as manifesting its neurotoxic effects from outside the cell. In part, this is because amyloid plaques are localized extracellularly in the AD brain and because administration of Aβ to cultured cells is known to be toxic *(38)*. However, it is uncertain whether the results of the in vitro studies accurately mimic the means by which Aβ is pathogenic in AD or whether it is the exclusive mechanism by which Aβ can be neurotoxic.

Evidence supporting a role for intracellular Aβ in the pathogenesis of AD includes in vitro work with synthetic Aβ peptide *(54,55)* and with mutant genes linked to FAD *(56,57)*, and findings from transgenic mice and AD brains *(40,41,50,58)*. It is our hypothesis that perturbations of the normal cellular processing of APP or selective reuptake of Aβ would cause the peptide to accumulate intracellularly. As nucleation is concentration dependent, intracellular Aβ accumulation would facilitate the peptide's formation of neurotoxic aggregates, ultimately causing cell death. The Aβ released following cell death would form an extracellular nidus for neuritic plaque formation, leading to secondary cellular damage by glial activation or other inflammatory responses. Despite evidence supporting this hypothesis, the role of intracellular Aβ in AD remains controversial. However, another age-related disorder called inclusion body myositis (IBM), is characterized by skeletal muscle fiber degeneration associated with accumulation of intracellular Aβ *(59)*.

2.6. *APP* TRANSGENIC AND KNOCKOUT MICE

A frequent approach used to elucidate the function of a protein in an organism is to knock out the encoding gene. Genetically modified mice have been generated that contain a functionally inactive *APP* gene *(60,60a)*; these mice exhibited behavioral deficits, showing slight decreases in motor activity and forelimb grip strength when compared to age-matched controls, which indicates that *APP* is necessary for optimal cellular functioning. Marked reactive gliosis was also observed in the brains of young mice, consistent with absence of *APP* leading to altered neural functioning and the activation of secondary processes. In addition, the *APP* null mice exhibited deficits in spatial learning, impaired long-term potentiation and a reduction in synaptic density, suggesting that *APP* may be necessary for maintenance of synaptic function during aging *(61)*.

Neuroanatomical studies of the brain did not reveal significant differences in the knock out mice as compared to the wild-type controls. Consequently, either *APP* is not essential for mouse embryonic and early neuronal development or the absence of *APP* was compensated by the highly homologous *APP*-like proteins 1 and 2 (*APLP1*, *APLP2*) *(62,63)*. Genetically modified mice have been derived that lack both *APP* and *APLP2 (64)*. The absence of both genes results in early lethality as 80% of double knock out mice die within the first week after birth. Those mice that survive beyond this time-point are reduced in body weight and show several postural and motor abnormalities. Thus, it appears that *APLP2* and *APP* are required for early postnatal development and that *APLP2* and *APP* can compensate for each other functionally. In sum, the results of these single and double knockout mice show that loss of *APP* function does not lead to AD neuropathology in mice, which indicates that *APP* mutations lead to AD by a mechanism other than loss of *APP* function.

To address the pathophysiological role of *APP* in transgenic mice, generally one of two experimental strategies have been utilized: either expression of the carboxyl fragment of *APP* termed C100 (which consists of 100–104 amino acids from the C' terminus of *APP* including Aβ) or expression of the full-length *APP* molecule, encoded by either genomic or cDNA sequences. Earlier approaches focused on overexpressing wild-type *APP* (APP_{WT}), which resulted in some pathological or behavioral alterations *(65–68)*. Following the identification of *APP* mutations, the consequences of their expression could be measured in the CNS of transgenic mice, particularly in regard to whether they were sufficient to induce AD-like pathology. The most dramatic successes have centered on transgenic models harboring APP molecules that either contained missense mutations near the γ-secretase site *(69)* or the β-secretase site *(70)*. Since then, numerous other laboratories have used comparable approaches. In short, although neither model fully mimics the complete spectrum of AD pathology, the robust Aβ deposits observed in the CNS of these transgenic mice provided confirmatory evidence that mutations within the *APP* gene are, in fact, pathological mutations responsible for some cases of early-onset FAD.

An in depth analysis of all the *APP* transgenic models is beyond the scope of this chapter, therefore we focus on two of the more prominent lines of transgenic mice that developed certain key pathological features, notably extensive extracellular Aβ deposits. Although the transcriptional promoter and *APP* mutant differed in each construct, there were, nevertheless, several histopathological commonalties observed in the brains of both transgenic lines. One common feature shared by the PD-APP *(69)* and prion protein (PrP)–APP_{SW} (PrP-APP_{SW}) *(70)* mice was the age-related increase in Aβ production in the brain. Relative levels of $Aβ_{40}$ and $Aβ_{42(43)}$ were measured using enzyme-linked immunosorbent assay (ELISA). Analysis of Aβ concentrations within the hippocampus of PD–APP mice revealed a 17-fold increase between the ages of 4 and 8 mo and a 500-fold increase from 4 to 18 mo of age *(71)*. Likewise, ELISA analysis of *PrP–APP_{SW}* brain showed 5 and 14 times the amount of $Aβ_{40}$ and $Aβ_{42(43)}$, respectively, in 11- to 13-mo- old transgenic mice when compared to 2- to 5-mo- old transgenic mice. This dramatic age-dependent increase in overall Aβ levels as well as in the $Aβ_{42(43)/40}$ ratio clearly mimics an important characteristic of the human disorder.

Despite the use of different neuronal promoters, both groups observed extracellular deposition of Aβ that was distributed in a region-specific fashion in the brain. This is true, for example, despite the use of the platelet-derived growth factor-beta (PDGF) promoter, which results in widespread neuronal expression throughout the CNS *(72)*. Aβ deposits in the *PD–APP* brain were localized within the hippocampus, corpus callosum, and cerebral cortex but absent in other brain structures, thus paralleling the regional distribution that occurs in human AD brains. In *PrP–APP*$_{SW}$ mice, Aβ deposits were found in frontal, temporal, and entorhinal cortex, hippocampus, presubiculum, subiculum, and cerebellum. The regional specificity of Aβ accumulation is intriguing because the promoters used in the transgene constructs expressed APP globally in neurons throughout the CNS, suggesting that region-specific factors may influence the accumulation and/or clearance of Aβ.

In addition to widespread Aβ deposition, other pathological markers characteristic of the AD brain such as reactive gliosis and synaptic loss were observed in both mice. Histopathologically, in the *PrP–APP*$_{SW}$ brains, amyloid cores were surrounded by glial fibriallary acidic protein (GFAP) immunoreactive astrocytes and dystrophic neurites. In *PD–APP* brain, the majority of amyloid plaques were also surrounded by extensive GFAP immunoreactive astrocytes and usually associated with dystrophic neurites from surrounding cells *(73)*. In addition, in *PD–APP* transgenic brains, the synaptic density in the molecular layer of the hippocampal dendate gyrus was markedly reduced following visualization with synaptophysin, a presynaptic marker, and MAP2, a dendritic marker *(69)*. This loss of synaptic density within the hippocampus, a structure important for learning and memory, is a major histopathological feature of the AD brain *(74)* and may be a causal factor in the memory loss associated with the disease.

Because an important clinical manifestation of AD is the loss of short-term memory, a comprehensive transgenic model would not only have to contain the major hallmarks of AD pathology (i.e., plaques and tangles) but would also have to exhibit the accompanying behavioral deficits. Several of the *APP* transgenic models do show deficits in certain behavioral tests designed to evaluate learning and memory performance, such as spatial reference and alternation tasks *(67,70,75,76)*. In addition, these transgenic mice also exhibit changes in synaptic plasticity such as induction of long-term potentiation (LTP) *(61,76,77)*.

In sum, transgenic mice expressing mutant *APP* at high levels were successful at producing amyloid deposits distributed in a region-specific manner similar to that observed in the AD brain, along with varying degrees of learning and memory deficits. One of the key features that distinguishes the APP models mentioned earlier from the other models is the high level of expression that is required for plaque formation (*see*, for example, ref. *78*). *PD–APP* transgenic mice expressed human APP to over 10 times the level of endogenous mouse APP at the protein level. The *PrP–APP*$_{SW}$ mice expressed the *APP* transgene mRNA in brain over fivefold the endogenous

APP levels in 14-mo-old transgenic mice. Notably, despite this high level of transgene APP expression and resultant prominent Aβ deposition, there is a relative paucity of neuronal cell death. The implications for the relationship of the genesis of Aβ plaques and neuronal cell death are not clear.

2.7. PRESENILIN TRANSGENIC AND KNOCKOUT MICE

Given the earlier age of disease onset in *PS1* versus *APP* FAD pedigrees (30–50 yr versus 50–60 yr) and the rapid clinical demise that occurs in *PS1* families, one might likely predict that overexpression of *PS1* mutations in transgenic mice would induce more severe pathology than *APP* transgenic mice. Surprisingly, these mice did not develop amyloid plaques. There is no obvious theoretical explanation to account for this observation and it may simply be a reflection of not achieving adequate levels of *PS1* overexpression in the appropriate cell types.

Several groups characterized transgenic mice overexpressing human mutant PS1 molecules *(29,32,33)*. Although the promoter, *PS1* mutation, and background strain of mice differed, two consistent findings emerged. First, overexpression of mutant human *PS1* in transgenic mice results in increased levels of $A\beta_{42(43)}$. Although overexpression of human wild-type PS1 increases $A\beta_{42(43)}$ levels in transgenic mice, these levels are dramatically increased in mice harboring *PS1* missense mutations. Not all *PS1* mutations elevate $A\beta_{42(43)}$ to comparable levels; for example, transgenic mice harboring the methionine-146-leucine (M146V) mutation had a greater increase in $A\beta_{42(43)}$ than did mice overexpressing the leucine-286-valine mutation *(32)*. This disparity may indicate that certain regions of the PS1 protein are more critical, at least with regards to modulating APP processing.

Both the PS1 and PS2 proteins are subject to endoproteolytic processing in vivo; the net effect is that it can be difficult to detect the holoproteins in vivo and that the major detectable species in brain are an approx 27-kDa N-terminal and approx 19-kDa C-terminal fragments *(79)*. The second finding to emerge was that PS1 and PS2 appear to compete for common proteolytic factor(s), as it was observed that saturable levels of the N- and C-terminal fragments accumulate at approximately 1 : 1 stoichiometry in transgenic mice, an effect independent of transgene-derived mRNA levels. These studies reveal that compromised accumulation of murine *PS1* derivatives resulting from overexpression of human *PS1* occurs in a manner independent of endoproteolytic cleavage, consistent with a model in which the abundance of PS1 fragments is regulated coordinately by competition for limiting cellular factors *(79,80)*.

The pathogenic means by which mutations in the presenilin genes lead to AD may involve three mechanisms. One likely mechanism clearly involves elevation of $A\beta_{42(43)}$ levels, as the transgenic models described earlier clearly illustrate; it is still not established whether this effect is primary or whether it lies downstream of other molecular processes. The second mechanism may involve enhanced sensitivity to apoptosis (*see* ref. *44* for review). Recently, *PS1* mutant knock-in mice have been developed that express the human PS1 M146V mutation at normal physiological levels *(81)*; primary hippocampal neurons from these *PS1* mutant knock-in mice exhibit increased vulnerability to Aβ toxicity. The third mechanism may involve disruption of calcium homeostasis *(81,82)*. For example, primary cells from the *PS1* mutant knock-in mice contain elevated calcium stores in the endoplasmic reticulum and deficits in capacitative calcium entry *(82a)*. Thus, although Aβ may be the most obvious readout of mutations in the presenilin genes, it may not necessarily be the primary effect. Recent transgenic data support this hypothesis. Chui et al. *(58)* developed mutant PS1 transgenic mice and found that neurodegeneration was significantly accelerated in mice older than 13 mo, without amyloid plaque formation. However, they reported significantly more neurons containing intracellularly deposited $A\beta_{42}$ in aged mutant transgenic mice, indicating that the pathogenic role of the *PS1* mutation is upstream of the amyloid cascade *(58)*.

To elucidate the physiological role of the PS1 molecule in vivo during development, *PS1*-deficient mice were created by effectively disrupting the murine gene in mouse embryos *(83,84)*. Knocking out the *PS1* gene at this early stage of development has a lethal effect; null mutants have abnormal skeletal deformities and hemorrhages in the central nervous system. These physical characteristics are similar to mice with inactivated Notch 1 *(85,86)*, which is not surprising given the homology between the presenilins and sel-12 *(87)*. These findings may indicate that PS1 expression is required for neuronal survival and normal neurogenesis *(83)*. The PS1 null mice can be "rescued" by being crossed to transgenic mice harboring either wild-type or mutant *PS1 (88,89)*, demonstrating that the PS1 mutation does not lead to a total loss of function during development.

Although overexpression of mutant *PS1* in transgenic mice increases $A\beta_{42(43)}$ levels, Aβ levels are decreased fivefold in the *PS1* knockout embryos *(90)*. The turnover of the membrane-associated fragments of APP was specifically decreased in the null mice. Therefore, it appears that PS1 mediates a proteolytic activity that cleaves the integral membrane domain of APP; simply stated, either PS1 may modulate γ-secretase cleavage of APP or it may even be the γ-secretase molecule *(91)*.

2.8. *APP* AND *PS1* DOUBLY TRANSGENIC MICE

The single transgenic models of AD, defined as carrying one FAD-linked mutation, demonstrate either a limited pathology, as in the case of mice overexpressing *PS1* mutations or a late onset of pathology, as in the case of mice overexpressing *APP* mutations. To create an animal model that develops more severe AD-like neuropathology, including perhaps an earlier age of onset, several laboratories have focused their attention on developing transgenic mice that harbor two or more FAD-linked mutations *(32,33,92)*. Consequently, transgenic mice have been generated that carry a mutant *APP* gene and a mutant *PS1* gene.

Alzheimer's disease-like pathology was accelerated in the double transgenic mice that carried either the A246E or M146L mutation in the *PS1* gene and the APP_{SW} gene *(32,33,92)*. Amyloid deposits were abundant at 6 mo of age and distributed in a region-specific manner in the cerebral cortex and hippocampus. By contrast, similar pathology was not evident in the single APP_{SW} transgenic mice until 9–12 mo of age *(70)*. As mentioned earlier, amyloid deposition was not present in the single *PS1* mutant transgenic mice (see, for example ref *58*). Taken together, these observations suggest that the mutant PS1 acts synergistically with APP_{SW} to accelerate APP processing and amyloid deposition. Because these mice develop plaques at a relatively early age, they may prove to be a more efficient model system to evaluate the usefulness of potential AD therapeutics and, moreover, certainly indicate that overexpression of more than one FAD-linked gene may be essential to develop an animal model that contains all of the hallmark pathological features of AD.

2.9. *ApoE* TRANSGENIC AND KNOCKOUT MICE

ApoE has been identified as a major risk factor that modifies the age of onset for AD *(93)*. Although apoE can bind to and stabilize Aβ *(23)*, its precise physiological role in the CNS or in the pathogenesis of AD remains to be established. To address its role during development, genetically modified mice have been derived in which the gene was effectively knocked out *(94,95)*. In short, no obvious phenotypic alterations were evident in *ApoE* null mice, which appeared to be relatively healthy when compared to wild-type controls; thus, *ApoE* is not essential for development. The *ApoE*-deficient mice, however, had significantly higher levels of serum cholesterol than age-matched controls receiving the same diet, consistent with a known role for apoE in the transport of cholesterol *(96)*.

To elucidate the role of human apoE in brain, the three different protein isoforms (ε2, ε3, or ε4) were individually overexpressed in transgenic mice on a null murine *ApoE* background *(97–99)*. This approach allowed for the characterization of the effects of human ApoE in mice without the confounding influence of the endogenous *ApoE* gene. Several approaches to express human *ApoE* in null mice have used neuronal specific promoters *(97,99,100)*. *ApoE* is normally expressed to relatively high levels in glial cells, although recent evidence for expression in neurons has also been provided *(101)*. Another approach utilized the GFAP promoter to direct expression to astrocytes, because ApoE in the CNS is primarily found in astrocytes. Immunohistochemical analysis of these transgenic mice at 14 mo of age failed to show any evidence of amyloid deposition or increase in Aβ levels.

To study the effects of human apoE isoforms on Aβ depostion in transgenic mice, *ApoE* transgenic mice (on a mouse null ApoE background) were crossed with *APP* mice containing the valine-717-phenylalanine mutation *(102)*. Aβ deposition was significantly less in ApoE 3 and 4 mice crossed with mutant *APP* mice compared to mutant *APP* mice alone. These findings are somewhat counterintuitive given the strong association between Aβ deposition and apoE isoform *(23)*. If true, these results implicate a potential role for apoE 3 and 4 in increasing clearance and/or decreasing aggregation of Aβ.

2.10. CONCLUDING REMARKS

The goal of these transgenic endeavors is to create an animal model which faithfully mimics the major histopathological and behavioral features of AD. It is expected that such an animal model would be an invaluable tool in the development of treatments to prevent or halt the progression of disease. Currently, none of the transgenic models expressing a single *FAD* gene meets this criterion, but the development of transgenic models that express more than one AD-associated gene may be the key to overcoming this inadequacy. Nevertheless, the single transgenic models have provided novel insights into the pathogenesis of AD. For instance, one consistent theme that has emerged from the genetic studies is that mutations in all of the genes linked to autosomal-dominant AD affect production or accumulation of Aβ. Certainly, transgenic models have been key in providing some of this supporting evidence, which clearly underscores the important pathological role of Aβ in the genesis of AD.

The AD transgenic mice are starting to pave the way for potential and novel therapeutic approaches toward the treatment of this insidious neurodegenerative disorder. One of the most promising of these therapies involves vaccination of transgenic mice with Aβ *(103)*. Schenk et al. showed that immunization of the PD-APP transgenic mice $A\beta_{42}$ at an early age (prior to onset of AD pathology/plaque formation) essentially prevented the development of Aβ deposition and other neuropathological changes such as neuritic dystrophy and astrogliosis. Likewise, immunization of older mice with well-established neuropathologies also was efficacious in reducing the extent and progression of the pathology. Whether this treatment will be effective (or even safe) in human patients awaits results from clinical trials.

Unexpected advances toward the development of a comprehensive transgenic model of AD occasionally emerge from previously unpredictable avenues of research. One recently described and very exciting model was reported by Capsoni et al. *(104)*. These authors created transgenic mice in which the CMV promoter was used to overexpress a neutralizing antibody directed against nerve growth factor (NGF). Levels of free NGF in the brains of transgenic mice were 53% less than in control mice. Intriguingly, the aged 15- to 17-mo old anti-NGF transgenic mice exhibited AD-type phenotypic changes including β-amyloid plaques, neurofibrillary tangles, tau hyperphosphorylation, neuronal death, and selective behavioral deficits. The mechanism by which neutralization of NGF in the brain leads to the hallmark features of AD neuropathology remains to be determined. Regardless of the mechanism, the bottom line is that this model currently represents the most comprehensive model of AD.

ACKNOWLEDGMENTS

The authors thank Dr. Malcolm Leissring for critical reading of the manuscript. This work has been supported by grants from the NIA (AG15409) and the State of California (98-15717).

REFERENCES

1. Yankner, B. A. (1996) Mechanisms of neuronal degeneration in Alzheimer's disease. *Neuron* **16,** 921–932.
2. Price, D. L. and Sisosida, S. S. (1998) Mutant genes in familial Alzheimer's disease and transgenic models. *Annu. Rev. Neurosci.* **21,** 479–505.
3. Cruts, M., van Duijn, C. M., Backhovens, H., Van den Broeck, M., Wehnert, A., Serneels, S., et al. (1998) Estimation of the genetic contribution of presenilin-1 and -2 mutations in a population-based study of presenile Alzheimer disease. *Hum. Mol. Genet.* **7,** 43–51.
4. Goate, A. M. (1998) Monogenetic determinants of Alzheimer's disease: APP mutations. *Cell Mol. Life Sci.* **54,** 897–901.
5. Levy, E., Carman, M. D., Fernandez-Madrid, I. J., Power, M. D., Lieberburg, I., van Duinen, S. G., et al. (1990) Mutation of the Alzheimer's disease amyloid gene in hereditary cerebral hemorrhage, Dutch type. *Science* **248,** 1124–1126.
6. Van Broeckhoven, C., Haan, J., Bakker, E., Hardy, J. A., Van Hul, W., Wehnert, A., et al. (1990) Amyloid β protein precursor gene and hereditary cerebral hemorrhage with amyloidosis (Dutch). *Science* **248,** 1120–1122.

7. Goate, A., Chartier-Harlin, M. C., Mullan, M., Brown, J., Crawford, F., Fidani, L., et al. (1991) Segregation of a missense mutation in the amyloid precursor protein gene with familial Alzheimer's disease. *Nature* **349,** 704–706.
8. Naruse, S., Igarashi, S., Kobayashi, H., Aoki, K., Inuzuka, T., Kaneko, K., et al. (1991) Mis-sense mutation Val—Ile in exon 17 of amyloid precursor protein gene in Japanese familial Alzheimer's disease. *Lancet* **337,** 978–979.
9. Murrell, J., Farlow, M., Ghetti, B., and Benson, M. D. (1991) A mutation in the amyloid precursor protein associated with hereditary Alzheimer's disease. *Science* **254,** 97–99.
10. Chartier-Harlin, M. C., Crawford, F., Houlden, H., Warren, A., Hughes, D., Fidani, L., et al. (1991) Early-onset Alzheimer's disease caused by mutations at codon 717 of the β-amyloid precursor protein gene. *Nature* **353,** 844–846.
11. Ancolio, K., Dumanchin, C., Barelli, H., Warter, J. M., Brice, A., Campion, D., et al. (1999) Unusual phenotypic alteration of β amyloid precursor protein (βAPP) maturation by a new val-715 —> met βAPP-770 mutation responsible for probable early-onset Alzheimer's disease. *Proc. Natl. Acad. Sci. USA* **96,** 4119–4124.
12. Mullan, M., Houlden, H., Windelspecht, M., Fidani, L., Lombardi, C., Diaz, P., et al. (1992) A locus for familial early-onset Alzheimer's disease on the long arm of chromosome 14, proximal to the alpha 1-antichymotrypsin gene. *Nat. Genet.* **2,** 340–342.
13. Hendriks, L., van Duijn, C. M., Cras, P., Cruts, M., Van Hul, W., van Harskamp, et al. (1992) Presenile dementia and cerebral haemorrhage linked to a mutation at codon 692 of the β-amyloid precursor protein gene. *Nat. Genet.* **1,** 218–221.
14. Sherrington, R., Rogaev, E. I., Liang, Y., Rogaeva, E. A., Levesque, G., Ikeda, M., et al. (1995) Cloning of a gene bearing missense mutations in early-onset familial Alzheimer's disease. *Nature* **375,** 754–760.
15. Levy-Lahad, E., Wasco, W., Poorkaj, P., Romano, D. M., Oshima, J., Pettingell, W. H., et al. (1995) Candidate gene for the chromosome 1 familial Alzheimer's disease locus. *Science* **269,** 973–977.
16. Li, J., Ma, J., and Potter, H. (1995) Identification and expression analysis of a potential familial Alzheimer disease gene on chromosome 1 related to AD3. *Proc. Natl. Acad. Sci. USA* **92,** 12,180–12,184.
17. Rogaev, E. I., Sherrington, R., Rogaeva, E. A., Levesque, G., Ikeda, M., Liang, Y., et al. (1995) Familial Alzheimer's disease in kindreds with missense mutations in a gene on chromosome 1 related to the Alzheimer's disease type 3 gene. *Nature* **376,** 775–778.
18. Steiner, H., Romig, H., Grim, M. G., Philipp, U., Pesold, B., Citron, M., et al. (1999) The biological and pathological function of the presenilin-1 Deltaexon 9 mutation is independent of its defect to undergo proteolytic processing. *J. Biol. Chem.* **274,** 7615–7618.
19. Lao, J. I., Beyer, K., Fernandez-Novoa, L., and Cacabelos, R. (1998) A novel mutation in the predicted TM2 domain of the presenilin 2 gene in a Spanish patient with late-onset Alzheimer's disease. *Neurogenetics* **1,** 293–296.

20. Corder, E. H., Saunders, A. M., Strittmatter, W. J., Schmechel, D. E., Gaskell, P. C., Small, G. W., et al. (1993) Gene dose of apolipoprotein E type 4 allele and the risk of Alzheimer's disease in late onset families. *Science* **261,** 921–923.
21. Saunders, A. M., and Roses, A. D. (1993) Apolipoprotein E4 allele frequency, ischemic cerebrovascular disease, and Alzheimer's disease. *Stroke* **24,** 1416–1417.
22. Strittmatter, W. J., Saunders, A. M., Schmechel, D., Pericak-Vance, M., Enghild, J., Salvesen, G. S., et al. (1993) Apolipoprotein E: high-avidity binding to β-amyloid and increased frequency of type 4 allele in late-onset familial Alzheimer disease. *Proc. Natl. Acad. Sci. USA* **90,** 1977–1981.
23. Strittmatter, W. J., Weisgraber, K. H., Huang, D. Y., Dong, L. M., Salvesen, G. S., Pericak-Vance, M., et al. (1993) Binding of human apolipoprotein E to synthetic amyloid β peptide: isoform-specific effects and implications for late-onset Alzheimer disease. *Proc. Natl. Acad. Sci. USA* **90,** 8098–8102.
24. Corder, E. H., Saunders, A. M., Risch, N. J., Strittmatter, W. J., Schmechel, D. E., Gaskell, P. C., Jr., et al. (1994) Protective effect of apolipoprotein E type 2 allele for late onset Alzheimer disease. *Nat. Genet.* **7,** 180–184.
25. Roses, A. D. (1996) Apolipoprotein E alleles as risk factors in Alzheimer's disease. *Annu. Rev. Med.* **47,** 387–400.
26. Blacker, D., Wilcox, M. A., Laird, N. M., Rodes, L., Horvath, S. M., Go, R. C., et al. (1998) Alpha-2 macroglobulin is genetically associated with Alzheimer disease. *Nat. Genet.* **19,** 357–360.
27. Selkoe, D. J. (1998) The cell biology of β-amyloid precursor protein and presenilin in Alzheimer's disease. *Trends Cell Biol.* **8,** 447–453.
28. Jarrett, J. T., Berger, E. P., and Lansbury, P. T., Jr. (1993. The carboxy terminus of the ß amyloid protein is critical for the seeding of amyloid formation: implications for the pathogenesis of Alzheimer's disease. *Biochemistry* **32,** 4693–4697.
29. Borchelt, D. R., Thinakaran, G., Eckman, C. B., Lee, M. K., Davenport, F., Ratovitsky, T., et al. (1996) Familial Alzheimer's disease-linked presenilin 1 variants elevate Aß1-42/1-40 ratio *in vitro* and *in vivo*. *Neuron* **17,** 1005–1013.
30. Cai, X. D., Golde, T. E., and Younkin, S. G. (1993) Release of excess amyloid β protein from a mutant amyloid β protein precursor. *Science* **259,** 514–516.
31. Citron, M., Oltersdorf, T., Haass, C., McConlogue, L., Hung, A. Y., Seubert, P., et al. (1992) Mutation of the ß-amyloid precursor protein in familial Alzheimer's disease increases β-protein production. *Nature* **360,** 672–674.
32. Citron, M., Westaway, D., Xia, W., Carlson, G., Diehl, T., Levesque, G., et al. (1997) Mutant presenilins of Alzheimer's disease increase production of 42- residue amyloid β-protein in both transfected cells and transgenic mice. *Nat. Med.* **3,** 67–72.
33. Duff, K., Eckman, C., Zehr, C., Yu, X., Prada, C. M., Perez-tur, J., et al. (1996) Increased amyloid-β42(43) in brains of mice expressing mutant presenilin 1. *Nature* **383,** 710–713.
34. Mehta, N. D., Refolo, L. M., Eckman, C., Sanders, S., Yager, D., Perez-Tur, J., et al. (1998) Increased Aβ42(43) from cell lines expressing presenilin 1 mutations. *Ann. Neurol.* **43,** 256–258.

35. Scheuner, D., Eckman, C., Jensen, M., Song, X., Citron, M., Suzuki, N., et al. (1996) Secreted amyloid β-protein similar to that in the senile plaques of Alzheimer's disease is increased *in vivo* by the presenilin 1 and 2 and APP mutations linked to familial Alzheimer's disease. *Nat. Med.* **2,** 864–870.
36. Suzuki, N., Cheung, T. T., Cai, X. D., Odaka, A., Otvos, L., Jr., Eckman, C., et al. (1994) An increased percentage of long amyloid β protein secreted by familial amyloid β protein precursor (βAPP717) mutants. *Science* **264,** 1336–1340.
37. Xia, W., Zhang, J., Kholodenko, D., Citron, M., Podlisny, M. B., Teplow, D. B., et al. (1997) Enhanced production and oligomerization of the 42-residue amyloid β-protein by Chinese hamster ovary cells stably expressing mutant presenilins. *J. Biol. Chem.* **272,** 7977-7982.
38. Cotman, C. W., Pike, C. J., and Copani, A. (1992) β-Amyloid neurotoxicity: a discussion of *in vitro* findings. *Neurobiol. Aging* **13,** 587–590.
39. Podlisny, M. B., Stephenson, D. T., Frosch, M. P., Lieberburg, I., Clemens, J. A., and Selkoe, D. J. (1992) Synthetic amyloid β-protein fails to produce specific neurotoxicity in monkey cerebral cortex. *Neurobiol. Aging* **13,** 561–567.
40. LaFerla, F. M., Tinkle, B. T., Bieberich, C. J., Haudenschild, C. C., and Jay, G. (1995) The Alzheimer's Aβ peptide induces neurodegeneration and apoptotic cell death in transgenic mice. *Nat. Genet.* **9,** 21–30.
41. LaFerla, F. M., Hall, C. K., Ngo, L., and Jay, G. (1996) Extracellular deposition of β-amyloid upon p53-dependent neuronal cell death in transgenic mice. *J. Clin. Invest.* **98,** 1626–1632.
42. Charriaut-Marlangue, C. and Ben-Ari, Y. (1995) A cautionary note on the use of the TUNEL stain to determine apoptosis. *Neuroreport* **7,** 61–64.
43. Loo, D. T., Copani, A., Pike, C. J., Whittemore, E. R., Walencewicz, A. J., and Cotman, C. W. (1993) Apoptosis is induced by β-amyloid in cultured central nervous system neurons. *Proc. Natl. Acad. Sci. USA* **90,** 7951–7955.
44. Mattson, M. P., Guo, Q., Furukawa, K., and Pedersen, W. A. (1998) Presenilins, the endoplasmic reticulum, and neuronal apoptosis in Alzheimer's disease. *J. Neurochem.* **70,** 1–14.
45. Unger, J. W. (1998) Glial reaction in aging and Alzheimer's disease. *Microsc. Res. Tech.* **43,** 24–28.
46. Gottlieb, T. M. and Oren, M. (1998) p53 and apoptosis. *Semin. Cancer Biol.* **8,** 359–368.
47. de la Monte, S. M., Sohn, Y. K., Ganju, N., and Wands, J. R. (1998) P53- and CD95-associated apoptosis in neurodegenerative diseases. *Lab. Invest.* **78,** 401–411.
48. de la Monte, S. M., Sohn, Y. K., and Wands, J. R. (1997) Correlates of p53- and Fas (CD95)-mediated apoptosis in Alzheimer's disease. *J. Neurol. Sci.* **152,** 73–83.
49. Kitamura, Y., Shimohama, S., Kamoshima, W., Matsuoka, Y., Nomura, Y., and Taniguchi, T. (1997) Changes of p53 in the brains of patients with Alzheimer's disease. *Biochem. Biophys. Res. Commun.* **232,** 418–421.
50. LaFerla, F. M., Troncoso, J. C., Strickland, D. K., Kawas, C. H., and Jay, G. (1997) Neuronal cell death in Alzheimer's disease correlates with apoE uptake and intracellular Aβ stabilization. *J. Clin. Invest.* **100,** 310–320.

51. Lassmann, H., Bancher, C., Breitschopf, H., Wegiel, J., Bobinski, M., Jellinger, K., et al. (1995) Cell death in Alzheimer's disease evaluated by DNA fragmentation in situ. *Acta Neuropathol. (Berl.)* **89,** 35–41.
52. Su, J. H., Anderson, A. J., Cummings, B. J., and Cotman, C. W. (1994) Immunohistochemical evidence for apoptosis in Alzheimer's disease. *NeuroReport* **5,** 2529–2533.
53. Su, J. H., Satou, T., Anderson, A. J., and Cotman, C. W. (1996) Up-regulation of Bcl-2 is associated with neuronal DNA damage in Alzheimer's disease. *NeuroReport* **7,** 437–440.
54. Knauer, M. F., Soreghan, B., Burdick, D., Kosmoski, J., and Glabe, C. G. (1992) Intracellular accumulation and resistance to degradation of the Alzheimer amyloid A4/β protein. *Proc. Natl. Acad. Sci. USA* **89,** 7437–7441.
55. Yang, A. J., Knauer, M., Burdick, D. A., and Glabe, C. (1995) Intracellular Aβ 1–42 aggregates stimulate the accumulation of stable, insoluble amyloidogenic fragments of the amyloid precursor protein in transfected cells. *J. Biol. Chem.* **270,** 14,786–14,792.
56. Martin, B. L., Schrader-Fischer, G., Busciglio, J., Duke, M., Paganetti, P., and Yankner, B. A. (1995) Intracellular accumulation of β-amyloid in cells expressing the Swedish mutant amyloid precursor protein. *J. Biol. Chem.* **270,** 26,727–26,730.
57. Wild-Bode, C., Yamazaki, T., Capell, A., Leimer, U., Steiner, H., Ihara, Y., et al. (1997) Intracellular generation and accumulation of amyloid β-peptide terminating at amino acid 42. *J. Biol. Chem.* **272,** 16,085–16,088.
58. Chui, D. H., Tanahashi, H., Ozawa, K., Ikeda, S., Checler, F., Ueda, O., et al. (1999) Transgenic mice with Alzheimer presenilin 1 mutations show accelerated neurodegeneration without amyloid plaque formation. *Nat. Med.* **5,** 560–564.
59. Askanas, V., Engel, W. K., Alvarez, R. B., and Glenner, G. G. (1992) β-Amyloid protein immunoreactivity in muscle of patients with inclusion-body myositis. *Lancet* **339,** 560–561.
60. Zheng, H., Jiang, M., Trumbauer, M. E., Sirinathsinghji, D. J., Hopkins, R., Smith, D. W., et al. (1995) β-Amyloid precursor protein-deficient mice show reactive gliosis and decreased locomotor activity. *Cell* **81,** 525–531.
60a. Muller, U., Cristina, N., Li, Z. W., Wolfer, D. P., Lipp, H. P., Rulicke, T., Brandner, S., Aguzzi, A., and Weissmann, C. (1994) Behavioral and anatomical deficits in mice homozygous for a modified beta-amyloid precursor protein gene. *Cell* **79,** 755–765.
61. Dawson, G. R., Seabrook, G. R., Zheng, H., Smith, D. W., Graham, S., O'Dowd, et al. (1999) Age-related cognitive deficits, impaired long-term potentiation and reduction in synaptic marker density in mice lacking the β-amyloid precursor protein. *Neuroscience* **90,** 1–13.
62. Wasco, W., Bupp, K., Magendantz, M., Gusella, J. F., Tanzi, R. E., and Solomon, F. (1992) Identification of a mouse brain cDNA that encodes a protein related to the Alzheimer disease-associated amyloid β protein precursor. *Proc. Natl. Acad. Sci. USA* **89,** 10,758–10762.
63. Wasco, W., Gurubhagavatula, S., Paradis, M. D., Romano, D. M., Sisodia, S. S., Hyman, B. T., et al. (1993) Isolation and characterization of APLP2

encoding a homologue of the Alzheimer's associated amyloid β protein precursor. *Nat. Genet.* **5,** 95–100.
64. von Koch, C. S., Zheng, H., Chen, H., Trumbauer, M., Thinakaran, G., van der Ploeg, L. H., et al. (1997) Generation of APLP2 KO mice and early postnatal lethality in APLP2/APP double KO mice. *Neurobiol. Aging* **18,** 661–669.
65. Higgins, L. S., Catalano, R., Quon, D., and Cordell, B. (1993) Transgenic mice expressing human β-APP751, but not mice expressing β-APP695, display early Alzheimer's disease-like histopathology. *Ann. NY Acad. Sci.* **695,** 224–227.
66. Lamb, B. T., Sisodia, S. S., Lawler, A. M., Slunt, H. H., Kitt, C. A., Kearns, W. G., et al. (1993) Introduction and expression of the 400 kilobase amyloid precursor protein gene in transgenic mice. *Nat. Genet.* **5,** 22–30.
67. Moran, P. M., Higgins, L. S., Cordell, B., and Moser, P. C. (1995) Age-related learning deficits in transgenic mice expressing the 751- amino acid isoform of human β-amyloid precursor protein. *Proc. Natl. Acad. Sci. USA* **92,** 5341–5345.
68. Quon, D., Wang, Y., Catalano, R., Scardina, J. M., Murakami, K., and Cordell, B. (1991) Formation of β-amyloid protein deposits in brains of transgenic mice. *Nature* **352,** 239–241.
69. Games, D., Adams, D., Alessandrini, R., Barbour, R., Berthelette, P., Blackwell, C., et al. (1995) Alzheimer-type neuropathology in transgenic mice overexpressing V717F β-amyloid precursor protein. *Nature* **373,** 523–527.
70. Hsiao, K., Chapman, P., Nilsen, S., Eckman, C., Harigaya, Y., Younkin, S., et al. (1996) Correlative memory deficits, Aβ elevation, and amyloid plaques in transgenic mice. *Science* **274,** 99–102.
71. Johnson-Wood, K., Lee, M., Motter, R., Hu, K., Gordon, G., Barbour, R., et al. (1997) Amyloid precursor protein processing and Aβ42 deposition in a transgenic mouse model of Alzheimer disease. *Proc. Natl. Acad. Sci. USA* **94,** 1550–1555.
72. Sasahara, M., Fries, J. W., Raines, E. W., Gown, A. M., Westrum, L. E., Frosch, M. P., et al. (1991) PDGF B-chain in neurons of the central nervous system, posterior pituitary, and in a transgenic model. *Cell* **64,** 217–227.
73. Masliah, E., Sisk, A., Mallory, M., Mucke, L., Schenk, D., and Games, D. (1996) Comparison of neurodegenerative pathology in transgenic mice overexpressing V717F β-amyloid precursor protein and Alzheimer's disease. *J. Neurosci.* **16,** 5795–5811.
74. Masliah, E. (1995) Mechanisms of synaptic dysfunction in Alzheimer's disease. *Histol. Histopathol.* **10,** 509–519.
75. Moechars, D., Gilis, M., Kuiperi, C., Laenen, I., and Van Leuven, F. (1998) Aggressive behaviour in transgenic mice expressing APP is alleviated by serotonergic drugs. *NeuroReport* **9,** 3561–3564.
76. Nalbantoglu, J., Tirado-Santiago, G., Lahsaini, A., Poirier, J., Goncalves, O., Verge, G., et al. (1997) Impaired learning and LTP in mice expressing the carboxy terminus of the Alzheimer amyloid precursor protein. *Nature* **387,** 500–505.
77. Chapman, P. F., White, G. L., Jones, M. W., Cooper-Blacketer, D., Marshall, V. J., Irizarry, M., et al. (1999) Impaired synaptic plasticity and learning in aged amyloid precursor protein transgenic mice. *Nat. Neurosci.* **2,** 271–276.

78. Malherbe, P., Richards, J. G., Martin, J. R., Bluethmann, H., Maggio, J., and Huber, G. (1996) Lack of β-amyloidosis in transgenic mice expressing low levels of familial Alzheimer's disease missense mutations. *Neurobiol. Aging* **17,** 205–214.
79. Thinakaran, G., Borchelt, D. R., Lee, M. K., Slunt, H. H., Spitzer, L., Kim, G., et al. (1996) Endoproteolysis of presenilin 1 and accumulation of processed derivatives *in vivo*. *Neuron* **17,** 181–190.
80. Thinakaran, G., Harris, C. L., Ratovitski, T., Davenport, F., Slunt, H. H., Price, D. L., et al. (1997) Evidence that levels of presenilins (PS1 and PS2) are coordinately regulated by competition for limiting cellular factors. *J. Biol. Chem.* **272,** 28,415–28,422.
81. Guo, Q., Sebastian, L., Sopher, B. L., Miller, M. W., Ware, C. B., Martin, G. M., et al. (1999) Increased vulnerability of hippocampal neurons from presenilin-1 mutant knock-in mice to amyloid β-peptide toxicity: central roles of superoxide production and caspase activation. *J. Neurochem.* **72,** 1019–1029.
82. Leissring, M. A., Paul, B. A., Parker, I., Cotman, C. W., and LaFerla, F. M. (1999) Alzheimer's presenilin-1 mutation potentiates inositol 1,4,5-trisphosphate-mediated calcium signaling in Xenopus oocytes. *J. Neurochem.* **72,** 1061–1068.
82a. Leissring, M. A., Akbari, Y., Fanger, C. M., Cahalan, M. D., Mattson, M. P., and LaFerla, F. M. (2000) Capacitative calcium entry deficits and elevated luminal calcium content in mutant presenilin-1 knockin mice. *J. Cell Biol.* **149,** 793–798.
83. Shen, J., Bronson, R. T., Chen, D. F., Xia, W., Selkoe, D. J., and Tonegawa, S. (1997) Skeletal and CNS defects in Presenilin-1-deficient mice. *Cell* **89,** 629–639.
84. Wong, P. C., Zheng, H., Chen, H., Becher, M. W., Sirinathsinghji, D. J., Trumbauer, M. E.,et al. (1997) Presenilin 1 is required for Notch1 and DII1 expression in the paraxial mesoderm. *Nature* **387,** 288–292.
85. Conlon, R. A., Reaume, A. G., and Rossant, J. (1995. Notch1 is required for the coordinate segmentation of somites. *Development* **121,** 1533–1545.
86. Hrabe de Angelis, M., McIntyre, J., 2nd, and Gossler, A. (1997) Maintenance of somite borders in mice requires the Delta homologue DII1. *Nature* **386,** 717–721.
87. Levitan, D. and Greenwald, I. (1995) Facilitation of lin-12-mediated signalling by sel-12, a Caenorhabditis elegans S182 Alzheimer's disease gene. *Nature* **377,** 351–354.
88. Davis, J. A., Naruse, S., Chen, H., Eckman, C., Younkin, S., Price, D. L., et al. (1998) An Alzheimer's disease-linked PS1 variant rescues the developmental abnormalities of PS1-deficient embryos. *Neuron* **20,** 603–609.
89. Qian, S., Jiang, P., Guan, X. M., Singh, G., Trumbauer, M. E., Yu, H., et al. (1998) Mutant human presenilin 1 protects presenilin 1 null mouse against embryonic lethality and elevates Aβ1-42/43 expression. *Neuron* **20,** 611–617.
90. De Strooper, B., Saftig, P., Craessaerts, K., Vanderstichele, H., Guhde, G., Annaert, W., et al. (1998) Deficiency of presenilin-1 inhibits the normal cleavage of amyloid precursor protein. *Nature* **391,** 387–390.

91. Wolfe, M. S., Xia, W., Ostaszewski, B. L., Diehl, T. S., Kimberly, W. T., and Selkoe, D. J. (1999) Two transmembrane asparates in presenilin-1 required for presenilin endoproteolysis and gamma-secretase activity. *Nature* **398,** 513–517.
92. Borchelt, D. R., Ratovitski, T., van Lare, J., Lee, M. K., Gonzales, V., Jenkins, N. A., et al. (1997) Accelerated amyloid deposition in the brains of transgenic mice coexpressing mutant presenilin 1 and amyloid precursor proteins. *Neuron* **19,** 939–945.
93. Roses, A. D. (1998) Apolipoprotein E and Alzheimer's disease. The tip of the susceptibility iceberg. *Ann. NY Acad. Sci.* **855,** 738–743.
94. Piedrahita, J. A., Zhang, S. H., Hagaman, J. R., Oliver, P. M., and Maeda, N. (1992) Generation of mice carrying a mutant apolipoprotein E gene inactivated by gene targeting in embryonic stem cells. *Proc. Natl. Acad. Sci. USA* **89,** 4471–4475.
95. Plump, A. S., Smith, J. D., Hayek, T., Aalto-Setala, K., Walsh, A., Verstuyft, J. G., et al. (1992) Severe hypercholesterolemia and atherosclerosis in apolipoprotein E- deficient mice created by homologous recombination in ES cells. *Cell* **71,** 343–353.
96. Mahley, R. W. (1988) Apolipoprotein E: cholesterol transport protein with expanding role in cell biology. *Science* **240,** 622–630.
97. Raber, J., Wong, D., Buttini, M., Orth, M., Bellosta, S., Pitas, R. E., et al. (1998) Isoform-specific effects of human apolipoprotein E on brain function revealed in ApoE knockout mice: increased susceptibility of females. *Proc. Natl. Acad. Sci. USA* **95,** 10,914–10919.
98. Smith, J. D., Sikes, J., and Levin, J. A. (1998) Human apolipoprotein E allele-specific brain expressing transgenic mice. *Neurobiol. Aging* **19,** 407–413.
99. Xu, P. T., Schmechel, D., Rothrock-Christian, T., Burkhart, D. S., Qiu, H. L., Popko, B., et al. (1996) Human apolipoprotein E2, E3, and E4 isoform-specific transgenic mice: human-like pattern of glial and neuronal immunoreactivity in central nervous system not observed in wild-type mice. *Neurobiol. Dis.* **3,** 229–245.
100. Bowman, B. H., Jansen, L., Yang, F., Adrian, G. S., Zhao, M., Atherton, S. S., et al. (1995) Discovery of a brain promoter from the human transferrin gene and its ulitization for development of transgenic mice that express human apolipoprotein E alleles. *Proc. Natl. Acad. Sci. USA* **92,** 12,115–12,119.
101. Xu, P. T., Gilbert, J. R., Qiu, H. L., Ervin, J., Rothrock-Christian, T. R., Hulette, C., et al. (1999) Specific regional transcription of apolipoprotein E in human brain neurons. *Am. J. Pathol.* **154,** 601-611.
102. Holtzman, D. M., Bales, K. R., Wu, S., Bhat, P., Parsadanian, M., Fagan, A. M., et al. (1999) Expression of human apolipoprotein E reduces amyloid-β deposition in a mouse model of Alzheimer's disease. *J. Clin. Invest.* **103,** R15–R21.
103. Schenk, D., Barbour, R., Dunn, W., Gordon, G., Grajeda, H., Guido, T., et al. (1999) Immunization with amyloid-beta attenuates Alzheimer-disease-like pathology in the PDAPP mouse. *Nature* **400,** 173–177.
104. Capsoni, S., Ugolini, G., Comparini, A., Ruberti, F., Berardi, N., and Cattaneo, A. (2000) Alzheimer-like neurodegeneration in aged antinerve growth factor transgenic mice. *Proc. Natl. Acad. Sci. USA* **97,** 6826–6831.

3
Glial Cells in Alzheimer's Disease

Robert E. Mrak and W. Sue T. Griffin

3.1. INTRODUCTION

Alzheimer's disease is characterized clinically by progressive and inevitable decline and loss of all higher cognitive functions over a period of years. This clinical decline is accompanied by the spread across cerebral cortical and subcortical regions of two salient neuropathological features: intraneuronal neurofibrillary tangles and complex neuritic β-amyloid-containing plaques *(1,2)*. These plaques contain extracellular deposits of β-amyloid and a number of other proteins *(3–6)*, as well as degenerating (dystrophic) neuritic processes and — importantly — activated glia elaborating a number of neurotrophic and immunomodulatory cytokines that drive and orchestrate the inception and evolution of these plaques *(7–10)*. These cardinal neuropathological features are, in turn, accompanied by progressive neuronal loss and decreased density of synaptic elements within the cerebral cortical neuropil *(11)*.

The spread of neurofibrillary tangles across cerebral cortical and subcortical regions follows a reasonably predictable pattern, to the extent that the cerebral cortical distribution pattern of these structure is the basis for a six-part pathological staging system that extends from early, subclinical involvement to end-stage disease *(12)*. The spread of neuritic plaques also shows progressive involvement of different cerebral cortical regions, but there is somewhat greater variability in the pattern of spread from patient to patient *(12)*. Patterns of neuronal cell loss associated with disease progression are not as well characterized, in part because such determinations are inherently more difficult. Neuronal cell loss is extensive, however, even in early, mild stages of the disease *(13)*, and such loss far exceeds that attributable to degeneration of neurofibrillary tangle-bearing neurons *(14)*.

Our understanding of disease progression and of mechanisms of neuronal loss in Alzheimer's disease has been advanced by the recent elucidation of glial mechanisms contributing to the development of Alzheimer-type

From: *Contemporary Clinical Neuroscience: Molecular Mechanisms of Neurodegenerative Diseases*
Edited by: M.-F. Chesselet © Humana Press Inc., Totowa, NJ

neuropathological changes *(15,16)* and by the introduction of *in situ* techniques for the identification of neuronal cell injury (manifested as DNA breaks) in paraffin-embedded tissue from autopsied patients *(17)*. In particular, the recent application of the TUNEL technique (Tdt-mediated dUTP-X-nick end labeling) to postmortem tissue has revealed considerable labelling of cerebral cortical neurons in Alzheimer's disease *(18–23)*. This technique detects late stages of apoptosis *(24)*, but also appears to identify nonapoptotic cell necrosis, and perhaps even damaged, but non-necrotic, cells that are vulnerable to metabolic disturbances *(25)*. We have shown that a significant portion of the TUNEL-positive neurons in Alzheimer's disease are associated with the two key histopathologic features of this disease [neurofibrillary tangles *(26)* and β-amyloid plaques *(27)*] and that the patterns of TUNEL positivity among different stages of tangle and plaque formation correlate with patterns of glial activation and glial association with these structures.

These lines of work suggest an important role for activated glia in the neuronal injury of Alzheimer's disease, and further suggest novel mechanisms for the spread of neuronal injury and neurodegeneration across cerebral regions in Alzheimer's disease. Of particular note are the roles of two key glia-derived cytokines : microglia-derived interleukin-1 and astrocyte-derived S100β. Interleukin-1 (IL-1) is a pleiotropic cytokine that orchestrates immunological responses in both peripheral tissues *(28)* and in the central nervous system *(29)*. In addition to trophic and potentially toxic effects on neurons, described in Subheading 3.2., IL-1 is known to activate astrocytes *(30)* and to induce astrocytic expression of the neuritotrophic cytokine S100β *(31)*. As will be discussed, S100β itself has a number of neurotrophic and gliotrophic actions, including promotion of neurite outgrowth *(32)* and of elevated intraneuronal free calcium levels *(33)*. The role of these cytokines, and of the activated glia that produce them, in the inception and spread of neuronal injury and loss in Alzheimer's disease is the subject of this review.

3.2. GLIAL ACTIVATION AND NEURONAL INJURY ASSOCIATED WITH NEUROFIBRILLARY TANGLE FORMATION

Intraneuronal neurofibrillary tangles are a major histopathological feature of Alzheimer's disease and have long been accepted as a histological hallmark for neuronal injury in this disease. Tangles correlate closely with degree of clinical impairment in Alzheimer patients *(34)* and anatomical patterns of tangle distribution are sufficiently predictable to serve as the basis for pathological staging of Alzheimer's disease *(12)*. We have recently been able to provide the first direct evidence of progressive neuronal injury associated with the appearance and evolution of neurofibrillary tangles in Alzheimer's disease, using the TUNEL technique. We found that the frequency of TUNEL positivity among neurons bearing neurofibrillary tangles increases progressively with stage of tangle formation, from 21% of neurons bearing early stages of neurofibrillary tangles to 87% of neurons bearing late stages *(26)*. This first demonstration of a progressive association between stage of neurofibrillary tangle formation and a molecular *in situ* index of neuronal cell injury not only confirms that tangle formation is associated with neuronal cell injury but also validates the TUNEL technique as an index of such injury.

Concomitant with this progressive neuronal injury, there is a progressive association of activated glia with neurons bearing neurofibrillary tangles *(35)*. Activated microglia, overexpressing IL-1, are found in close association with 48% of neurons bearing early stages of neurofibrillary tangles, and this frequency increases to 92% of neurons bearing late stages of tangles. A similar pattern of progressive association is seen between activated astrocytes, overexpressing the neurotrophic and potentially neurotoxic cytokine S100β, and tangle-bearing neurons. Activated astrocytes, overexpressing S100β, are found in association with 21% of neurons bearing early stages of neurofibrillary tangles, and this figure increases to 91% of neurons bearing late stages of tangles. This progressive association of activated, cytokine-elaborating glia with neurons bearing successive stages of neurofibrillary tangle formation suggests an important role for glial–neuronal interactions in the progression of neurofibrillary degeneration and in the associated neuronal injury in tangle-bearing neurons. However, most neuronal loss in Alzheimer's disease is not attributable to neurofibrillary tangle formation, as the extent of neuronal loss in Alzheimer's disease greatly exceeds the numbers of neurons undergoing neurofibrillary changes *(18)*.

3.3. GLIAL ACTIVATION AND NEURONAL INJURY ASSOCIATED WITH NEURITIC PLAQUE FORMATION

Neuritic plaques are found throughout the cerebral cortex, and in great numbers in end-stage Alzheimer's disease. Despite long-standing suspicions of neuronal injury associated with these plaques, evidence for such an effect — or even for postulated toxic mechanisms — has proven elusive. A great deal of attention has focused on the potential neurotoxicity of β-amyloid, but experimental attempts to demonstrate such β-amyloid-associated neurotoxicity have yielded equivocal results *(36–38)*. In vivo intracerebral injections of β-amyloid have been shown to result in neurodegeneration and neuronal loss, but only in primates and only in old age, suggesting that additional, possibly age-related factors are necessary for β-amyloid-associated neurotoxicity *(39)*. There is also the well-recognized observation that occasional elderly patients without discernible cognitive impairment manifest abundant extracellular deposits of amyloid peptide *(40,41)*, suggesting both that the amyloid peptide itself is not neurotoxic and that aging alone is insufficient to initiate β-amyloid-associated neurotoxicity. Indeed, the early "diffuse" amyloid peptide deposits of Alzheimer's disease appear to acquire neuritotoxic characteristics not seen in those benign, diffuse amyloid deposits of nondemented elderly patients *(42)*.

We have shown, using the TUNEL technique, that the extent of neuronal injury in plaque-associated neurons increases progressively with plaque stages, representing a hypothesized sequence of plaque evolution *(27)*. The frequency of TUNEL positivity in plaque-associated neurons increases from 40% in early, diffuse amyloid deposits (compared to a frequency of TUNEL positivity of 20% for neurons not associated with plaques) to 70–80% and 100% for neurons associated with either neuritic or dense-core non-neuritic plaques, respectively. This finding suggests that as plaques mature from diffuse amyloid deposits to neuritic plaques, there is progressive damage to associated neurons *(27)*. As a further finding, the total numbers of plaque-associated neurons (normalized for plaque size) showed a dramatic 70% decrease for dense-core, non-neuritic plaques, the so-called end-stage, or "burnt-out" plaques. Taken together, these findings show progressive neuronal injury and loss associated with the evolution of β-amyloid plaques in Alzheimer's disease and thus provide the first direct evidence that the appearance and progression of β-amyloid plaques is a major cause of neuronal injury and loss in Alzheimer's disease.

A key difference between the early, diffuse amyloid deposits found in patients with Alzheimer's disease and the diffuse amyloid deposits sometimes found in nondemented elderly patients is the presence of activated glia, overexpressing cytokines, in the diffuse plaques of Alzheimer patients *(8,10)* but not in the diffuse plaques of the nondemented elderly *(43)*. We have shown that most (78%) diffuse amyloid deposits in Alzheimer's disease contain activated IL-1-immunoreactive microglia *(8)* in contrast to the "benign" diffuse plaques of the nondemented elderly, which are not associated with activated microglia *(43)*. As these early amyloid plaques of Alzheimer's disease evolve into the destructive neuritic forms, there is an increase in the number of plaque-associated microglia, from an average of two microglia per plaque (in 10-μm-thick sections) in diffuse deposits to four to seven microglia per plaque in neuritic forms. Moreover, the microglia associated with the later, neuritic plaque forms are larger and more intensely immunoreactive than those associated with diffuse deposits *(8)*.

There is also evidence that activated astrocytes, overexpressing S100β, are involved in driving plaque progression in Alzheimer's disease. Tissue levels of biologically active S100β are elevated in the brain of Alzheimer patients *(44)* and activated astrocytes, overexpressing S100β, show a pattern of progressive association with evolving β-amyloid plaques similar to that seen for activated microglia overexpressing IL-1 *(10)*. Such activated astrocytes are found associated with most (80%) diffuse amyloid deposits in Alzheimer's disease, in small numbers (one per plaque) and are found in virtually all neuritic plaque forms, in greater numbers (two to four astrocytes per plaque) and with greater degrees of activation *(10)*. Even more striking is the finding that the numbers of activated, S100β-immunoreactive astrocytes in cerebral cortical tissue sections in Alzheimer's disease correlate with the extent of dystrophic neurite formation and the extent of neuritic expression of the β-amyloid precursor protein in Alzheimer's disease. Indeed, even *within individual neuritic plaques*, the numbers of these activated astrocytes, overexpressing S100β, correlate with the extent of dystrophic neurite formation and with the extent of neuritic expression of the β-amyloid precursor protein *(10)*. These results collectively indicate that amyloid deposits in Alzheimer's disease are foci of immunological activity, in contrast to the relative inertness of those diffuse amyloid deposits found in the nondemented elderly, and that this immunological activity correlates closely with neuronal injury and loss.

3.4. ACTIVATED GLIA AS ORGANIZING AND DRIVING ELEMENTS IN PROGRESSION OF NEUROPATHOLOGICAL CHANGES IN ALZHEIMER'S DISEASE: THE CYTOKINE CYCLE

In contrast to the paucity of evidence supporting a direct neurotoxic effect for β-amyloid in the plaques of Alzheimer's disease, there is ample evidence for potentially neurotoxic effects engendered by chronic activation of immunological mechanisms. We have proposed that chronic, sustained microglial overexpression of IL-1 initiates a cascade of cellular and molecular events — the *cytokine cycle* — with neurotoxic effects that culminate in the progressive neuronal injury and loss (and the progressive neurological decline) of Alzheimer's disease *(16)*. The consequences of this chronic IL-1 overexpression include (1) astrocyte activation and upregulation of astrocytic expression of S100β *(31)* and α_1-antichymotrypsin *(45)*, (2) stimulation of neuronal synthesis *(46,47)* and processing *(48)* of β-amyloid precursor protein (βAPP), thus favoring the release of both amyloid peptide fragments (and further deposition of β-amyloid) and gliotrophic *(49)* secreted β-APP fragments, and (3) autocrine effects with further activation of microglia and further IL-1 expression *(50–52)*. Astrocytic S100β overexpression, in turn, may (1) promote increases in intracellular free calcium concentrations in neurons *(33)*, (2) promote growth of neuronal processes *(53)*, (3) induce astrocytic nitric oxide synthase activity *(54)* with release of potentially neurotoxic nitric oxide, and (4) induce neuronal β-APP and interleukin-6 production *(55)*. This cascade includes several potentially neurotoxic steps, including raised intraneuronal free-calcium concentrations, overstimulation of neuritic outgrowth, and increased tissue levels of nitric oxide. Feedback mechanisms, with further activation of microglia and promotion of interleukin-1 overexpression, both sustain the immunological process and promote continuing neuronal injury. The demonstrated, progressive prominence of activated glia, overexpressing potent neurotrophic factors, both in β-amyloid plaques and in association with neurofibrillary tangle-bearing neurons, suggests that neurotoxicity associated with activation of the cytokine cycle drives the demonstrated neuronal injury and loss associated with the progression of β-amyloid plaque and neurofibrillary tangle pathology in Alzheimer's disease.

3.5. GLIAL ACTIVATION AS A COMMON RISK FACTOR FOR ALZHEIMER'S DISEASE

One approach to Alzheimer pathogenesis is the investigation of conditions known to predispose to Alzheimer's disease or to accelerated appearance of age-associated Alzheimer-type (senile) neuropathological changes. Known predisposing conditions for Alzheimer's disease, in addition to aging, include Down's syndrome and head injury. Patients with chronic, intractable epilepsy show accelerated appearance of Alzheimer-type "senile" changes. We have shown early and striking activation of microglia and astrocytes with overexpression of IL-1 and S100β, respectively, in all of these conditions.

Normal aging is characterized by progressive increases in the numbers of activated astrocytes overexpressing S100β in the brain *(56)*, and experimental animals with accelerated senescence also show acceleration of this astrocytic S100β overexpression *(57)*. There are also increases in tissue levels of IL-1 mRNA in aged brain *(58)*. We have provided evidence that the distribution of IL-1$^+$ microglia determines, in part, the distribution of neuritic plaques in Alzheimer's disease. In Alzheimer's disease, the distribution of IL-1$^+$ microglia correlates with that of neuritic plaques both across brain regions *(59)* and across cerebral cortical layers *(60)*. Even more striking, however, is the additional finding that the cortical laminar distribution of IL-1⁺ microglia in age-matched control patients correlates highly with the cortical laminar distribution of neuritic plaques found in Alzheimer patients *(60)*. This latter finding suggests that pre-existing laminar distribution patterns of IL-1⁺ microglia (i.e., those seen in control patients) are important in determining the observed laminar distribution of neuritic plaques in Alzheimer patients. This result, together with the observed increase in numbers of activated IL-1⁺ microglia in aged brain, may explain the experimental observation that intracerebral injections of purified β-amyloid is neurotoxic in aged, but not young, primates *(39)*.

Down's syndrome is a condition wherein the entire array of Alzheimer-type neuropathological changes accumulate gradually and virtually inevitably over the course of several decades *(61)*. Adult Down's syndrome patients show cognitive impairment that is similar to early cognitive changes in Alzheimer's disease *(62,63)*, and the time-course of neurofibrillary tangle formation in Down's syndrome displays regional patterns comparable to those observed in aging and Alzheimer's disease *(64)*. We have shown glial inflammatory changes, including a profusion of activated astrocytes and microglia overexpressing S100β and IL-1, as well as overexpression of neuronal β-APP and astrocytic S100β, in young and even fetal Down's tissue *(7,65,66)*. These changes precede by decades the appearance of classic Alzheimer-type neuropathological changes.

Head injury is now recognized as an important risk factors for the later development of Alzheimer's disease *(67,68)*. Following severe, acute head injury, there is increased neuronal expression of β-APP *(69)* and there is cerebral cortical deposition of diffuse β-amyloid protein in approximately one-third of patients *(70)*. These changes are accompanied by dramatic increases in the numbers of activated, IL-1⁺ microglia, which correlate with the numbers of neurons overexpressing β-APP. Plaquelike clustering of dystrophic βAPP⁺ neurites, invariably associated with activated microglia overexpressing IL-1, can also be seen *(69)*. These findings suggest that early glial inflammatory and neuronal acute-phase responses are important factors underlying the increased risk of Alzheimer's disease that follows head injury.

In addition to these recognized risk factors for the development of Alzheimer's disease, there are other conditions in which the incidence of Alzheimer-type neuropathological changes has been shown to be increased. Epilepsy, for example, is not a recognized risk factor for Alzheimer's disease, although there is a small but significant increased risk of dementia in these patients *(71)*. Patients with chronic intractable epilepsy show an accelerated appearance of Alzheimer-type "senile" neuropathological changes *(72)*, and this is most striking in patients who carry the Alzheimer-associated ApoE ε4 allele *(73)*. We have shown that neurons in the temporal lobe from patients with chronic intractable epilepsy overexpress β-APP *(74)*, perhaps in response to neuronal injury resulting from hyperexcitability, and that this β-APP overexpression correlates with the appearance of increased numbers of activated microglia overexpressing IL-1 *(74)* and astrocytes overexpressing S100β *(75)*. Another example is human immunodeficiency virus (HIV) infection. Patients with HIV infection also show accelerated appearance of Alzheimer-type "senile" neuropathological changes *(76)*, and in these patients there is also neuronal overexpression of β-APP and glial activation with overexpression of IL-1 and of S100β *(77,78)*.

3.6. NEURONAL INJURY, GLIAL ACTIVATION, AND THE PROGRESSION OF ALZHEIMER LESIONS

Neurofibrillary tangles appear early in the course of Alzheimer's disease, and the appearance and evolution of neurofibrillary tangles within neurons are accompanied by progressive associations of these neurons with activated, cytokine-elaborating glia *(35)*, as well as by progressively more frequent evidence of neuronal DNA damage in these neurons *(26)*. The extent and pattern of neurofibrillary tangle distribution appear to be relatively predictable *(12)*, and these tangles preferentially affect a subpopulation of cortical neurons with long corticocortical projections *(79)*. These corticocortical projections have been implicated in the distinct laminar patterns of neuritic plaque distribution within brain regions *(12,80,81)*. Indeed, these corticocortical projection patterns suggest that transcortical spread of neuronal damage and loss in Alzheimer's disease may be engendered in remote target regions via corticocortical projections from damaged or dying neurons. In support of this idea, we have found evidence of neuronal injury in the form of overexpression of the neuronal acute-phase reactant, β-APP, and activation of glia in target regions of ablated brainstem nuclei in rats *(82)*. These result suggests that overexpression of β-APP and β-amyloid deposition in corticocortical target regions are potential downstream consequences of injury and death of neurons with corticocortical projections. Our results, showing progressive neuronal cell damage and eventual neuronal loss as plaques evolve from diffuse to more complex forms, suggest that plaque-associated neuronal injury is a major cause of neuronal cell injury and loss in Alzheimer's disease. The lack of strong evidence for direct β-amyloid neurotoxicity suggests that other plaque-associated elements contribute to neuronal dysfunction or loss and thus to the transcortical spread of neuronal damage via subsequent dysfunction or loss of corticocortical projections. The combination of plaque-associated neuronal injury and corticocortical projection-associated neuronal injury provides an obvious mechanism for propagating β-amyloid deposition resulting from neuronal injury-induced overexpression of β-APP. This overexpression of β-APP favors synthesis and release of β-amyloid and of secreted APP (sAPP) fragments *(48)*. The latter fragments, in turn, activate microglia and increase synthesis and release of IL-1 *(49)*, initiating a cascade of amyloidogenic and neurotoxic effects *(15,16)* that propagates the entire complex of β-amyloid-associated neuropathological changes.

REFERENCES

1. Alzheimer, A. (1907) Ueber eine eigenartige Erkrankung der Hirnrinde. *Allgemeine. Z. Psychiatrie* **64,** 146–148.
2. Alzheimer, A. (1907) Ueber eine eigenartige Erkrankung der Hirnrinde. *Zentralbl. Gsante. Neurol. Psychiatrie* **18,** 177–179.
3. Masters, C. L., Simms, G., Weinman, N. A., Multhaup, G., McDonald, B. L., and Beyreuther, K. (1985) Amyloid plaque core protein in Alzheimer disease and Down syndrome. *Proc. Natl. Acad. Sci. USA* **82,** 4245–4249.
4. Abraham, C. R., Selkoe, D. J., and Potter, H. (1988) Immunochemical identification of the serine protease inhibitor α_1-antichymotrypsin in the brain amyloid deposits of Alzheimer's disease. *Cell* **52,** 487–501.
5. McComb, R. D., Miller, K. A., and Carson, S. D. (1991) Tissue factor antigen in senile plaques of Alzheimer's disease. *Am. J. Pathol.* **139,** 491–494.
6. McGeer, P. L., Akiyama, H., Itagaki, S., and McGeer, E. G. (1989) Activation of the classical complement pathway in brain tissue of Alzheimer patients. *Neurosci. Lett.* **107,** 341–346.
7. Griffin, W. S. T., Stanley, L. C., Ling, C., White, L., MacLeod, V., Perrot, L. J., White, C. L., III, and Araoz, C. (1989) Brain interleukin 1 and S-100 immunoreactivity are elevated in Down syndrome and Alzheimer disease. *Proc. Natl. Acad. Sci. USA* **86,** 7611–7615.
8. Griffin, W. S. T., Sheng, J. G., Roberts, G. W., and Mrak, R. E. (1995) Interleukin-1 expression in different plaque types in Alzheimer's disease: significance in plaque evolution. *J. Neuropathol. Exp. Neurol.* **54,** 276–281.
9. Sheng, J. G., Mrak, R. E., and Griffin, W. S. T. (1994) S100β protein expression in Alzheimer disease: potential role in the pathogenesis of neuritic plaques. *J. Neurosci. Res.* **39,** 398–404.
10. Mrak, R. E., Sheng, J. G., and Griffin, W. S. T. (1996) Correlation of astrocytic S100β expression with dystrophic neurites in amyloid plaques of Alzheimer's disease. *J. Neuropathol. Exp. Neurol.* **55,** 273–279.
11. Terry, R. D., Masliah, E., Salmon, D. P., et al. (1991) Physical basis of cognitive alterations in Alzheimer's disease: synapse loss is the major correlate of cognitive impairment. *Ann. Neurol.* **30,** 572–580.
12. Braak, H. and Braak, E. (1991) Neuropathological stageing of Alzheimer-related changes. *Acta Neuropathol.* **82,** 239–259.

13. Gomez-Isla, T., Price, J., McKeel, D., Morris, J., Greenberg, S., Petersen, R., et al. (1998) Profound loss of layer II entorhinal cortex neurons occurs in very mild Alzheimer's disease. *J. Neurosci.* **16,** 4491–4500.
14. Gomez-Isla, T., Hollister, R., West, H., Mui, S., Growdon, J., Petersen, R., et al. (1997) Neuronal loss correlates with but exceeds neurofibrillary tangles in Alzheimer's disease. *Ann. Neurol.* **41,** 17–24.
15. Mrak, R. E., Sheng, J. G., and Griffin, W. S. T. (1995) Glial cytokines in Alzheimer's disease: review and pathogenic implications. *Hum. Pathol.* **26,** 816–823.
16. Griffin, W. S. T., Sheng, J. G., Royston, M. C., Gentleman, S. M., McKenzie, J. E., Graham, D. I., et al. (1998) Glial–neuronal interactions in Alzheimer's disease: the potential role of a 'cytokine cycle' in disease progression. *Brain Pathol.* **8,** 65–72.
17. Gavrieli, Y., Sherman, Y., and Ben-Sasson, S. A. (1992) Identification of program cell death in situ via specific labeling of nuclear DNA fragmentation. *J. Cell Biol.* **119,** 493–501.
18. Troncoso, J. C., Sukhov, R. R., Kawas, C. H., and Koliatsos, V. E. (1996) In situ labeling of dying cortical neurons in normal aging and in Alzheimer's disease: correlations with senile plaques and disease progression. *J. Neuropathol. Exp. Neurol.* **55,** 1134–1142.
19. Su, J. H., Anderson, A. J., Cummings, B. J., and Cotman, C. W. (1994) Immunohistochemical evidence for apoptosis in Alzheimer's disease. *NeuroReport* **5,** 2533.
20. Lassmann, H., Bancher, C., Breitschopf, H., Wegiel, J., Bobinski, M., Jellinger, K., and Wisniewski, H. M. (1995) Cell death in Alzheimer's disease evaluated by DNA fragmentation in situ. *Acta Neuropathol.* **89,** 35–41.
21. Smale, G., Nichols, N. R., Brady, D. R., Finch, C. E., and Horton, W. E., Jr. (1995) Evidence for apoptotic cell death in Alzheimer's disease. *Exp. Neurol.* **133,** 225–230.
22. Anderson, A., Su, J. H., and Cotman, C. W. (1996) DNA damage and apoptosis in Alzheimer's disease colocalization with *cjun* immunoreactivity, relationship to brain area and effect of postmortem delay. *J. Neurosci.* **16,** 1710–1719.
23. Mullaart, E., Boerrigter, M. E., Ravid, R., Swaab, D. F., and Vijg, J. (1990) Increased levels of DNA breaks in cerebral cortex of Alzheimer's disease patients. *Neurobiol. Aging* **11,** 169–173.
24. Frankfurt, O. S., Robb, J. A., Sugarbaker, E. V., and Villa, L. (1996) Monoclonal antibody to single-stranded DNA is a specific and sensitive cellular marker of apoptosis. *Exp. Cell Res.* **226,** 387–397.
25. Stadelmann, C., Bruck, W., Bancher, C., Jellinger, K., and Lassman, H. (1998) Alzheimer disease: DNA fragmentation indicates increased neuronal vulnerability, but not apoptosis. *J. Neuropathol. Exp. Neurol.* **57,** 456–464.
26. Sheng, J. G., Mrak, R. E., and Griffin, W. S. T. (1998) Progressive neuronal DNA damage associated with neurofibrillary tangle formation in Alzheimer disease. *J. Neuropathol. Exp. Neurol.* **57,** 323–328.
27. Sheng, J. G., Zhou, X. Q., Mrak, R. E., and Griffin, W. S. T. (1998) Progressive neuronal injury associated with amyloid plaque formation in Alzheimer's disease. *J. Neuropathol. Exp. Neurol.* **57,** 714–717.

28. Dinarello, C. A. and Wolff, S. M. (1993) The role of Interleukin-1 in disease. *N. Engl. J. Med.* **328,** 106–113.
29. Rothwell, N. J. (1991) Functions and mechanisms of interleukin 1 in the brain. *TiPS* **12,** 430–436.
30. Giulian, D., Woodward, J., Young, D. G., Krebs, J. F., and Lachman, L. B. (1988) Interleukin-1 injected into mammalian brain stimulates astrogliosis and neovascularization. *J. Neurosci.* **8,** 2485–2490.
31. Sheng, J. G., Ito, K., Skinner, R. D., Mrak, R. E., Rovnaghi, C. R., Van, Eldik, L. J., et al. (1996) In vivo and in vitro evidence supporting a role for the inflammatory cytokine interleukin-1 as a driving force in Alzheimer pathogenesis. *Neurobiol. Aging* **17,** 761–766.
32. Kligman, D. and Marshak, D. R. (1985) Purification and characterization of a neurite extension factor from bovine brain. *Proc. Natl. Acad. Sci. USA* **82,** 7136–7139.
33. Barger, S. W. and Van Eldik, L. J. (1992) S100 stimulates calcium fluxes in glial and neuronal cells. *J. Biol. Chem.* **267,** 9689–9694.
34. Arriagada, P. V., Growden, J. H., Hedley-White, E. T., and Hyman, B. T. (1992) Neurofibrillary tangles but not senile plaques parallel duration and severity in Alzheimer's disease. *Neurology* **42,** 639.
35. Sheng, J. G., Mrak, R. E., and Griffin, W. S. T. (1997) Glial-neuronal interactions in Alzheimer disease: progressive association of IL-1α^+ microglia and S100β^+ astrocytes with neurofibrillary tangle stages. *J. Neuropathol. Exp. Neurol.* **56,** 285–290.
36. Cotman, C. W., Pike, C. J., and Copani, A. (1992) β-Amyloid neurotoxicity: a discussion of in vitro findings. *Neurobiol. Aging* **13,** 587–590.
37. Manelli, A. M. and Puttfarcken, P. S. (1995) β-Amyloid-induced toxicity in rat hippocampal cells: in vitro evidence for the involvement of free radicals. *Brain Res. Bull.* **38,** 569–576.
38. Stein-Behrens, B., Adams, K., Yeh, M., and Sapolsky, R. (1992) Failure of β-amyloid protein fragment 25–35 to cause hippocampal damage in the rat. *Neurobiol. Aging* **13,** 577–579.
39. McKee, A. C., Kowall, N. W., Schumacher, J. S., and Beal, M. F. (1998) The neurotoxicity of amyloid β protein in aged primates. *Amyloid Int. J. Exp. Clin. Invest.* **5,** 1–9.
40. Crystal, H., Dickson, D., Fuld, P., Masur, D., Scott, R., Mehler, M., et al. (1988) Clinico-pathologic studies in dementia: nondemented subjects with pathologically confirmed Alzheimer's disease. *Neurology* **38,** 1682–1168.
41. Katzman, R., Terry, R., DeTeresa, R., Brown, T., Davies, P., Fuld, P., et al. (1988) Clinical, pathological, and neurochemical changes in dementia: a subgroup with preserved mental status and numerous neocortical plaques. *Ann. Neurol.* **23,** 138–144.
42. Knowles, R. B., Gomez-Isla, T., and Hyman, B. T. (1998) Aβ associated neuropil changes: correlation with neuronal loss and dementia. *J. Neuropathol. Exp. Neurol.* **57,** 1122–1130.
43. Mackenzie, I. R., Hao, C., and Munoz, D. G. (1995) Role of microglia in senile plaque formation. *Neurobiol. Aging* **16,** 797–804.

44. Marshak, D. R., Pesce, S. A., Stanley, L. C., and Griffin, W. S. T. (1992) Increased S100 neurotrophic activity in Alzheimer disease temporal lobe. *Neurobiol. Aging* **13,** 1–7.
45. Das, S. and Potter, H. (1995) Expression of the Alzheimer amyloid-promoting factor antichymotrypsin is induced in human astrocytes by IL-1. *Neuron* **14,** 447–456.
46. Forloni, G., Demicheli, F., Giorgi, S., Bendotti, C., and Angaretti, N. (1992) Expression of amyloid precursor protein mRNAs in endothelial, neuronal and glial cells: modulation by interleukin1. *Brain Res. Mol. Brain Res.* **16,** 128–134.
47. Goldgaber, D., Harris, H. W., Hla, T., Maciag, T., Donnelly, R. G., Jacobsen, J. S., et al. (1989) Interleukin 1 regulates synthesis of amyloid B-protein precursor mRNA in human endothelial cells. *Proc. Natl. Acad. Sci. USA* **86,** 7606–7610.
48. Buxbaum, J. D., Oishi, M., Chen, H. I., Pinkas-Kramarski, R., Jaffe, E. A., Gandy, S. E., et al. (1992) Cholinergic agonists and interleukin 1 regulate processing and secretion of the Alzheimer β/A4 amyloid protein precursor. *Proc. Natl. Acad. Sci. USA* **89,** 10,075–10,078.
49. Barger, S. W. and Harmon, A. D. (1997) Microglial activation by Alzheimer amyloid precursor protein and modulation by apolipoprotein E. *Nature* **388,** 878–881.
50. Ganter, S., Northoff, H., Mannel, D., and Gebicke-Harter, P. J. (1992) Growth control of cultured microglia. *J. Neurosci. Res.* **33,** 218–230.
51. Lee, S. C., Liu, W., Dickson, D. W., Brosnan, C. F., and Berman, J. W. (1993) Cytokine production by human fetal microglia and astrocytes. Differential induction by lipopolysaccharide and IL1β. *J. Immunol.* **150,** 2659–2667.
52. Sebire, G., Emilie, D., Wallon, C., Hery, C., Devergne, O., Delfraissy, J. F., et al. (1993) In vitro production of IL6, IL1β, and tumor necrosis factor α by human embryonic microglial and neural cells. *J. Immunol.* **150,** 1517–1523.
53. Marshak, D. R. (1990) S100 as a neurotrophic factor. *Prog. Brain Res.* **86,** 169–181.
54. Hu, J., Castets, F., Guevara, J. L., and Van Eldik, L. J. (1996) S100β stimulates inducible nitric oxide synthase activity and mRNA levels in rat cortical astrocytes. *J. Biol. Chem.* **271,** 2543–2547.
55. Li, Y., Barger, S. W., Liu, L., Mrak, R. E., and Griffin, W. S. T. (1999) S100 induction of the pro-inflammatory cytokine interleukin-6 in neurons: implications for Alzheimer pathogenesis. *Neurochem.* **74,** 143–150.
56. Sheng, J. G., Mrak, R. E., Rovnaghi, C. R., Kozlowska, E., Van, E., LJ, and Griffin, W. S. T. (1996) Human brain S100β and S100β mRNA expression increases with age: pathogenic implications for Alzheimer's disease. *Neurobiol. Aging* **17,** 359–363.
57. Griffin, W. S. T., Sheng, J. G., and Mrak, R. E. (1998) Senescence-accelerated overexpression of S100β in brain of SAMP6 mice. *Neurobiol. Aging* **19,** 71–76.
58. Sheng, J. G., Mrak, R. E., and Griffin, W. S. T. (1998) Enlarged and phagocytic, but not primed, interleukin-1α-immunoreactive microglia increase with age in normal human brain. *Acta Neuropathol.* **95,** 229–234.

59. Sheng, J. G., Mrak, R. E., and Griffin, W. S. T. (1995) Microglial interleukin-1α expression in brain regions in Alzheimer's disease: correlation with neuritic plaque distribution. *Neuropathol. Applied Neurobiol.* **21,** 290–301.
60. Sheng, J. G., Griffin, W. S. T., Royston, M. C., and Mrak, R. E. (1998) Distribution of IL-1-immunoreactive microglia in cerebral cortical layers: implications for neuritic plaque formation in Alzheimer's disease. *Neuropathol. Appl. Neurobiol.* **24,** 278–283.
61. Wisniewski, K. E., Wisniewski, H. M., and Wen, G. Y. (1985) Occurrence of neuropathological changes and dementia of Alzheimer's disease in Down's syndrome. *Ann. Neurol.* **17,** 278–282.
62. Brugge, K. L., Nichols, S. L., Salmon, D. P., Hill, L. R., Delis, D. C., Aaron, L., et al. (1994) Cognitive impairment in adults with Down's syndrome: similarities to early cognitive changes in Alzheimer's disease. *Neurology* **44,** 232–238.
63. Soininen, H., Partanen, J., Jousmaki, V., Helkala, E. L., Vanhanen, M., Majuri, S., et al. (1993) Age-related cognitive decline and electroencephalogram slowing in Down's syndrome as a model of Alzheimer's disease. *Neuroscience* **53,** 57–63.
64. Hof, P. R., Bouras, C., Perl, D. P., Sparks, D. L., Mehta, N., and Morrison, J. H. (1995) Age-related distribution of neuropathologic changes in the cerebral cortex of patients with Down's syndrome. Quantitative regional analysis and comparison with Alzheimer's disease. *Arch. Neurol.* **52,** 379–391.
65. Griffin, W. S. T., Sheng, J. G., McKenzie, J., Royston, M. C., Gentleman, S. M., Brumback, R. A.,et al. (1998) Life-long overexpression of S100 in Down's syndrome: implications for Alzheimer pathogenesis. *Neurobiol. Aging* **2,** 35–42.
66. Royston, M. C., McKenzie, J. E., Gentleman, S. M., Sheng, J. G., Mann, D. M. A., Griffin, W. S. T., et al. (1999) Overexpression of the neuritotrophic cytokine S100β in Down's syndrome: correlation with patient age and β-amyloid deposition. *Neuropathol. Appl. Neurobiol.* **25,** 387–393.
67. Gautrin, D. and Gauthier, S. (1989) Alzheimer's disease: environmental factors and etiologic hypotheses. *Can. J. Neurol. Sci.* **16,** 375–387.
68. Gentleman, S. M. and Roberts, G. W. (1991) Risk factors in Alzheimer's disease. *Br. Med. J.* **304,** 118–119.
69. Griffin, W. S. T., Sheng, J. G., Gentleman, S. M., Graham, D. I., Mrak, R. E., and Roberts, G. W. (1994) Microglial interleukin-1α expression in human head injury: correlations with neuronal and neuritic β-amyloid precursor protein expression. *Neurosci. Lett.* **176,** 133–136.
70. Gentleman, S. M., Graham, D. I., and Roberts, G. W. (1993) Molecular pathology of head injury: altered β-APP metabolism and the aetiology of Alzheimer's disease. *Prog. Brain Res.* **96,** 237–246.
71. Breteler, M. M., de Groot, R. R., van Romunde, L. K., and Hofman, A. (1994) Risk of dementia in patients with Parkinson's disease epilepsy and severe head trauma: a register-based followup study. *Am. J. Epidemiol.* **142,** 1300–1305.
72. Mackenzie, I. R. and Miller, L. A. (1994) Senile plaques in temporal lobe epilepsy. *Acta Neuropathol.* **87,** 504–510.
73. Gouras, G. K., Relkin, N. R., Sweeney, D., Munoz, D. G., Mackenzie, I. R., and Gandy, S. (1997) Increased apolipoprotein E epsilon 4 in epilepsy with senile plaques. *Ann. Neurol.* **41,** 402–404.

74. Sheng, J. G., Boop, F. A., Mrak, R. E., and Griffin, W. S. T. (1994) Increased neuronal β-amyloid precursor protein expression in human temporal lobe epilepsy: association with interleukin-1α immunoreactivity. *J. Neurochem.* **63,** 1872–1879.
75. Griffin, W. S. T., Yeralan, O., Sheng, J. G., Boop, F. A., Mrak, R. E., Rovnaghi, C. R., et al. (1995) Overexpression of the neurotrophic cytokine S100β in human temporal lobe epilepsy. *J. Neurochem.* **65,** 228–233.
76. Esiri, M. M., Biddolph, S. C., and Morris, C. S. (1998) Prevalence of Alzheimer plaques in AIDS. *J. Neurol. Neurosurg. Psychiatry* **65,** 29–33.
77. Stanley, L. C., Mrak, R. E., Woody, R. C., Perrot, L. J., Zhang, S., Marshak, D. R., et al. (1994) Glial cytokines as neuropathogenic factors in HIV infection: pathogenic similarities to Alzheimer's disease. *J. Neuropathol. Exp. Neurol.* **53,** 231–238.
78. Mrak, R. E. and Griffin, W. S. T. (1997) The role of chronic self-propagating glial responses in neurodegeneration: implications for long-lived survivors of human immunodeficiency virus. *J. NeuroVirol.* **3,** 241–246.
79. Lewis, D. A., Campbell, M. J., Terry, R. D., and Morrison, J. H. (1987) Laminar and regional distributions of neurofibrillary tangles and neuritic plaques in Alzheimer's disease: a quantitative study of visual and auditory cortices. *J. Neurosci.* **7,** 1799–1808.
80. Clinton, J., Roberts, G. W., Gentleman, S. M., and Royston, M. C. (1993) Differential pattern of β-amyloid protein deposition within cortical sulci and gyri in Alzheimer's disease. *Neuropathol. Applied Neurobiol.* **19,** 277–281.
81. Rogers, J. and Morrison, J. H. (1985) Quantitative morphology and regional and laminar distributions of senile plaques in Alzheimer's disease. *J. Neurosci.* **5,** 2801–2808.
82. Ito, K., Ishikawa, Y., Skinner, R. D., Mrak, R. E., Morrison-Bogorad, M., Mukawa, J., et al. (1997) Lesioning of the inferior olive using a ventral surgical approach: characterization of temporal and spatial responses at the lesion site and in cerebellum. *Mol. Chem. Neuropathol.* **31,** 245–264.

4
Inflammation in Alzheimer's Disease

Caleb E. Finch, Valter Longo, Aya Miyao,
Todd E. Morgan, Irina Rozovsky, Yubei Soong,
Min Wei, Zhong Xie, and Hadi Zanjani

4.1. INTRODUCTION

We develop a new perspective on the interactions of amyloids and Alzheimer's disease (AD), in which molecular and cellular changes of aging have a key role. According to our view, a subset of basic mechanisms in aging is equivalent to chronic inflammatory processes, which predispose to the deposition of amyloids in the brain and other organs.

The term *amyloid* was introduced by Rudolf Virchow in the 1850s to describe ''starchy'' inclusion bodies in animals; for historical perspectives, *see* Schwartz *(98)*, Pepys *(85)*, and Sipe *(104)*. Amyloids are most commonly characterized as fibrillar aggregates, which can be formed from diverse proteins and which have extensive β-sheet interactions as detected histochemically by the binding of the dyes Congo red or thioflavin-S *(85,104)*. Some aggregated forms of the same protein are not recognized as amyloids because of the lack of histochemical signals for bound Congo red or thioflavin-S, e.g. the diffuse deposits of the amyloid β-peptide amyloid (Aβ) as described in Subheading 4.2. Moreover, we have observed Congo red binding with a hyperchromic shift that is characteristic of β-sheet structures in oligomeric, slowly sedimenting aggregates of Aβ *(74)*. These examples indicate that the archaic term *amyloid* requires cautious use in its application to molecular structure and biological activity, because it may exclude many states in amyloid-forming proteins that are biologically interesting.

A major goal is to identify the causal chains in AD, which we suggest are best understood as multiple concurrent inflammatory processes during aging. These complex processes are subject to many modulations by genetic variations at multiple loci. They may also be sensitive to endogenous levels

From: *Contemporary Clinical Neuroscience: Molecular Mechanisms of Neurodegenerative Diseases*
Edited by: M.-F. Chesselet © Humana Press Inc., Totowa, NJ

of sex steroids, to steroid replacements and other drugs taken to optimize outcomes of aging, and to poorly defined environmental factors.

The clinical symptoms of AD are rare before the age of 60, although subclinical deterioration may exist for several decades before impairments are obvious. After 60, the risk of AD then increases exponentially and then doubles every 5 yr *(52)*. Although the incidence reaches 30–50% by the ninth decade *(52,70,89)*, nonetheless some individuals reach 100 yr or more, despite carrying the strong heritable risk factor of homozygotic apoE ε4/ ε4, without clinical dementia *(70,106)*. These multiple outcomes of aging give a strong basis for ultimate optimism, as we identify segments of these complex inflammatory processes that are ongoing during life in multiple organ systems.

4.2. ALZHEIMER'S DISEASE

Brain Amyloids and Other Aggregates of Aβ Peptides

A diagnostic characteristic of AD is a sufficient density of extracellular senile plaques (SPs) that contain the amyloid β peptide (Aβ) in certain brain regions. The Aβ peptides of up to 43 amino acid residues are endoproteolytically derived from the amyloid precursor protein (APP). Senile plaque amyloids consist mainly of Aβ 1–42 but with some longer and shorter peptides, whereas cerebral blood vessels accumulate amyloid containing the slightly shorter Aβ 1–40. Another diagnostic of AD is the accumulation of intraneuronal aggregates of hyperphosphorylated tau, which are described as neurofibrillary tangles (NFTs). APP, Aβ, and tau are widely produced by cells throughout the body. From one perspective, the puzzle of AD is to understand why proteins that are present throughout life in body fluids (Aβ) or within neurons (tau) aggregate during aging. We note briefly that the accumulations of aggregated Aβ and hyperphosphorylated tau are extremely common during aging of primates and most other mammals that live longer than 10 yr *(26,88)*. These and other species-generalized aging changes define a canonical pattern of aging in mammals *(24)*.

We emphasize that many other forms of aggregated Aβ 1–43 peptides are found widely during aging in many brain regions. These heterogeneous *extracellular* materials range widely in morphology and binding of Congo red, which is a required criterion for designation as an "amyloid." At one extreme may be oligomeric forms of Aβ ("ADDLs"), which would not be detected by the usual aqueous immunocytochemisty because of their solubility *(60)*. A higher level of Aβ aggregation is represented by the amorphous, or diffuse, Aβ deposits detected by immunohistochemistry to Aβ peptides, but do not distinctly bind Congo red, and hence, are not called amyloids *(3,136)*. The highly compact, Congo red binding Aβ-containing deposits of senile plaques and cerebral vessels are the classic amyloid of AD brains. Because of neurons with abnormal dystrophic neurites (swollen, twisted) that are nearby or growing though their matrix, senile plaques are also called neuritic plaques. Another type of deposit in AD brains is described as a "fleecy amyloid" *(114)*. Although the amorphous deposits may arise before the senile plaques, there is no information on causality. The apparent rapidity with which brain amyloids can form after head trauma *(30)* indicates the need for detailed studies on the time-course of amyloidogenesis, which may be forthcoming from mouse models engineered with human AD genes.

The significance of these diverse Aβ-containing deposits to AD pathogenesis is highly controversial. On one hand, Terry, Masliah, and colleagues of the UC San Diego Alzheimer Center have emphasized that in the cerebral cortex, the amyloid load is less strongly correlated with the degree of clinical dementia than the synaptic loss *(112,113)*. During the course of clinically demonstrated AD, the total amyloid load does not change *(46)*. On the other hand, Cotman and colleagues of the UCIrvine Alzheimer Center have shown strong correlations of cognitive functions with the total amyloid load as determined in AD brains at various stages *(16)* and in dogs *(17)*. Moreover, careful examination of the neocortex at very early stages of AD from the Washington University (St. Louis) Alzheimer Center showed all those with "minimal cognitive dysfunction" had many neuritic plaques, whereas cognitively normal individuals of the same age had a much lower density of amyloid deposits *(76)*. Diffuse plaques were found in all brains and, even in the early AD, they were fivefold more common than neuritic plaques. Although some authors emphasize that neurons tend to have normal morphology around diffuse plaques with loss of synapses *(112)*, others have observed a smaller cholinergic neuron fiber density in nondemented elderly with diffuse Aβ-containing deposits, also consistent with early pathogenesis *(7)*.

A large body of work on the role of Aβ in AD has focused on the neurotoxicity of fibrillar Aβ. Complex aggregates form rapidly during incubation of various Aβ peptides (Aβ1–40, Aβ1–42, Aβ25–35) at ambient temperatures; these high-molecular-weight aggregates have widely varying toxicity *(102)*. However, work from this laboratory in collaboration with Klein and Krafft of Northwestern University has demonstrated that oligomeric (soluble) Aβ aggregates are highly toxic to neurons *(60,81)*. Transgenic mice that overexpress human AD genes and have increased production of Aβ peptides also show neuronal dysfunctions in the absence of Aβ deposits *(41,42,43,89,107)*, which implies a role for small Aβ aggregates. The neurotoxic pathways involve oxidative stress *(8,60,81)* and appear to be mediated by signaling systems with Fyn and Rac 1 (Longo and Finch, unpublished). Superoxide and redox-active iron are both implicated as mediators of Aβ neurotoxicity (*8*; Longo and Finch, unpublished).

Inflammatory Processes in Amyloid Aggregates

Besides the neurotoxicity of various forms of aggregated Aβ peptides, they may participate in inflammatory mechanisms at many levels (Table 1). By inflammatory mechanisms, we mean to include several subsets of the cellular and molecular changes observed in injured peripheral tissues. We note several major exceptions to inflammatory processes in the brain during AD from those of peripheral inflammation: (1) the absence of swelling; (2) the absence of pain (the brain parenchyma is unique in its paucity of nociception); and (3) the scarcity of B- and T-cells, thereby sharply distinguishing AD from multiple sclerosis in which autoreactive T-cells have the major role in pathogenesis. With its ''cold'' variety of inflammation, the brain may offer a unique opportunity to study complement functions independently of B- and T-cells. Finch and Marchalonis *(25)* proposed that AD is a model for the evolutionarily early stages of inflammatory mechanisms that preceded combinatorial cellular mechanisms in immune responses.

A prominent cellular change during AD is the activation of microglia *(15,19,22,72,86,91)*. Microglia are bone-marrow-derived cells of the monocyte lineages that, like peripheral tissue macrophages, become phagocytic and produce reactive oxygen species. Glial activation has been recognized from the beginning of AD science: activated glia were described in "presenile dementia'' brains by Alzheimer in 1907 *(4)*, whereas microglia were observed near fibrous Aβ in senile plaques by Terry et al. in 1964 *(111)*. *See* Chapter 3 for more details of glia in senile plaques. In general, fewer activated microglia are associated with diffuse Aβ deposits *(15,87,93)*. A continuum of aggregation states is indicated, in which increased microglial and astrocyte activation parallels the intensity of thioflavin-S staining *(15,87)*. Of particular interest, the activation of microglia appears to preceed that of astrocytes. During AD, astrocytes also become activated, as generally evaluated by the increase of cellular extensions containing GFAP, the intermediate filament. It is widely recognized that GFAP expression increases in response to local brain injury *(64)*. Moreover, we observed that systemic pathophysiology can stimulate GFAP expression (e.g., in association with wasting diseases and pathology of non-neural organs) *(34)*.

It is now clear that Aβ peptides can activate microglia/monocytes and astrocytes (Table 1), the latter including oligomeric Aβ forms (Aβ–derived diffusible ligands [ADDLs]) *(45)*. Aβ also stimulates astrocyte production of interleukin-1 (IL-1) *(31,45)*. Moreover, Aβ can directly activate the

Table 1
Pro-inflammatory Activities of Aβ Peptides

Astrocytes
NF-κB, NO, d IL-1
GFAP fibril thickening (stellation) *(45)*
Microglia
Immunoepitopes of activation
Respiratory bursts *(6,121)*
Interactions with complement
Binds and activates C1q *(122,125)*

classical complement cascade by binding to C1q, which is the initial component of the classical complement cascade (Table 1). A general hypothesis is being considered in AD research, by which aggregated Aβ initiates inflammatory responses *(2,15,22,25,91)*.

Many inflammatory proteins are detected in senile plaques including cytokines, complement factors, and acute-phase proteins (Table 2). A general note of caution is that these histochemical observations are semiquantitative at best and are sensitive to fixation and to the specificity of the antibodies. Of great interest to inflammatory mechanisms, C1q shows strong immunostaining in senile plaques *(1,22,91)*. Activation of C1q can produce the anaphylactic peptides (C3a, C4a, C5a), which are chemoattractants and which, like C1q itself, can stimulate oxygen bursts *(20)*. The complement cascade can culminate in production of the cytocidal membrane attack complex (MAC), which contains C5b–C9. Although MAC components and MAC inhibitors are found in AD brains *(137,141)*, there is no information on their role in neuron death during AD.

The amorphous/diffuse deposits of Aβ appear to have fewer inflammatory components. C1q immunohistochemical signals are markedly less in diffuse plaques of the same brains, which show strong signals in neuritic plaques *(1,22)*. However, C3d, apoE, and apoJ are regularly detected in diffuse/amorphous plaques and neuritic plaques of affected brain regions *(142)*. Ongoing work from this lab indicates the presence of inflammatory markers in very early stages of AD *(140)*, the CDR of 0.5 or minimal cognitive impairment *(76)*.

In contrast to the robust indications of inflammatory processes in the AD brain, several cytokines in the cerebrospinal fluid (IL-1 , IL-1ra, IL-6, tumor necrosis factor [TNF]) did not change during the dramatic and rapid brain atrophy observed by brain imaging during longitudinal studies of the same patients from the Oxford Aging and Alzheimer Center *(63)*. These findings

Table 2
Neuroinflammatory Changes in Alzheimer and Aging Brain

	Alzheimer's disease senile plaques[a]	Normal aging rodent[b]	Normal aging human
Astrocytes	X	X(GFAP)	X (GFAP)[c]
Microglia	X	X (OX-6, -42)	X
Neurite abnormalities	X (NFT)	X (no NFT)	
Aβ	X		X
APP	X		
α_1-antichymotrypsin	X		
α_2-macroglobulin	X		
apoE	X	X (mRNA)	
apoJ (clusterin)	X	X (mRNA)	Corpora amylacea[e]
CRP	X		
Heme oxygenase-1	X	X (ICC)	
Complement factors			
C1q	X, decrease in CSF[f]	X (mRNA)	
C3	X		Increase in CSF[d]
C9	X		Corpora amylaceae[e]
Cytokines			
IL-1	X		
IL-6	X		Plasma[g]
TGF-β1	X	X (mRNA)	

[a] References *22*, *25*, and *137*, and Chapter 3.
[b] References *75* and *86*.
[c] Reference *79*.
[d] Reference *67*.
[e] Reference *103*.
[f] Reference *105*.
[g] References *13*, *18*, and *92*.

agree with studies of these cytokines, which were based on single-time cerebrospinal fluid CSF sampling *(29,68)*. However, other CSF inflammatory markers may change markedly during AD: on one hand, the level of C1q varied inversely with the clinical rating, which implies its consumption by complement activation *(105)*. However, IL-6 and its soluble receptor in CSF do not change consistently during AD *(68)*. We also note the increased expression of complement genes in other neurological diseases independently of AD-like amyloid deposits. For example, we observed increased C1q and apoJ in sporadic amyotrophic lateral sclerosis *(35)*. Because C1q may be activated by many components of neurodegeneration, including myelin and DNA released from dying cells, there may be multiple steps in AD which involve inflammatory mechanisms.

The sources of inflammatory proteins in AD brains are poorly understood. The skeptic would simply dismiss these findings as a postmortem artifact of blood-brain barrier breakdown during death. However, immunoglobulins are not found in the same senile plaques, which present certain other serum proteins *(22,91)*. Local brain cells are a major potential source of inflammatory proteins in association with Aβ aggregates. Our laboratory was the first to show by *in situ* hybridization that C1q mRNA is relatively abundant in neurons and microglia of human and rodent brains *(61,62,83,94)*. Moreover, we and others (Patrick and Edith McGeer, University of British Columbia; Joseph Rogers, Sun Health Research Insititute, Sun City AZ) in 1991–1993 reported that C1qB mRNA increases several fold in frontal cortex of AD brains *(49,61,62,83,123)*. Consistent with these findings, we also detected C1q immunoreactivity in surviving hippocampal CA1 pyramidal neurons after excitotoxin lesions *(94)*. At the time, these findings were viewed very skeptically. However, others soon showed C1q immunostaining in neurons of AD brains *(1)* and increased C1q mRNA in AD brains *(28)*. Recently, we showed that rat brain can synthesize *de novo* bioactive C1q during responses to lesions *(32)*. Evidence from many sources indicates that brain contains mRNA and proteins that represent most, if not all, classical and alternate path complement components, including the C9 of the membrane attack complex (e.g., ref. *137*). The multiple functions of C1q include intracellular activities (binding to calreticulin) as well as interactions with a wide range of other systems that mediate normal tissue renewal *(20)*. Moreover, we observed that C1q mRNA is expressed during brain development in association with regional synaptogenesis *(50)*. We also briefly note that some of the inflammatory proteins associated with aggregated Aβ can also modify the activities of glial cells (e.g., apoE attenuates activation of astrocytes by Aβ) *(44)*, whereas apoJ (clusterin) activates microglia *(135)*.

4.3. INFLAMMATORY MECHANISMS IN BRAIN AGING IN THE ABSENCE OF AD

Microglia and astrocytes become activated during aging in rodents, but without *any conventional* indications of neurodegeneration. For example, we observed a threefold increase in the numbers of microglia in corticostriatal bundles of 24-mo-old rats (Fig. 1) and in corpus callosum (Fig. 2) *(75)*. The rats of this study were male F_1(F344 × BN) hybrids, which are in excellent health at 24 mo and have mean life spans of 33 mo, as discussed in ref. *75*. Many other molecular indices of activated microglial and astrocytes

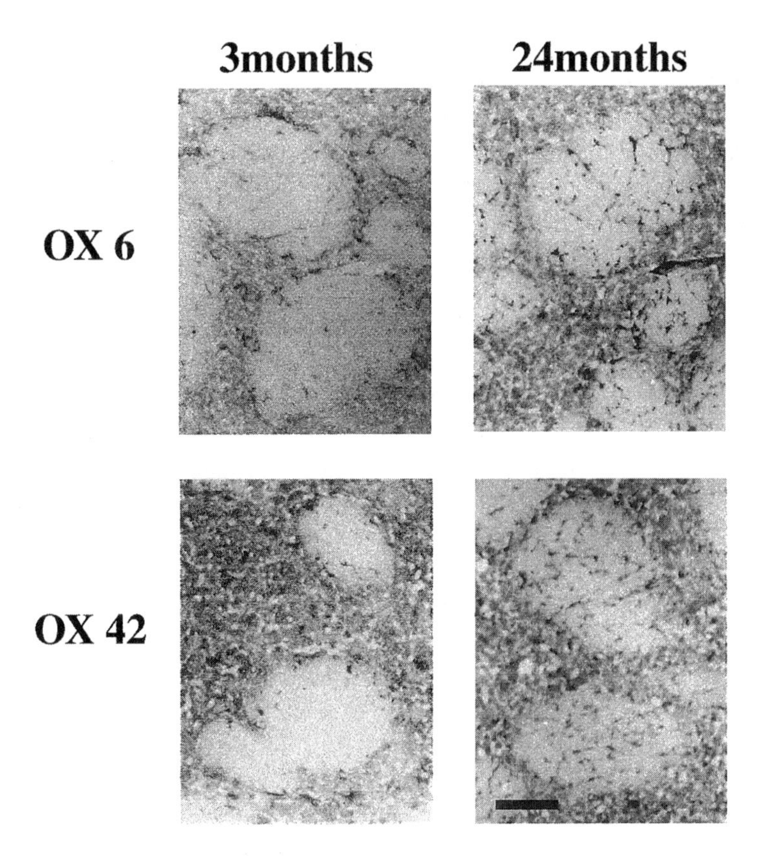

Fig. 1. Corticostriatal bundles within the caudate-putamen of 3- versus 24-mo-old male rats (Fisher–Brown Norway F1 hybrids) processed for OX6 (anti-major histocompatibility complex class 2) and OX42 (anti-complement type 3 receptor) immunoreactivity. Increased OX6 and OX42 immunoreactivity (indicative of activated microglia) is evident by 24 mo, an age considered as middle-aged for this long-lived genotype. Scale bar=50 µm. (Adapted from ref. *75*).

are found throughout the aging brain in this study and in others, and in several other rodent genotypes *(33,79,80,84,95,99,139)*.

GFAP, the intermediate filament of astrocytes, shows robust progressive increases in expression during normal aging in rodents and humans. By *in situ* hybridization, GFAP mRNA increases progressively per astrocyte during aging (Fig. 2) *(73,75)*. The increased expression of GFAP in rodent aging represents mainly increased activation of astrocytes, with small to negligible increases in the total numbers of astrocytes (e.g., in the hippocampus) *(66)*. We calculate that GFAP mRNA increases from puberty onward at a rate of about five mRNA copies per astrocyte per month *(84)*.

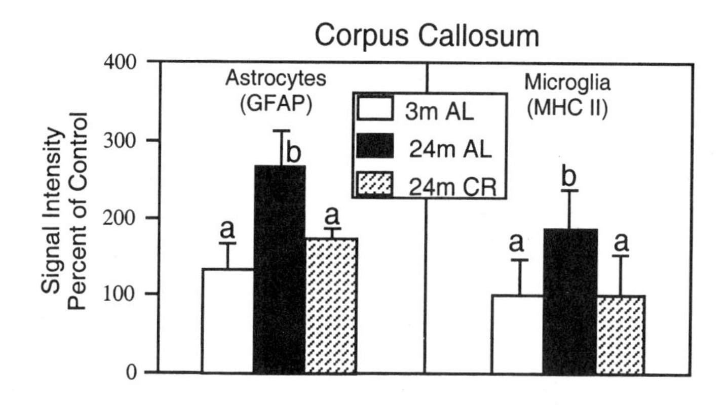

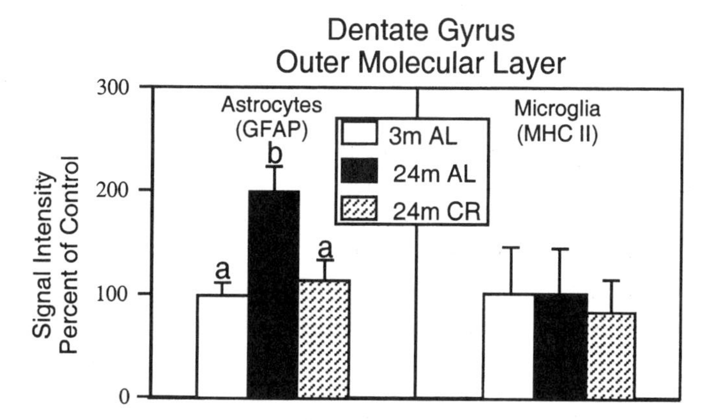

Fig. 2. Effect of age and caloric restriction (CR) on astrocytic and microglial activation, as shown with GFAP and MHCII immunoreactivity in the corpus callosum and outer molecular layer of dentate gyrus; 3- versus 24-mo-old male rats (Fisher–Brown Norway F1 hybrids). GFAP or major histocompatibility complex class II (MHCII) levels were assessed by quantification of respective immunocytochemical signals, expressed as a percentage of 3-mo-old *ad libitum* (3mAL). Bars represent mean ± SEM; eight rats per group; $p < 0.05$. **a** is relative to 24-mo-ad libitum (24mAL) and **b** is relative to 24-mo-old caloric restricted (24mCR). (From *ref.* 75).

Human brains also show increased levels of GFAP and fibrous astrocytes during aging in the absence of pathology *(36,79)*.

The increase of GFAP during aging is associated with increased transcription, as shown by *in situ* hybridization with intron–RNA probes *(75,139)* and by nuclear run-on *(56,65)*. GFAP is the first example of a gene in the brain or other tissues that becomes progressively activated during aging in the absence of specific pathological changes. The GFAP increase during aging is thus a candidate for designation as a canonical change of aging *(24,27)*. Because GFAP transcription is also activated by oxidative stress *(73)*, we hypothesize that the age-related increases of GFAP transcription are mediated by generalized inflammatory background during aging.

The rat GFAP promoter contains an NFκB element, which mediates responses to oxidative stress *(128)*. This complex transcriptional control locus also regulates GFAP induction by IL-1 and repression by transforming growth factor-β1 (TGF-β) *(58)*. Moreover, Aβ and hydrogen peroxide treatment of astrocytes increase the concentrations of nuclear proteins that bind to this element *(128)*. As noted earlier, in diffuse amyloid deposits, microglia appear to become activated before astrocytes *(87)*, which suggests that, during AD, astrocyte activation in nascent amyloids may be secondary to oxidative stress from activated microglia. However, during aging, we observe that GFAP mRNA increases in some rat brain regions in the absence of microglial activation (Mhc II epitopes) in those same regions (e.g., the outer molecular layer of the dentate gyrus) (Fig. 2). Further work is needed with more markers of glial activation to define the interrelationships among astrocytes, microglia, and oligodendroglial changes during aging.

There are multiple sources of reactive oxygen species and oxidative stress during aging. Oxidized groups of proteins increase during aging in rodents and human brains (reviewed in ref. *75*). In the case of rodents, we can absolutely rule out Aβ amyloids as a factor, because aging laboratory rodent brains do not accumulate Aβ peptides. (Some types of fibrils are found in granules within astrocytes, which are immunoreactive for proteoglycans and laminin, but not for those Aβ peptides tested *[51])* A clue to the mechanisms involved in glial activation during aging is its attenuation by modest caloric restriction (Fig. 2) *(75)*. Because caloric restriction decreases the amount of oxidized proteins in the brain and other organs *(75,100)*, we hypothesize that this is a factor in the attenuation of glial activation by caloric restriction.

Although astrocytes are widely considered as supportive or trophic cells for neurons, there is chemical evidence from primary cultures that astrocytes can produce and release superoxide and other reactive oxygen species *(10,116)*, which, in turn, can inhibit gap-junction permeability *(10)*. Examination of effects of aging on astrocyte production of reactive oxygen species might be very informative as a mechanism favoring subsequent neurodegenerative changes.

Besides these indications of activated glia during aging, we note that a subset of the same inflammatory factors associated with AD also increase during brain aging. In aging rat striatum, we find modest increases of mRNAs for C1q, apoJ (clusterin), and TGF-β1 *(84)*. For example, in rat neostriatum C1qB mRNA increased about 50% in the F344 rat by its mean life span, whereas the trend for increase in the longer-lived hybrid (F344 × BN)F_1 was not significant *(84)*. In rat hippocampus, IL-1β is also increased, as is IL-1β binding *(109)*. Pending mRNA studies on human brains, we note that aging human brains commonly have extracellular bodies, the corpora amylacea, which are 2- to 20-mm in diameter in which polysaccharides are a major component *(12,77)*. However, corpora amylacea also immunoreactive for many complement factors *(103)*. In AD, the corpora amylacea increase in numbers further above control aged brains, but also in Parkinson disease and multiple sclerosis *(103)*. These somewhat scattered observations give a rationale for an in-depth analysis of how and why normal aging promotes increased expression of cytokines, complement factors, and other inflammatory mediators.

4.4. AMYLOIDS AND AGING IN NON-NEURAL TISSUES

We now extend our discussion to tissues outside of the brain, which show many parallel changes during aging, although different amyloids are involved. Accumulations in non-neural tissues of extracellular amyloids during aging ("senile amyloids") are very common in human populations *(14,55,85,98)* (Table 3). By amyloids, as noted earlier, we mean fibrillar proteinaceous materials that bind Congo red or thioflavin-S. About 20 different proteins are found to form tissue amyloids *(55,85,104)*. Some of these proteins are pentraxins that form aggregates with a pentameric organization (e.g., C-reactive protein [CRP] and serum amyloid]SAA]), which are evolutionarily ancient components of host defense mechanisms with roles in antimicrobial defense and tissue repair *(25)*.

However, many other aggregated proteins do not meet the standard criteria for tissue amyloids, as noted in Subheading 4.1., which merit more consideration if we are to to understand the causes of amyloid deposits in Alzheimer disease (AD) and other age-related diseases. Recall that in the earliest stages of AD, the most abundant form is nonfibrillar, diffuse deposits of the amyloid-β peptide (Aβ) which do not significantly bind Congo red or thioflavin-S. Moreover, the skin of those with clinical AD has nonfibrillar Aβ deposits, which are detected by immunohistochemistry at higher frequency than in age-matched controls *(48,129)*.

Here we note terminology used by the general field of amyloidologists, which recognizes amyloidosis syndromes in three general categories: *primary* (idiopathic); *secondary* (associated with chronic inflammation, e.g., rheumatoid arthritis or tuberculosis); and *familial (47,85,104)*. The brain amyloid of AD would appear to fit all three categories of the traditional amyloidoses. Some amyloidoses are associated with abnormal depos-

Table 3
Senile Amyloid in Human Organs

Organ	Type of amyloid	Incidence	Ref.
Brain: senile plaques and cerebral vessels	Aβ peptide	High	*89*
Heart:			
Aorta	Apolipoproten A1	High	*132*
Atrium	α-ANF	High	*53*
Myocardium	Transthyretin	Moderate	*12a,47*
Lumbar disks	Unknown	Low–high	*138*
Lung	Unknown	Moderate	*59,120*
Pituitary	Prolactin	Unknown	*131*

its of elevated serum amyloids, e.g., as in the amyloid A of familial Mediterranean fever *(12a,47)*. This may be pertinent to those familial forms of AD that are associated with chronic, albeit modest, increased production and body fluid levels of the Aβ peptide *(89)*.

In the heart and aorta, several types of amyloids increase after 50 yr *(69)*. Amyloids in the myocardium include atrial natriuretic peptide (α-ANP) *(53)* and transthyretin, particularly in African-Americans who carry a transthyretin mutation (isoleu 122) *(12a,47)*. Myocardial amyloids can accumulate sufficiently to modify heart structure and function, causing arrhythmias and conduction disturbances and they may be a significant cause of heart failure in the elderly *(9,47)*. The aorta accumulates different (and unidentified) amyloids, particularly in the medial layer *(78)*. "Senile" amyloids accumulate in other vital organs to varying degrees. We note the potential value of a thorough study of non-neural amyloids in individuals whose brains are characterized for the neuropathology of AD. This effort might identify a new relationship between peripheral and central inflammatory processes of aging, in which amyloid depositions could be a variable outcome.

Senile amyloids are also widely reported in domestic animals. Aging dogs have well-characterized accumulations of the Aβ peptide in cerebral vessels and as senile plaques *(11,37,110,126)*. However, aging dogs also commonly have other (not identified) amyloids in the heart, lung, and intestine *(118)*. Of great interest, the accumulation of amyloids during aging in different tissues varies widely between individual dogs *(37,126)* as it does in humans (Table 3).

Laboratory rodents are well known for the lack of brain amyloid during aging *(51)*, unless engineered with certain human familial Alzheimer transgenes (ref. *88,107* and Chapter 2). However, in kidney and other non-neural tissues, senile amyloid deposits commonly increase during aging in widely used strains of mice *(38,96,115,117,130,133)*. Particularly early and extensive amyloid deposits arise in strains of the senescence accelerated (SAM) strains of mice, which accumulate amyloid fibrils containing a lipoprotein homolog of human apoA-II *(38,39,40)*. Remarkably, mice engineered to over-express TGFβ1 showed age-related deposits of the Aβ peptide in cerebral vessels *(134)*, which suggests the importance of TGF-β1 and possibly other cytokines in tissue amyloid deposits. Next, we consider evidence for age-related increases in cytokines and other inflammatory regulators.

4.5. INFLAMMATION AND AGING IN NON-NEURAL TISSUES

In asking how the inflammatory features of AD emerge in relation to the course of the disease, we must consider the evidence that many inflammatory markers also emerge during brain aging in many species of mammals. Much data indicate a progressive increase in inflammatory markers in peripheral blood during aging in the general human population (e.g., elevated blood levels of IL-6 *(23,82)*. Of particular interest are the increases of IL-6 in community-dwelling elderly of two large samplings from well characterized populations: the Duke Established Populations for Epidemiological Studies of the Elderly (EPESE) *(13)* and the Framingham Study *(92)*. In the Duke EPESE sample, plasma IL-6 showed progressive average elevations from 70 to 99+ years, with the strongest upward trend in white males. The subgroup of those with very high IL-6 levels (> 5 pg/mL) doubled at later ages, from 10% (70–79 yr) to 20% (90–99+) *(13)*. Another indication is the increased frequency of apparently healthy elderly with modestly elevated plasma CRP *(5,92)*. In some conditions (e.g., chronic hemodialysis), elevations of CRP are a strong predictor of cardiovascular disease and mortality risk *(143)*. These and other peripheral markers suggest that inflammatory degenerative processes may be ongoing in many organs during aging. Consistent with this possibility, the Duke EPESE sample showed a highly significant correlation between high IL-6 and poor self-rated health *(13)*. However, a caveat that shows the complexity of the issues that must be considered is the similar correlation between high IL-6 and depression, because depression is also associated with poor self-ratings in many cognitive domains *(18)*.

At tissue levels, there are also many indications of inflammatory processes during aging, which extend the findings on brain aging (Subheading 4.3.). The liver, which is a producer of CRP and other acute-phase inflammatory proteins, manifests inflammatory mechanisms during aging.

For example, aging mice have an increased basal level of a transcription factor, C/EBPδ, which regulates many acute-phase genes and other genes that mediate responses to oxidative stress *(90)*. The C/EBPδ mRNA was about fivefold higher in control mice aged 24 versus 3 mo; after injection of the inflammatory stimulus lipopolysaccharide, the return to baseline was much slower *(90)*. T-kininogen, another acute-phase protein, shows spontaneous elevations during aging in rats that predict death within 4 mo *(101,124)*; this observation is consistent with the association of elevated CRP with increased mortality in hemodialysis patients noted above.

4.6. SYNTHESIS AND CONCLUSIONS

We have shown that inflammatory processes associated with Alzheimer's disease, an age-related condition, also develop during "normal" aging to a lesser extent, not only in the brain but in many other tissues. We propose that a major feature of aging is the development of a general inflammatory tone, which, in turn, is a precondition for other specific pathogenic processes. It is worth serious thought that macrophage/monocytes may be a crucial determinant of the outcomes of aging in a wide range of tissues. For example, macrophage monocytes are prominent in brain aging (microglia), in vascular aging (foam cells in the arterial wall), and in the bones (osteoclasts) and arthritic conditions of joints. Macrophages also produce estradiol in breast tissues and may thus influence breast cancer *(71)*; this observation suggests that activation of macrophagelike cells during aging could have many other consequences to sex steroid sensitive cells in the environment. The slow accumulation of oxidized epitopes in long-lived proteins could be a fundamental background factor in these inflammatory processes. Among the mechanisms that cause protein oxidation is the nonenzymatic reaction of blood glucose with ε-amino groups *(100)*. In turn, glycoxidized proteins can propagate free-radical reactions leading to crosslinking and the attraction of tissue macrophages *(57,97)*. Of course, many other mechanisms can lead to protein oxidation.

Inflammation is also known to promote amyloid formation in non-neural tissues *(12a,25,54,55,104)*. For example, tuberculosis with major host inflammatory responses frequently leads to systemic amyloidosis *(104,119)*. Renal dialysis, through little understood processes that lead to the accumulation of inflammatory cells, is also associated with tissue amyloids *(21,55)*. Thus, we may consider a global hypothesis of aging, in which chronic, initially low-grade inflammatory processes progress during aging to become proamyloidogenic in different tissues. Individual outcomes of aging may depend on how smouldering inflammatory processes are fanned by the external environment, according to the proclivities of the genotype and species, and by chance variations in hemopoeiesis that may stochastically modify the reactivity of macrophage/monocyte cells *(27)*.

REFERENCES

1. Afagh, A., Cummings, B. J., Cribbs, D. H., Cotman, C. W., and Tenner, A. J. (1996) Localization and cell association of C1q in Alzheimer's disease brain. *Exp. Neurol.* **138,** 22–32.
2. Aisen, P. S. and Pasinetti, G. M. (1998) Glucocorticoids in Alzheimer's disease. The story so far. *Drugs Aging* **12,** 1–6.
3. Akiyama, H., Mori, H., Saido, T., Kondo, H., Ikeda, K., and McGeer, P. L. (1999) Occurrence of the diffuse amyloid beta-protein (Abeta) deposits with numerous Abeta-containing glial cells in the cerebral cortex of patients with Alzheimer's disease. *Glia* **25,** 324–331.
4. Alzheimer, A. (1907) Uber eine einartige Erkrankung der Hindrinde. *Allg. Zeitschr. Psychiatrie Psych.-Gerichtl. Med.* **64,** 146–148. Translated as "A characteristic disease of the cerebral cortex," in *The Early Story of Alzheimer's Disease* (Bick K., Amaducci, L., and Pepeu, G. eds.), Livonia Press, Padova, distributed through Raven Press, New York.
5. Ballou, S. P. and Kushner, I. (1997) Chronic inflammation in older people: recognition, consequences, and potential intervention. *Clin. Geriatr. Med.* **13,** 653–659.
6. Barger, S. W. and Harmon A. D. (1997) Microglial activation by Alzheimer amyloid precursor protein and modulation by apolipoprotein E. *Nature.* **388,** 878–881.
7. Beach, T. G. and McGeer, E. G. (1992) Senile plaques, amyloid beta-protein, and acetylcholinesterase fibres: laminar distributions in Alzheimer's disease striate cortex. *Acta Neuropathol. (Berl.)* **93,** 146–153.
8. Behl, C., Davis, J.B., Lesley, R., and Schubert, D. (1994) Hydrogen peroxide mediates amyloid beta protein toxicity. *Cell* **77,** 817–827.
9. Benson, M. D. (1997) Aging, amyloid, and cardiomyopathy. *N. Engl J. Med.* **336,** 502–504.
10. Bolaños, J. P. and Medina, J. M. (1996) Induction of nitric oxide synthase inhibits gap junction permeability in cultured rat astrocytes. *J. Neurochem.* **66,** 2091–2099.
11. Borras, D., Ferrer, I., and Pumarola, M . (1999) Age-related changes in the brain of the dog. *Vet. Pathol.* **36,** 202–211.
12. Cavanagh, J. B. (1999) Corpora-amylacea and the family of polyglucosan diseases. *Brain Res. Brain Res. Rev.* **29,** 265–295.

12a. Buxbaum, J. N. and Tagoe, C. E. (2000) The genetics of the amyloidoses. *Annu. Rev. Med.* **51,** 543–569.

13. Cohen, H. J., Pieper, C. F., Harris, T., Rao, K. M., and Currie, M. S. (1997) The association of plasma IL-6 levels with functional disability in community-dwelling elderly. *J. Gerontol. A: Biol. Sci. Med. Sci.* **52,** M201–M208.

14. Cornwell, G. G., 3rd, Johnson, K. H., and Westermark, P. (1995) The age related amyloids: a growing family of unique biochemical substances. *J. Clin. Pathol.* **48,** 984–989.
15. Cotman, C. W., Tenner, A. J., and Cummings, B. J. (1996) β-Amyloid converts an acute phase injury response into chronic injury responses. *Neurobiol. Aging* **17,** 723–731.
16. Cummings, B. J., Pike, C. J., Shankle, R., and Cotman, C. W. (1996) Beta-amyloid deposition and other measures of neuropathology predict cognitive status in Alzheimer's disease. *Neurobiol Aging* **17,** 921–933.
17. Cummings, B. J., Head, E., Afagh, A. J., Milgram, N. W., and Cotman, C. W.(1996) Beta-amyloid accumulation correlates with cognitive dysfunction in the aged canine. *Neurobiol Learn Mem.* **66,** 11–23.
18. Dentino, A. N., Pieper, C. F., Rao, M. K., Currie, M. S., Harris, T., Blazer, D. G., et al. (1999) Association of interleukin-6 and other biologic variables with depression in older people living in the community. *J. Am. Geriatr. Soc.* **47,** 6–11.
19. Dickson, D. W., Lee, S. C., Mattiace, L. A., Yen, S. H. C., and Brosnan, C. (1993) Microglia and cytokines in neurological disease with special reference to AIDS and Alzheimer's disease. *Glia* **7,** 75–83.
20. Eggleton, P., Reid, K. B., and Tenner, A. J. (1998) C1q–how many functions? How many receptors? *Trends Cell Biol.* **8,** 428–431.
21. Ehlerding, G., Schaeffer, J., Drommer, W., Miyata, T., Koch, K. M., and Floege, J. (1998) Alterations of synovial tissue and their potential role in the deposition of beta2-microglobulin-associated amyloid. *Nephrol. Dial. Transplant.* **13,** 1465–1475.
22. Eikelenboom, P. and Veerhuis, R. (1996) The role of complement and activated microglia in the pathogenesis of Alzheimer's disease. *Neurobiol. Aging* **17,** 673–680.
23. Ershler, W. B. (1993) Interleukin-6: a cytokine for gerontologists. *J. Am. Geriatr. Soc.* **41,** 176–181.
24. Finch, C. E.(1993) Neuron atrophy during aging: programmed or sporadic? *Trends Neurosci.* **16,** 104–110.
25. Finch, C. E. and Marchalonis, J. (1996) An evolutionary perspective on amyloid and inflammatory features of Alzheimer disease. *Neurobiol. Aging* **17,** 809–815.
26. Finch, C.E. and Sapolsky, R.M. (1999). The evolution of Alzheimer disease, the reproductive schedule, and apoE isoforms. *Neurobiol. Aging* **20,** 427–428.
27. Finch, C. E. and Kirkwood, T. B. L. (1999) *Chance, Development, and Aging.* Oxford Univerity Press, Oxford.
28. Fischer, B., Schmoll, H., Riederer, P., Bauer, J., Platt, D., and Popa-Wagner, A. (1996) Complement C1q and C3 mRNA expression in the frontal cortex of Alzheimer's patients. *J. Mol. Med.* **73,** 465–471.
29. Garlind, A., Brauner, A., Hojeberg, B., Basun, H., and Schultzberg, M. (1999) Soluble interleukin-1 receptor type II levels are elevated in cerebrospinal fluid in Alzheimer's disease patients. *Brain Res.* **826**, 112–116.

30. Gentleman, S. M., Greenberg, B. D., Savage, M. J., Noori, M., Newman, S. J., Roberts, G. W., et al. (1997) Abeta 42 is the predominant form of amyloid beta-protein in the brains of short-term survivors of head injury. *NeuroReport* **8,** 1519–1522.
31. Gitter, B. D., Cox, L. M., Rydel, R. E., and May, P. C. (1995) Amyloid beta peptide potentiates cytokine secretion by interleukin-1 beta-activated human astrocytoma cells. *Proc. Natl. Acad. Sci. USA.* **92,** 10,738–10,741.
32. Goldsmith, S. K., Wals, P., Rozovsky, I., Morgan, T. E., and Finch, C. E. (1997) Kainic acid and decorticating lesions stimulate the synthesis of C1q protein in adult rat brain. *J. Neurochem* **68,** 2046–2052.
33. Gordon, M. N., Schreier, W. A., Ou, X., Holcomb, L. A., and Morgan, D. G. (1997) Exaggerated astrocyte reactivity after nigrostriatal deafferentation in the aged rat. *J. Comp. Neurol.* **388,** 106–119.
34. Goss, J. R., Finch, C. E., and Morgan, D. G. (1990) GFAP RNA prevalence is increased in aging and in wasting mice. *Exp. Neurol.* **108,** 266–268.
35. Grewal, R. P., Morgan, T. E., and Finch, C. E. (1999) C1qB and clusterin mRNA are increased in association with sporadic ALS. *Neurosci. Lett.* **271,** 65–67.
36. Hansen, L. A., Armstrong, D. M., and Terry, R. D. (1987) An immunohistochemical quantification of fibrous astrocytes in the aging human cerebral cortex. *Neurobiol. Aging* **8,**1–6.
37. Head, E., Callahan, H., Muggenburg, B. A., Cotman, C. W., and Milgram, N. W. (1998) Visual-discrimination learning ability and beta-amyloid accumulation in the dog. *Neurobiol. Aging* **19,** 415–425.
38. Higuchi, K., Kitagawa, K., Naiki, H., Hanada, K., Hosokawa, M., and Takeda, T. (1991) Polymorphism of apolipoprotein A-II (apoA-II) among inbred strains of mice. Relationship between the molecular type of apoA-II and mouse senile amyloidosis. *Biochem. J.* **279,** 427–433.
39. Higuchi, K., Kogishi, K., Wang, J., Chen, X., Chiba, T., Matsushita, T., et al. (1998) Fibrilization in mouse senile amyloidosis is fibril conformation-dependent. *Lab. Invest.* **78,** 1535–1542.
40. Higuchi, K., Matsumura, A., Honma, A., Takeshita, S., Hashimoto, K., Hosokawa, M., et al. (1983) Systemic senile amyloid in senescence-accelerated mice. A unique fibril protein demonstrated in tissues from various organs by the unlabeled immunoperoxidase method. *Lab. Invest.* **48,** 231–240.
41. Holcomb, L., Gordon, M. N., McGowan, E., Yu, X., Benkovic, S., Jantzen, P., et al. (1998) Accelerated Alzheimer-type phenotype in transgenic mice carrying both mutant amyloid precursor protein and presenilin 1 transgenes. *Nat. Med.* **4,** 97–100.
42. Hsia, A. Y., Masliah, E., McConlogue, L., Yu, G. Q., Tatsuno, G., Hu, K., et al. (1999) Plaque-independent disruption of neural circuits in Alzheimer's disease mouse models. *Proc. Natl. Acad. Sci. USA* **96,** 3228–3233.
43. Hsiao, K. K., Borchelt, D. R., Olson, K., Johannsdottir, R., Kitt, C., Yunis, W., et al. (1995) Age-related CNS disorder and early death in transgenic FVB/N mice overexpressing Alzheimer amyloid precursor proteins. *Neuron* **15,** 1203–1218.

44. Hu, J., LaDu, M. J., and Van Eldki, L. J. (1998) Apolipoprotein E attenuates β-amyloid-induced astrocyte activation. *J. Neurochem.* **71,** 1626–1634.
45. Hu, J., Akama, K. T., Krafft, G. A, Chromy, B. A., and Van Eldik, L. J. (1998) Amyloid-beta peptide activates cultured astrocytes: morphological alterations, cytokine induction and nitric oxide release. *Brain Res.* **785,** 195–206.
46. Hyman, B. T., Marzloff, K., and Arriagada, P. V. (1993) The lack of accumulation of senile plaques or amyloid burden in Alzheimer's disease suggests a dynamic balance between amyloid deposition and resolution. *J. Neuropathol. Exp. Neurol.* **52,** 594–600.
47. Jacobson, D. R., Pastore, R. D., Yaghoubian, R., Kane, I., Gallo, G., Buck, F. S., et al. (1997) Variant-sequence transthyretin (isoleucine 122) in late-onset cardiac amyloidosis in black Americans. *N. Engl. J. Med.* **336,** 466–473.
48. Joachim, C. L., Mori, H., and Selkoe, D. J. (1989) Amyloid beta-protein deposition in tissues other than brain in Alzheimer's disease. *Nature* **341,** 226–230.
49. Johnson, S. A., Lampert-Etchells, M., Rozovsky, I., Pasinetti, G., and Finch, C. E. (1992) Complement mRNA in the mammalian brain: responses to Alzheimer's disease and experimental lesions. *Neurobiol. Aging* **13,** 641–648.
50. Johnson, S. A., Pasinetti, G. M., and Finch, C. E. (1994) Expression of complement C1qB and C4 mRNAs during rat brain development. *Dev. Brain Res.* **80,** 163–174.
51. Jucker, M., Walker, L. C., Kuo, H., Tian, M., and Ingram, D. K. (1994) Age-related fibrillar deposits in brains of C57BL/6 mice. A review of localization, staining characteristics, and strain specificity. *Mol. Neurobiol.* **9,** 125–133.
52. Katzman, R. and Kawas, C.H. (1994) The epidemiology of Alzheimer disease, in *Alzheimer Disease* (Terry, R. D., Katzman, R., and Bick, K. L., eds.) Raven, New York, pp. 105–122.
53. Kawamura, S., Takahashi, M., Ishihara, T., and Uchino, F. (1995) Incidence and distribution of isolated atrial amyloid: histologic and immunohistochemical studies of 100 aging hearts. *Pathol. Int.* **45,** 335–342.
54. Kisilevsky, R. (1994) Inflammation-associated amyloidogenesis: lessons for Alzheimer's amyloidogenesis. *Mol. Neurobiol.* **8,** 65–66.
55. Kisilevsky, R. and Fraser, P. E. (1997) A beta amyloidogenesis: unique, or variation on a systemic theme? *Crit. Rev. Biochem. Mol. Biol.* **32,** 361–404.
56. Krekoski, C. A., Parhad, I. M., Fung, T. S., and Clark, A. W. (1996) Aging is associated with divergent effects on Nf-L and GFAP transcription in rat brain. *Neurobiol. Aging* **17,** 833–841.
57. Kristal, B. S. and Yu, B. P. (1992) An emerging hypothesis: synergistic induction of aging by free radicals and Maillard reactions. *J Gerontol.* **47,** B107–B114.
58. Krohn, K., Rozovsky, I., Wals, P., Teter, B., Anderson, C. P., and Finch, C. E. (1999) Glial fibrillary acidic protein (GFAP) transcription responses to TGF-β1 and IL-1β are mediated by an NF-1 like site in the near-upstream promoter. *J. Neurochem.* **72,** 1353–1361.
59. Kunze, W. P. (1979) Senile pulmonary amyloidosis. *Pathol. Res. Pract.* **164,** 413–422.

60. Lambert, M. P., Barlow, A. K., Chromy, B., Edwards, C., Freed, R., Liosatos, M., et al. (1998) Diffusible, non-fibrillar ligands derived from Ab $_{1\text{-}42}$. *Proc. Natl. Acad. Sci. USA* **95,** 6448–6453.
61. Lampert-Etchells, M., Johnson, S. A., and Finch, C. E. (1991) Alzheimer's and normal brain contain mRNA for complement components C1qB, C3, and C4. *Soc. Neurosci. Abstr.* **17,** 196.
62. Lampert-Etchells, M., Pasinetti, G. M., Finch, C. E., and Johnson, S. A. (1993) Regional localization of cells containing C1qb and C4 mRNAs in the frontal cortex during Alzheimer disease. *Neurodegeneration* **2,** 111–121.
63. Lanzrein, A. S., Johnston, C. M., Perry, V. H., Jobst, K. A., King, E. M., and Smith, A. D. (1998) Longitudinal study of inflammatory factors in serum, cerebrospinal fluid, and brain tissue in Alzheimer disease: interleukin-1beta, interleukin-6, interleukin-1 receptor antagonist, tumor necrosis factor-alpha, the soluble tumor necrosis factor receptors I and II, and alpha1-antichymotrypsin. *Alzheimer Dis. Assoc. Disord.* **12,** 215–227.
64. Laping, N. J., Teter, B., Nichols, N. R., Rozovsky, I., and Finch, C. E. (1994) Glial fibrillary acidic protein: regulation by hormones, cytokines, and growth factors. *Brain Pathol.* **4,** 259–274.
65. Laping, N. J., Teter, B., Anderson, C., O'Callaghan, J. P., Johnson, S. A., and Finch, C. E. (1994) Age-related increases in glial fibrillary acidic protein are not associated with proportionate changes in transcription rates or DNA methylation in the cerebral cortex and hippocampus of male rats. *J. Neurosci. Res.* **39,** 710–717.
66. Lindsay, J. D., Landfield, P. W., and Lynch, G. (1979) Early onset and topographical distribution of hypertrophied astrocytes in hippocampus of aging rats: a quantitative study. *J. Gerontol.* **34,** 661–671.
67. Loeffler, D.A., Brickman, C.M., Juneau, P.L., Perry, M.F,, Pomara, N., and Lewitt, P.A. (1997) Cerebrospinal fluid C3a increases with age, but does not increase further in Alzheimer's disease. *Neurobiol. Aging* **18,** 555–557.
68. Marz, P., Heese, K., Hock, C., Golombowski, S., Muller-Spahn, F., Rose-John, S., et al. (1997) Interleukin-6 (IL-6) and soluble forms of IL-6 receptors are not altered in cerebrospinal fluid of Alzheimer's disease patients. *Neurosci. Lett.* **239,** 29–32.
69. McCarthy, R. E., 3rd, and Kasper, E. K. (1998) A review of the amyloidoses that infiltrate the heart. *Clin. Cardiol.* **21,** 547–552.
70. Meyer, M. R., Tschanz, J. T., Norton, M. C., Welsh-Bohmer, K. A., Steffens, D. C., Wyse B. W., et al. (1998) APOE genotype predicts when—not whether—one is predisposed to develop Alzheimer disease. *Nat. Genet.* **19,** 321–322.
71. Mor, G.,Yue, W., Santen, R. J., Gutierrez, L., Eliza, M., Berstein, L. M., et al. (1998) Macrophages, estrogen and the microenvironment of breast cancer. *J. Steroid Biochem. Mol. Biol.* **67,** 403–411.
72. Morgan, T. E., Nichols, N. R., Pasinetti, G. M., and Finch, C. E. (1993) TGF-β1 mRNA increases in macrophage/microglia cells of the hippocampus in response to deafferentation and kainic acid-induced in neurodegeneration. *Exp. Neurol.* **120,** 291–301.

73. Morgan, T. E., Rozovsky, I., Goldsmith, S. K., Stone, D. J., Yoshida, T., and Finch, C. E. (1997) Increased transcription of the astrocyte gene GFAP during middle-age is attenuated by food restriction: implications for the role of oxidative stress. *Free Radical Biol. Med.* **23,** 524–528.
74. Morgan, T. E., Wals, P. A., Mohtashemi, I., Stine, W. B., Klein, W. L., Krafft, G. A., et al. (1999) Aβ-derived diffusible ligands (ADDLs): clusterin (apo J), Congo red binding and toxicity. *Soc. Neurosci. Abstr.***25,** 2130.
75. Morgan, T. E., Xie, Z., Goldsmith, S., Yoshida, T., Lanzrein, A.-S., Stone, D., et al. (1999) The mosaic of brain glial hyperactivity during normal aging and its attenuation by food restriction. *Neuroscience* **89,** 687–699.
76. Morris, J. C., Storandt, M., McKeel, D. W., Jr, Rubin, E. H., Price, J. L., Grant, E. A., et al. (1996) Cerebral amyloid deposition and diffuse plaques in "normal" aging: evidence for presymptomatic and very mild Alzheimer's disease. *Neurology* **46,** 707–719.
77. Mrak, R. E., Groffom, S. T., and Graham D. I. (1997) Aging-associated changes in human brain. *J. Neuropathol. Exp. Neurol.* **56,** 1269–1275.
78. Mucchiano, G., Cornwell, G. G., 3d., and Westermark, P. (1992) Senile aortic amyloid. Evidence for two distinct forms of localized deposits. *Am. J. Pathol.* **140,** 871–877.
79. Nichols, N. R., Day, J. R., Laping, N. J., Johnson, S. A., and Finch, C. E. (1993) GFAP mRNA increases with age in rat and human brain. *Neurobiol. Aging* **14,** 421–429.
80. O'Callaghan, J. P. and Miller, D. B. (1991) The concentration of GFAP increases with age in the mouse and rat brain. *Neurobiol. Aging* **12,** 171–174.
81. Oda, T., Wals, P., Osterburg, H. H., Johnson, S. A., Pasinetti, G. M., Morgan, T. E., et al. (1995) Clusterin (apoJ) alters the aggregation of amyloid β-peptide (Aβ_{1-42}) and forms slowly sedimenting A complexes that cause oxidative stress. *Exp. Neurol.* **136,** 22–31.
82. Papanicolaou, D. A., Wilder, R. L., Manolagas, S. C., and Chrousos, G. P. (1998) The pathophysiologic roles of interleukin-6 in human disease. *Ann. Intern. Med.* **128,** 127–137.
83. Pasinetti, G. M., Johnson, S. A., Rozovsky, I., Lampert-Etchells, M., Morgan, D. G., Gordon, M. N., et al. (1992) Complement C1qB and C4 mRNA responses to lesioning in rat brain (1992) *Exp. Neurol.* **118,** 117–125.
84. Pasinetti, G. M., Hassler, M., Stone, D., and Finch, C. E. (1999) Glial gene expression during aging in rat striatum and in long-term responses to 6-OHDA lesions. *Synapse* **31,** 278–284.
85. Pepys, M. B. (1988) Amyloidosis, in *Immunological Diseases*, 4th ed. (Samter, M., ed.) Little, Brown, and Co., Boston, Vol I, pp. 631–674.
86. Perry, V. H., Matyszak, M., and Fearn, S. (1993) Altered antigen expression of microglia in aged rodent CNS. *Glia* **7,** 60–67.
87. Pike, C. J., Cummings, B. J., Monzavi, R., and Cotman, C. W. (1994) Beta-amyloid-induced changes in cultured astrocytes parallel reactive astrocytosis associated with senile plaques in Alzheimer's disease. *Neuroscience* **63,** 517–531.

88. Price, D. L., Martin, L. J., Sisodia, S. S., Walker, L. C., and Cork, L. C. (1992) Alzheimer's disease-type brain abnormalities in animal models. *Prog. Clin. Biol. Res.* **379,** 271–287.
89. Price, D. L., Tanzi, R. E., Borchelt, D. R., and Sisodia, S. S. (1998) Alzheimer's disease: genetic studies and transgenic models. *Annu. Rev. Genet.* **32,** 461–493.
90. Rabek, J. P., Scott, S., Hsieh, C. C., Reisner, P. D., and Papaconstantinou, J. (1998) Regulation of LPS-mediated induction of C/EBP delta gene expression in livers of young and aged mice. *Biochim. Biophys. Acta* **1398,** 137–247.
91. Rogers, J., Webster, S., Lue, L. F., Brachova, L., Civin, W. H., Emmerling, M., et al. (1996) Inflammation and Alzheimer's disease pathogenesis. *Neurobiol. Aging* **17,** 681–686.
92. Roubenoff, R., Harris, T. B., Abad, L. W., Wilson, P. W., Dallal, G. E., and Dinarello, C. A. (1998) Monocyte cytokine production in an elderly population: effect of age and inflammation. *J. Gerontol. A. Biol. Sci. Med. Sci.* **53,** M20–M26.
93. Rozemuller, J. M., van der Valk, P., and Eikelenboom, P. (1992) Activated microglia and cerebral amyloid deposits in Alzheimer's disease. *Res. Immunol.* **143,** 646–649.
94. Rozovsky, I., Morgan, T. E., Willoughby, D. A., Dugich-Djordevich, M. N., Pasinetti, G. M., Johnson, S. A., et al. (1994) Selective expression of clusterin (SGP-2) and complement C1q and C4 during responses to neurotoxins *in vivo* and *in vitro*. *Neuroscience* **62,** 741–758.
95. Rozovsky, I., Finch, C. E., and Morgan, T. E. (1998) Age-related activation of microglia and astrocytes: in vitro studies show persistence of phenotypes of aging, increased proliferation, and resistance to down-regulation. *Neurobiol. Aging* **19,** 97–103.
96. Scheinberg, M. A., Cathcart, E. S., Eastcott, J. W., Skinner, M., Benson, M., and Shirahama, T. (1976) The SJL/J mouse: a new model for spontaneous age-associated amyloidosis. I. Morphologic and immunochemical aspects. *Lab. Invest.* **35,** 47–54.
97. Schmidt, A. M., Yan, S. D., Wautier, J. L., and Stern, D. (1999) Activation of receptor for advanced glycation end products: a mechanism for chronic vascular dysfunction in diabetic vasculopathy and atherosclerosis. *Circ. Res.* **84,** 489–497.
98. Schwartz, P. (1970) *Amyloidosis: Cause and Manifestations of Senile Deterioration*. Charles C C Thomas, Springfield, IL.
99. Scott, S. A. and Mandybur, T. I. (1996) Astrocytic and microglial alterations in the aged mouse brain, in *Pathobiology of the Aging Mouse* (Mohr, U., ed.), ISLI, Washington, DC, Vol. 2, pp. 39–52.
100. Sell, D. R., Lane, M. A., Johnson, W. A., Masoro, E. J., Mock, O. B., Reiser, K. M., et al. (1996) Longevity and the genetic determination of collagen glycoxidation kinetics in mammalian senescence. *Proc. Natl. Acad. Sci. USA.* **93,** 485–490.
101. Sierra, F., Coeytaux, S., Juillerat, M., Ruffieux, C., Gauldie, J., and Guigoz, Y. (1992) Serum T-kininogen levels increase two to four months before death. *J. Biol. Chem.* **267,** 10,665–10,669.

102. Simmons, L. K., May, P. C., Tomaselli, K. J., Rydel, R. E., Fuson, K. S., Brigham, E. F., et al. (1994) Secondary structure of amyloid beta peptide correlates with neurotoxic activity in vitro. *Mol. Pharmacol.* **45,** 373–379.
103. Singhrao, S. K., Morgan, B. P., Neal, J. W., and Newman, G. R. (1995) A functional role for corpora amylacae based on evidence from complement studies. *Neurodegeneration* **4,** 335–345.
104. Sipe, J. D. (1994) Amyloidosis. *Crit. Rev. Clin. Lab. Sci.* **31,** 325–354.
105. Smyth, M. D., Cribbs, D. H., Tenner, A. J., Shankle, W. R., Dick, M., Kesslak, J. P., et al. (1994) Decreased levels of C1q in cerebrospinal fluid of living Alzheimer patients correlate with disease state. *Neurobiol. Aging* **15,** 609–614.
106. Sobel, E., Louhija, J., Sulkava, R., Davanipour, Z., Kontula, K., Miettinen, H., et al. (1995) Lack of association of apolipoprotein E allele epsilon 4 with late-onset Alzheimer's disease among Finnish centenarians. *Neurology* **45,** 903–907.
107. Price, D. L., Tanzi, R. E., Borchelt, D. R., and Sisodia, S. S. (1998) Alzheimer's disease: genetic studies and transgenic models. *Annu. Rev. Genet.* **32,** 461–493.
108. Rabek, J. P., Scott, S., Hsieh, C. C., Reisner, P. D., and Papaconstantinou, J. (1998) Regulation of LPS-mediated induction of C/EBP delta gene expression in livers of young and aged mice. *Biochim. Biophys. Acta* **1398,** 137–247.
109. Takao, T., Nagano, I., Tojo, C., Takemura, T., Makino, S., Hashimoto, K., et al. (1996) Age-related reciprocal modulation of interleukin-1beta and interleukin-1 receptors in the mouse brain–endocrine–immune axis. *Neuroimmunomodulation.* **3,** 205–212.
110. Tekirian, T. L., Cole, G. M., Russell, M. J., Yang, F., Wekstein, D. R., Patel, E., et al. (1996) Carboxy terminal of beta-amyloid deposits in aged human, canine, and polar bear brains. *Neurobiol. Aging* **17,** 249–257.
111. Terry, R. D., Gonatas, N. K., and Weiss, M. (1964) Ultrastructural studies in Alzheimer's presenile dementia. *Am. J. Pathol.* **44,** 269–297.
112. Terry, R. D., Masliah, E., Salmon, D. P., Butters, N., DeTeresa, R., Hill, R., et al. (1991) Physical basis of cognitive alterations in Alzheimer's disease: synapse loss is the major correlate of cognitive impairment. *Ann. Neurol.* **30,** 572–580.
113. Terry, R. D., Masliah, E., and Hansen, L. A. (1994) Structural basis of the cognitive alterations in Alzheimer disease, in *Alzheimer Disease* (Terry, R. D., Katzman, R., and Bick, K. L., eds.) Raven, New York, pp. 179–196.
114. Thal, D. R., Sassin, I., Schultz, C., Haass, C., Braak, E., and Braak, H. (1999) Fleecy amyloid deposits in the internal layers of the human entorhinal cortex are comprised of N-terminal truncated fragments of Abeta. *J. Neuropathol. Exp. Neurol.* **58,** 210–216.
115. Thung, P. J. (1957) The relation between amyloid and ageing in comparative pathology. *Gerontology* **1,** 234–254.
116. Tolias, C.M., McNeil, C.J., Kazlauskaite, J., and Hillhouse, E.W. (1999) Superoxide generation from constitutive nitric oxide synthase in astrocytes in vitro regulates extracellular nitric oxide availability. *Free Radical Biol. Med.* **26,** 99–106.
117. Thung, P. J. (1957) The relation between amyloid and ageing in comparative pathology. *Gerontology* **1,** 234–254.

118. Uchida, K., Okuda, R., Yamaguchi, R., Tateyama, S., Nakayama, H., and Goto, N. (1993) Double-labeling immunohistochemical studies on canine senile plaques and cerebral amyloid angiopathy. *J. Vet. Med. Sci.* **55,** 637–642.
119. Urban, B. A., Fishman, E. K., Goldman, S. M., Scott, W. W., Jr, Jones, B., Humphrey, R. L., et al. (1993) CT evaluation of amyloidosis: spectrum of disease. *Radiographics* **13,** 1295–1308.
120. Utz, J. P., Swensen, S. J., and Gertz, M. A. (1996) Pulmonary amyloidosis. The Mayo Clinic experience from 1980 to 1993. *Ann. Intern. Med.* **124,** 407–413.
121. Van Muiswinkel, F. L., Raupp, S. F., de Vos, N. M., Smits, H. A., Verhoef, J., Eikelenboom, P., et al. (1999) The amino-terminus of the amyloid-beta protein is critical for the cellular binding and consequent activation of the respiratory burst of human macrophages. *J. Neuroimmunol.* **96,** 121–130.
122. Velazquez, P., Cribbs, D. H., Poulos, T. L., and Tenner, A. J. (1997) Aspartate residue 7 in amyloid beta-protein is critical for classical complement pathway activation: implications for Alzheimer's disease pathogenesis. *Nat. Med.* **3,** 77–79.
123. Walker, D. G. and McGeer, P. L. (1992) Complement gene expression in human brain: comparison between normal and Alzheimer disease cases. *Brain Res. Mol. Brain. Res.* **14,** 109–116.
124. Walter, R., Murasko, D. M., and Sierra, F. (1998) T-kininogen is a biomarker of senescence in rats. *Mech. Ageing Dev.* **106,** 129–144.
125. Webster, S., Bonnell, B., and Rogers, J. (1997) Charge-based binding of complement component C1q to the Alzheimer amyloid beta-peptide. *Am. J. Pathol.* **150,** 1531–1536.
126. Wegiel, J., Wisniewski, H. M., Dziewiatkowski, J., Tarnawski, M., Dziewiatkowska, A., Morys, J., et al. (1996) Subpopulation of dogs with severe brain parenchymal beta amyloidosis distinguished with cluster analysis. *Brain Res.* **728,** 20–26.
128. Wei, M., Rozovsky, I., Lopez, L. M., Morgan, T. E., and Finch, C. E. (1999) Oxidative stress and the GFAP promoter. *Soc. Neurosci. Abstr.* **25,** 1317.
129. Wen, G. Y., Wisniewski, H. M., Blondal, H., Benedikz, E., Frey, H., Pirttila, T., et al. (1994) Presence of non-fibrillar amyloid beta protein in skin biopsies of Alzheimer's disease (AD), Down's syndrome and non-AD normal persons. *Acta Neuropathol (Berl.)* **88,** 201–206.
130. West, W. T. and Murphy, E. D. (1965) Sequence of deposition of amyloid in strain A mice and relationship to renal disease. *J. Natl. Cancer Inst.* **35,** 167–174.
131. Westermark, P., Eriksson, L., Engstrom, U., Enestrom, S., and Sletten, K. (1997) Prolactin-derived amyloid in the aging pituitary gland. *Am J Pathol.* **150,** 67–73.
132. Westermark, P., Mucchiano, G., Marthin, T., Johnson, K. H., and Sletten, K. (1995) Apolipoprotein A1-derived amyloid in human aortic atherosclerotic plaques. *Am. J. Pathol.* **147,** 1186–1192.
133. Westermark, P., Sletten, K., Naeser, P., and Natvig, J. B. (1979) Characterization of amyloid of ageing obese-hyperglycaemic mice and their lean littermates. *Scand. J. Immunol.* **9,** 193–196.

134. Wyss-Coray, T., Masliah, E., Mallory, M., McConlogue, L., Johnson-Wood, K., Lin, C., et al. (1997)Amyloidogenic role of cytokine TGF-beta1 in transgenic mice and in Alzheimer's disease. *Nature* **389,** 603–606.
135. Xie, Z., Wals, P. A., Walsh, J. P., Finch, C. E., and Morgan, T. E. (1998) Characterization of clusterin-induced microglial activation. *Soc. Neurosci.* **24,** 1944.
136. Yamaguchi, H., Hirai, S., Morimatsu, M., Shoji, M., and Ihara, Y. (1988) A variety of cerebral amyloid deposits in the brains of the Alzheimer-type dementia demonstrated by beta protein immunostaining. *Acta Neuropathol. (Berl.)* **76,** 541–549.
137. Yasojima, K., Schwab, C., McGeer, E. G., and McGeer, P. L. (1999) Up-regulated production and activation of the complement system in Alzheimer's disease brain. *Am. J. Pathol.* **154,** 927–936.
138. Yasuma, T., Arai, K., and Suzuki, F. (1992) Age-related phenomena in the lumbar intervertebral discs. Lipofuscin and amyloid deposition. *Spine* **17,** 1194–1198.
139. Yoshida, T., Goldsmith, S., Morgan, T. E., Stone, D., and Finch, C. E. (1996) Transcription supports age-related increases of GFAP gene expression in the male rat brain. *Neurosci. Lett.* **215,** 107–110.
140. Zanjani, H., Lanzrein, A. S., McKeel, D. W., Morris, J. C., and Finch, C. E. (1999) clusterin and complement components C1q and C3 deposits increased at very early stages of Alzheimer disease (AD). *Soc. Neurosci. Abstr.* **25,** 1102.
141. Zhan, S. S., Veerhuis, R., Janssen, I., Kamphorst, W., and Eikelenboom, P. (1994) Immunohistochemical distribution of the inhibitors of the terminal complement complex in Alzheimer's disease. *Neurodegeneration* **3,** 111–117.
142. Zhan, S. S., Veerhuis, R., Kamphorst, W., and Eikelenboom, P. (1995) Distribution of beta amyloid associated proteins in plaques in Alzhemer's disease and in the non-demented elderly. *Neurodegeneration* **4,** 291–297.
143. Zimmermann, J., Herrlinger, S., Pruy, A., Metzger, T., and Wanner, C. (1999) Inflammation enhances cardiovascular risk and mortality in hemodialysis patients. *Kidney Int.* **55,** 648–658.

5

Proteolysis in Neurodegenerative Diseases

Vivian Y. H. Hook and Liane Mende-Mueller

5.1. INTRODUCTION

It has become increasingly evident that proteolytic events are primary mechanisms in the development of major neurodegenerative diseases that include Alzheimer's disease (AD), Huntington's disease (HD), and Parkinson's disease (PD). This chapter will illustrate the theme that proteolytic mechanisms are primary contributors to the pathogenesis of AD, HD, and PD (Fig. 1). In each of these diseases, genetic mutations result in expression of protein precursors that undergo limited proteolysis to result in the formation of neurotoxic peptides. Of paramount importance is the deposition of each of these toxic peptide fragments as protein aggregates in the brain, which are manifested as specific neuropathologies. These gene mutations and resultant peptide fragments eventually contribute to the behavioral abnormalities that are characteristic of each of these neurodegenerative diseases — AD, HD, and PD (Table 1).

Specifically in Alzheimer's disease (AD), genetic mutations in the amyloid precursor protein (APP) gene and the presenilin 1 and 2 genes (reviewed in refs. *1–3*) result in enhanced conversion of APP into the smaller, neurotoxic β-peptide (Aβ) in the brain. Aβ becomes deposited in extracellular brain amyloid plaques, the hallmark of AD neuropathology. Evidence from studies of transgenic mice expressing these mutant genes suggests that elevated Aβ leads to cognitive deficits in AD *(1–3)*.

In Huntington's disease (HD), the *IT15* gene undergoes expansion in CAG trinucleotide repeats, resulting in an expanded polyglutamine domain in the huntingtin protein *(4,5)*. Of particular interest is the finding that an NH_2-terminal fragment(s) of huntingtin becomes deposited in nuclear inclusions in the brains of HD patients *(6)* and transgenic mice *(7,8)*. These neuropathologic inclusions are implicated in the pathogenesis of HD.

With respect to Parkinson's disease (PD), genetic mutations in the *α-synuclein* gene have recently been found to be linked to PD *(9,10)*.

From: *Contemporary Clinical Neuroscience: Molecular Mechanisms of Neurodegenerative Diseases*
Edited by: M.-F. Chesselet © Humana Press Inc., Totowa, NJ

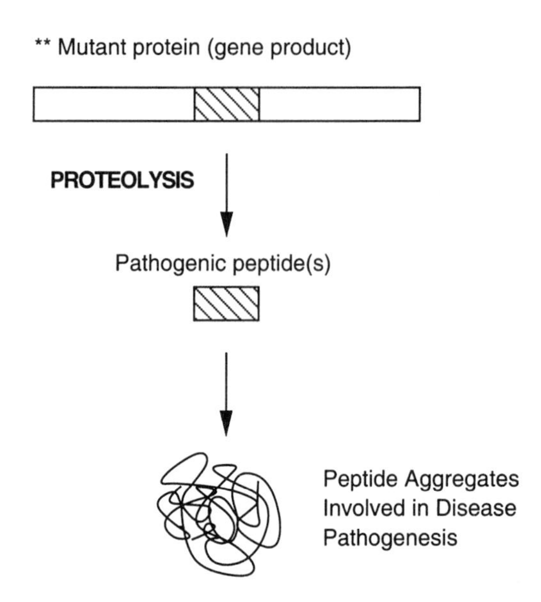

Fig. 1. Pathogenesis of mutant disease products by proteolysis. A common feature of mechanisms involved in AD, HD, and PD neurodegenerative diseases involve proteolysis of the mutant protein gene product. Proteolysis results in peptide fragment(s) derived from the protein precursor. These neurotoxic peptide fragments become incorporated into protein aggregrates that are involved in the pathogenesis of neurodegenerative diseases.

Table 1
Neurodegenerative Disease Gene Products in Pathological Brain Deposits

Neurodegenerative Disease gene	Pathogenic peptide generated by proteolysis of gene product	Neuropathological deposits
Amyloid precursor protein (APP) in Alzheimer's disease	Aβ peptides	Amyloid plaques
Huntingtin protein in Huntington's disease	NH_2-terminal huntingtin protein fragments	Nuclear inclusions
α-Synuclein in Parkinson's disease	NAC (nonamyloid component)	Lewy bodies

The α-synuclein gene product is represented by the NCAP protein (non-Aβ component precursor) *(11–13)* that is proteolytically cleaved to form the NAC peptide that possesses amyloidogenic properties *(14–16)*. Moreover, α-synuclein is present within Lewy bodies that are characteristic of PD brain neuropathology.

It is clear that proteases are important in AD, HD, and PD neurode-generative mechanisms. However, the specific brain proteases responsible for these proteolytic events have not been identified. In this chapter, the proteolytic processing of APP, the huntingtin protein, and α-synuclein will be discussed as a means to understanding the properties of the proteases that contribute to these neurodegenerative diseases.

5.2. ALZHEIMER'S DISEASE

Clinical Features

Alzheimer's disease is a progressive cognitive disorder that generally appears in the sixth or seventh decade of life and results in a gradual degeneration of memory and cognitive processes. The cognitive deficits incapacitate the AD patients to the point of total dependency on supplemental health care for survival. Although several palliative drug therapies are available for treating AD, none are effective for any length of time *(17)*. Ten percent of people over 65 yr of age are afflicted with AD. With the gradual aging of the American population, it is predicted that a larger fraction of the population will be affected by this disease.

Elevated Aβ Peptides in Amyloid Plaques of AD Brains

Alzheimer's disease is diagnosed postmortem by typical neuropathology illustrated by the presence of amyloid plaques, especially in hippocampal and cortical brain regions *(1–3)*. The primary component of amyloid plaques is the β-peptide (or Aβ peptide) that consists of three peptide forms that differ in their COOH-termini (Fig. 2). These β peptides are comprised of the Aβ1–40, Aβ1–42, and Aβ1–43 forms that possess 40, 42, or 43 residues of the same primary sequence differing only at their COOH-termini. The Aβ1–42 form contains two additional amino acids at the COOH-terminus of the Aβ1–40 peptide form, and the Aβ1–43 form contains one additional amino acid at its COOH-terminus compared to the Aβ1–42 form. The three forms of Aβ are first synthesized as the larger APPs *(18–22)*. Clearly, proteolytic processing of APP is required to generate the smaller Aβ peptides.

All forms of Aβ accumulate in AD brains and have been shown in numerous studies to be neurotoxic *(23–25)*. Moreover, there appears to be a preferential elevation of the Aβ1–42 and Aβ1–43 peptide forms in AD. Genetic mutations in the APP gene and in the presenilin 1 and 2 genes have been characterized in transgenic mice and in AD patients *(1–3)*; these studies show a strong relationship between APP and presenilin gene mutations with the observed increases in levels of Aβ1–42 and Aβ1–43, amyloid plaque

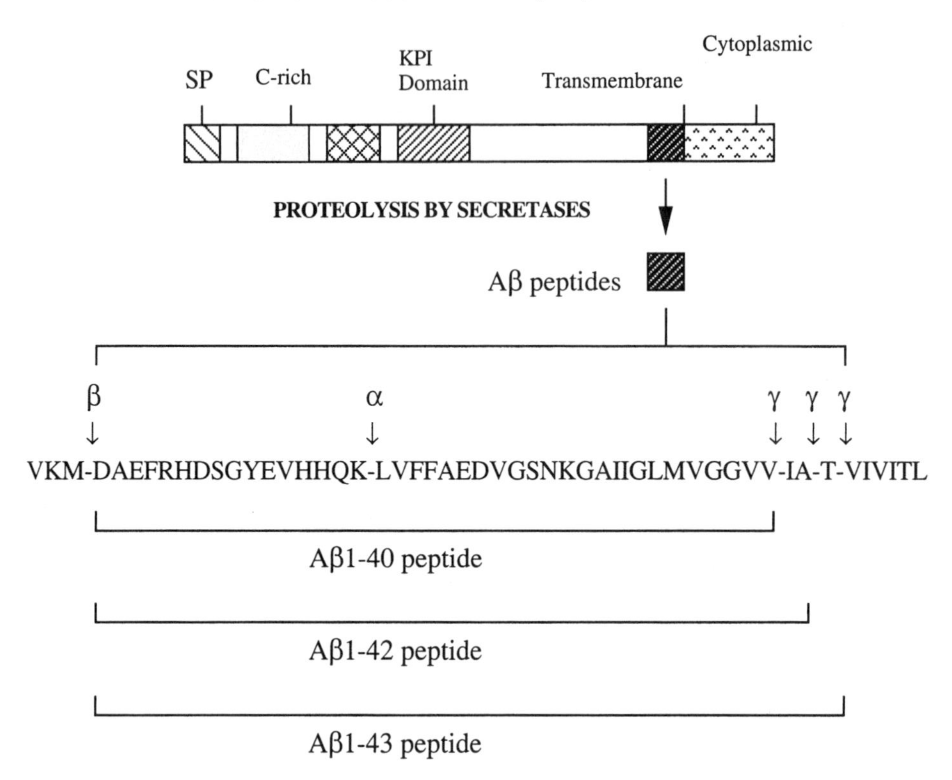

Fig. 2. Amyloid precursor protein (APP), Aβ peptides, and secretases. The structure of APP is schematically illustrated. APP contains a signal peptide (SP) sequence, a cysteine-rich region (C-rich), a KPI (kunitz protease inhibitor) domain for APP-751 and APP-770 forms (APP-695 lacks the KPI domain), a transmembrane domain, and a cytoplasmic COOH-terminal domain. Most importantly, Aβ peptides are present within APP near the transmembrane domain. The APP precursor undergoes proteolysis by secretases to generate the Aβ peptides that are known to be neurotoxic, and accumulate in amyloid plaques in AD brains. Proteolytic processing of APP generates three forms of Aβ peptides of 40, 42, and 43 amino acids in length. These peptide forms possess the same NH_2-terminus beginning with Asp; they differ in their COOH-termini, as illustrated. Proteolysis at the β-secretase site generates the NH_2-terminus of Aβ peptides. Proteolysis at the COOH-termini of Aβ peptides occurs at γ-secretase sites; it is noted that there are three different γ-secretase sites. Proteolysis may also occur within the Aβ peptide at the α-secretase site, which precludes formation of Aβ peptides. Although the APP processing sites have been named the β-, γ,- and α-secretases, the protease responsible for cleavage at these sites have not yet been definitively identified.

formation, and cognitive deficits. These data support the hypothesis of a causal relationship between enhanced Aβ production and amyloid plaques in AD brains associated with the cognitive deficits that are characteristic of AD.

APP Processing by Secretase Enzymes

All forms of Aβ peptides are derived from a larger precursor protein, the amyloid precursor protein (APP) (Fig. 2). There are three major forms of APP consisting of 695, 751, and 771 amino acids, which result from alternative splicing of the APP gene product. Each form of APP contains the Aβ peptides *(18–22)*. The APP-751 and APP-771 forms include a kunitz protease inhibitor domain.

Proteases, known as ''secretases,'' produce Aβ peptides by cleaving APP at specific peptide bonds at or near the NH_2- and COOH-termini of the Aβ peptide sequences within APP *(1,2,18–22)*. The secretases are categorized according to their specific cleavage sites within APP, which are related to the production of Aβ peptides. The secretase that cleaves at the NH_2-terminal end of Aβ is known as the β-secretase. The β-secretase is predicted to cleave between Met-↓Asp to generate the NH_2-terminus of Aβ. The protease(s) that cleaves at the COOH-termini of Aβ are known as γ-secretase(s), which determine whether Aβ1–40, Aβ1–42, or Aβ1–43 are produced. Production of Aβ1–40 would require γ-secretase cleavage between Val-↓Ile. Aβ1–42 and Aβ1–43 production would require γ-secretase cleavages between Ala-↓Thr and Thr-↓Val, respectively. It is not known whether different γ-secretases produce the three different forms of Aβ peptides. However, because specific increases in Aβ1–42 and Aβ1–43 occur in AD, compared to lesser changes in Aβ1–40, it is likely that several γ-secretases exist to generate the COOH-termini of the different Aβ peptides.

In addition to β- and γ-secretases, normal cleavage within the Aβ sequence occurs between Lys-↓Leu by α-secretase *(1,2)*. Therefore, α-secretase cleavage of APP precludes formation of Aβ peptides.

Genetic studies point to a critical role of the β- and γ–secretases to increase the production of Aβ peptides in AD, especially the extended Aβ1–42 and Aβ1–43 peptide forms. Mutations in the APP gene, which are located near secretase processing sites within APP, are genetically linked to AD in certain families *(1–3,26–29)*. Transgenic mice that overexpress mutant APPs develop brain amyloid plaques, show elevated Aβ peptide levels in the brain, and display deficits in cognition and memory *(1–3,30–32)*. Moreover, numerous AD-linked genetic mutations in the presenilin 1 and 2 genes enhance the production of Aβ and favor the elevation of Aβ1–42 and Aβ1–43 over Aβ1–40 in transgenic mice *(33,34)* and tranfected cell lines *(34–36)*. The selective increase in Aβ1–42 and Aβ1–43 by mutant presenilins suggests that different γ-secretases may be responsible for producing the Aβ peptide forms.

APP Trafficking and Processing in the Secretory Pathway

Neuronal peptides destined for secretion are typically routed to the secretory pathway to allow for release of these peptides into the extracellular environment. Studies of the cellular trafficking of APP and its processing are important to define the possible locations of secretases within the cell. Thus, although the secretases themselves have not been found, numerous studies have established that APP subcellular trafficking and processing occur in the secretory pathway *(1–3)*.

The deduced primary sequence of the human APP cDNA indicates that it possesses an NH_2-terminal signal sequence, which serves as a mechanism to route translated proteins to the secretory pathway. The secretion of peptides routed to the secretory pathway are typically stimulated by neuronal receptor activation; indeed, muscarinic receptor stimulation of hippocampal neurons releases Aβ peptides *(37)*. In addition, APP undergoes axonal transport to nerve terminals *(38,39)*, which is consistent with trafficking of vesicles to axon terminals for secretion. Many in vivo studies have provided ample evidence for the trafficking and processing of APP in the secretory pathway.

In vitro studies of APP transfected into cell lines have provided valuable information concerning the subcellular compartments involved in APP trafficking and processing. Investigations of Aβ peptides, detected by sensitive sandwich enzyme-linked immunosorbent assays (ELISAs), show trafficking of APP in the secretory pathway, where Aβ peptide production occurs. Evidence sugggests that APP processing occurs in the early secretory pathway, including the RER (rough endoplasmic reticulum) and Golgi apparatus, and in post-Golgi vesicles *(1,2,40,41)*. In addition, a high proportion of APP exists in a membrane-bound form and becomes incorporated into the cell membrane *(1–3)*. The APP protein can be internalized from the cell surface to endosomes, where some APP processing may also occur.

These findings predict that APP processing into Aβ peptides may occur at several locations within the secretory pathway. It is, therefore, logical to predict that the corresponding secretases are present with APP in the secretory pathway. This knowledge is important for consideration of candidate secretases. For example, recent in vitro studies in transfected cells have suggested that caspases may cleave APP to Aβ, and, thus, caspases have been proposed as candidate secretases *(42)*. However, caspases are present in the cytosol of cells, whereas APP is inaccessible to cytosolic components, because APP is contained within the subcellular organelles of the secretory pathway. Therefore, caspases are not considered to be candidates *(43)*.

Several recent studies reported the identification of a candidate aspartyl protease known as BACE (β-site APP-cleaving enzyme) or Asp2 *(44–47)* for β-secretase processing of APP. This novel aspartyl protease cDNA clone was obtained by expression cloning of a human embryonic kidney cell cDNA library expressed in HEK293 cells *(44)*, purification and cloning of the human aspartyl protease *(45)*, and bioinformatic approaches to identify aspartyl proteases based on their predicted conserved active site residues *(46,47)*. BACE, or Asp2, increases Aβ formation when cotransfected with APP in cell lines. Recombinant BACE, or Asp2, has been shown to cleave at the β-secretase site. This enzyme is expressed in the brain, with the highest expression in the pancreas, as well as in the kidney and other tissues. Studies have not yet tested for colocalization of the aspartyl protease with APP, APP-derived intermediates, and Aβ within the identical cell type and subcellular compartment in vivo. Moreover, it will be important to test these candidate β-secretase enzymes in knockout mice to assess their likelihood as proteases involved in Aβ formation.

It is predicted that the secretases should be colocalized with APP and Aβ peptides in the secretory pathway. It will be most exciting when authentic secretases are established, which is now an area of intense investigation. Knowledge of the secretases is essential for understanding the proteolytic mechanisms underlying the development of Alzheimer's disease.

5.3. HUNTINGTON'S DISEASE

Clinical Features

Huntington's disease (HD) is a neurodegenerative hereditary disorder, characterized by neurological signs usually including chorea, personality change, and ultimately dementia *(5,48)*. The onset of the disease occurs in mid-life in approximately 90% of HD patients. In some cases, HD occurs in juveniles; sporadic cases of HD also occur *(49,50)*.

Huntington's disease is characterized by neuronal loss, especially of striatal neurons. Such neuronal loss may result in modified activity of the nigrostriatal dopamine pathway and lead to chorea *(51)*. Degeneration of the striatum in HD brains occurs in a gradient, with degeneration beginning dorsomedially and extending ventrolaterally. The severity of the disease is graded 0 to 4. In grade 1, 50% of neurons in the caudate nucleus are lost, and the putamen and ventral striatum are intact. However, in grade 4, almost all neurons in the dorsal striatum have been destroyed, and ventral neurons are spared; grade 4 represents the end stage of the disease *(52)*. Some affected neurons contain nuclear inclusions of protein aggregates *(6)*. The formation of nuclear inclusions has been implicated in the pathogenesis of HD because formation of neuronal inclusions precedes onset of disease symptoms in transgenic mouse models of the disease *(7)*.

Proteolysis of the Trinucleotide Repeat Expanded IT15 Gene Product: Deposition of NH_2- Terminal Fragments of the Huntingtin Protein in Nuclear Inclusions

Huntington's disease is inherited in an autosomal dominant fashion. The *IT15* gene, which carries the mutation causing HD, maps to chromosome 4 and encodes the huntingtin protein of 3144 amino acids *(4)*. Importantly, the *IT15* gene in HD contains an expanded CAG trinucleotide repeat region, which is part of the first exon. The CAG repeats result in a polyglutamine domain near the NH_2-terminus of the huntingtin protein. The repeat is polymorphic in normal brain with 8–39 repeats. In HD patients, expansions of 36–121 repeats have been reported *(53)*. The length of the repeated polyglutamine expansion is inversely correlated with the age of onset of the disease. In addition, the primary sequence of the huntingtin protein is unique *(4)* and possesses no significant homology with known proteins, except for a single leucine zipper motif *(54)*.

Importantly, NH_2-terminal fragments of the huntingtin protein are contained within nuclear inclusions of HD brains *(6)*. Using an antibody generated against 17 residues of the NH_2-terminus of the huntingtin protein, anti-1-17 immunoreactivity was detected in nuclear inclusions by immunocytochemistry and Western blots. However, nuclear inclusions were not recognized by antibodies specific for the COOH-terminal regions of the huntingtin protein. These data indicate cleavage of the intact 350-kDa (approximately) huntingtin protein, such that the NH_2-terminal fragment(s) of the huntingtin protein, containing the polyglutamine expansion, becomes deposited as protein aggregates within nuclear inclusions. Nuclear inclusions also stain positively for ubiquitin, suggesting possible ubiquitin-mediated proteosome degradation of the huntingtin protein; however, this possibility has not yet been definitively determined.

The intact huntingtin protein is normally located in the cytoplasm of cells. Thus, proteolysis of the huntingtin protein is presumably followed by nuclear translocation of NH_2-terminal fragments containing the polyglutamine region, and incorporation of these peptide fragments within nuclear inclusions (Fig. 3). In our preliminary studies, several NH_2-terminal fragments are detected by the anti-1-17 serum (Hook et al., unpublished data). However, the precise cleavage sites of the huntingtin protein have not yet been determined. Knowledge of the proteolytic cleavage sites that result in the huntingtin protein NH_2-terminal fragments, which become deposited

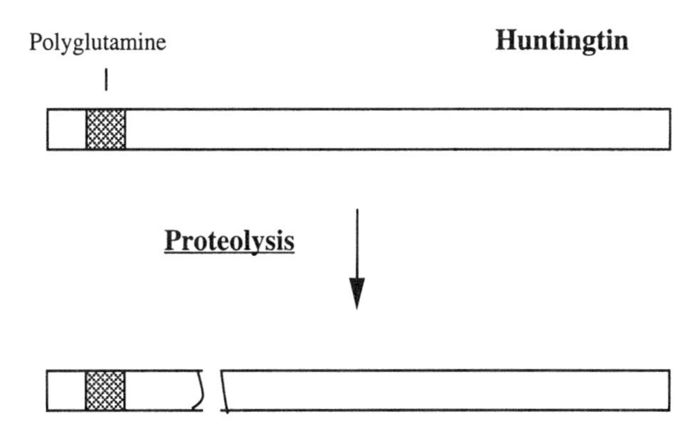

Fig. 3. Huntingtin protein: proteolysis to generate NH_2-terminal fragments. The huntingtin protein represents the gene product of the *IT15* gene that is genetically linked to Huntington's disease. The mutant gene product contains expansion of a polyglutamine region near the NH_2-terminus of huntingtin. Proteolysis generates NH_2-terminal fragment(s) that become incoporated into nuclear inclusions in affected brain neurons. The precise proteolytic cleavage sites of huntingtin have not been determined, as indicated by the ragged peptide fragments. It will be important to define the cleavage sites that will allow identification of proteases involved in huntingtin protein processing.

in nuclear inclusions, will be essential for future identification of proteases that process the huntingtin protein.

Expression of mutant NH_2-terminal fragments in transgenic mice have confirmed the role of these peptide fragments in the pathogenesis of HD. Mice expressing huntingtin protein fragments with 115–156 repeats developed brain nuclear inclusions, and mice showed behavioral symptoms of the disease *(7,8)*. These nuclear inclusions in mice were also stained by ubiquitin antibodies, suggesting involvement of a ubiquitin/proteosome system.

In *Drosophila*, expression of a huntingtin protein NH_2-terminal fragment with 120 repeats resulted in rapid degeneration of photoreceptor cells. Lower degrees of degeneration were produced when a fragment containing a fewer number of 75 repeats was expressed in *Drosophila (55)*. These results support the hypothesis that huntingtin-derived peptide fragments, containing expanded polyglutamine repeats, are involved in neurotoxicity and neurodegeneration in HD. *See* Chapter 14 for further discussion of proteolysis in HD.

Cell Biology of Normal Huntingtin Protein

Huntingtin is a cytoplasmic protein expressed in many tissues, yet the mutation of the protein only affects neuronal cells. Huntingtin protein expression is high in medium-sized striatal neurons that contain GABA and enkephalin, or GABA and substance P *(51)*. The normal protein is thought to function in vesicle trafficking *(56–58)*. The protein was localized by immunoelectron microscopy to microtubules and vesicle membranes, and western blots detected the huntingtin protein in synaptosomal fractions *(59)*. In subcellular fractionations of fibroblasts, huntingtin colocalized to clathrin-coated vesicles and with plasma membranes *(58)*. Clathrin-coated vesicles are part of the trans-Golgi network and secretory system *(60)*.

Proteolytic Mechanisms in Other Trinucleotide Repeat Neurodegenerative Diseases

Huntington's disease represents one of several neurodegenerative diseases that are the result of expansions of trinucleotide repeats in the disease gene. The spinocerebellar ataxias (SCA) types 1, 2, 3, 6, and 7 involve CAG repeats in affected genes *(61–65)* SCA3 is also known as Machado Joseph disease *(63)*. Expansions of trinucleotide repeats in affected genes are involved in spinal and bulbar muscular atrophy (SBMA, or Kennedy disease) *(66)* and dentatorubral–pallidoluysian atrophy (DRPLA) *(67)*. Among these diseases, expansions of CAG trinucleotide repeats are present in unrelated genes on different chromosomes. In each case, the expanded polyglutamine region produces toxic effects on vulnerable neurons. Notably, SCA1 (*see* Chapter 11) and SCA3 (*68; see* also Chapter 12) transgenic mouse models expressing polyglutamine expanded regions of the respective mutant genes show the formation of nuclear inclusions. These triplet repeat neurodegenerative diseases involve CAG repeats in the coding region of the mutant disease gene, resulting in a polyglutamine expansion in the mutant gene product. In contrast, several triplet repeat diseases contain expansions in noncoding sequences of the gene, which includes fragile X, myotonic dystrophy, and Friedrich's ataxia *(69)*; the roles of these gene mutations in noncoding regions in disease pathogenesis is unknown. Please see Chapter 14 for further details.

Clearly, proteolysis of proteins encoded by mutant genes containing expansions of trinucleotide repeats may represent similar molecular mechansims responsible for neurodegeneration involving mutant genes containing expansions of trinucleotide repeats.

5.4. PARKINSON'S DISEASE

Clinical Features

Parkinson's disease (PD) is the second most common neurodegenerative disease, following Alzheimer's disease. PD is a movement disorder in which affected individuals display resting tremor, rigidity, and bradykinesia (slowness in initiating movements), and it is sometimes associated with difficulty in maintaining posture. The prevalence of PD increases with aging to approximately 3.4% among those above 75 yr old *(70–72)*.

Parkinson's disease involves neuronal degeneration that results in the loss of dopaminergic cells in the substantia nigra of PD brains. The degree of dopamine depletion in the caudate nucleus and putamen correlates with loss of cells in the substantia nigra. Other dopaminergic systems in the brain are also affected, but to a lesser degree than nigrostriatal projections.

Neuropathologic characterization of PD brains indicates the presence of intracytoplasmic inclusion bodies, known as Lewy bodies (LBs) that represent a significant marker for PD. LBs are found in several brain regions including substantia nigra, locus coeruleus, hypothalamus, cerebral cortex, and other regions *(71–73)*. LBs contain neurofilaments and ubiquitin, as well as other protein components. LBs in brain are also a characteristic of certain dementia *(72–73)*.

Molecular mechanisms involved in the development of PD are thought to be multifactorial, with environmental factors influencing genetically predisposed individuals as they age. Moreover, PD most likely involves polygenic inheritance.

Genetic Mutations in the α-Synuclein and Parkin Genes in PD

Mutant genes involved in PD have only recently been identified. Two mutations in the *α-synuclein* gene have been identified in PD (autosomal dominant), consisting of missense mutations resulting in an Ala to Thr substitution at position 53 (Ala53Thr) *(9)* and an Ala to Pro substitution at position 30 (Ala30Pro) *(10)*. Importantly, α-synuclein has been identified as a major component of LBs in PD *(72,73)*, as well as in LBs in dementia and AD *(11,14–16)*. α-Synuclein belongs to a family of related proteins, which includes β- and γ-synuclein *(74,75)*; however, β- and γ-synuclein are not found in LBs *(72)*. Possible relationships between LB pathology and neuronal loss are not clear at the present time.

α-Synuclein was first isolated from vesicles from the electric organ of *Torpedo californica*; in addition, rat homologs have been described *(74,76)*. Human α-synuclein is homologous to zebra finch synelfin *(77)*. Phosphoneuroprotein 14 (PNP14) represents another member of the synuclein family and is homologous to β-synuclein *(13)*. β-Synuclein lacks the NAC (nonamyloid component peptide) peptide sequence present within α-synuclein. γ-Synuclein represents a third member of the synuclein family, which is expressed in numerous breast tumors *(74)*, suggesting a role in cancer.

Because genetic mutations of the *α-synuclein* gene involved in PD have only recently been discovered, transgenic mice expressing mutant α-synucleins have not yet been developed. Expression of the human α-synuclein Ala30Pro substitution could be expressed in transgenic mice, because the mouse *α-synuclein* gene sequence is identical to human at position 30. However, expression of the human Ala53Thr mutation in transgenic mice is not predicted to produce phenotypic effects, as the normal mouse α-synuclein gene possesses Thr at position 53, which represents the human mutation in PD. Even more recently, mutations in the *parkin* gene *(78,79)* have been identified in juvenile parkinsonism *(78)*. The newly identified parkin protein consists of 465 amino acids, with a segment possessing some homology to ubiquitin. Resemblance to ubiquitin suggests a role for yet another possible proteolytic component, parkin, in PD. Characterization of the parkin protein in brain and its possible presence in LBs has not yet been determined. Thus, it is not currently known whether the parkin protein undergoes proteolysis.

Proteolytic Processing of the α-Synuclein Gene Product

Prior to the discovery of the genetic mutations in the *α-synuclein* gene in PD, a proteolytic fragment of *α-synuclein* was originally isolated and identified from amyloid plaques of Alzheimer's disease (AD). This peptide fragment was known as the ''nonamyloid component'' (NAC) of AD amyloid plaques *(11,12)*. NAC is a 35-amino-acid peptide that is acidic and it has been demonstrated to be highly amyloidogenic *(14,15)*. Molecular cloning of NAC revealed that it is first synthesized as a precursor protein, NACP, of 19 kDa *(11)*. NACP and α-synuclein are the same protein; this protein is currently more commonly referred to as α-synuclein. LBs contain α-synuclein, it is unclear if they contain the 35-residue NAC peptide.

Clearly, α-synuclein (NACP) undergoes proteolytic processing to generate the smaller amyloidogenic peptide NAC (Fig. 4). It is of interest to note that NAC is flanked by monobasic or dibasic lysines at its NH_2- and COOH-termini within the α-synuclein precursor. These basic residues suggest proteolytic processing at these sites to generate the NAC peptide. Specific proteases cleaving at these basic lysine residue processing sites within α-synuclein would be essential in the production and deposition of NAC in amyloid plaques of AD. It will also be important to assess whether the NAC peptide is also a component of LBs in PD.

α-Synuclein (NACP) is colocalized with synaptic vesicles, as determined by the colocalization with subcellular organelles containing the vesicle protein synaptophysin *(12,80)*. Comparisons of the subcellular localization of α-synuclein (NACP) and its proteolytic peptide product NAC have not yet been determined. Such knowledge will be helpful in predicitng the location of α-synuclein-cleaving protease(s).

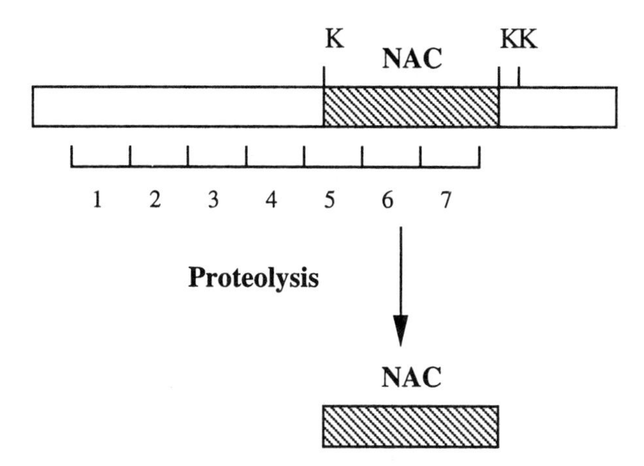

Fig. 4. Proteolysis of α-synuclein to generate the NAC peptide. The structure of the α-synuclein precursor contains the NAC peptide. α-synuclein also contains seven repetitive degenerate sequences of the consensus sequence KTKEGV (indicated by the numbered regions 1–7). Proteolysis of α-synuclein (also known as NACP, for NAC precursor) is required to generate the NAC peptide product. Proteolytic processing of the precursor at single and paired lysine residues that flank NAC at its NH_2- and COOH-termini is required to generate NAC. Proteases responsible for generating NAC have not been identified.

5.5. FUTURE POTENTIAL OF PROTEASE INHIBITORS FOR THERAPEUTIC TREATMENT OF NEURODEGENERATIVE DISEASES

It is clear from the discussions presented in this chapter that proteolytic processing of mutant gene products into potentially neurotoxic peptide fragments, which are deposited in neuropathological inclusions and aggregates, represent significant mechanisms responsible for the development of AD, HD, and PD neurodegenerative diseases. Identification of the specific proteases that mediate proteolytic processing of APP in AD, the huntingtin protein in HD, and α-synuclein in PD and AD are critical to the future discovery of drug inhibitors that block proteolysis of these mutant gene products, thereby preventing the neurotoxic effects of neurodegenerative disease peptides.

It will, therefore, be important to find the authentic brain proteases that are responsible for the development of these devastating neurodegenerative diseases. These protease enzymes will provide logical drug targets for inhibition by chemical molecules as therapeutic agents for the treatment of these neurodegenerative diseases.

ACKNOWLEDGMENTS

This work was supported by grants from the Hereditary Disease Foundation and the National Institutes of Health.

REFERENCES

1. Selkoe, D. J. (1998) The cell biology of β-amyloid precursor protein and presenilin in Alzheimer's disease. *Trends Cell Biol.* **8,** 447–453.
2. Price, D. L., Tanzi, R. E., Boirchelt, D. R., and Sisodia, S.S. (1998) Alzheimer's disease: genetic studies and transgenic models. *Annu. Rev. Genet.* **32,** 461–493.
3. Price, D. L., Sisodia, S. S., and Borchelt, D. R. (1998) Genetic neurodegenerative diseases; the human illness and transgenic models. *Science* **282,** 1079–1083.
4. The Huntington's Disease Collaborative Research Group (1993) A novel gene containing a trinucleotide repeat that is expanded and unstable on Huntington's Disease chromosomes. *Cell* **72,** 971–983.
5. Vonsattel, J. P. G. and DiFiglia, M. (1998) Huntington disease. *J. Neuropathol. Exp. Neurol.* **57,** 369–384.
6. DiFiglia, M., Sapp, E., Chase, K. O., Davies, S. W., Bates, G. P., Vonsattel, J. P., et al. (1997) Aggregation of huntingtin in neuronal intranuclear inclusions and dystrophic neurites in brain. *Science* **277,** 1990–1993.
7. Davies, S. W., Turmaine, M., Cozens, B. A., DiFiglia, M., Sharp, A. H., Ross, C. A., et al. (1997) Formation of neuronal intranuclear inclusions underlies the neurological dysfunction in mice transgenic for the HD mutation. *Cell* **90,** 537–548.
8. Scherzinger, E., Lurz, R., Turmaine, M., Mangiarini, L., Hollenbach, B., Hasenbank, R., et al. (1997) Huntingtin-encoded polyglutamine expansions form amyloid-like protein aggregates in vitro and in vivo. *Cell* **90,** 549–558.
9. Polymeropoulos, M. H., Lavedan, C., Leroy, E., Ide, S. E., Dehejia, A., Dutra, A., et al. (1997) Mutation in the alpha-synuclein gene identified in families with Parkinson's disease. *Science* **276,** 2045–2047.
10. Kruger, R., Kuhn, W., Muller, T., Woitalla, D., Graeber, M., Kosel, S., et al. (1998) Ala30Pro mutation in the gene encoding alpha-synuclein in Parkinson's disease. *Nat. Genet.* **18,** 106–108.
11. Ueda, K., Fukushima, H. J., Masliah, E., Xia, Y., Iwai, A., Yoshimoto, M., et al. (1993) Molecular cloning of cDNA encoding an unrecognized component of amyloid in Alzheimer's disease. *Proc. Natl. Acad. Sci. USA* **90,** 11,282–11,286.
12. Iwai, A., Masliah, E., Yoshimoto, M., Ge, N., Flanagan, L., Rohan de Silva, H. A., et al. (1995) The precursor protein of non-Aβ component of Alzheimer's disease amyloid is a presynaptic protein of the central nervous system. *Neuron* **14,** 467–475.
13. Jakes, R., Spillantini, M. G., and Goedert, M. (1994) Identification of two distinct synucleins from human brain. *FEBS Lett.* **345,** 27–32.
14. Iwai, A., Yoshimoto, M., Masliah, E., and Saitoh, T. (1995) Non-Aβ component of Alzheimer's disease amyloid (NAC) is amyloidogenic. *Biochemistry* **34,** 10,139–10,145.

15. Yoshimoto, M., Iwai, A., Kang, D., Otero, D. A. C., Xia, Y., and Saitoh, T. (1995) NACP, the precursor protein of the non-amyloid β/A4 protein (Aβ) component of Alzheimer's disease amyloid, binds Aβ and stimulates Aβ aggregation. *Proc. Natl. Acad. Sci. USA* **92,** 9141–9145.
16. Masliah, E., Iwai, A., Mallory, M., Ueda, K., and Saitoh, T. (1996) Altered presynaptic protein NACP is associated with plaque formation and neurodegeneration in Alzheimer's disease. *Am. J. Pathol.* **148,** 201–210.
17. Standaert, D. G. and Young, A. B. (1996) in *The Pharmacological Basis of Therapeutics* (Goodman and Gilman,eds.), McGraw-Hill, New York, p. 514.
18. Kang, J., Lemaire, H. G., Unterbeck, A., Salbaum, J. M., Masters, C. L., et al. (1987) The precursor of Alzheimer's disease amyloid A4 protein resembles as cell-surface receptor. *Nature* **325,** 733–736.
19. Robakis, N. K., Ramakrishna, N., Wolfe, G., and Wisniewski, H. M. (1987) Molecular cloning and characterization of a cDNA encoding the cerebrovascular and the neuritic plaque amyloid peptides. *Proc. Natl. Acad. Sci. USA* **84,** 4190–4194.
20. Goldgaber, D., Lermena, M. I., McBride, O. W., Saffiotti, U., and Gajdusek, D. C. (1987) Characterization and chromosomal localization of a cDNA encoding brain amyloid of Alzheimer disease. *Science* **235,** 877–880.
21. Tanzi, R. E., Gusella, J. F., Watkins, P. C., Bruns, G. A. P., St. George-Hyslop, P., et al. (1987) Amyloid β protein gene: cDNA, mRNA distribution, and genetic linkage near the Alzheimer locus. *Science* **235,** 880–884.
22. Tanzi, R. E., McClatchey, A. I., Lampert, E. D., Villa-Komaroff, L., Gusella, J. F., et al. (1988) Protease inhibitor domain encoded by an amyloid protein precursor mRNA associated with Alzheimer's disease. *Nature* **331,** 528–530.
23. Iversen, L. L. (1995) The toxicity in vitro of beta-amyloid protein. *Biochem. J.* **311,** 1–16.
24. Selkoe, D. (1996) Amyloid beta-protein and the genetics of Alzheimer's disease. *J. Biol. Chem.* **271,** 18,295–18,298.
25. Kuo, K. M., Emmerling, M. R., Vigo-Pelfrey, C., Kasunic, T. C., Kirkpatrick, J. B., Murdock, G. H., Ball, M. J., and Roher, A. E. (1996) Water soluble Aβ(N-40, N-42) oligomers in normal and Alzheimer's disease brains. *J. Biol. Chem.* **271,** 4077–4081.
26. Levy, E., Carman, M. D., Fernandez-Madrid, I. J., Power, M. D., Lieberburg, I., et al. (1990) Mutation of the Alzheimer's disease amyloid gene in hereditary cerebral hemorrage, Dutch type. *Science* **248,** 1124–1126.
27. Goate, A., Chartier-Harlin, M. C., Mullan, M., Braown, J. Crawford, F., et al. (1991) Segregation of a missense mutation in the amyloid precursor protein gene with familial Alzheimer's disease. *Nature* **349,** 704–706.
28. Hendricks, L., van Duijn, C. M., Cras, P., Cruts, M., Van Hul, W., et al. (1992) Presenile dementia and cerebral haemorrhage linked to a mutation at codon 692 of the β-amyloid precursor protein gene. *Nat. Genet.* **1,** 218–221.
29. Murrell, J., Farlow, M., Ghetti, B., and Benson, M.D. (1991) A mutation in the amyloid precursor protein associated with hereditary Alzheimer's disease. *Science* **254,** 97–99.

30. Lamb, B. T., Call, L. M., Slunt, H. H., Bardel, K. A., Lawler, A. M., et al. (1997) Altered metabolism of familial Alzheimer's disease-linked amyloid precursor protein variants in yeast artificial chromosome transgenic mice. *Hum. Mol. Genet.* **6,** 1535–1541.
31. Hsiao, K., Chapman, P., Nilsen, S., Eckman, C., Harigaya, Y., et al. (1996) Correlative memory deficits, Aβ elevation and amyloid plaques in transgenic mice. *Science* **274,** 99–102.
32. Sturchler-Pierrat, C., Abramowski, D., Duke, M., Wiederhold, K. H., Mistl, C., et al. (1997) Two amyloid precursor protein transgenic mouse models with Alzheimer disease-like pathology. *Proc. Natl. Acad. Sci. USA* **94,** 13,287–13,292.
33. Borchelt, D. R., Thinakaran, G., Eckman, C. B., Lee, M. K., Davenport, F., et al. (1996) Familial Alzheimer's disease-linked presenilin 1 variants elevate Aβ1–42/1–40 ratio in vitro and in vivo. *Neuron* **17,** 1005–1013.
34. Borchelt, D. R., Ratovitski, T., Van Lare, J., Lee, M. K., Gonzales, V. B., et al. (1997) Accelerated amyloid deposition in the brains of transgenic mice co-expressing mutant presenilin 1 and amyloid precursor proteins. *Neuron* **19,** 939–945.
35. Citron, M., Westaway, D., Xia, W., Carlson, G., Kiehl, T., et al. (1997) Mutant presenilins of Alzheimer's disease increase production of 42-residue amyloid β-protein in both transfected cells and transgenic mice. *Nat. Med.* **3,** 67–72.
36. Tomita, T., Tokuhiro, S., Hashimoto, T., Aiba, K., Saido, T. C., Maruyama, K., and Iwatsubo, T. (1998) Molecular dissection of domains in mutant presenilin 2 that mediate overproduction fo amyloidogenic forms of amyloid beta peptides. *J. Biol. Chem.* **273,** 6277–6284.
37. Nitsch, R. M., Flack, B. E., Wurtman, R. J., and Growdon, J. H. (1992) Release of Alzheimer's amyloid precursor derivatives stimulated by activation of muscarinic acetylcholine receptors. *Science* **258,** 304–307.
38. Koo, E. H., Sisodia, S. S., Archer, D. R., Martin, L. J., Weidemann, A., et al. (1990) Precursor of amyloid protein in Alzheimer disease undergoes fast anterograde axonal transport. *Proc. Natl. Acad. Sci. USA* **87,** 1561–1565.
39. Sisodia, S. S., Koo, E. H., Hoffman, P. N., Perry, G., and Price, D. L. (1993) Identfication and transport of full-length amyloid precursor proteins in rat peripheral nervous system. *J. Neurosci.* **13,** 3136–3142.
40. Hartmann, T., Bieger, S. C., Bruhl, B., Tienari, P., Ida, N., Allsop, D., et al. (1997) Distinct sites of intracellular production of Alzheimer's disease Aβ40/42 amyloid protein. *Nat. Med.* **3,** 1016–1020.
41. Haass, C., Lemere, C. A., Capell, A., Citron, M., Seubert, P., et al. (1995) The Swedish mutation causes early-onset Alzheimer's disease by β-secretase cleavage within the secretory pathway. *Nat. Med.* **1,** 291–296.
42. Gervais, F. G., Xu, D., Robertson, G. S., Vaillancourt, J. P., Zhu, Y., Huang, J., et al. (1999) Involvement of caspases in proteolytic cleavage of Alzheimer's amyloid-β precursor protein and amyloidogenic Aβ peptide formation. *Cell* **97,** 395–406.
43. Haass, C. (1999) Dead end for neurodegeneration. *Nature* **399,** 204–206.
44. Vassar, R., Bennet, B. D., Babu-Khan, S., Kahn, S., Mendiaz, E. A., Denis, P., et al. (1999) β-Secretase clevage of Alzheimer's amyloid precursor protein by the transmembrane aspartic protease BACE. *Science* **286,** 735–741.

45. Sinha, S., Anderson, J. P., Barbour, R., Basi, G. S., Caccavello, R., Davis, D., et al. (1999) Purication and cloning of amyloid precursor protein β-secretase from human brain. *Nature* **402,** 537–540.
46. Hussain, I., Powell, D., Howlett, D. R., Tew, D. G., Meek, T. D., Chapman, C., et al. (1999) Identification of a novel aspartic protease (Asp2) as β-secretase. *Mol. Cell. Neurosci.* **14,** 419–427.
47. Yan, R., Bienkowski, M. J., Shuck, M. E., Miao, H., Tory, M. C., Pauley, A. M., et al. (1999) Membrane-anchored aspartyl protease with Alzheimer's disease beta-secretase activity. *Nature* **402,** 533–537.
48. Harper, P. S. (ed.) (1996) *Huntington's Disease* 2nd ed. W.B. Saunders, London.
49. Goldberg, Y. P., Kremer, B., Andrew, S. E., Theilmann, J., Graham, R. K., Squitieri, F., et al. (1993) Molecular analysis of new mutations for Huntington's disease: intermediate alleles and sex of origin effects. *Nat. Gen.* **5,** 174–179.
50. Vogel, F. and Motuksky, A. G. (1986) *Human Genetics* 2nd ed. Springer-Verlag, New York.
51. Albin, R. L., Young, A. B., and Penney, J. B. (1989) The functional anatomy of basal ganglia disorders. *Trends Neurosci.* **12,** 366–375.
52. Vonsattel, J. P., Myers, R. H., Stevens, T. J., Ferrante, R. J., Bird, E. D., and Richardson, E. P. (1985) Neuropathological classification of huntington's disease. *J. Neuropathol. Exp. Neurol.* **44,** 559–577.
53. Rubinsztein, D. C., Leggo, J., Coles, R., Almqvist, E., Biancalana, V., Cassiman, J. J., et al. (1996) Phenotypic characterization of individuals with 30–40 CAG repeats in the huntington disease (HD) gene reveals HD cases with 36 repeats and apparently normal elderly individuals with 36–39 repeats. *Am. J. Hum. Genet.* **59,** 16–22.
54. Perutz, M. F., Johnson, M., Suzuki, M., and Finch, J. T. (1994) Glutamine repeats as polar zippers: their possible role in inherited neurodegenerative diseases. *Proc. Natl. Acad. Sci. USA* **91,** 5355–5358.
55. Jackson, G., Salecker, I., Dong, X., Yao, X., Arnheim, N., Faber, P. W., et al. (1998) Polyglutamine-expanded human huntingtin transgenes induce degeneration of Drosophila photoreceptor neurons. *Neuron* **21,** 633–642.
56. Gutekunst, C. A., Levey, A. I., Heilman, C. J., Whaley, W. L., Yi, H., Nash, N. R., et al. (1995) Identficationa nd localization of huntingtin in brain and human lymphoblastoid cell lines with anti-fusion protein antibodies. *Proc. Natl. Acad. Sci. USA* **92,** 8710–8714.
57. Tukamoto, T., Ukina, N., Ide, K., and Kanazawa, I. (1997) Huntington's disease gene product, huntingtin, associates with microtubules in vitro. *Mol. Brain Res.* **51,** 8–14.
58. Velier, J., Kim, M., Schwarz, C., Kim, T. W., Sapp, E., Chase, K., et al. (1998) Wild-type and mutant huntingtins function in vesicle trafficking in the secretory and endocytic pathways. *Exp. Neurol.* **152,** 34–40.
59. DiFiglia, M., Sapp, E., Chase, K., Schwarz, C., Meloni, A., Young, C., et al. (1995) Huntingtin is a cytoplasmic protein associated with vesicle in human and rat brain neurons. *Neuron* **14,** 1075–1081.
60. Pearse, B. M. F. and Robinson, M. S. (1990) Clathrin, adaptors, and sorting. *Annu. Rev. Cell Biol.* **6,** 151–171.

61. Orr, H. T., Chung, M. Y., Banfi, S., Kwiatkowski, T. J., Servadio, A., Beaudet, A. L., et al. (1993) Expansion of an unstable trinucleotide CAG repeat in spinocerebellar ataxia type 1. *Nat. Genet.* **4,** 221–226.
62. Sanpei, K., Takano, H., Igarashi, S., Sato, T., Oyake, M., Sasaki, H., et al. (1996) Identification of the spinocerebellar ataxia type 2 gene using a direct identification of repeat expansion and cloning technique, DIRECT. *Nat. Genet.* **14,** 277–284.
63. Kwaguchi, Y., Okamoto, T., Taniwaki, M., Aizawa, M., Inoue, M., et al. (1994) CAG expansions in a novel gene for Machado-Joseph disease at chromosome 14q32.1. *Nat. Genet.* **8,** 221–228.
64. Zhuchenko, O. et al. (1997) Autosomal dominant cerebellar ataxia (SCA6) associated with small polyglutamine expansions in the alpha 1A-voltage dependent calcium channel. *Nat. Genet.* **15,** 62–69.
65. David, G., Abbas, N., Stevanin, G., Durr, A., Yvert, G., et al. (1997) Cloning of the SCA7 gene reveals a highly unstable CAG repeat expansion. *Nat. Genet.* **17,** 65–70.
66. Spada, A. R., Wilson, E. M., Lubahan, D. B., Harding, A. E., and Fischbeck, K. H. (1991) Androgen receptor gene mutations in X-linked spinal and bulbar muscular atrophy. *Nature* **352,** 77–79.
67. Koide, R., Ikeuchi, T., Onodera, O., Tanaka, H., Igarashi, S., Endo, K., et al. (1994) Unstable expansion of CAG repeat in hereditary dentatorubral-pallidoluysian atrophy (DRPLA). *Nat. Genet.* **6,** 9–13.
68. Paulson, H. L., Perez, M. K., Tkrottier, Y., Trojanowski, J. Q., Subramony, S. H., Das, S. S., et al. (1997) Intranuclear inclusions of expanded polyglutamine protein in spinocerebellar ataxia type 3. *Neuron* **19,** 333–344.
69. Petersen, A., Mani, K., and Brundin, P. (1999) Recent advances on the pathogenesis of Huntington's disease. *Exp. Neurol.* **157,** 1–18.
70. Wood, N. (1997) Genes and parkinsonism. *J. Neurol. Neurosurg. Psychiatry* **62,** 305–309.
71. Riess, O., Jakes, R., and Kruger, R. (1998) Genetic dissection of familial Parkinson's disease. *Mol. Med. Today* **4,** 438–444.
72. Spillantini, M. G., Crowther, R. A., Jakes, R., Hasegawa, M., and Goedert, M. (1998) α-synuclein in filamentous inclusions of Lewy bodies from parkinson's disease and dementia with Lewy bodies. *Proc. Natl. Acad. Sci. USA* **95,** 6469–6473.
73. Iseki, E., Marui, W., Kosaka, K., Akiyama, H., Ueda, K., and Iwatsubo, T. (1998) Degenerative terminals of the perforant pathway are human α-synuclein-immunoreactive in the hippocampus of patients with diffuse Lewy body disease. *Neuroci. Lett.* **258,** 81–84.
74. Clayton, D.F. and George, J.M. (1998) The synucleins: a family of proteins involved in synaptic function, plasticity, neurodegeneration and disease. *TINS* **21,** 249–254.
75. Lavedan, C., Leroy, E., Dehejia, A., Buchholtz, S., Dutra, A., Nussbaum, R. L., et al. (1998) Identification, localization, and characterization of the human γ-synuclein gene. *Hum. Genet.* **103,** 106–112.

76. Martoeaux, L., Campanelli, J. T., and Scheller, R. H. (1988) Synuclein: a neuron-specific protein localized to the nucleus and presynaptic nerve terminal. *J. Neurosci.* **8,** 2804–2815.
77. George, J. M., Woods, J. H., and Clayton, D. F. (1995) Characterization of a novel protein regulated during the critical period for song learning in the zebra finch. *Neuron* **15,** 248–254.
78. Kitada, T., Asakawa, S., Hattori, N., Matsumine, H., Yamamura, Y., et al. (1998) Mutations in the parkin gene cause autosomal recessive juvenile parkinsonism. *Nature* **392,** 605–608.
79. Abbas, N., Lucking, C. B., Ricard, S., Durr, A., Bonifati, V., et al. (1999) A wide variety of mutations in the *parkin* gene are responsible for autosomal recessive parkinsonism in Europe. *Hum. Mol. Genet.* **8,** 567–574.
80. Hsu, L., Mallory, M., Xia, Y., Veinbergs, I., Hashimoto, M., Yoshimoto, M., et al. (1998) Expression pattern of synucleins (non-Aβ component of Alzheimer's disease amyloid precursor protein/α-synuclein) during murine brain development. *J. Neurochem.* **71,** 338–344.

6

Treatment Approaches for Alzheimer's Disease

Michael Gold, Kevin M. Felsenstein, and Perry Molinoff

Alzheimer's disease (AD) is the most common cause of dementia and can be thought of as a prototype of a primary neurodegenerative disorder *(1,2)*. Studies of the epidemiology of AD have demonstrated an age-related increase in prevalence from approximately 10% in persons 65 yr old to as high as 50% in persons reaching 85 yr of age *(3)*. In addition to age, family history, head trauma, general anesthesia, and a poor education have been implicated as risk factors for AD *(4–6)*. The "graying" of our society as well as successes in developing treatments for other chronic disorders implies that unless ways are found to reduce the incidence of AD, the societal costs attributable to AD will increase in the coming years. The public health problem posed by AD is succintly illustrated by the fact that the current annual cost associated with AD in the United States is approximately US $100 billion. The projected quadupling of the affected population in the next 20–30 yr *(7)* serves to underscore the scope of the challenge.

Because AD is predominantly a disease of the elderly, treatments that delay the onset or delay the progression of the disease can have a significant impact over time. It has been estimated that delaying the onset of AD by 5–7 yr can lead to a 50% reduction in the prevalence of this disease in the span of one generation *(8)*. Clinically, AD is recognized as a heterogeneous disorder with typical *(9)* as well as atypical presentations *(10,11)* and a variable natural history *(12,13)*. The typical clinical presentation of AD is that of a person with progressive memory loss (a requisite deficit) and a functional deficit that often serves as the trigger for families to seek evaluations of the person affected. Although plateaus in decline are commonly described, patients with AD invariably decline and require increasing amounts of assistance.

The views expressed in this chapter are those of the authors and do not imply any endorsement or approval from Bristol-Myers Squibb.

From: *Contemporary Clinical Neuroscience: Molecular Mechanisms of Neurodegenerative Diseases*
Edited by: M.-F. Chesselet © Humana Press Inc., Totowa, NJ

Clinico-pathologically validated *(14)* diagnostic criteria for AD have been available for almost 20 yr *(15)*. Consequently, there has been significant progress in separating those patients who are thought to have a relatively isolated degenerative process (probable AD) from those who have additional pathology superimposed on the primary degenerative process (possible AD). This separation has allowed the investigation of the natural history and phenotypic variations of patients with a wide spectrum of pathology. One of the major benefits of the standardization of definitions and criteria is that diagnostic accuracy has increased dramatically such that, in expert hands, there is approximately an 80–90% accuracy in premortem diagnosis *(16)*. In the absence of noninvasive, inexpensive diagnostic tests, the clinical diagnosis has been robust enough that clinical trials in this population have been successfully executed, leading to the approval of several medications for the symptomatic treatment of AD. Currently approved treatments for AD using inhibitors of acetylcholinsterase are symptomatic, have very modest effect sizes, have significant sideeffects problems, and are effective in a minority of patients who take them. There are no approved disease-modifying treatments as of this time.

The identification and development of treatments for patients with AD presents several difficult challenges to the pharmaceutical industry. From the perspective of pharmaceutical development, the lack of a clear etiology for the sporadic form of AD makes target selection of a molecular target problematic. In addition to the diagnostic difficulties mentioned earlier, the chronic progressive nature of the disease requires large, prolonged, and costly clinical trials. Generic problems of oral bioavailability, the need for compounds that cross the blood-brain barrier, central nervous system (CNS) selectivity, and avoidance of systemic adverse events all apply in this area.

Treating physicians are faced with problems related to efficacy and compliance *(17)*. It is entirely plausible that with AD, like Parkinson's disease, significant amounts of pathology accrue before symptoms become apparent. Physicians are now faced with the fact that several medications that have been approved for the treatment of AD have effect sizes that are modest at best *(18)*. Because these medications are associated with significant side effects *(19)*, patient compliance is a significant issue. Physicians are also faced with the problems of maintaining patient compliance in a disease where progression continues and the best one can currently hope for is a brief delay in disease progression. In the context of a disease process that can take 7–10 yr, currently achievable delays of 6 mo do not amount to a dramatic improvement in patient care.

Patients and caregivers are faced with many difficult choices in deciding how to engage the treatment of AD. Although many patients and caregivers are eager to try new medications, the lack of broad efficacy and the very modest effect on those who do respond means that many patients and their families will derive no benefit from exposure to medications yet will be incur all of the liabilities of exposure to medications with nontrivial side effects. It should also be made clear that patients with AD and their caregivers are in a particularly vulnerable economic position. Because the annual cost of symptomatic treatment is relatively small compared to the annual cost of institutionalization, pharmacoeconomic assessments have concentated on patients who transition from a community-dwelling state into some form of institutionalized care *(20)*. The pharmacoeconomic assessments of treating mildly affected patients with purely symptomatic drugs remains to be conducted *(21)*.

Advances in the treatment of AD have proceeded along several avenues and are the product of major advances in the pathological and molecular biological understanding of AD. In the area of molecular biology, the discovery of mutations associated with early-onset familial forms of AD has served to focus the attention of industry and academia on the role of β-amyloid as the intial step and/or progression of AD.

A comprehensive review of amyloid cascade hypothesis is beyond the scope of this chapter; however, ref. *22* is suggested. The core arguments in support of the amyloid cascade hypothesis, are as follows:

- The deposition of β-amyloid in the form of plaques is a *sine qua non* of AD *(23)*.
- β-Amyloid in its various forms is neurotoxic *(24)*.
- β-Amyloid overproduction resulting from mutations in the amyloid precursor protein (APP) and *PS-1* and *PS-2* genes *(25)* produces a dementia that meets the pathological criteria for AD although there are phenotypic differences (age of onset, myoclonus) *(26)*.
- β-Amyloid activates or modulates many of the other systems or mechanisms that are implicated in AD, such as inflammation *(27)*, free-radical formation *(28)*, hyperphosphorylation of tau *(29)*,cholinergic neurotransmission *(30)*, and apoptosis *(31)*.

As a consequence of the identification of the APP and *PS-1* and *PS-2* genes, the mutations associated with these genes and their effect on amyloid processing, a consensus has emerged that abnormal processing of the APP molecule, resulting in elevated levels of β-amyloid, is the most likely cause of AD in these familial forms. Because β-amyloid deposition is a necessary, but not sufficient, condition for developing the sporadic form of AD, treatments aimed at the biology of amyloid may be efficacious in the sporadic form of AD and will be the most likely candidates to emerge in the near future. The recent reports detailing the cloning and identification of the putative β-secretase *(32–35)* as well as the recent reports of the use of fibrillar β-amyloid as a "vaccine" *(36)* should accelerate development of compounds and techniques for interfering with β-amyloid deposition. The inhibition of amyloid synthesis and/or deposition would be expected to slow or halt the progression of the disease, and depending on when the treatment is initiated, such treatments could lead to some functional improvement.

For purposes of this chapter, the discussion of the treatment of AD is divided into treatments aimed at specific and potentially causal mechanisms. This is followed by a discussion of treaments aimed at the various pathological changes noted in AD, a discussion of systemic treatments that may affect AD indirectly, and, finally, by a discussion of approached aimed at disease management.

β-Amyloid is currently the leading candidate as the most probable culprit in AD. Whether it is directly causal as in cases of early-onset familial AD or part of a neurodegenerative cascade that results in the dysregulation of β-amyloid, the data regarding its neurotoxicity and ability to interact with other systems involved in AD support targeting β-amyloid as a high-priority issue. To discuss potential targets and interventions for the treatment of AD, a model of β-amyloid processing is presented in Fig. 1 and a model of β-amyloid deposition is presented in Fig. 2.

In the context of these models, two general types of interventions may be defined:

- Inhibition of systems that:
  - Proteolytically process APP into amyloidogenic fragments (i.e., inhibition of the β- and/or γ-secretases) *(37)*
  - Facilitate the aggregation of β-amyloid fragments into proto-fibrils *(38,39)*
  - Facilitate crosslinkage, posttranslational modifications and stabilization of proto-fibrils into full-fledged β-amyloid fibrils *(40)*
  - Facilitate the extracellular deposition of β-amyloid *(41)*
- Enhancement of systems that:
  - Process APP into nonamyloidogenic fragments (i.e., ehancement of the α-secretase) *(42)*
  - Solubilize β-amyloid fibrils
  - Clear β-amyloid from the extracellular space [amino-peptidases *(43)*]

This list identifies several specific enzymatic targets (i.e., α-, β-, and γ-secretases) that are currently in the process of being isolated and for which specific inhibitors or agonists are being developed. This figure also suggests that compounds such as inhibitors of β-amyloid polymerization, inhibitors of β-amyloid crosslinkage or the induction of immune responses to the various forms of β-amyloid may be viable techniques for reducing the neurotoxic effect of β-amyloid. Conversely, compounds that solubilize β-amyloid fibrils in various stages of assembly may work by enhancing the clearance of β-amyloid from the CNS. Augmentation of endogenous

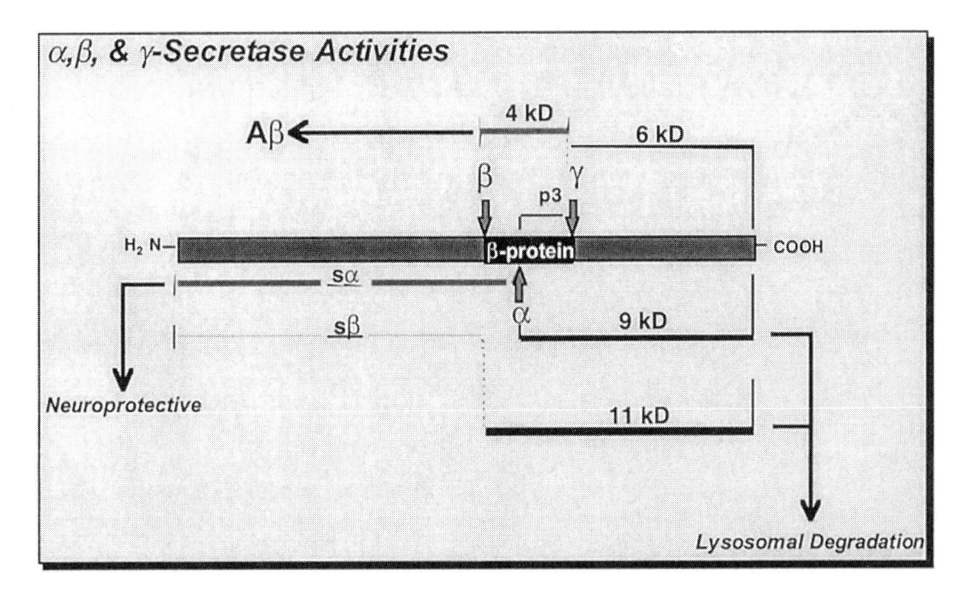

Fig. 1. Proteolytic processing of β-APP. The relative positions of the α-, β-, and γ-secretase cleavage sites are indicated along with the products resulting from these proteolytic activities. The 4-kDa β-protein domain is indicated by the white type.

mechanisms for clearing β-amyloid such as amino-peptidases may also have a role.

Transgenic species can be used to test large number of compounds in a relatively short period of time. There are unresolved issues related to the exact nature of the pathological changes, strain effects, and behavioral changes seen in transgenic mice and their relevance to the pathology seen in man *(44)*. Although there is no confirmation that the cleavage of APP in TG mice is carried out by enzymes identical to those of man, analysis of the end products suggest that the systems for processing APP in TG mice are the same as man. Therefore, it must be acknowledged that barring the development of noninvasive techniques for assesssing the various pools of APP and its products in vivo *(45)*, transgenic animals are likely to remain our best models for testing novel compounds directed at amyloid processing but are far from being true models of AD *(46)*.

There are several caveats related to the treatment of AD on the basis of inducing a reduction in the levels of β-amyloid:

1. Although, by definition, the brains of patients with AD have β-amyoid deposits, it is not clear that there is an overproduction of β-amyloid in the sporadic form of the disease. Pathological data demonstrating that total β-amy-

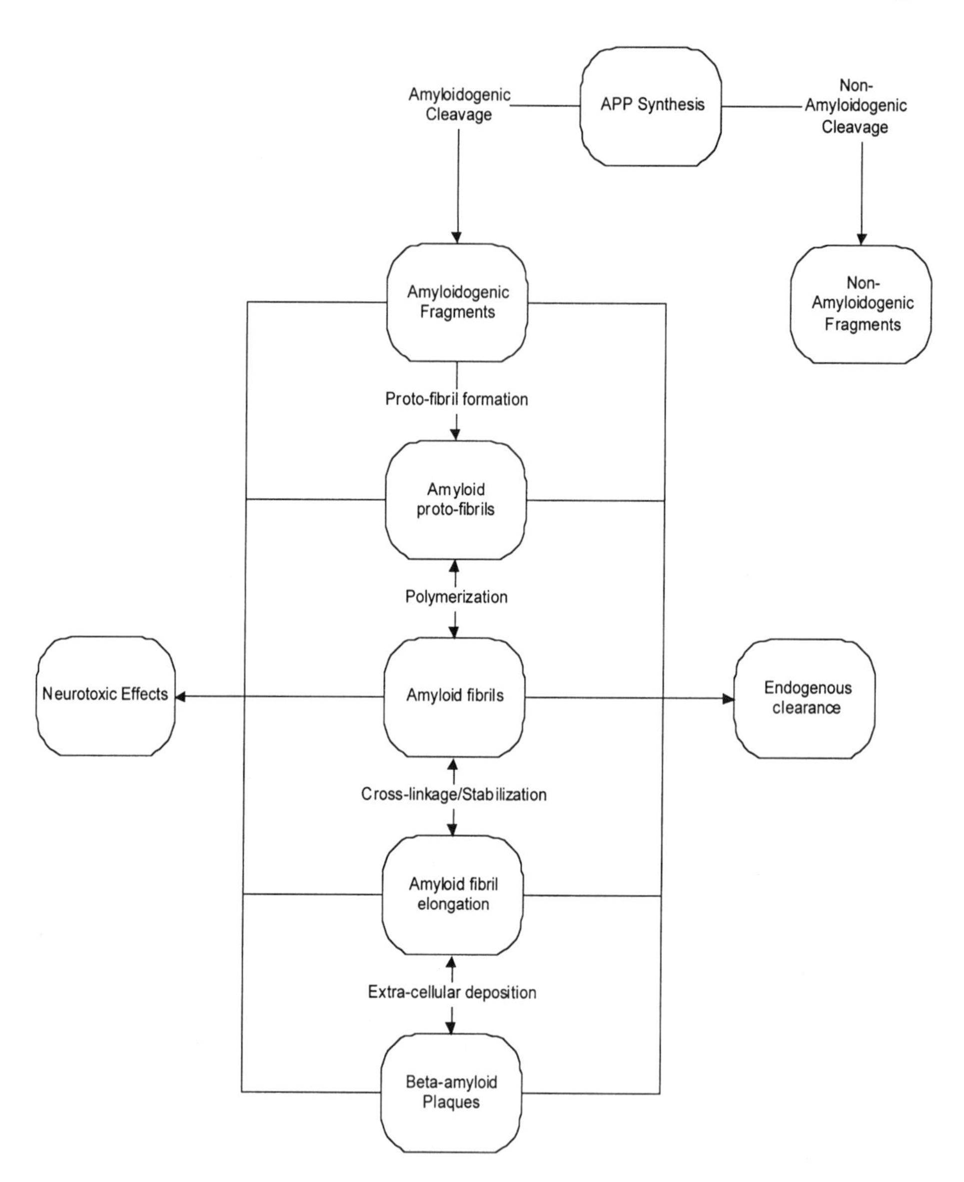

Fig. 2. Model of β-amyloid deposition.

loid levels tend to level off as the disease progresses has been taken as evidence of continuous clerance of β-amyloid *(47)*.

2. The long-term safety effects of supressing β-amyloid production have not been defined. This is a highly conserved system whose functions are just now beginning to be understood.

3. It is not clear when β-amyloid begins to be deposited in human beings and when β-amyloid-reducing treatments should be instituted. There are data that β-amyloid levels in the plasma begin to rise in the fourth decade of life. Furthermore, there is a hypothesis that β-amyloid deposition may begin to accelerate around the time of menopause *(48)*.
4. It is not clear what constitutes a pathological burden of β-amyloid, as there are persons who have pathological burdens of β-amyloid but are cognitively normal *(49)*. This underscores the idea that β-amyloid deposition is necessary, but insufficient, to cause AD.
5. It is not clear how long or how far levels of β-amyloid need to be reduced in order to have a clinically detectable effect.
6. It is not clear if supression of the total β-amyloid load or only of the soluble pool is necessary *(50)*.
7. The lack of adequate animal models and the lack of surrogates for clinical end points requires that clinical trials approach these questions empirically *(51)*.

An alternative approach to identifying targets for the treatment of AD is bases on the hypothesis that the deposition of β-amyloid, while not necessarily causal, is pathogenic and initiates a broad-based pathological cascade as illustrated in Fig. 3.

The following potential targets/mechanisms also can be considered:

- Free-radical scavengers *(28)*
  - Vitamin E *(52)*, HWA-285 *(53)*
  - Spin-trapping nitrones *(54)*
  - Idebenone *(55)*
- Inhibitors of free-radical production
  - MAO inhibitors
    - Selegeline *(52)*, Lazabemide *(56)*
    - Rasagiline *(57)*
  - Anti-inflammatory agents *(58,59)*
  - Nonsteroidal anti-inflammatory drugs (NSAIDS) (i.e., Indocin, COX-2 inhibitors) *(60)*
  - Prednisone *(61)*
  - Interleukin blockers
  - Hydroxycholoroquine/colchicine *(27)*
  - Immune modulation *(62)*
- Cytoskeletal/synaptic stabilizers
  - Sabeluzole *(63)*
  - Tau phosphorylases (kinases and phosphatases) *(64)*
  - Synaptophysin/synaptotagmin
  - apoE-related compounds
- Apoptosis
  - Caspase inhibitors *(65)*
  - Neurotrophic agents *(66)*
    - AIT-082 *(67)*

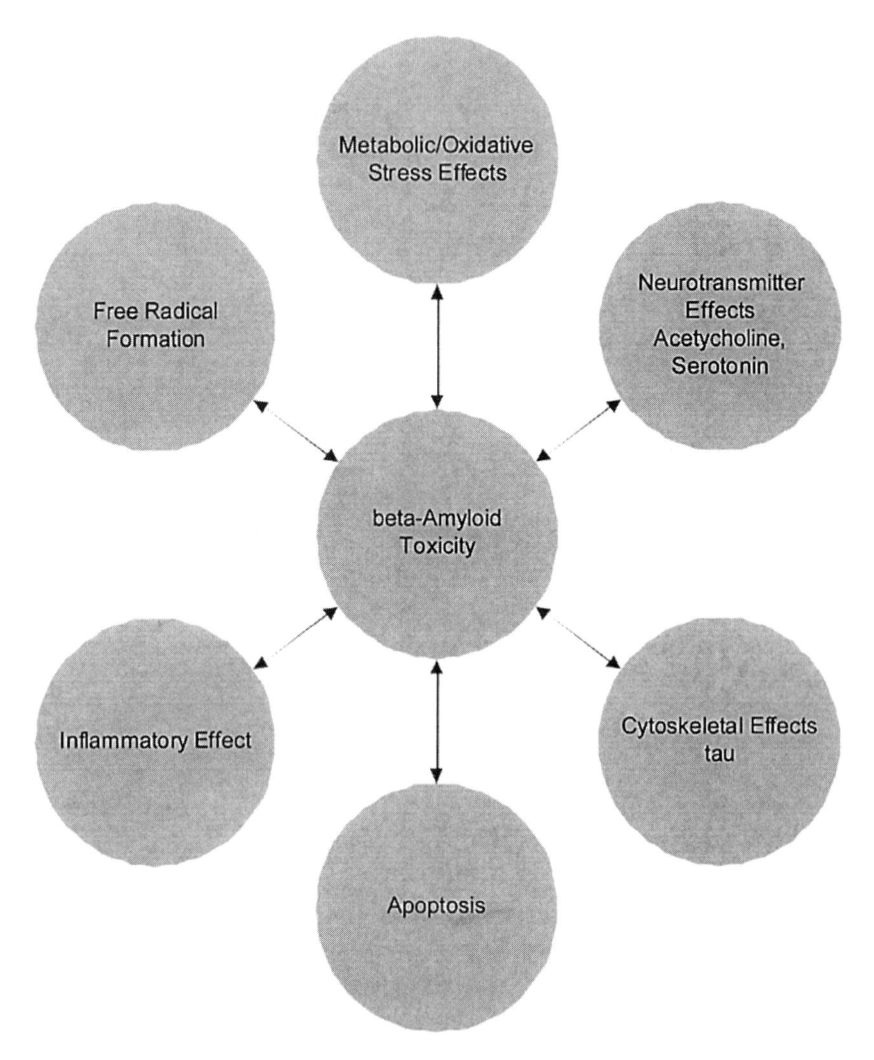

Fig. 3. Pathological results of β-amyloid.

- Neurochemical effects
  - Augmentation of acetylcholine
    - Acetylcholinesterase inhibitors *(68)*
    - Muscarinic receptor based
      - M1 agonists *(69)* [i.e., Milameline *(70)*, Xanolemine *(71)*, AF-102B *(72)*]
      - M2 antagonists *(73)*
      - Mixed agonists/antagonists *(74)*
    - Nicotinic Agonists (i.e., ABT-41875, SIB-1553A76, GTS-2177)
    - High-affinity choline uptake *(78)*

- Hormonal releasing factors
  - TRH79 or TRH analogues *(80)*
  - Corticotropin *(81)*
  - JTP-4819 *(82)*
- Augmentation of long-term potentiation (LTP)
  - Ampakines (CX516) *(83)*
  - Cycloserine *(84)*
  - Somatostatin *(85)*
- Bezodiazepine inverse agonists [i.e., S-8510 *(86)*, flumazenil *(87)*]
- Multiple effects
  - Estrogen *(88)*
  - HWA-285 *(53)*, Denbufylline *(89)*
  - Posatirelin *(90)*

As with the previous hypothesis, there are many caveats, including the following:

1. Neurons may be affected by pathological changes along multiple systems simultaneously. This would suggest that combination therapy will be required.
2. It is not known whether subsets of patients have one or another pathological change as the predominant expression of β-amyloid toxicity. Until AD patients can be phenotyped in terms of their pathology, selective treatments will not be available.
3. Because the relative contribution of each of these pathological changes remains unknown as do the time frames in which they occur, treatments aimed at the down stream consequences of β-amyloid toxicity are likeky to be palliative, at best.
4. All the liabilities associated with combination therapies (i.e., side effects, tolerability, drug interactions, induction, etc.) will likely limit the efficacy of combination therapies.
5. Because there are so many parallel pathological pathway, combination therapies that block only one or two of these paths will likely be ineffective. This is analogous to the experience with neuroprotectants in the treatment of ischemic stroke.

If one believes the hypothesis that the deposition of β-amyloid in AD is an epiphenomenon and that the real genesis of the disease is a failure of cellular metabolism *(91)* which, in turn, causes the same kinds of pathological changes as β-amyloid, as illustrated by Fig. 3, then the following targets are suggested:

- Enhancers of oxidative phosphorylation
  - Metabolic cofactors [i.e., thiamine *(92)*]
  - Facilitators of mitochondrial function [i.e., ALCAR *(93)*]
- Detoxification of mitochondrial by-products
- Protectants from glutamatergic neurotoxicity
  - *N*-Methyl-D-Aspartate (NMDA) antagonists [i.e., Memantine *(94)*]

Another alternative hypothesis [tau hypothesis *(95)*] posits that AD is primarily a disorder of microtubules. Failure of neurons to maintain their internal scaffolding because of the hyperphosphorylation of tau leads to a variety of cellular dysfunction, neuronal death, and the deposition of neurofibrillary tangles with their characteristic paired-helical filaments. The tau hypothesis is bolstered by the closer relationship between the other *sine qua non* of AD, neurofibrillary tangles, and the severity of dementia. The recent identification of mutations of the *tau* gene on chromosome 17 and their association with fronto-temporal dementias indicates that abnormal tau is sufficient to produce a dementing disorder *(96)*. However, the dementias caused by chromosome 17 mutations are phenotypically quite distinct from AD, suggesting that a tauopathy is unlikely to be the singular cause of AD. In order to address this question, transgenic models incorporating β-amyloid overproduction and abnormal tau production should be very helpful in sorting out the relative contributions of each to the overall development of pathology. Figure 4 illustrates a model of tau polymerization and suggests some potential targets as well.

- Inhibitors of tau hyperphophorylations *(97)*
- Selective tau dephosphorylation *(98)*
- Inhibition of tau dimerization or polymerization
- Solubilization of tau polymers

The last category of potential targets are related to the comorbid disorders that are also associated with age and AD. Cerebrovascular diseases and/or metabolic disorders that adversely affect cerebral blood flow can accelerate the onset of AD and can also accelerate its progression once established *(99)*. As such, risk factors for cerebrovascular disease *(100)* such as hypertension, diabetes, hyperlipidemia, hypercholesterolemia, head trauma, depression, and smoking are all modifiable risk factors for AD. The recent demonstration that reductions in blood pressure can either slow the development of cognitive decline *(101)* or even reverse it *(102)*, suggests that, in some cases, improvements in cerebral blood flow allows viable but dysfunctional neurons to recuperate to some extent and normalize their functions. Effective treatments exist for all these common disorders. On the positive side, the protective effect of education and good nutrition as the foundation for the "added reserve" hypothesis *(103)* suggests that good education and nutrition allow the optimization of neural development of young children and may delay the onset of AD by up to several years.

Finally, there are treatments aimed at the behavioral manifestations of AD. Longitudinal studies of patients with AD have demonstrated that the decision to institutionalize a person with dementia is not predicated on the severity of their cognitive impairment, but rather on the development of behavioral problems and the loss of continence. Behavioral problems with AD run the gamut of psychopathology, including anxiety, apathy, depression, obsessive-

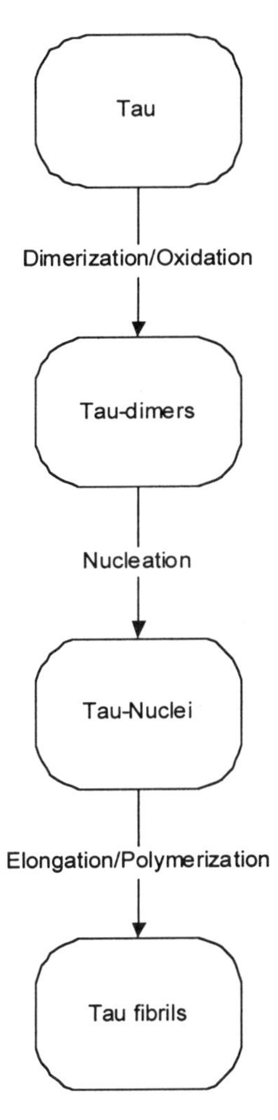

Fig. 4. Model of tau polymerization.

compulsive behaviors, dis-inhibition leading to verbal and physical agression, hallucinations, delusions, paranoia, and psychosis. It should come as no surprise that the entire psychiatric armamentarium has been used for the management of these patients in an attempt to make their behaviors "manageable." In what has to be one of the most troubling gaps in evidence-based medicine, the treatment of these behavioral problems is often based on anecdoctal reports, reports of small case series or underpowered clinical trials. Commonly used psychoactive compounds for the management of behavioral disturbances in AD include neuroleptics, both typical (i.e., haloperidol) and atypical (risperidone), benzodiazepines (diazepam, lorazepam), antidepressants (fluoxitine, sertraline), anxiolytics (midazolam, buspirone), mood-stabilizing agents (phenytoin, carbamazepine, valproic acid), β-blockers, nonbenzodiazepine hypnotics (chloral hydrate, zaleplon), and stimulants (cycloserine). Clinicians should be particularly cautious with the use of psychoactive compounds because patients with AD may require much smaller doses and because caregivers may also be receiving psychoactive medications at the same time.

In summary, there are many potential targets for the treatment of Alzheimer's disease at various stages of disease progression. It is abundantly clear that treatments aimed at the symptomatology of AD are going to be limited in efficacy and duration of effect as the pathological cascade continues unabated. Treatments aimed at the various pathological derangements that have been described may serve to delay disease progression. However, because there are several independent pathological mechanisms at play, it is unlikely that these treatments will have significant effects on their own. Because of their closer links to the cause of AD, treatments aimed at reducing the β-amyloid burden present the best opportunity for arresting or potentially regressing the pathological changes in AD in a parsimonious fashion. The combination of β-amyloid modulators with drugs that can enhance cognition and/or reduce the downstream pathological cascade may be the best option for patients until such time as the etiologies for AD are clarified and improved phenotyping of patients allows specific treatments to be developed.

REFERENCES

1. Treves, T. A. (1991) Epidemiology of Alzheimer's disease. *Psychiatr. Clin. North Am.* **14,** 251–265.
2. Fratiglioni, L. (1996) Epidemiology of Alzheimer's disease and current possibilities for prevention. *Acta Neurol. Scand.* **165(Suppl.),** 33–40.
3. Fratiglioni, L., Forsell, Y., Aguero, T. H., and Winblad, B. (1994) Severity of dementia and institutionalization in the elderly, prevalence data from an urban area in Sweden. *Neuroepidemiology* **13,** 79–88.
4. Bachman, D. L., Wolf, P. A., Linn, R., Knoefel, J. E., Cobb, J., Belanger, A., et al. (1992) Prevalence of dementia and probable senile dementia of the Alzheimer type in the Framingham Study. *Neurology* **42,** 115–119.
5. Hendrie, H. C. (1997) Epidemiology of Alzheimer's disease. *Geriatrics* **52(Suppl. 2),** S4–S8
6. Slooter, A. J. and van Duijn, C. M. (1997) Genetic epidemiology of Alzheimer disease. *Epidemiol. Rev.* **19,** 107–119.
7. Whitehouse, P. J. (1997) Pharmacoeconomics of dementia. *Alzheimer Dis. Associat. Dis.* **11(Suppl 5),** S22–S32.

8. Brookmeye, R., Gray, S., and Kawas, C. (1998) Projections of Alzheimer's disease in the United States and the public health impact of delaying disease onset. *Am. J. Public Health* **88,**1337–1342.
9. Morris, J. C. (1997) Alzheimer's disease, a review of clinical assessment and management issues. *Geriatrics* **52(Suppl. 2),** S22–S25
10. Hansen, L. A. (1997) The Lewy body variant of Alzheimer disease. *J. Neural Transm.* **51(Suppl.),** 83–93.
11. Johnson, J. K., Head, E., Kim, R., Starr, A., and Cotman, C. W. (1999) Clinical and pathological evidence for a frontal variant of Alzheimer disease. *Arch. Neurol.* **56,** 1233–1239.
12. Brooks, J. O. and Yesavage, J. A. Identification of fast and slow decliners in Alzheimer disease, a different approach. *Alzheimer Dis. Associat. Dis.* **9(Suppl. 1),** S19–S25
13. Piccini, C., Bracco, L., Falcini, M., Pracucci, G., and Amaducci, L. (1995) Natural history of Alzheimer's disease, prognostic value of plateaux. *J. Neurol. Sci.* **131,** 177–182.
14. Rasmusson, D. X., Brandt, J., Steele, C., Hedreen, J. C., Troncoso, J. C., and Folstein, M. F. (1996) Accuracy of clinical diagnosis of Alzheimer disease and clinical features of patients with non–Alzheimer disease neuropathology. *Alzheimer Dis. Associat. Disord.* **10,** 180–188.
15. McKhann, G., Drachman, D., Folstein, M., Katzman, R., Price, D., and Stadlan, E. M. (1984) Clinical diagnosis of Alzheimer's disease, report of the NINCDS-ADRDA Work Group under the auspices of Department of Health and Human Services Task Force on Alzheimer's Disease. *Neurology* **34,** 939–944.
16. Galasko, D., Hansen, L. A., Katzman, R., Wiederholt, W., Masliah, E., Terry, R., et al. (1994) Clinical–neuropathological correlations in Alzheimer's disease and related dementias. *Arch. Neurol.* **51,** 888–895.
17. Filley, C. M., Chapman, M. M., and Dubovsky, S. L. (1996) Ethical concerns in the use of palliative drug treatment for Alzheimer's Disease. *J. Neuropsychiatry Clin. Neurosci.* **8,** 202–205.
18. McLendon, B. M. and Doraiswamy, P. M. (1999) Defining meaningful change in Alzheimer's disease trials, the donepezil experience. *J. Geriatr. Psychiatry Neurol.* **12,** 39–48.
19. Domingo, J. L. (1995) Adverse effects of potential agents for the treatment of Alzheimer's disease, a review. *Adverse Drug Reacti. Toxicol. Rev.* **14,** 101–115.
20. Fenn, P. and Gray, A. (1999) Estimating long-term cost savings from treatment of Alzheimer's disease. A modelling approach. *Pharmacoeconomics* **16,** 165–174.
21. Neumann, P. J., Hermann, R. C., Kuntz, K. M., Araki, S. S., Duff, S. B., Leon, J., et al. (1999) Cost-effectiveness of donepezil in the treatment of mild or moderate Alzheimer's disease. *Neurology* **52,** 1138–1145.
22. Hardy, J. (1997) Amyloid, the presenilins and Alzheimer's disease. *Trends Neurosci.* **20,**154–159.
23. The National Institute on Aging, and Reagan Institute Working Group on Diagnostic Criteria for the Neuropathological Assessment of Alzheimer's

Disease. (1997) Consensus recommendations for the postmortem diagnosis of Alzheimer's disease. *Neurobiol. Aging* **18,** S1–S2
24. Small, D. H. (1998) The role of the amyloid protein precursor (APP) in Alzheimer's disease, does the normal function of APP explain the topography of neurodegeneration? *Neurochem. Res.* **23,** 795–806.
25. Selkoe, D. J. (1999) Translating cell biology into therapeutic advances in Alzheimer's disease. *Nature* **399,** A23–A31
26. Tilley, L., Morgan, K., and Kalsheker, N. (1998) Genetic risk factors in Alzheimer's disease. *Mol. Pathol.* **51,** 293–304.
27. Aisen, P. S. (1997) Inflammation and Alzheimer's disease, mechanisms and therapeutic strategies. *Gerontology* **43,** 143–149.
28. Frolich, L. and Riederer, P. (1995) Free radical mechanisms in dementia of Alzheimer type and the potential for antioxidative treatment. *Arzneimittel-Forschung* **45,** 443–446.
29. Alvarez, A., Toro, R., Caceres, A., and Maccioni, R. B. (1999) Inhibition of tau phosphorylating protein kinase cdk5 prevents beta- amyloid-induced neuronal death. *FEBS Lett.* **459,** 21–426.
30. Ehrenstein, G., Galdzicki, Z., and Lange, G. D. (1997) The choline-leakage hypothesis for the loss of acetylcholine in Alzheimer's disease. *Biophys. J.* **73,** 1276–1280.
31. Kaltschmidt, B., Uherek, M., Wellmann, H., Volk, B., and Kaltschmidt, C. (1999) Inhibition of NF-kappaB potentiates amyloid beta-mediated neuronal apoptosis. *Proc. Natl. Acad. Sci. USA* **96,** 9409–9414.
32. Hussain, I., Powell, D., Howlett, D. R., Tew, D. G., Meek, T. D., Chapman, C., et al. (1999) Identification of a novel aspartic protease (Asp 2) as beta-secretase. *Mol. Cell Neurosci.* **14,**419–427.
33. Sinha, S., Anderson, J. P., Barbour, R., Basi, G. S., Caccavello, R., Davis, D., et al. (1999) Purification and cloning of amyloid precursor protein beta-secretase from human brain. *Nature* **402,** 537–540.
34. Vassar, R., Bennett, B. D., Babu-Khan, S., Kahn, S., Mendiaz, E. A., Denis, P., et al. (1999) Beta-secretase cleavage of Alzheimer's amyloid precursor protein by the transmembrane aspartic protease BACE. *Science* **286,** 735–741.
35. Yan, R., Bienkowski, M. J., Shuck, M. E., Miao, H., Tory, M. C., Pauley, A. M., et al. (1999) Membrane-anchored aspartyl protease with Alzheimer's disease beta- secretase activity. *Nature* **402,** 533–537.
36. Schenk, D., Barbour, R., Dunn, W., Gordon, G., Grajeda, H., Guido, T., et al. (1999) Immunization with amyloid-beta attenuates Alzheimer-disease-like pathology in the PDAPP mouse. *Nature* **400,**173–177.
37. Durkin, J. T., Murthy, S., Husten, E. J., Trusko, S. P., Savage, M. J., Rotella, D. P., et al. Rank-order of potencies for inhibition of the secretion of abeta40 and abeta42 suggests that both are generated by a single gamma-secretase. *J. Biol. Chem.* **274,** 20,499–20,504.
38. Janciauskiene, S., Garcia, D. F., Carlemalm, E., Dahlback, B., and Eriksson, S. (1995) Inhibition of Alzheimer beta-peptide fibril formation by serum amyloid P component. *J. Biol. Chem.* **270,** 26,041–26,044.

39. Soto, C. (1999) Plaque busters, strategies to inhibit amyloid formation in Alzheimer's disease. *Mol. Med. Today* **5,** 343–350.
40. Zhang, W., Johnson, B. R., and Bjornsson, T. D. (1997) Pharmacologic inhibition of transglutaminase-induced cross–linking of Alzheimer's amyloid beta-peptide. *Life Sci.* **60,** 2323–2332.
41. Howlett, D. R., Perry, A. E., Godfrey, F., Swatton, J. E., Jennings, K. H., Spitzfaden, C., et al. (1999) Inhibition of fibril formation in beta-amyloid peptide by a novel series of benzofurans. *Biochem. J.* **340,** 283–289.
42. Racchi, M., Solano, D. C., Sironi, M., and Govoni, S. (1999) Activity of alpha-secretase as the common final effector of protein kinase C-dependent and -independent modulation of amyloid precursor protein metabolism. *J. Neurochem.* **72,** 2464–2470.
43. Kuda, T., Shoji, M., Arai, H., Kawashima, S., and Saido, T. C. (1997) Reduction of plasma glutamyl aminopeptidase activity in sporadic Alzheimer's disease. *Biochem. Biophys. Res. Commun.* **231,** 526–530.
44. Price, D. L., Becher, M. W., Wong, P. C., Borchelt, D. R., Lee, M. K., and Sisodia, S. S. (1996) Inherited neurodegenerative diseases and transgenic models. *Brain Pathol.* **6,** 467–480.
45. Saito, Y., Buciak, J., Yang, J., and Pardridge, W. M. (1995) Vector-mediated delivery of 125I-labeled beta–amyloid peptide A beta 1–40 through the blood-brain barrier and binding to Alzheimer disease amyloid of the A beta 1–40/vector complex. *Proc. Natl. Acad. Sci. USA* **92,** 10,227–10,231.
46. Holcomb, L., Gordon, M. N., McGowan, E., Yu, X., Benkovic, S., Jantzen, P., et al. (1998) Accelerated Alzheimer-type phenotype in transgenic mice carrying both mutant amyloid precursor protein and presenilin 1 transgenes. *Nature Med.* **4,** 97–100.
47. Hyman, B. T., Marzloff, K., and Arriagada, P. V. (1993) The lack of accumulation of senile plaques or amyloid burden in Alzheimer's disease suggests a dynamic balance between amyloid deposition and resolution. *J. Neuropathol. Exp. Neurol.* **52,** 594–600.
48. Finch, C. E. and Sapolsky, R. M. (1999) The evolution of Alzheimer disease, the reproductive schedule, and apoE isoforms. *Neurobiol. Aging* **20,** 407–428.
49. Dickson, D. W. (1997) Neuropathological diagnosis of Alzheimer's disease, a perspective from longitudinal clinicopathological studies. *Neurobiol. Aging* **18,** S21–S26
50. Moir, R. D., Lynch, T., Bush, A. I., Whyte, S., Henry, A., Portbury, S., et al. (1998) Relative increase in Alzheimer's disease of soluble forms of cerebral Abeta amyloid protein precursor containing the Kunitz protease inhibitory domain. *J. Biol. Chem.* **273,** 5013–5019.
51. The Ronald and Nancy Reagan Research Institute of the Alzheimer's Association and the National Institute on Aging Working Group. Consensus report of the Working Group on "Molecular and Biochemical Markers of Alzheimer's Disease." (1998) *Neurobiol. Aging* **19,** 109–116.
52. Sano, M., Ernesto, C., Thomas, R. G., Klauber, M. R., Schafer, K., Grundman, M., et al. (1997) A controlled trial of selegiline, alpha-tocopherol, or both as

treatment for Alzheimer's disease. The Alzheimer's Disease Cooperative Study. *N. Engl. J. Med.* **336,** 1216–1222.
53. Rother, M., Erkinjuntti, T., Roessner, M., Kittner, B., Marcusson, J., and Karlsson, I. (1998) Propentofylline in the treatment of Alzheimer's disease and vascular dementia, a review of phase III trials. *Dement. Geriatr. Cogn. Disord.* **9(Suppl. 1),** 36–43.
54. Zhong, B., Lu, X., and Silverman, R. B. (1998) Syntheses of amino nitrones. Potential intramolecular traps for radical intermediates in monoamine oxidase-catalyzed reactions. *Bioorg. Med. Chem.* **6,** 2405–2419.
55. Gutzmann, H. and Hadler, D. (1998) Sustained efficacy and safety of idebenone in the treatment of Alzheimer's disease, update on a 2-year double-blind multicentre study. *J. Neural Transm.* **54(Suppl.),** 301–310.
56. Cesura, A. M., Borroni, E., Gottowik, J., Kuhn, C., Malherbe, P., Martin, J., et al. (1999) Lazabemide for the treatment of Alzheimer's disease, rationale and therapeutic perspectives. *Adv. Neurol.* **80,** 521–528.
57. Finberg, J. P., Lamensdorf, I., Commissiong, J. W., and Youdim, M. B. (1996) Pharmacology and neuroprotective properties of rasagiline. *J. Neural. Transm.* **48(Suppl.),** 95–101.
58. McGeer, P. L. and McGeer, E. G. (1999) Inflammation of the brain in Alzheimer's disease, implications for therapy. *J. Leukocyte Biol.* **65,** 409–415.
59. Breitner, J. C. (1996) The role of anti-inflammatory drugs in the prevention and treatment of Alzheimer's disease. *Annu. Rev. Med.* **47,** 401–411.
60. Scharf, S., Mander, A., Ugoni, A., Vajda, F., and Christophidis, N. (1999) A double-blind, placebo-controlled trial of diclofenac/misoprostol in Alzheimer's disease. *Neurology* **53,** 197–201.
61. Aisen, P. S., Marin, D., Altstiel, L., Goodwin, C., Baruch, B., Jacobson, R., et al. (1996) A pilot study of prednisone in Alzheimer's disease. *Dementia* **7,** 201–206.
62. Tan, J., Town, T., Paris, D., Mori, T., Suo, Z., Crawford, F., et al. (1998) Microglial Activation resulting from CD40–CD40L interaction after beta-amyloid stimulation. *Science* **286,** 2352–2355.
63. Mohr, E., Nair, N. P., Sampson, M., Murtha, S., Belanger, G., Pappas, B., et al. (1997) Treatment of Alzheimer's disease with sabeluzole, functional and structural correlates. *Clin. Neuropharmacol.* **20,** 338–345.
64. Hoshi, M., Takashima, A., Noguchi, K., Murayama, M., Sato, M., Kondo, S., et al. (1996) Regulation of mitochondrial pyruvate dehydrogenase activity by tau protein kinase I/glycogen synthase kinase 3beta in brain. *Proc. Natl. Acad. Sci. USA* **93,** 2719–2723.
65. Michel, P., Lambeng, N., and Ruberg, M. (1999) Neuropharmacologic Aspects of Apoptosis, Significance for Neurodegenerative Diseases. *Clin. Neuropharmacol.* **22,** 137–150.
66. Hefti, F. (1997) Pharmacology of neurotrophic factors. *Annu. Rev. Pharmacol. Toxicol.* **37,** 239–267.
67. Grundman, M., Corey-Bloom, J., and Thal, L. J. (1998) Perspectives in clinical Alzheimer's disease research and the development of antidementia drugs. *J. Neural Transm.* **53(Suppl.),** 255–275.

68. Krall, W. J., Sramek, J. J., and Cutler, N. R. (1999) Cholinesterase inhibitors: a therapeutic strategy for Alzheimer disease. *Ann. Pharmacother.* **33,** 441–450.
69. Avery, E. E., Baker, L. D., and Asthana, S. (1997) Potential role of muscarinic agonists in Alzheimer's disease. *Drugs Aging* **11,** 450–459.
70. Sedman, A. J., Bockbrader, H., and Schwarz, R. D. (1995) Preclinical and phase 1 clinical characterization of CI-979/RU35926, a novel muscarinic agonist for the treatment of Alzheimer's disease. *Life Sci.* **56,** 877–882.
71. Bodick, N. C., Offen, W. W., Levey, A. I., Cutler, N. R., Gauthier, S. G., Satlin, A., et al. (1997) Effects of xanomeline, a selective muscarinic receptor agonist, on cognitive function and behavioral symptoms in Alzheimer disease. *Arch. Neurol.* **54,** 465–473.
72. Fisher, A., Heldman, E., Gurwitz, D., Haring, R., Karton, Y., Meshulam, H., et al. (1996) M1 agonists for the treatment of Alzheimer's disease. Novel properties and clinical update. *Ann. NY Acad. Sci.* **777,** 189–196.
73. Lachowicz, J. E., Lowe, D., Duffy, R. A., Ruperto, V., Taylor, L. A., Guzik, H., et al. (1999) SCH 57790: a novel M2 receptor selective antagonist. *Life Sci.* **64,** 535–539.
74. Muller, D., Wiegmann, H., Langer, U., Moltzen-Lenz, S., and Nitsch, R. M. Lu (1998) 25-109, a combined m1 agonist and m2 antagonist, modulates regulated processing of the amyloid precursor protein of Alzheimer's disease. *J. Neural. Transm.* **105,** 1029–1043.
75. Potter, A., Corwin, J., Lang, J., Piasecki, M., Lenox, R., and Newhouse, P. A. (1999) Acute effects of the selective cholinergic channel activator (nicotinic agonist) ABT-418 in Alzheimer's disease. *Psychopharmacology (Berl.)* **142,** 334–342.
76. Vernier, J. M., El-Abdellaoui, H., Holsenback, H., Cosford, N. D., Bleicher, L., Barker, G., et al. (1999) 4-[[2-(1-Methyl-2-pyrrolidinyl)ethyl]thio]phenol hydrochloride (SIB-1553A), a novel cognitive enhancer with selectivity for neuronal nicotinic acetylcholine receptors. *J. Med. Chem.* **42,** 1684–1686.
77. Arendash, G. W., Sengstock, G. J., Sanberg, P. R., and Kem, W. R. (1995) Improved learning and memory in aged rats with chronic administration of the nicotinic receptor agonist GTS-21. *Brain Res.* **674,** 252–259.
78. Murai, S., Saito, H., Abe, E., Masuda, Y., Odashima, J., and Itoh, T. (1994) MKC-231, a choline uptake enhancer, ameliorates working memory deficits and decreased hippocampal acetylcholine induced by ethylcholine aziridinium ion in mice. *J. Neural. Transm. Gen. Sect.* **98,** 1–13.
79. Mellow, A. M., Aronson, S. M., Giordani, B., and Berent, S. (1993) A peptide enhancement strategy in Alzheimer's disease, pilot study with TRH–physostigmine infusions. *Biol. Psychiatry* **34,** 271–273.
80. Horita, A., Carino, M. A., Zabawska, J., and Lai, H. (1989) TRH analog MK-771 reverses neurochemical and learning deficits in medial septal-lesioned rats. *Peptides* **10,** 121–124.
81. Behan, D. P., Heinrichs, S. C., Troncoso, J. C., Liu, X. J., Kawas, C. H., Ling, N., et al. (1995) Displacement of corticotropin releasing factor from its binding protein as a possible treatment for Alzheimer's disease. *Nature* **378,** 284–287.

82. Toide, K., Shinoda, M., and Miyazaki, A. (1998) A novel prolyl endopeptidase inhibitor, JTP-4819—its behavioral and neurochemical properties for the treatment of Alzheimer's disease. *Rev. Neurosci.* **9,** 17–29.
83. Hampson, R. E., Rogers, G., Lynch, G., and Deadwyler, S. A. (1998) Facilitative effects of the ampakine CX516 on short-term memory in rats: enhancement of delayed-nonmatch-to-sample performance. *J. Neurosci.* **18,** 2740–2747.
84. Fakouhi, T. D., Jhee, S. S., Sramek, J. J., Benes, C., Schwartz, P., Hantsburger, G., et al. (1995) Evaluation of cycloserine in the treatment of Alzheimer's disease. *J. Geriatr. Psychiatry Neurol.* **8,** 226–230.
85. Craft, S. (1999) Enhancement of memory in Alzheimer disease with insulin and somatostatin, but not glucose. *Arch. Gen. Psychiatry.* **56,** 1135–1140.
86. Abe, K., Takeyama, C., and Yoshimura, K. (1998) Effects of S-8510, a novel benzodiazepine receptor partial inverse agonist, on basal forebrain lesioning-induced dysfunction in rats. *Eur. J. Pharmacol.* **347,** 145–152.
87. Ott, B. R., Thompson, J. A., and Whelihan, W. M. (1996) Cognitive effects of flumazenil in patients with Alzheimer's disease. *J. Clin. Psychopharmacol.* **16,** 400–402.
88. van Duijn, C. M. (1999) Hormone replacement therapy and Alzheimer's disease. *Maturitas* **31,** 201–205.
89. Treves, T. A. and Korczyn, A. D. (1999) Denbufylline in dementia, A double-blind controlled study. *Dement. Geriatr. Cogn. Disord.* **10,** 505–510.
90. Parnetti, L., Ambrosoli, L., Abate, G., Azzini, C., Balestreri, R., Bartorelli, L., et al. (1995) Posatirelin for the treatment of late-onset Alzheimer's disease, a double–blind multicentre study vs citicoline and ascorbic acid. *Acta Neurolog. Scand.* **92,** 135–140.
91. Markesbery, W. R. (1997) Oxidative stress hypothesis in Alzheimer's disease. *Free Radical Biol. Med.* **23,** 134–147.
92. Mimori, Y., Katsuoka, H., and Nakamura, S. (1996) Thiamine therapy in Alzheimer's disease. *Metab. Brain Dis.* **11,** 89–94.
93. Thal, L. J., Carta, A., Clarke, W. R., Ferris, S. H., Friedland, R. P., Petersen, R. C., et al. A 1-year multicenter placebo-controlled study of acetyl-L-carnitine in patients with Alzheimer's disease. *Neurology* **47,** 705–711.
94. Fleischhacker, W. W., Buchgeher, A., and Schubert, H. (1986) Memantine in the treatment of senile dementia of the Alzheimer type. *Prog. Neuro-psychopharmacol. Biol. Psychiatry* **10,** 87–93.
95. Goedert, M. (1996) Tau protein and the neurofibrillary pathology of Alzheimer's disease. *Ann. NY Acad. Sci.* **777,** 121–131.
96. Sperfeld, A. D., Collatz, M. B., Baier, H., Palmbach, M., Storch, A., Schwarz, J., et al. (1999) FTDP-17, an early-onset phenotype with parkinsonism and epileptic seizures caused by a novel mutation. *Ann. Neurol.* **46,** 708–715.
97. Imahori, K., Hoshi, M., Ishiguro, K., Sato, K., Takahashi, M., Shiurba, R., et al. (1998) Possible role of tau protein kinases in pathogenesis of Alzheimer's disease. *Neurobiol. Aging* **19,** S93–S98
98. Trojanowski, J. Q. and Lee, V. M. (1995) Phosphorylation of paired helical filament tau in Alzheimer's disease neurofibrillary lesions, focusing on phosphatases. *FASEB J.* **9,** 1570–1576.

99. Snowdon, D. A., Greiner, L. H., Mortimer, J. A., Riley, K. P., Greiner, P. A., and Markesbery, W. R. (1997) Brain infarction and the clinical expression of Alzheimer disease. The Nun Study. *JAMA* **277,** 813–817.
100. Sparks, D. L. (1997) Coronary artery disease, hypertension, ApoE, and cholesterol, a link to Alzheimer's disease? *Ann. NY Acad. Sci.* **826,** 128–146.
101. Forette, F., Seux, M. L., Staessen, J. A., Thijs, L., Birkenhager, W. H., Babarskiene, M. R., et al. (1998) Prevention of dementia in randomised double-blind placebo-controlled Systolic Hypertension in Europe (Syst-Eur) trial. *Lancet* **352,** 1347–1351.
102. Tedesco, M. A., Ratti, G., Mennella, S., Manzo, G., Grieco, M., Rainone, A. C., et al. (1999) Comparison of losartan and hydrochlorothiazide on cognitive function and quality of life in hypertensive patients. *Am. J. Hypertens.* **12,** 1130–1134.
103. Alexander, G. E., Furey, M. L., Grady, C. L., Pietrini, P., Brady, D. R., Mentis, M. J., et al. (1997) Association of premorbid intellectual function with cerebral metabolism in Alzheimer's disease, implications for the cognitive reserve hypothesis. *Am. J. Psychiatry* **154,** 165–172.

7
Tau and α-Synuclein in Neurodegenerative Diseases

Benoit I. Giasson, Christina A. Wilson, John Q. Trojanowski, and Virginia M. Y. Lee

7.1. INTRODUCTION

The past 2 yr have been extremely prolific in the area of neurodegenerative research, particularly with regard to diseases involving the proteins tau and synuclein. Tau aggregation in the form of filaments has long been implicated in diseases such as Alzheimer's disease (AD), progressive supranuclear palsy (PSP), and corticobasal degeneration (CBD), as well as others. The recent discovery of *tau* gene mutations in patients afflicted by a heterogeneous disease entity termed fronto-temporal dementia and parkinsonism linked to chromosome 17 (FTDP-17) has provided genetic corroboration for the importance of tau in disease and opens novel avenues of investigation into the nature of tau dysfunctions that lead to the demise of neurons. The discovery of mutations in α-synuclein in familial cases of Parkinson's disease (PD) has led to the revelation that this protein likely plays a prominent role in the etiology of several sporadic neurodegenerative disorders including PD, dementia with Lewy body (DLB) and multiple system atrophy (MSA), collectively grouped as synucleinopathies. In common with the subset of neurodegenerative diseases known as tauopathies because they are characterized by prominent filamentous tau aggregates in neurons and glia, similar fibrillary inclusions also accumulate in the brains of patients with synucleinopathies, but these inclusions are comprised predominantly of α-synuclein aggregates. In this chapter, the current knowledge of synuclein and tau proteins and their possible aberrant, malevolent role(s) in the onset and/or progression of brain diseases is reviewed.

From: *Contemporary Clinical Neuroscience: Molecular Mechanisms of Neurodegenerative Diseases*
Edited by: M.-F. Chesselet © Humana Press Inc., Totowa, NJ

7.2. THE SYNUCLEIN PROTEINS

Synucleins are small proteins (123–143 amino acids) characterized by repetitive imperfect repeats (KTKEGV) distributed throughout most of the amino-terminal half of the polypeptide and an acidic carboxyl-terminal region (*see* Fig. 1). The first synuclein was cloned from the electric ray, *Torpedo california*, by screening an expression library with an antiserum raised against cholinergic vesicles *(1)*. This protein was named synuclein because of its initial localization within the neuronal nuclei and presynaptic nerve terminals; however, localization of mammalian synucleins to the nucleus was not confirmed by subsequent studies. Three human synuclein proteins, termed α, β, and γ, are encoded by separate genes mapped to chromosomes 4q21.3–q22 *(2–4)*, 5q23 *(4,5)*, and 10q23.2–q23.3 *(6,7)*, respectively. The most recently cloned synuclein protein, synoretin, has a close homology to γ-synuclein and is predominantly expressed within the retina *(8)*.

α-Synuclein, also referred to as the nonamyloid component of senile plaques precursor protein (NACP) *(9)*, SYN1 *(10)* or synelfin *(11)*, is a heat-stable, "natively unfolded" protein *(12,13)* of poorly defined function. It is predominantly expressed in central nervous system (CNS) neurons where it is localized to presynaptic terminals *(11,14–16)*. Although highly overexposed Northern blots suggest that α-synuclein also may be expressed at low levels in many peripheral organs, these data must be interpreted with caution *(9)*. Expression of α-synuclein has also been demonstrated in a megakaryocyte cell line and in platelets, where it is loosely associated with organelles such as the endoplasmic reticulum *(17)*. Electron microscopy studies have localized α-synuclein in close proximity to synaptic vesicles at axonal termini *(11,16)*, suggesting a role for α-synuclein in neurotransmission or synaptic organization, and biochemical analysis has revealed that a small fraction of α-synuclein may be associated with vesicular membranes, but most of α-synuclein is cytosolic *(11,18)*. Further supporting the notion that it may have a vesicular function, α-synuclein can bind to rat brain vesicles in vitro *(19)*. Structurally, α-synuclein is predicted to form amphipathic helixes that can associate with phospholipid bilayers *(11)*, and an increase in α-helical secondary structure correlates with the binding of α-synuclein to small synthetic acidic unilamellar vesicles *(13)*. The expression pattern of α-synuclein is altered in subsets of neurons that form the brain nuclei involved in male zebra finch song learning during the critical developmental period when singing is acquired *(11)*. This suggests that α-synuclein may be involved in neuronal plasticity, although it does not seem to play a role in initial synaptic formation because it localizes to synapses after they are formed in cultured rat hippocampal neurons *(14)*. Interestingly, α- and

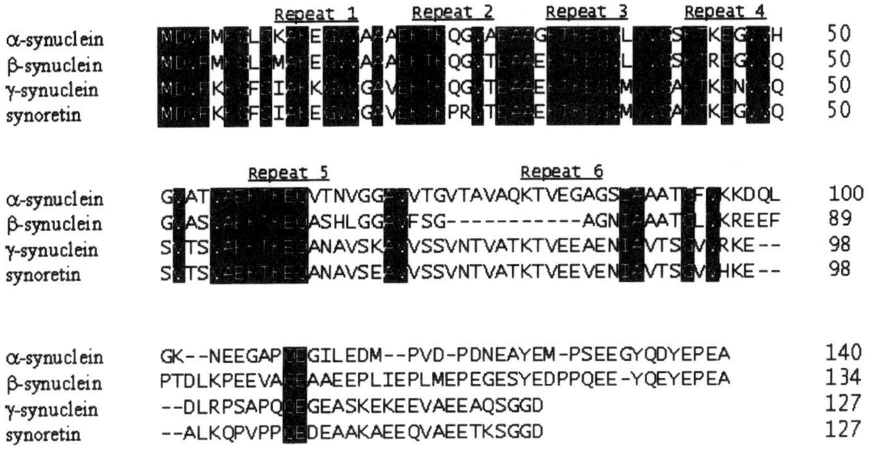

```
                          Repeat 1      Repeat 2     Repeat 3      Repeat 4
α-synuclein   MDVFMKGLSKAKEGVVAAAEKTKQGVAEAAGKTKEGVLYVGSKTKEGVVH   50
β-synuclein   MDVFMKGLSMAKEGVVAAAEKTKQGVTEAAEKTKEGVLYVGSKTREGVVQ   50
γ-synuclein   MDVFKKGFSIAKEGVVGAVEKTKQGVTEAAEKTKEGVMYVGAKTKENVVQ   50
synoretin     MDVFKKGFSIAKEGVVGAVEKTKPRVTEAAEKTKEGVMYVGAKTKEGVVQ   50

                       Repeat 5              Repeat 6
α-synuclein   GVATVAEKTKEQVTNVGGAVVTGVTAVAQKTVEGAGSIAAATGFVKKDQL   100
β-synuclein   GVASVAEKTKEQASHLGGAVFSG-----------AGNIAAATGLVKREEF   89
γ-synuclein   SVTSVAEKTKEQANAVSKAVVSSVNTVATKTVEEAENIAVTSGVVRKE--   98
synoretin     SVTSVAEKTKEQANAVSEAVVSSVNTVATKTVEEVENIAVTSGVVHKE--   98

α-synuclein   GK--NEEGAPQEGILEDM--PVD-PDNEAYEM-PSEEGYQDYEPEA   140
β-synuclein   PTDLKPEEVAQEAAEEPLIEPLMEPEGESYEDPPQEE-YQEYEPEA   134
γ-synuclein   --DLRPSAPQQEGEASKEKEEVAEEAQSGGD                127
synoretin     --ALKQPVPPQEDEAAKAEEQVAEETKSGGD                127
```

Fig. 1. Amino acid sequence alignment of the human synuclein proteins. The imperfect repeats of the type KTKEGV are identified. The black background highlights amino acid residues conserved between all four proteins. The sequences of α– and β–synucleins were obtained from Jakes et al. *(15)*, and γ–synuclein and synoretin were obtained from Ji et al. *(26)* and Surguchov et al. *(8)*, respectively.

β-synucleins are selective inhibitors of mammalian phospholipase D2 and, hence, may play a role in the control of synaptic vesicle cycling *(20)*.

The second member of the synuclein family that was identified is known as β-synuclein, initially named phosphoneuroprotein-14 or PNP-14, and it is a heat-stable protein predominantly expressed in neuronal axon termini of CNS neurons *(15,21–23)*, although it has also been localized to testicular Sertoli cells *(24)*. Although it is the least studied synuclein, it is highly homologous at the amino acid sequence level to α-synuclein and the localization of both proteins overlaps extensively in neurons, suggesting that the functions of α- and β-synuclein may be similar.

The third protein discovered to be homologous to α-synuclein is γ-synuclein (also termed persyn or breast-specific cancer gene-1; BSCG-1) *(25,26)* and it also is expressed in the brain and as well as in the spinal cord, but it is most abundant in the peripheral nervous system (PNS) including neurons of the dorsal root ganglia and trigeminal ganglia *(7,25)* (*see* clone D3 in ref. *27*). In addition, γ-synuclein is highly expressed in the stratum granulosum of the epidermis *(28)* and at low levels in several other organs *(7,26)*. Unlike α- and β-synucleins, γ-synuclein is distributed throughout the neuronal cytosol *(25)*, where it may alter the metabolism of the neuronal cytoskeleton *(29)*. Interestingly, γ-synuclein expression is upregulated in advanced infiltrating breast carcinoma *(6,26)*, and overexpression in breast cancer cells augments cell motility, invasiveness, and metastasis *(30)*. Furthermore, synuclein proteins may be involved in signaling, as the expression of synoretin affects the regulation of signal transduction pathways by activating Elk-1 *(8)*.

α-Synuclein Mutations and Aggregation in Neuronal Diseases

α-Synuclein was first associated with a neurodegenerative disease when a fragment thereof corresponding to amino acids 61–95, referred to as the nonamyloid component of senile plaques (NAC), was isolated from proteolytically digested sodium dodecyl sulfate-insoluble fractions of Alzheimer's brain *(9)*. In addition, antibodies to NAC recognize a significant percentage of diffuse and mature plaques *(9,31–33)*. Although NAC may extend beyond amino acids 61–95 *(9)*, the full-length protein is not a component of plaques because plaques are not labeled by antibodies recognizing either ends of α-synuclein. NAC may be involved in the formation of amyloid senile plaques, as it can form amyloidogenic aggregates *(32)* and can stimulate the aggregation of β-amyloid, the major component of plaques *(34)*. These findings notwithstanding, the role of NAC in AD pathogenesis still remains to be evaluated. Interestingly, one study has suggested that a dinucleotide repeat polymorphism in the promoter of α-synuclein may confer protection against the apolipoprotein E ε4 allele associated with AD *(35)*; however, this finding was not confirmed in a later study *(36)*.

Genetic and histopathological findings have illuminated the significant contribution of α-synuclein to the etiology of PD. The identification of point mutations in α-synuclein in a small number of familial cases of PD strongly supports the notion that synuclein may play a causal role in this disease. Autosomal dominant mutations in α-synuclein were identified in a German kindred harboring an A30P mutation resulting from G to C transversion at position 88 *(37)* and in a large Italian family (the Contorsi kindred) and five Greek families with a A53T mutation resulting from an G to A transition at position 209 *(38,39)*. Families harboring the A53T mutation may have existed in close contact, suggesting a possible common ancestor *(40)*. Thus far, α-synuclein mutations resulting in disease have only been found in a small number of kindreds afflicted with familial PD. Analysis of a large number of patients with sporadic or familial cases of either PD or DLB, as well as a small number of patients with MSA, failed to reveal mutations in α-synuclein *(41–50)*. Furthermore, no mutations in the γ-synuclein gene have been demonstrated in families with PD *(51)*.

Mounting evidence supports the idea that α-synuclein is the major component of several proteinaceous inclusions characteristic of specific neurodegenerative diseases. Pathological synuclein aggregations are restricted to the α-synuclein isoforms as β- and γ-synucleins have not been detected in these inclusions. The presence of α-synuclein positive aggregates is disease-specific. Lewy bodies, neuronal fibrous cytoplasmic inclusions that are histopathological hallmarks of PD and DLB *(52,53)*, are strongly labeled with antibodies to α-synuclein *(54–61)*. Dystrophic ubiquitin-positive neurites associated with PD pathology, termed Lewy neurites (LN) *(62)*, and CA2/CA3 ubiquitin neurites are also α-synuclein positive *(54,56,58–60)*. Furthermore, pale bodies, putative precursors of LBs *(58,60,61)*, thread-like structures in the perikarya of slightly swollen neurons *(61)*, and glial silver-positive inclusions in the midbrains of patients with LB diseases *(63)* are also immunoreactive for α-synuclein. α-Synuclein is likely the major component of glial cell inclusions (GCIs) and neuronal cytoplasmic inclusions in MSA and Hallervorden–Spatz disease (brain iron accumulation type 1) *(60,64–68)*. α-Synuclein immunoreactivity is present in some dystrophic neurites in senile plaques in AD *(33)*, but it is not detected in Pick bodies, neurofibrillary tangles (NFTs), neuropil threads, or in the neuronal or glial inclusion characteristic of PSP, CBD, motor neuron disease, and trinucleotide-repeat diseases *(59,60)*.

Further evidence supports the notion that α-synuclein is the actual building block of the fibrillary components of LBs, LNs, and GCIs. Immunoelectron microscopic studies have demonstrated that these fibrils are intensely labeled with α-synuclein antibodies *in situ (31,57,60,61,65,68)*. Sarcosyl-insoluble α-synuclein filaments with straight and twisted morphologies can also be observed in extracts of DLB and MSA brains *(54,67)*. Moreover, α-synuclein can assemble in vitro into elongated homopolymers with similar widths as sarcosyl-insoluble fibrils or filaments visualized *in situ (69–73)*. Polymerization is associated with a concomitant change in secondary structure from random coil to anti-parallel β-sheet structure *(73)* consistent with the Thioflavine-S reactivity of these filaments *(72,73)*. Furthermore, the PD-associated α-synuclein mutation, A53T, may accelerate this process, as recombinant A53T α-synuclein has a greater propensity to polymerize than wild-type α-synuclein *(69,71,73)*. This mutation also affects the ultrastructure of the polymers; the filaments are slightly wider and are more twisted in appearance, as if assembled from two protofilaments *(69–71)*. The A30P mutation may also modestly increase the propensity of α-synuclein to polymerize *(73)*, but the pathological effects of this mutation also may be related to its reduced binding to vesicles *(19)*. Interestingly, carboxyl-terminally truncated α-synuclein may be more prone to form filaments than the full-length protein *(74)*. Although the pathological implications of the latter finding is still unclear, it is possible that aberrant proteolysis of α-synuclein may form "seeds" that could initiate α-synuclein filament assembly.

7.3. TAU EXPRESSION AND FUNCTION

Tau is a collection of microtubule (MT)-associated proteins (MAPs) *(75)* expressed from a single gene on chromosome 17 *(76,77)*. In the adult human brain, 6 isoforms ranging between 352 and 441 amino acids in length are produced as a result of alternative RNA splicing *(78,79)* (Fig. 2). The incorporation or exclusion of exon 2 or exons 2 and 3 results in proteins with 0 (0N), 29 (1N), or 58 (2N) amino acid inserts in the amino-terminal region. Similarly, exon 10 can be alternatively spliced to yield products containing either three (3R) or four (4R) tandem repeats of 31 or 32 amino acids. In the adult brain, 3R and 4R tau are present at approximately equal amounts and 2N tau isoforms are significantly underrepresented relative to 0N or 1N isoforms *(80,81)*. The expression of tau isoforms is developmentally regulated, as only the smallest tau polypeptide (0N, 3R) is expressed in fetal brain *(78,80)*. Furthermore, alternative splicing and inclusion of exon 4A *(77,82)* yields a group of higher-molecular-weight tau proteins, termed "big tau", which is expressed predominantly in the peripheral nervous system *(82–84)*. Tau is preferentially found in neurons *(85,86)* but can also be detected in some oligodendrocytes and astrocytes *(86–89)*.

The ability of tau to induce MT assembly, nucleation, and bundling is well documented *(75,90–96)*. The MT binding domain of tau resides within the carboxyl-terminal region containing the three or four MT-binding repeats (3R, 4R) and 13 or 14 amino acid interrepeats *(97–99)*. Individual repeats are capable of binding MTs, albeit with weaker affinities than the full-length protein *(100)*. Although each repeat makes a significant contribution to the overall MT affinity *(101)*, MT binding is more complex than a simple linear array of binding sites *(94,95)*. The spacer between MT-binding repeats one and two (R1–R2) can also contribute to MT binding, as it has more than twice the binding affinity than any individual repeat *(101)*. Furthermore, a proline-rich region upstream of the repeat region *(94)*, more precisely the sequence KKVAVVR (amino acids 215–221), exerts a strong positive influence on MT binding and assembly *(102)*. The molecular details governing the tau–MT interaction remains incompletely defined and controversial; nevertheless, it appears that tau can bind to at least two regions in either α- and β- tubulin [*(103)*, and references herein]. Four-repeat tau has a greater MT polymerization and binding capacity than 3R tau *(80,98)*. The amino-terminal inserts do not significantly contribute to the binding affinity of tau; however, they may induce MT bundling *(94)*. The ability of

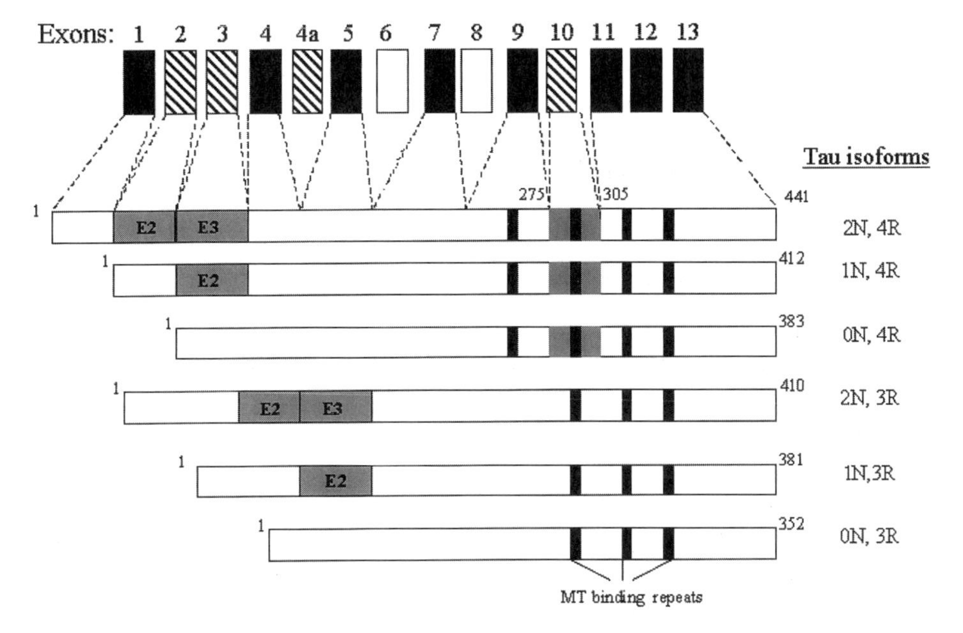

Fig. 2. Schematic of exon organization and the six brain tau isoforms generated by alternative splicing. Alternative splicing of exons 2, 3, and 10 produce the six alternative products. Putative exons 6 and 8 are not used in brain. Exon 4A, which is also not used in the brain, is included in the PNS leading to the translation of larger tau isoforms, termed "big tau" (*see* text). Black bars depict the 18–amino-acid MT–binding repeats.

tau to bind and modulate MT assembly is negatively regulated by phosphorylation *(90,93,104–106)*.

Surprisingly, tau is not essential for MT function because disruption of expression by a genetically engineered null mutation does not result in an overt phenotype *(107)*. Moreover, depletion of axonal tau in rat sympathetic neurons by microinjection of anti-tau antibodies had no detectable effects on the dynamics of axonal MTs *(108)*. Thus, it seems likely that the ability of tau to modulate MT assembly can be compensated for by other MAPs.

The distribution of tau in cultured rat sympathetic and hippocampal neurons suggests that it may serve functions other than the stabilization of MTs. In these cultures, tau is more concentrated at the distal end than the proximal end of axons even though axonal MTs in the distal end are less stable and turn over more rapidly *(109,110)*. It is possible that these apparent discrepancies may be the result of differential phosphorylation of tau within these axonal regions, but these observations also suggest that the abundance of tau at the growth cone neck may reflect an alternative role for tau. Tau expression can inhibit kinesin-dependent trafficking of organelles such as mitochondria and vesicles *(111)*. Additionally, the amino-terminal projection domain of tau interacts with the plasma membrane, although the importance of this observation is still unknown *(112)*. Furthermore, tau has been shown to exist in complex with phospholipase C-γ *(113)* and to increase the activity of this enzyme *(114)*.

Tau Pathogenesis in Alzheimer's Disease

In AD, tau aggregates into cytoplasmic inclusions in the form of neurofibrillary tangles (NFTs) in neuronal cell bodies, and neuropil threads and dystrophic neurites of senile plaques in neuronal processes *(115)*. Ultrastructurally, these aberrant structures are comprised of 8- to 20-nm twisted double-helical ribbons, referred to as paired helical filaments (PHFs) *(116,117)* and the less abundant 15-nm-wide straight filaments (SFs) *(118,119)*. Compelling biochemical and immuno-electron microscopic studies have demonstrated that PHFs are comprised of tau *(120–124)*. SFs are a structural variant of PHFs and are likely entirely composed of tau *(118)*. Abnormally aggregated tau isolated from AD brain, referred to as PHF-tau (or A68), contains all six CNS tau isoforms *(125)* aberrantly hyperphosphorylated at >25 Ser or Thr residues *(126–128)*. It is still unclear which enzymes are responsible for this hyperphosphorylation of tau, as numerous kinases and phosphatases can modulate tau phosphorylation in vivo and/or in vitro *(129,130)*. It is likely that a change in a combination of enzymatic activities is involved in generating hyperphosphorylated PHF-tau.

The mechanism of PHF-tau formation in neurons remains enigmatic. Because hyperphosphorylation is the most prominent difference between PHF-tau and normal tau, it would be reasonable to hypothesize that phosphorylation may induce tau filament formation. However, there is no direct evidence to support this model, and nonphosphorylated, recombinant tau can assemble into filaments in vitro *(131)*. It is more likely that abnormal phosphorylation increases the pool of MT-unbound tau, which then becomes available for PHF formation. Supporting this notion are the findings that (1) hyperphosphorylation of tau precedes PHF formation *(132,133)*, (2) phosphorylation inhibits MT binding *(90,93,106,134)*, and (3) the ability of PHF-tau to bind to MTs is greatly impaired, but this loss of function can be overcome by dephosphorylation *(93,105,135)*.

Interestingly, tau filament assembly in vitro can be facilitated by long polyanionic molecules such as strongly or moderately sulfated glycosaminoglycans and nucleic acids *(136–140)*. Moreover, sulfated glycosaminoglycans and nucleic acids has been shown to prevent tau MT-binding, and heparin sulfate, chondroitin sulfate, and dermatan sulfate proteoglycans have been colocalized with NFTs of AD brains *(136,141,142)*. It is unclear how sulfated glycosaminoglycans appear within the cytoplasm, although a likely explanation would involve leakage from membrane-bound organelles.

PHF-tau also is modified by ubiquitination *(143)*, glycation *(144,145)*, and N-linked glycosylation *(146)*. Ubiquitination occurs after aggregation, probably as an attempt by the cellular machinery to degrade these protein aggregates, and it is unlikely to contribute to PHF formation *(133,147)*. Glycation is a nonenzymatic addition of reduced carbohydrates, and the presence of this modification is likely to result from the slow turnover of PHFs. The importance of N-linked gycosylation is undetermined, although it may contribute to the maintenance of PHF structure *(146)*.

FTDP-17: Direct Genetic Evidence for the Importance of Tau in Disease

FTDP-17 refers to a group of autosomal dominant hereditary neurodegenerative disorders characterized by behavioral changes with subsequent cognitive disturbance and, in some cases, parkinsonism *(148)*. Most, if not all, FTDP-17 families show tau deposits either in neurons or in both neurons and glia without accompanying amyloid deposition *(149–157)*. Genetic analysis has revealed 12 different mutations in the *tau* gene in at least 26 FTDP-17 families, establishing that in FTDP-17 kindreds, tau mutations are pathogenic for the disease (Table 1). The mutations can be divided into two functional groups: missense mutations that impair the ability of tau to bind to MTs and promote MT assembly, and exonic or intronic mutations that alter the inclusion of exon 10 during splicing. Missense mutations G272V, Δ280, P301L, V337M, and R406W belong to the former category *(81,155,158,159)*. These mutations may lead to pathogenesis through an initial loss of function, followed by a gain of toxic effect. The reduced capacity of these mutants to stabilize MTs may lead to a loss of MT function, such as fast axonal transport. Pathology may subsequently be compounded by a progressive accumulation of tau in the cytoplasm and eventual aggregation into insoluble filaments. Moreover, mutations P301L, V337M, and R406W may accelerate tau filament formation *(160,161)*. Consistent with the location of these mutations within tau, aggregated tau from cases with mutation V337M or R406W is predominantly comprised of all six CNS tau isoforms, whereas only 4R-tau is present in the case of P301L *(81,162)*.

Some pathogenic missense mutations and silent mutations at or close to exon 10 can alter the splicing efficiency of this exon, as demonstrated by exon-trapping analysis *(155,163,164)*. The mutations may affect splicing via three different mechanisms. First, in cases of known intronic mutations

Table 1
Tau Mutations in FTDP–17

Mutations	Exon/intron location	Protein domain	Functional impact	Ref.
G272V	Exon 9	Repeat 1	Reduced MT binding	*159,163*
N279K	Exon 10	Interrepeat 1–2	Altered splicing	*162*
Δ280	Exon 10	Interrepeat 1–2	Altered splicing/reduced MT binding	*159*
L284L (T to C)	Exon 10	Interrepeat 1–2	Altered splicing	*155*
P301L	Exon 10	Repeat 2	Reduced MT binding	*159,162,163,183*
S305N	Exon 10	Interrepeat 2–3	Altered splicing	*149*
V337M	Exon 12	Interrepeat 3–4	Reduced MT binding	*184*
R406W	Exon 13	C–terminus	Reduced MT binding	*163*
E10+3 (G to A)	Intron 10	N/A	Altered splicing	*171*
E10+13 (A to G)	Intron 10	N/A	Altered splicing	*163*
E10+14 (C to T)	Intron 10	N/A	Altered splicing	*163*
E10+16 (C to T)	Intron 10	N/A	Altered splicing	*154,163,185*

Note: The positions of the mutations are assigned according to the longest brain tau isoform (441 amino acid long).

and the missense mutation S305N, it has been proposed that altered splicing efficiency may be due to the disruption of a putative inhibitory RNA stem loop structure at the 5' boundary of the intron following exon 10 (Fig. 3) *(155,163)*. This secondary structure may compete with the U1 snRNP or other splicing factors for the binding of the splice donor site, and its destabilization leads to the increased inclusion of exon 10. However, attempts to rescue the putative function of the stem-loop structure with compensatory double mutants were not successful, suggesting that other elements beside the secondary structure are involved *(155)*. The S305N mutation also changes the 5' splice site of intron 10 to a stronger splice site (GUguga to AUguga) *(165)*, which likely also contributes to the effect on splicing by this mutation. Consistent with this pathogenic mechanism, an increased ratio of exon 10+/exon 10– tau RNA in the brains of patients with intronic mutations has been reported *(163)*. A second mechanism by which splicing is affected is demonstrated by the N279K mutation, which may enhance the insertion of exon 10 by improving an exon-splicing enhancer. At the RNA level, this mutation changes a nucleotide stretch from TAAGAA to GAAGAA. The latter sequence is a repeat of GARs (where R is a purine), which can act as an exon-splice enhancer *(166–169)*. The notion that this nucleotide stretch is a splicing enhancer is supported by the finding that the deletion of nucleotides AAG by the Δ280 mutation obliterates exon 10 inclusion *(155)*. Finally, a third mechanism is demonstrated by the silent mutation L284L (CTT→CTC), which likely affects splicing because it disrupts the sequence UUAG that can act as a putative exon-splicing silencer *(170)*, and thereby increases the ratio of exon 10+/exon 10– tau mRNA *(155)*. Consistent with the notion that an alteration in RNA splicing is the cause of pathogenesis, biochemical postmortem analysis of the brains of affected patients with mutations predicted to increase exon 10 splicing (i.e., E10+14, N279K, E10+3 mutations) showed an increase in the abundance of 4R-tau over 3R-tau and the specific accumulation of aggregated 4R-tau *(81,150,154,162,171)*.

The mechanism by which changes in the 3R/4R-tau ratio lead to neuronal and, in some cases, glial dysfunction and death is still nebulous. Four repeat-tau and 3R-tau may bind to distinct sites on MTs *(101)*, and the overproduction of one group of isoforms may result in a pool of MT-unbound tau that may polymerize into filaments over time. It is also possible that a specific ratio of tau isoforms is required for the normal maintenance and function of MTs. Although speculative, the possibility that specific isoforms might have other, undetermined functions should not be overlooked.

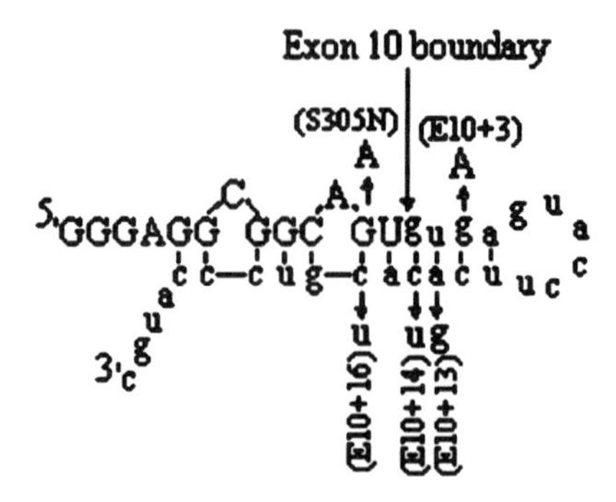

Fig. 3. Structure of the putative inhibitory RNA stem loop structure at the 5' boundary of the intron following exon 10. Pathogenic mutations that can affect the stability of this secondary structure are depicted.

Other Tauopathies Involving Specific Isoforms of Tau

Pick's disease is a fronto-temporal-type dementia characterized by the presence of Pick bodies, round-shaped neuronal inclusions composed of granular material together with 10- to 20-nm diameter filaments *(172)*. These disease specific filamentous tau inclusions contain 3R-tau isoforms exclusively *(173,174)*. The reasons for this selective aggregation of 3R-tau isoforms is unknown, but a possible explanation is that neurons expressing specifically these forms of tau are more vulnerable in Pick's disease. The restricted expression of 3R-tau in the granule cell layer of the dentate gyrus demonstrates that expression of tau isoforms can be cell-type specific *(79)*. This concept has not been extensively studied and further evaluation is certainly warranted.

Progressive supranuclear palsy (PSP) and corticobasal degeneration (CBD) are late onset neurodegenerative disorders characterized by both neuronal and glial tau inclusions. Aggregated tau in these diseases is predominantly comprised of 4R-tau isoforms *(175)*. In CBD, tau precipitates in the form of astrocytic plaques, oligodendroglial "coil bodies," and neuronal inclusions sometimes termed corticobasal bodies *(176,177)*. Neuronal tau in PSP brains aggregates in the form of classical flame-shaped NFTs or globose NFTs *(176)*. PSP also features distinctive glial tau inclusions termed oligodendroglial "coiled bodies," tufted astrocytes, and thorn-shaped astrocytes *(177)*.

Genetic changes in the *tau* gene may contribute to the risk of developing PSP. Conrad et al. *(178)* reported a link between PSP and a polymorphic dinucleotide repeat region found between exons 9 and 10 of the *tau* gene. Subsequent studies confirmed this correlation *(179–181)*, and it was recently demonstrated that this association is the result of a specific haplotype that also contains at least eight single nucleotide polymorphisms *(182)*.

7.4. CONCLUSIONS AND FUTURE DIRECTIONS

Two major groups of neurodegenerative diseases, synucleopathies and tauopathies, exhibit aberrant proteinaceous inclusions that result in cellular dysfunction. There may be multiple mechanisms by which these aggregates mediate their destructive consequences. First, the accumulation of either synuclein or tau in inclusions may reduce the levels of functional molecules, which alone may be detrimental to the cell. However, the presence of inclusions may also act as a barrier that interferes with overall cellular functions such as axonal transport or cellular morphology. In the end, it is likely that both the depletion of functional protein and the presence of cytoplasmic obstacles formed by aggregated filaments are instrumental in the ultimate demise of neurons. Further investigation, including the development of transgenic mouse models, is warranted to enhance the current understanding of normal synuclein and tau functions as well as the mechanism(s) involved in the intracellular aggregation of these proteins in order to improve preventative and therapeutic strategies.

ACKNOWLEDGMENTS

B.I.G. is a recipient of a fellowship from the Human Frontier Science Program Organization. C.A.W. is a Howard Hughes predoctoral fellow. This work was supported by grants from the National Institute on Aging, and the Dana Foundation and a Pioneer Award from the Alzheimer's Association. The authors want to thank Dr. S. Rueter for her critical reading of this manuscript. We also thank our colleagues in the Center for Neurodegenerative Disease Research, Departments of Pathology and Laboratory Medicine, Neurology, Psychiatry, and the Penn Alzheimer Center for their assistance, and the families of patients who made this research possible.

REFERENCES

1. Maroteaux, L., Campanelli, J. T., and Scheller, R. H. (1988) Synuclein: a neuron-specific protein localized to the nucleus and presynaptic nerve terminal. *J. Neurosci.* **8,** 2804–2815.
2. Shibasaki, Y., Baillie, D. A. M., St.Clair, D., and Brookes, A. J. (1995) High-resolution mapping of SNCA encoding α-synuclein, the non-Aβ component of Alzheimer's disease amyloid precursor, to human chromosome 4q21.3 to q22 by fluorescence in situ hybridization. *Cytogenet.Cell Genet.* **71,** 54–55.
3. Chen, X., Rohan de Silva, H. A., Pettenati, M. J., Rao, P. N., St.George-Hyslop, P., Roses, A. D., et al. (1995) The human NACP/α-synuclein gene: chromosome assignment to 4q21.3-q22 and Taq1 RFLP ananysis. *Genomics* **26,** 425–427.
4. Spillantini, M. G., Divane, A., and Goedert, M. (1995) Assignment of human α-synuclein(SNCA) and β-synuclein(SNCB) genes to chromosomes 4q21 and 5q35. *Genomics* **27,** 379–381.

5. Lavedan, C., Leroy, E., Torres, R., Dehejia, A., Dutra, A., Buchhlotz, S., et al. (1998) Genomic organization and expression of the human β-synuclein gene (SNCB). *Genomics* **54,** 173–175.
6. Ninkina, N. N., Alimova-Kost, M. V., Paterson, J. W. E., Delaney, L., Cohen, B. B., Imreh, S., et al. (1998) Organization, expression and polymorphism of the human persyn gene. *Hum. Mol. Genet.* **7,** 1417–1424.
7. Lavedan, C., Leroy, E., Dehejia, A., Buchhlotz, S., Dutra, A., Nussbaum, R. L., et al. (1998) Identification, location and characterization of the human γ-synuclein gene. *Hum. Genet.* **103,** 106–112.
8. Surguchov, A., Surgucheva, I., Solessio, E., and Baehr, W. (1999) Synoretin—a new protein belonging to the synuclein family. *Mol. Cell. Neurosci.* **13,** 95–103.
9. Uéda, K., Fukushima, H., Masliah, E., Xia, Y., Iwai, A., Yoshimoto, M., et al. (1993) Molecular cloning of cDNA encoding an unrecognized component of amyloid in Alzheimer's disease. *Proc. Natl. Acad. Sci. USA* **90,** 11,282–11,286.
10. Maroteaux, L. and Scheller, R. H. (1991) The rat brain synucleins; family of proteins transiently associated with neuronal membrane. *Mol. Brain Res.* **11,** 335–343.
11. George, J. M., Jin, H., Woods, W. S., and Clayton, D. F. (1995) Characterization of a novel protein regulated during the critical period for song learning in the zebra finch. *Neuron* **15,** 361–372.
12. Weinreb, P. H., Zhen, W., Poon, A. W., Conway, K. A., and Lansbury, P. T. (1996) NACP, a protein implicated in Alzheimer's disease and learning, is natively unfolded. *Biochemistry* **35,** 13,709–13,715.
13. Davidson, W. S., Jonas, A., Clayton, D. F., and George, J. M. (1998) Stabilization of α-synuclein secondary structure upon binding to synthetic membranes. *J. Biol. Chem.* **273,** 9443–9449.
14. Withers, G. S., George, J. M., Banker, G. A., and Clayton, D. F. (1997) Delayed localization of synelfin (synuclein, NACP) to presynaptic terminals in cultured rat hippocampal neurons. *Dev. Brain Res.* **99,** 87–94.
15. Jakes, R., Spillantini, M. G., and Goedert, M. (1994) Identification of two distinct synucleins from human brain. *FEBS Lett.* **345,** 27–32.
16. Iwai, A., Masliah, E., Yoshimoto, M., Ge, N., Flanagan, L., Rohan de Silva, H. A., et al. (1995) The precursor protein of non-Aβ component of Alzheimer's disease amyloid is a presynaptic protein of the central nervous system. *Neuron* **14,** 467–475.
17. Hashimoto, M., Yoshimoto, M., Sisk, A., Hsu, L. J., Sundsumo, M., Kittel, A., et al. (1997) NACP, a synaptic protein involved in Alzheimer's disease, is differentially regulated in during megakaryocyte differentiation. *Biochem. Biophys. Res. Commun.* **237,** 611–616.
18. Irizarry, M. C., Kim, T.-W., McNamara, M., Tanzi, R. E., George, J. M., Clayton, D. F., et al. (1996) Characterization of the precursor protein of the non-Aβ component of senile plaques (NACP) in the human central nervous system. *J. Neuropathol. Exp. Neurol.* **55,** 889–895.
19. Jensen, P. H., Nielsen, M. S., Jakes, R., Dotti, C. G., and Goedert, M. (1998) Binding of α-synuclein to brain vesicles is abolished by familial Parkinson's disease mutation. *J. Biol. Chem.* **273,** 26,292–29,294.

20. Jenco, J. M., Rawlingson, A., Daniels, B., and Morris, A. J. (1998) Regulation of phospholipase D2: selective inhibition of mammalian phospholipase D isoenzymes by α- and β -synucleins. *Biochemistry* **37,** 4901–4909.
21. Nakajo, S., Omata, K., Aiuchi, T., Shibayama, T., Okahashi, I., Ochiai, H., et al. (1990) Purification and characterization of a novel brain-specific 14-kDa protein. *J. Neurochem.* **55,** 2031–2038.
22. Shibayama-Imazu, T., Okahashi, I., Omata, K., Nakajo, S., Ochiai, H., Nakai, Y., et al. (1993) Cell and tissue distribution and development change of neuron specific 14 kDa protein (phosphoneuroprotein 14). *Brain Res.* **622,** 17–25.
23. Nakajo, S., Shioda, S., Nakai, Y., and Nakaya, K. (1994) Localization of phosphoneuroprotein 14 (PNP 14) and its mRNA expression in rat brain determined by immunocytochemistry and in situ hybridization. *Mol. Brain Res.* **27,** 81–86.
24. Shibayama-Imazu, T., Ogane, K., Hasegawa, Y., Nakajo, S., Shioda, S., Ochiai, H., et al. (1998) Distribution of PNP 14 (β-synuclein) in neuroendocrine tissues: localization in Sertoli cells. *Mol. Reprod. Dev.* **50,** 163–169.
25. Buchman, V. L., Hunter, H. J. A., Pinõn, L. G. P., Thompson, J., Privalova, E. M., Ninkina, N. N., et al. (1998) Persyn, a member of the synuclein family, has a distinct pattern of expression in the developing nervous system. *J. Neurosci.* **18,** 9335–9341.
26. Ji, H., Liu, Y. E., Jia, T., Wang, M., Liu, J., Xiao, G., et al. (1997) Identification of a breast cancer-specific gene, BCSG1, by differential cDNA sequencing. *Cancer Res.* **57,** 759–764.
27. Akopian, A. N. and Wood, J. N. (1995) Peripheral nervous system-specific genes identified by substractive cDNA cloning. *J. Biol. Chem.* **270,** 21,264–21,270.
28. Ninkina, N. N., Privalona, E. M., Pinõn, L. G. P., Davies, A. M., and Buchman, V. L. (1999) Developmentally regulated expression of persyn, a member of the synuclein family, in skin. *Exp. Cell. Res.* **246,** 308–311.
29. Buchman, V. L., Adu, J., Pinõn, L. G. P., Ninkina, N. N., and Davies, A. M. (1998) Persyn, a member of the synuclein family, influences neurofilament network integrity. *Nat. Neurosci.* **1,** 101–103.
30. Jia, T., Liu, Y. E., Liu, J., and Shi, Y. E. (1999) Stimulation of breast cancer invasion and metastasis by synuclein γ. *Cancer Res.* **59,** 742–747.
31. Takeda, A., Hashuimoto, M., Mallory, M., Sundsumo, M., Hansen, L., Sisk, A., et al. (1998) Abnormal distribution of the non-Aβ component of Alzheimer's disease amyloid precursor/α-synuclein in Lewy body disease as revealed by proteinase K and formic acid pretreatment. *Lab. Invest.* **78,** 1169–1177.
32. Iwai, A., Yoshimoto, M., Masliah, E., and Saitoh, T. (1995) Non-Aβ component of Alzheimer's disease amyloid (NAC) is amyloidogenic. *Biochemistry* **34,** 10,139–10,145.
33. Masliah, E., Iwai, A., Mallory, M., Uéda, K., and Saitoh, T. (1996) Altered presynaptic protein NACP is associated with plaque formation and neurodegeneration in Alzheimer's disease. *Am. J. Pathol.* **148,** 201–210.
34. Yoshimoto, M., Iwai, A., Kang, D., Otero, D. A. C., Xia, Y., and Saitoh, T. (1995) NACP, the precursor protein of the non-amyloid β/A4 protein (Aβ)

component of Alzheimer's disease amyloid, binds Aβ and stimulates Aβ aggregation. *Proc. Natl. Acad. Sci.USA* **92,** 9141–9145.
35. Xia, Y., Rohan de Silva, H. A., Rosi, B. L., Yamaoka, L. H., Rimmler, J. B., Pericak-Vance, M. A., et al. (1996) Genetic studies in Alzheimer's disease with an NACP/α-synuclein polymorphism. *Ann. Neurol.* **40,** 207–215.
36. Hellman, N. E., Grant, E. A., and Goate, A. M. (1998) Failure to replicate a protective effect of allele 2 of NACP/α -synuclein polymorphism in Alzheimer's disease: an association study. *Ann. Neurol.* **44,** 278–281.
37. Krüger, R., Kuhn, W. M. T., Woitalla, D., Graeber, M., Kösel, S., Przuntek, H., et al. (1998) Ala30Pro mutation in the gene encoding α-synuclein in Parkinson's disease. *Nat. Gen.* **18,** 106–108.
38. Polymeropoulos, M. H., Lavedan, C., Leroy, E., Ide, S. E., Dehejia, A., Dutra, A., et al. (1997) Mutations in the α-synuclein gene identified in families with Parkinson's disease. *Science* **276,** 2045–2047.
39. Papadimitriou, A., Veletza, V., Hadjigeorgiou, G.M., Patrikiou, A., Hirano, M., and Anastasopoulos, I. (1999) Mutated α-synuclein gene in two Greek kindreds with familial PD: incomplete penetrance. *Neurology* **52,** 651–654.
40. Golbe, L. I. (1999) Alpha-synuclein and Parkinson's disease. *Mov. Disorders* **14,** 6–9.
41. Chan, P., Tanner, C. M., Jiang, X., and Langston, J. W. (1998) Failure to find the α-synuclein gene missense mutation ($G^{209}A$) in 100 patients with younger onset Parkinson's disease. *Neurology* **50,** 513–514.
42. Parsian, A., Racette, B., Zhang, Z. H., Chakraverty, S., Rundle, M., Goate, A., et al. (1998) Mutation, sequence analysis, and association studies of α-synuclein in Parkinson's disease. *Neurology* **51,** 1757–1759.
43. Higuchi, S., Arai, H., Matsushita, S., Matsui, T., Kimpara, T., Takeda, A., et al. (1998) Mutation in the α-synuclein gene and sporadic Parkinson's disease, Alzheimer's disease, and dementia with Lwey bodies. *Exp. Neurol.* **153,** 164–166.
44. Chan, P., Jiang, X., Forno, L. S., Di Monto, D. A., Tanner, C. M., and Langston, J. W. (1999) Absence of mutation in the coding region of the α-synuclein gene in Parkinson's disease. *Neurology* **50,** 1136–1137.
45. Farrer, M., Wavrant-De Vrieze, F., Crook, R., Boles, L., Perez-Tur, J., Hardy, J., et al. (1998) Low frequency of α-synuclein mutations in familial Parkinson's disease. *Ann. Neurol.* **43,** 394–397.
46. Zareparsi, S., Kay, J., Camicioli, R., Kramer, P., Nutt, J., Bird, T., et al. (1998) Analysis of the α-synuclein G209A mutation in familial Parkinson's disease. *Lancet* **351,** 37–38.
47. Hu, C.-J., Sung, S.-M., Liu, H.-C., and Chang, J.-G. (1999) No mutation of G209A in the alpha-synuclein gene in sporadic Parkinson's disease among Taiwan Chinese. *Eur. Neurol.* **41,** 85–87.
48. Vaughan, J., Durr, A., Tassin, J., Bereznai, B., Gasser, T., Bonifati, V., et al. (1998) The α-synuclein Ala53Thr mutation is not a common cause of familial Parkinson's disease: a study of 230 European cases. *Ann. Neurol.* **44,** 270–273.
49. Vaughan, J. R., Farrer, M. J., Wszolek, Z. K., Gasser, T., Durr, A., Agid, Y., et al. (1999) Sequencing of the α-synuclein gene in a large series of cases of

familial Parkinson's disease fails to reveal any further mutations. *Hum. Mol. Genet.* **7,** 751–753.
50. El-Agnaf, O. M. A., Curran, M. D., Wallace, A., Middleton, D., Murgatroyd, C., Curtis A., et al. (1998) Mutation screening in exons 3 and 4 of α-synuclein in sporadic Parkinson's and sporadic and familial dementia with Lewy bodies cases. *NeuroReport* **9,** 3925–3927.
51. Lincoln, S., Gwinn-Hardy, K., Goudreau, J., Chartier-Harlin, M. C., Baker, M., Mouroux, V., et al. (1999) No pathogenic mutations in the persyn gene in Parkinson's disease. *Neurosci. Lett.* **259,** 65–66.
52. Forno, L. S. (1996) Neuropathology of Parkinson's disease. *J. Neuropathol. Exp. Neurol.* **55,** 259–272.
53. Pollanen, M. S., Dickson, D. W., and Bergeron, C. (1993) Pathology and biology of the Lewy body. *J. Neuropathol. Exp. Neurol.* **52,** 183–191.
54. Spillantini, M. G., Crowthier, R. A., Jakes, R., Hasegawa, M., and Goedert, M. (1998) α-Synuclein in filamentous inclusions of Lewy bodies from Parkinson's disease and dementia with Lewy bodies. *Proc. Natl. Acad. Sci.USA* **95,** 6469–6473.
55. Wakabayashi, K., Matsumoto, K., Takayama, K., Yoshimoto, M., and Takahashi, H. (1997) NACP, a presynaptic protein, immunoreactivity in Lewy bodies in Parkinson's disease. *Neurosci. Lett.* **239,** 45–48.
56. Spillantini, M. G., Schmidt, M. L., Lee, V. M. Y., Trojanowski, J. Q., Jakes, R., and Goedert, M. (1997) α-Synuclein in Lewy bodies. *Nature* **388,** 839–840.
57. Baba, M., Nakajo, S., Tu, P., Tomita, T., Nakaya, K., Lee, V. M. Y., et al. (1998) Aggregation of α-synuclein in Lewy bodies of sporadic Parkinson's disease amd dementia with Lewy bodies. *Am. J. Pathol.* **152,** 879–884.
58. Irizarry, M. C., Growdon, W., Gomez-Isla, T., Newell, K., George, J. M., Clayton, D. F., et al. (1998) Nigral and cortical Lewy bodies and dystrophic nigral neurites in Parkinson's disease and cortical Lewy body disease contain α-synuclein immunoreactivity. *J. Neuropathol. Exp. Neurol.* **57,** 334–337.
59. Takeda, A., Mallory, M., Sundsumo, M., Honer, W., Hansen, L., and Masliah, E. (1998) Abnormal accumulation of NACP/α-synuclein in neurodegenerative disorders. *Am. J. Pathol.* **152,** 367–372.
60. Wakabayashi, K., Hayashi, S., Kakita, A., Yamada, M., Toyoshima, Y., Yoshimoto, M., et al. (1998) Accumulation of α-synuclein is a cytopathological feature common to Lewy body disease and multiple system atropy. *Acta Neuropathol.* **96,** 445–452.
61. Arima, K., Uéda, K., Sunohara, N., Hirai, S., Izumiyama, Y., Tonozuka-Uehara, H., et al. (1998) Immunoelectron-microscopic demonstration of NACP/α-synuclein-epitopes on the filamentous component of Lewy bodies in Parkinson's disease and in dementia with Lewy bodies. *Brain Res.* **808,** 93–100.
62. Braak, H., Braak, E., Yilmazer, D., de Vos, R.A.I., Jansen, E. N. H., Bohl, J., et al. (1994) Amygdala pathology in Parkinson's disease. *Acta Neuropathol.* **88,** 493–500.
63. Arai, T., Uéda, K., Akiyama, H., Haga, C., Kondo, H., Kuroki, N., et al. (1999) Argyophilic glial inclusions in the midbrain of patients with Parkinson's disease

and diffuse Lewy body disease are immunopositive for NACP/α-synuclein. *Neurosci. Lett.* **259,** 83–86.

64. Arawaka, S., Saito, H., Murayama, S., and Mori, H. (1998) Lewy body in neurodegeneration with brain iron accumulation type 1 is immunoreactive for α-synuclein. *Neurology* **51,** 887–889.
65. Tu, P., Galvin, J. E., Baba, M., Giasson, B., Tomita, T., Leigth, S., et al. (1998) Glial cytoplasmic inclusions in white matter oligodendrocytes of multple system atrophy brain contain insoluble α-synuclein. *Ann. Neurol.* **44,** 415–422.
66. Wakabayashi, K., Yoshimoto, M., Tsuji, S., and Takahashi, H. (1998) α-Synuclein immunoreactivity in glial cytoplasmic inclusions in multiple system atrophy. *Neurosci. Lett.* **249,** 180–182.
67. Spillantini, M. G., Crowthier, R. A., Jakes, R., Cairns, N. J., Lantos, P. L., and Goedert, M. (1998) Filamentous α-synuclein inclusions link multiple system atrophy with Parkinson's disease and dementia with Lewy bodies. *Neurosci. Lett.* **251,** 205–208.
68. Arima, K., Uéda, K., Sunohara, N., Arakawa, K., Hirai, S., Nakamura, M., et al. (1998) NACP/α-synuclein immunoreactivity in fibrillary components of neuronal and oligodendroglial cytoplasmic inclusions in the pontine nuclei in multiple system atrophy. *Acta Neuropathol.* **96,** 439–444.
69. Giasson, B. I., Uryu, K., Trojanowski, J. Q., and Lee, V. M. Y. (1999) Mutant and wild type human α-synucleins assemble into elongated filaments with distinct morphologies *in vitro*. *J. Biol. Chem.* **274,** 7619–7622.
70. El-Agnaf, O. M. A., Jakes, R., Curran, M. D., and Wallace, A. (1998) Effect of the mutation Ala^{30} to Pro and Ala^{53} to Thr on the physical and morphological properties of α-synuclein protein implicated in Parkinson's disease. *FEBS Lett.* **440,** 67–70.
71. Conway, K. A., Harper, J. D., and Lansbury, P. T. (1998) Accelerated in vitro fibril formation by a mutant α-synuclein linked to early-onset Parkinson disease. *Nat.Med.* **11,** 1318–1320.
72. Hashimoto, M., Hsu, L. J., Sisk, A., Xia, Y., Takeda, A., Sundsmo, M., et al. (1998) Human recombinant NACP/α-synuclein is aggregated and fibrillated in vitro: relevance for Lewy body disease. *Brain Res.* **799,** 301–306.
73. Narhi, L., Wood, S. J., Steavenson, S., Jiang, Y., Wu, G. M., Anafi, D., et al. (1999) Both familial Parkinson's disease mutations accelerate α-synuclein aggregation. *J. Biol. Chem.* **274,** 9843–9846.
74. Crowthier, R. A., Jakes, R., Spillantini, M. G., and Goedert, M. (1998) Synthetic filaments assembled from C-terminally truncated α-synuclein. *FEBS Lett.* **436,** 309–312.
75. Cleveland, D. W., Hwo, S.-Y., and Kishimoto, T. (1977) Purification of Tau, a microtubule-associated protein that induces assembly of microtubules from purified tubulin. *J. Mol. Biol.* **116,** 207–225.
76. Neve, R. L., Harris, P., Kosik, K. S., Kurnit, D. M., and Donlon, T. A. (1986) Identification of cDNA clones for the human microtubule-associated protein tau and chromosomal localization of the genes for tau and microtubule-associated protein 2. *Brain Res.* **387,** 271–280.

77. Andreadis, A., Brown, W. M., and Kosik, K. S. (1992) Structure and novel exons of the human tau gene. *Biochemistry* **31,** 10,626–10,633.
78. Goedert, M., Spillantini, M. G., Jakes, R., Rutherford, D., and Crowther, R. A. (1989) Multiple isoforms of human microtubule-associated protein tau: sequences and localization in neurofibrillary tangles of Alzheimer's disease. *Neuron* **3,** 519–526.
79. Goedert, M., Spillantini, M. G., Potier, M. C., Ulrich, J., and Crowther, R. A. (1989) Cloning and sequencing of the cDNA encoding an isoform of microtubule- associated protein tau containing four tandem repeats: differential expression of tau protein mRNAs in human brain. *EMBO J.* **8,** 393–399.
80. Goedert, M. and Jakes, R. (1990) Expression of separate isoforms of human tau protein: correlation with the tau pattern in brain and effects on tubulin polymerization. *EMBO J.* **9,** 4225–4230.
81. Hong, M., Zhukareva, V., Vogelsberg-Ragaglia, V., Wszolek, Z., Reed, L., Miller, B. I., et al. (1998) Mutation-specific functional impairments in distinct tau isoforms of hereditary FTDP-17. *Science* **282,** 1914–1917.
82. Goedert, M., Spillantini, M. G., and Crowther, R. A. (1992) Cloning of a big tau microtubule-associated protein characteristic of the peripheral nervous system. *Proc. Natl. Acad. Sci. USA* **89,** 1983–1987.
83. Georgieff, I. S., Liem, R. K., Mellado, W., Nunez, J., and Shelanski, M. L. (1991) High molecular weight tau: preferential localization in the peripheral nervous system. *J. Cell Sci.* **100,** 55–60.
84. Taleghany, N. and Oblinger, M. M. (1992) Regional distribution and biochemical characteristics of high molecular weight tau in the nervous system. *J. Neurosci. Res.* **33,** 257–265.
85. Binder, L. I., Frankfurter, A., and Rebhun, L. I. (1985) The distribution of tau in the mammalian central nervous system. *J. Cell Biol.* **101,** 1371–1378.
86. Mighelli, A., Butler, M., Brown, K., and Shelanski, M. L. (1988) Light and electron microscope localization of the microtubule-associated tau protein in rat brain. *J. Neurosci.* **8,** 1846–1851.
87. LoPresti, P., Szuchet, S., Papasozomenos, S. C., Zinkowski, R. P., and Binder, L. I. (1995) Functional implications for the microtubule-associated protein tau: localization in oligodendrocytes. *Proc. Natl. Acad. Sci.USA* **92,** 10,369–10,373.
88. Tashiro, K., Hasegawa, M., Ihara, Y., and Iwatsubo, T. (1997) Somatodendritic localization of phosphorylated tau in neonatal and adult rat cerebral cortex. *NeuroReport* **8,** 2797–2801.
89. Papasozomenos, S. C. and Binder, L. I. (1987) Phosphorylation determines two distinct species of tau in the central nervous system. *Cell Motil. Cytoskeleton* **8,** 210–226.
90. Drechsel, D. N., Hyman, A. A., Cobb, M. H., and Kirschner, M. W. (1992) Modulation of the dynamic instability of tubulin assembly by the microtubule-associated protein tau. *Mol. Biol. Cell* **3,** 1141–1154.
91. Kanai, Y., Takemura, R., Oshima, T., Mori, H., Ihara, Y., Yanagisawa, M., et al. (1989) Expression of multiple tau isoforms and microtubule bundle formation in fibroblasts transfected with a single tau cDNA. *J. Cell Biol.* **109,** 1173–1184.

92. Drubin, D. G. and Kirschner, M. W. (1986) Tau protein function in living cells. *J. Cell Biol.* **103,** 2739–2746.
93. Bramblett, G. T., Goedert, M., Jakes, R., Merrick, S. E., Trojanowski, J. Q., and Lee, V. M. Y. (1993) Abnormal tau phosphorylation at Ser396 in Alzheimer's disease recapitulates development and contributes to reduced microtubule binding. *Neuron* **10,** 1089–1099.
94. Kanai, Y., Chen, J., and Hirokawa, N. (1992) Microtubule bundling by tau proteins in vivo: analysis of functional domains. *EMBO J.* **11,** 3953–3960.
95. Preuss, U., Biernat, J., Mandelkow, E.-M., and Mandelkow, E. (1997) The "jaws" of tau-microtubule interaction examined in CHO cells. *J. Cell Sci.* **110,** 789–800.
96. Brandt, R. and Lee, G. (1993) Functional organization of microtubule-associated protein tau. Identification of regions which affect microtubule growth, nucleation, and bundle formation in vitro. *J. Biol. Chem.* **268,** 3414–3419.
97. Himmler, A., Drechsel, D., Kirschner, M. W., and Martin, D. W. (1989) Tau consists of a set of proteins with repeated C-terminal microtubule-binding domains and variable N-terminal domains. *Mol. Cell. Biol.* **9,** 1381–1388.
98. Butner, K. A. and Kirschner, M. W. (1991) Tau protein binds to microtubules through a flexible array of distributed weak sites. *J. Cell Biol.* **115,** 717–730.
99. Lee, G., Neve, R. L., and Kosik, K. S. (1989) The microtubule binding domain of tau protein. *Neuron* **2,** 1615–1624.
100. Ennulat, D. J., Liem, R. K., Hashim, G. A., and Shelanski, M. L. (1989) Two separate 18-amino acid domains of tau promote the polymerization of tubulin. *J. Biol. Chem.* **264,** 5327–5330.
101. Goode, B. L. and Feinstein, S. C. (1994) Identification of a novel microtubule binding and assembly domain in the developmentally regulated inter-repeat region of tau. *J. Cell Biol.* **124,** 769–782.
102. Goode, B. L., Denis, P. E., Panda, D., Radeke, M. J., Miller, H. P., Wilson, L., et al. (1997) Functional interactions between the proline-rich and repeat regions of tau enhance microtubule binding and assembly. *Mol. Biol. Cell* **8,** 353–365.
103. Chau, M.-F., Radeke, M. J., de Inés, C., Barasoain, I., Kohlstaedt, L. A., and Feinstein, S. C. (1998) The microtubule-associated protein tau cross-links to two distinct sites on each α and β tubulin monomer via separate domains. *Biochemistry* **37,** 17,692–17,703.
104. Biernat, J., Gustke, N., Drewes, G., Mandelkow, E. M., and Mandelkow, E. (1993) Phosphorylation of Ser262 strongly reduces binding of tau to microtubules: distinction between PHF-like immunoreactivity and microtubule binding. *Neuron* **11,** 153–163.
105. Yoshida, H. and Ihara, Y. (1993) Tau in paired helical filaments is functionally distinct from fetal tau: assembly incompetence of paired helical filament-tau. *J. Neurochem.* **61,** 1183–1186.
106. Lindwall, G. and Cole, R. D. (1984) Phosphorylation affects the ability of tau protein to promote microtubule assembly. *J. Biol. Chem.* **259,** 5301–5305.
107. Harada, A., Oguchi, K., Okabe, S., Kuno, J., Terada, S., Ohshima, T.,et al. (1994) Altered microtubule organization in small-caliber axons of mice lacking tau protein. *Nature* **369,** 488–491.

108. Tint, I., Slaughter, T., Fischer, I., and Black, M. M. (1998) Acute inactivation of tau has no effect on dynamics of microtubules in growing axons of cultured sympathetic neurons. *J. Neurosci.* **18,** 8660–8673.
109. Mandell, J. W. and Banker, G. A. (1996) A spatial gradient of tau protein phosphorylation in nascent axons. *J. Neurosci.* **16,** 5727–5740.
110. Kempf, M., Clement, A., Faissner, A., Lee, G., and Brandt, R. (1996) Tau binds to the distal axon early in development of polarity in a microtubule- and microfilament-dependent manner. *J. Neurosci.* **16,** 5583–5592.
111. Ebneth, A., Godemann, R., Stammer, K., Illenberger, S., Trinczek, B., Mandelkow, E.-M., et al. (1998) Overexpression of tau protein inhibits kinesin-dependent trafficking of vesicles, mitochondria, and endoplasmic reticulum: implication for Alzheimer's disease. *J. Cell Biol.* **143,** 777–794.
112. Brandt, R., Léger, J., and Lee, G. (1995) Interaction of tau with the neuronal plasma membrane mediated by tau's amino-terminal projection domain. *J. Cell Biol.* **131,** 1327–1340.
113. Jenkins, S. M. and Johnson, G. V. W. (1999) Tau complexes with phospholipase C-γ in situ. *NeuroReport* **9,** 67–71.
114. Hwang, S. C., Jhon, D.-Y., Bae, Y. S., Kim, J. H., and Rhee, S. G. (1996) Activation of phospholipase C-γ by the concerted action of tau proteins and arachidonic acid. *J. Biol. Chem.* **271,** 18,342–18,349.
115. Trojanowski, J. Q. and Lee, V. M. Y. (1995) Phosphorylation of paired helical filament tau in Alzheimer's disease neurofibrillary lesions: focusing on phosphatases. *FASEB J.* **9,** 1570–1576.
116. Kidd, M. (1963) Paired helical filaments in electron microscopy of Alzheimer's disease. *Nature* **197,** 192–194.
117. Crowther, R. A. and Wischik, C. M. (1985) Image reconstruction of the Alzheimer paired helical filament. *EMBO J.* **4,** 3661–3665.
118. Crowther, R. A. (1991) Straight and paired helical filaments in Alzheimer disease have a common structural unit. *Proc. Natl. Acad. Sci.USA* **88,** 2288–2292.
119. Yagishita, S., Itoh, Y., Nan, W., and Amano, N. (1981) Reappraisal of the fine structure of Alzheimer's neurofibrillary tangles. *Acta Neuropathol. (Berl.)* **54,** 239–246.
120. Goedert, M., Wischik, C. M., Crowther, R. A., Walker, J. E., and Klug, A. (1988) Cloning and sequencing of the cDNA encoding a core protein of the paired helical filament of Alzheimer disease: identification as the microtubule-associated protein tau. *Proc. Natl. Acad. Sci. USA* **85,** 4051–4055.
121. Kondo, J., Honda, T., Mori, H., Hamada, Y., Miura, R., Ogawara, M., et al. (1988) The carboxyl third of tau is tightly bound to paired helical filaments. *Neuron* **1,** 827–834.
122. Wischik, C. M., Novak, M., Edwards, P. C., Klug, A., Tichelaar, W., and Crowther, R. A. (1988) Structural characterization of the core of the paired helical filament of Alzheimer disease. *Proc. Natl. Acad. Sci.USA* **85,** 4884–4888.
123. Kosik, K. S., Orecchio, L. D., Binder, L., Trojanowski, J. Q., Lee, V. M. Y., and Lee, G. (1988) Epitopes that span the tau molecule are shared with paired helical filaments. *Neuron* **1,** 817–825.

124. Lee, V. M. Y., Balin, B. J., Otvos, L. J., and Trojanowski, J. Q. (1991) A68: a major subunit of paired helical filaments and derivatized forms of normal Tau. *Science* **251,** 675–678.
125. Goedert, M., Spillantini, M. G., Cairns, N. J., and Crowther, R. A. (1992) Tau proteins of Alzheimer paired helical filaments: abnormal phosphorylation of all six brain isoforms. *Neuron* **8,** 159–168.
126. Morishima-Kawashima, M., Hasegawa, M., Takio, K., Suzuki, M., Yoshida, H., Titani, K., et al. (1995) Proline-directed and non-proline-directed phosphorylation of PHF-tau. *J. Biol. Chem.* **270,** 823–829.
127. Hanger, D. P., Betts, J. C., Loviny, T. L. F., Blackstock, W. P., and Anderton, B. H. (1998) New phosphorylation sites identified in hyperphosphorylated tau (paired helical filament-tau) from Alzheimer's disease brain using nanoelectrospray mass spectrometry. *J. Neurochem.* **71,** 2465–2476.
128. Matsuo, E. S., Shin, R. W., Billingsley, M. L., Van deVoorde, A., O'Connor, M., Trojanowski, J. Q., et al. (1994) Biopsy-derived adult human brain tau is phosphorylated at many of the same sites as Alzheimer's disease paired helical filament tau. *Neuron* **13,** 989–1002.
129. Billingsley, M. L. and Kincaid, R. L. (1997) Regulated phosphorylation and dephosphorylation of tau protein: effects on microtubule interaction, intracellular trafficking and neurodegeneration. *Biochem. J.* **323,** 577–591.
130. Johnson, G. V. W. and Hartigan, J. A. (1998) Tau protein in normal and Alzheimer's disease brain: an update. *Alzheimer's Dis. Rev.* **3,** 125–141.
131. Crowther, R. A., Olesen, O. F., Smith, M. J., Jakes, R., and Goedert, M. (1994) Assembly of Alzheimer-like filaments from full-length tau protein. *FEBS Lett.* **337,** 135–138.
132. Braak, E., Braak, H., and Mandelkow, E. M. (1994) A sequence of cytoskeleton changes related to the formation of neurofibrillary tangles and neuropil threads. *Acta Neuropathol. (Berl.)* **87,** 554–567.
133. Bancher, C., Grundke-Iqbal, I., Iqbal, K., Fried, V. A., Smith, H. T., and Wisniewski, H. M. (1991) Abnormal phosphorylation of tau precedes ubiquitination in neurofibrillary pathology of Alzheimer disease. *Brain Res.* **539,** 11–18.
134. Gustke, N., Steiner, B., Mandelkow, E. M., Biernat, J., Meyer, H. E., Goedert, M., et al. (1992) The Alzheimer-like phosphorylation of tau protein reduces microtubule binding and involves Ser-Pro and Thr-Pro motifs. *FEBS Lett.* **307,** 199–205.
135. Iqbal, K., Zaidi, T., Bancher, C., and Grundke-Iqbal, I. (1994) Alzheimer paired helical filaments. Restoration of the biological activity by dephosphorylation. *FEBS Lett.* **349,** 104–108.
136. Goedert, M., Jakes, R., Spillantini, M. G., Hasegawa, M., Smith, M. J., and Crowther, R. A. (1996) Assembly of microtubule-associated protein tau into Alzheimer-like filaments induced by sulphated glycosaminoglycans. *Nature* **383,** 550–553.
137. Kampers, T., Friedhoff, P., Biernat, J., Mandelkow, E. M., and Mandelkow, E. (1996) RNA stimulates aggregation of microtubule-associated protein tau into Alzheimer-like paired helical filaments. *FEBS Lett.* **399,** 344–349.

138. Perez, M., Valpuesta, J. M., Medina, M., de Garcini, E. M., and Avila, J. (1996) Polymerization of tau into filaments in the presence of heparin: the minimal sequence required for tau-tau interaction. *J. Neurochem.* **67,** 1183–1190.
139. Arrasate, M., Perez, M., Valpuesta, J. M., and Avila, J. (1997) Role of glycosaminoglycans in determining the helicity of paired helical filaments. *Am. J. Pathol.* **151,** 1115–1122.
140. Hasegawa, M., Crowther, R. A., Jakes, R., and Goedert, M. (1997) Alzheimer-like changes in microtubule-associated protein Tau induced by sulfated glycosaminoglycans. Inhibition of microtubule binding, stimulation of phosphorylation, and filament assembly depend on the degree of sulfation. *J. Biol. Chem.* **272,** 33,118–33,124.
141. DeWitt, D. A., Silver, J., Canning, D. R., and Perry, G. (1993) Chondroitin sulfate proteoglycans are associated with the lesions of Alzheimer's disease. *Exp. Neurol.* **121,** 149–152.
142. Snow, A. D., Mar, H., Nochlin, D., Kresse, H., and Wight, T. N. (1992) Peripheral distribution of dermatan sulfate proteoglycans (decorin) in amyloid-containing plaques and their presence in neurofibrillary tangles of Alzheimer's disease. *J. Histochem. Cytochem.* **40,** 105–113.
143. Morishima-Kawashima, M., Hasegawa, M., Takio, K., Suzuki, M., Titani, K., and Ihara, Y. (1993) Ubiquitin is conjugated with amino-terminally processed tau in paired helical filaments. *Neuron* **10,** 1151–1160.
144. Ledesma, M. D., Bonay, P., and Avila, J. (1995) tau protein from Alzheimer's disease is glycated at its tubulin-binding domain. *J. Neurochem.* **65,** 1658–1664.
145. Ledesma, M. D., Bonay, P., Colaço, C., and Avila, J. (1994) Analysis of microtubule-associated protein tau glycation in paired helical filaments. *J. Biol. Chem.* **269,** 21,614–21,619.
146. Wang, J. Z., Grundke-Iqbal, I., and Iqbal, K. (1996) Glycosylation of microtubule-associated protein tau: an abnormal posttranslational modification in Alzheimer's disease. *Nat. Med.* **2,** 871–875.
147. Iwatsubo, T., Hasegawa, M., Esaki, Y., and Ihara, Y. (1992) Lack of ubiquitin immunoreactivity at both ends of neuropil threads. *Am. J. Pathol.* **140,** 277–282.
148. Foster, N. L., Wilhelmsen, K., Sima, A. A., Jones, M. Z., D'Amato, C. J., and Gilman, S. (1997) Frontotemporal dementia and parkinsonism linked to chromosome 17: a consensus conference. *Ann. Neurol.* **41,** 706–715.
149. Iijima, M., Tabira, T., Poorkaj, P., Schellenberg, G. D., Trojanowski, J. Q., Lee, V. M. Y., et al. (1998) A distinct familial presenile dementia with a novel missense mutation in the tau gene. *NeuroReport* **10,** 497–501.
150. Spillantini, M. G., Goedert, M., Crowther, R. A., Murrell, J. R., Farlow, M. R., and Ghetti, B. (1997) Familial multiple system tauopathy with presenile dementia: a disease with abundant neuronal and glial tau filaments. *Proc. Natl. Acad. Sci.USA* **94,** 4113–4118.
151. Spillantini, M. G., Crowther, R. A., Kamphorst, W., Heutink, P., and van Swieten, J. C. (1998) Tau pathology in two Dutch families with mutations in the microtubule-binding region of tau. *Am. J. Pathol.* **153,** 1359–1363.

152. Sima, A. A., Defendini, R., Keohane, C., D'Amato, C., Foster, N. L., Parchi, P., et al. (1996) The neuropathology of chromosome 17-linked dementia. *Ann. Neurol.* **39,** 734–743.
153. Reed, L. A., Schmidt, M. L., Wszolek, Z. K., Balin, B. J., Soontornniyomkij, V., Lee, V. M. Y., et al. (1998) The neuropathology of a chromosome 17-linked autosomal dominant parkinsonism and dementia ("pallido-ponto-nigral degeneration"). *J. Neuropathol. Exp. Neurol.* **57,** 588–601.
154. Goedert, M., Spillantini, M. G., Crowther, R. A., Chen, S. G., Parchi, P., Tabaton, M., et al. (1999) Tau gene mutation in familial progressive subcortical gliosis. *Nat. Med.* **5,** 454–457.
155. D'Souza I., Poorkaj, P., Hong, M., Nochlin, D., Lee, V. M. Y., Bird, T. D., et al. (1999) Missense and silent tau gene mutations cause frontotemporal dementia with parkinsonism-chromosome 17 type by affecting multiple alternative RNA splicing regulatory elements. *Proc. Natl. Acad. Sci.USA* **96,** 5598–5603.
156. Spillantini, M. G., Crowther, R. A., and Goedert, M. (1996) Comparision of the neurofibrillary pathology in Alzheimer's disease and familial presenile dementia with tangles. *Acta Neuropathol.* **92,** 42–48.
157. Sumi, S. M., Bird, T. D., Nochlin, D., and Raskind, M. A. (1992) Familial presenile dementia with psychosis associated with cortical neurofibrillary tangles and degeneration of the amygdala. *Neurology* **42,** 120–127.
158. Hasegawa, M., Smith, M. J., and Goedert, M. (1998) Tau proteins with FTDP-17 mutations have a reduced ability to promote microtubule assembly. *FEBS Lett.* **437,** 207–210.
159. Rizzu, P., Van Swieten, J.C., Joosse, M., Hasegawa, M., Stevens, M., Tibben, A., et al. (1999) High prevalence of mutations in the microtubule-associated protein tau in a population study of frontotemporal dementia in the Netherlands. *Am. J. Hum. Genet.* **64,** 414–421.
160. Arrasate, M., Pérez, M., Armas-Portela, R., and Ávila, J. (1999) Polymerization of tau peptides into fibrillar structures. The effect of FTDP-17 mutations. *FEBS Lett.* **446,** 199–202.
161. Nacharaju, P., Lewis, J., Easson, C., Yen, S., Hackett, J., Hutton, M., et al. (1999) Accerated filament formation from tau protein with specific FTDP-17 missence mutations. *FEBS Lett.* **447,** 195–199.
162. Clark, L. N., Poorkaj, P., Wszolek, Z. K., Geschwind D. H., Nasreddine Z. S., Miller, B., et al. (1998) Pathogenic implications of mutations in the tau gene in pallido-ponto-nigral degeneration and related chromosome 17-linked neurodegenerative disorders. *Proc. Natl. Acad. Sci.USA* **95,** 13,103–13,107.
163. Hutton, M., Lendon, C. L., Rizzu, P., Baker, M., Froelich, S., Houlden, H., et al. (1998) Association of missense and 5'-splice-site mutations in tau with the inherited dementia FTDP-17. *Nature* **393,** 702–705.
164. Hasegawa, M., Smith, M. J., Iijima, M., Tabira, T., and Goedert, M. (1999) FTDP-17 mutations N279K and S305N in tau produce increased splicing of exon 10. *FEBS Lett.* **443,** 93–96.
165. Senapathy, P., Shapiro, M. B., and Harris, N. L. (1990) Splice junctions, branch point sites, and exons: sequence statistics, identification, and applications to genome project. *Methods Enzymol.* **183,** 252–278.

166. Xu, R., Teng, J., and Cooper, T. A. (1993) The cardiac troponin T alternative exon contains a novel purine-rich positive splicing element. *Mol. Cell. Biol.* **13,** 3660–3674.
167. Watakabe, A., Tanaka, K., and Shimura, Y. (1993) The role of exon sequences in splice site selection. *Genes Dev.* **7,** 407–418.
168. Lavigueur, A., La, B. H., Kornblihtt, A. R., and Chabot, B. (1993) A splicing enhancer in the human fibronectin alternate ED1 exon interacts with SR proteins and stimulates U2 snRNP binding. *Genes Dev.* **7,** 2405–2417.
169. Cooper, T. A. and Mattox, W. (1997) The regulation of splice-site selection, and its role in human disease. *Am. J. Hum. Genet.* **61,** 259–266.
170. Si, Z. H., Rauch, D., and Stoltzfus, C. M. (1998) The exon splicing silencer in human immunodeficiency virus type 1 Tat exon 3 is bipartite and acts early in spliceosome assembly. *Mol. Cell. Biol.* **18,** 5404–5413.
171. Spillantini, M. G., Murrell, J. R., Goedert, M., Farlow, M. R., Klug, A., and Ghetti, B. (1998) Mutation in the tau gene in familial multiple system tauopathy with presenile dementia. *Proc. Natl. Acad. Sci.USA* **95,** 7737–7741.
172. Rewcastle, N. B. and Ball, M. J. (1968) Electron microscopy of the inclusion bodies in Pick's disease. *Neurology* **18,** 1205–1213.
173. Delacourte, A., Sergeant, N., Wattez, A., Gauvreau, D., and Robitaille, Y. (1998) Vulnerable neuronal subsets in Alzheimer's and Pick's disease are distinguished by their tau isoform distribution and phosphorylation. *Ann. Neurol.* **43,** 193–204.
174. Sergeant, N., David, J. P., Lefranc, D., Vermersch, P., Wattez, A., and Delacourte, A. (1997) Different distribution of phosphorylated tau protein isoforms in Alzheimer's and Pick's diseases. *FEBS Lett.* **412,** 578–582.
175. Sergeant, N., Wattez, A., and Delacourte, A. (1999) Neurofibrillary degeneration in progressive supranuclear palsy and corticobasal degeneration: tau pathologies with exclusive "exon 10" isoforms. *J. Neurochem.* **72,** 1243–1249.
176. Feany, M. B., Mattiace, L. A., and Dickson, D. W. (1996) Neuropathologic overlap of progressive supranuclear palsy, Pick's disease and corticobasal degeneration. *J. Neuropathol. Exp. Neurol.* **55,** 53–67.
177. Chin, S. S. M. and Goldman, J. E. (1996) Glial inclusions in CNS degenerative disease. *J. Neuropathol. Exp. Neurol.* **55,** 499–508.
178. Conrad, C., Andreadis, A., Trojanowski, J. Q., Dickson, D. W., Kang, D., Chen, X., et al. (1997) Genetic evidence for the involvement of tau in progressive supranuclear palsy. *Ann. Neurol.* **41,** 277–281.
179. Bennett, P., Bonifati, V., Bonuccelli, U., Colosimo, C., De Mari, M., Fabbrini, G., et al. (1998) Direct genetic evidence for involvement of tau in progressive supranuclear plasy. *Neurology* **51,** 982–985.
180. Higgins, J. J., Litvan, I., Pho, L. T., Li, W., and Nee, L. E. (1998) Progressive supranuclear gaze palsy is in linkage disequilibrium with the τ and not the α-synuclein gene. *Neurology* **50,** 270–273.
181. Oliva, R., Tolosa, E., Ezquerra, M., Molinuevo, J. L., Valldeoriola, F., Burquera, J., et al. (1998) Significant changes in the tau A0 and A3 alleles in progressive supranuclear palsy and improved genotyping by silver detection. *Arch. Neurol.* **55,** 1122–1124.

182. Baker, M., Litvan, I., Houlden, H., Adamson, J., Dickson, D., Perez-Tur, J., et al. (1999) Association of an extended haplotype in the tau gene with progressive supranuclear palsy. *Hum. Mol. Genet.* **8,** 711–715.
183. Dumanchin, C., Camuzat, A., Campion, D., Verpillat, P., Hannequin, D., Dubois, B., et al. (1998) Segregation of a missense mutation in the microtubule-associated protein tau gene with familial frontotemporal dementia and parkinsonism. *Hum. Mol. Genet.* **7,** 1825–1829.
184. Poorkaj, P., Bird, T. D., Wijsman, E., Nemens, E., Garruto, R. M., Anderson, L., et al. (1998) Tau is a candidate gene for chromosome 17 frontotemporal dementia. *Ann. Neurol.* **43,** 815–825.
185. Morris, H. R., Perez-Tur, J., Janssen, J. C., Brown, J., Lees, A. J., Wood, N. W., et al. (1999) Mutation in the tau exon 10 splice site region in familial frontotemporal dementia. *Ann. Neurol.* **45,** 270–271.

8

PET Investigations of Parkinson's Disease

Tamara Hershey, Stephen M. Moerlein,
and Joel S. Perlmutter

8.1. INTRODUCTION

Degeneration of nigrostriatal neurons with subsequent striatal dopamine deficiency leads to Parkinson's disease (PD) with tremor, slowness, reduced spontaneous movement, stiffness, and postural instability (Hoehn, Yahr, 1967). The progression of PD is gradual but relentless; fortunately, symptomatic treatment provides substantial relief, especially early in the course of the disease. As yet, the pathogenesis of the substantia nigral degeneration remains unknown although a specific defect in the *α-synuclein* gene on chromosome 4 has been found in a few rare extended families with familial parkinsonism (Polymeropoulos et al., 1997). However, gene-sequencing analysis has not identified a defect in this gene in many other families with PD (Chan et al., 1998; Parsian et al., 1998), and the link between a genetic defect and the pathophysiology of the disease remains an area of intensive research. Furthermore, the functional consequences of the nigrostriatal degeneration on the development of the clinical manifestations of the disease and the response to pharmacotherapy are other areas of active investigation.

Positron emission tomography (PET) provides an in vivo measure of brain function and has been a useful tool for investigation of the pathophysiology and therapy of PD. This technique, depending on the type of radiopharmaceutical, experimental design, method of data collection, and data analysis can be used to measure different aspects of cerebral function. This review will focus on studies relating to PD that illustrate the utility of different types of PET-based measurements, including regional pathophysiology, pharmacologic activation, radioligand binding, and assessment of neurotransmitter function.

8.2. REGIONAL PATHOPHYSIOLOGY

Positron emission tomography can provide measurements of regional cerebral blood flow or metabolism either at rest or during activation. In general,

From: *Contemporary Clinical Neuroscience: Molecular Mechanisms of Neurodegenerative Diseases*
Edited by: M.-F. Chesselet © Humana Press Inc., Totowa, NJ

blood flow and metabolism reflect neuronal activity (Raichle, 1987; Perlmutter, Moerlein, 1999), but it is important to note that higher flow or metabolism may accompany either higher levels of excitation or inhibition because both may cost energy to maintain or change membrane gradients. Because the surface-to-volume ratio in terminal fields is high, regional flow or metabolism may preferentially reflect either activity of the neuronal input to a region or the activity of local interneurons (Jueptner, Weiller, 1995). Under pathological conditions, changes in flow may not coincide with changes in local metabolism or neuronal activity (Perlmutter, Raichle, 1984).

A critical feature of resting flow or metabolism experiments is the state of the subject during the scan. Subjects are commonly asked to lie still with their eyes closed. Although this condition is referred to as "resting" or "baseline," it is clear that subjects are engaging in some activity or experiencing some state during the scans, such as anxiety, boredom, discomfort, trying to stay awake, or keeping their eyes closed and body still. Recent studies have shown that the "baseline" state of the brain has a particular pattern of high activity that may reflect specific active internal cognitive or affective states (Shulman et al., 1997). This high level of activity, particularly in the posterior cingulate and parietal cortex, decreases when a defined task (e.g., reading words) needs to be performed and these internal states are temporarily suspended.

Given these caveats, regional PET measurements have been used to identify sites of abnormal function that occur in diseases such as PD. For example, abnormalities of local flow or metabolism found in basal ganglia in patients with PD reflect the known striatal dopamine deficiency, but PET studies have found other regional alterations such as changes in specific frontal regions subsequently confirmed to be part of the underlying pathophysiology (Martin et al., 1986; Perlmutter, Raichle, 1985). Application of sophisticated statistical techniques to determine the pattern of regional glucose metabolism permits differentiation of idiopathic PD patients from normals and those with a variant of parkinsonism called striatonigral degeneration (Eidelberg et al., 1994). However, this method does not provide information on the actual activity levels of different regions.

8.3. PHARMACOLOGIC ACTIVATION

Comparisons of PET scans collected during different physiological or behavioral states (e.g., saline vs drug infusion, or staring at fixation vs reading words) permit identification of normal and abnormal brain function not found in the resting state alone. Such activation studies can be a powerful way to determine the functional consequences of a neurological disorder, such as PD.

There are several important methodological issues critical for the interpretation of activation studies. First, qualitative measures of flow, using PET counts of regional radioactivity distribution, do not require arterial blood sampling and are much simpler studies than fully quantified regional cerebral blood flow (rCBF) measurements. Fortunately, PET counts are linearly related to rCBF, at least in the range of flows relevant for most brain studies, permitting good estimation of regional responses with normalized counts (Fox et al., 1984) assuming that there is no global blood flow shift produced by the activation. When making comparisons between conditions with PET counts, it is necessary to first determine that the condition (e.g., drug activation) does not produce changes in global blood flow. If it does, then regional changes in qualitative normalized flow may misrepresent the absolute change in local flow or neuronal activity. In other words, what seems to go up may have gone down!

There also is a substantial statistical challenge in analyzing activation studies. These studies typically involve large numbers of regional comparisons potentially leading to false-positive responses or Type 1 errors. Several sophisticated techniques have been developed that allow us to compare conditions or groups. These techniques often differ in the degree of conservatism with which they approach the problem of multiple comparisons. Some, such as the hypothesis generation and hypothesis-testing approach, are designed to minimize Type 1 errors and to ensure that each finding is reliable (Burton et al., 1997; Drevets et al., 1992). However, this strategy may have limited sensitivity for detection of low-level responses. Alternatives include the widely used Statistical Parametric Mapping analysis software (SPM96 [Friston et al., 1995]). This method examines the entire data set for voxels and clusters of voxels that have significant group or condition effects or interactions, using a multiple comparison correction, and recent versions of this software also appropriately correct for differences in regional variance. For small group sizes, the Statistical non-Parametric Mapping software (SnPM [Holmes, 1994; Holmes et al., 1996]) can be used, which is less sensitive than SPM to the problems of small sample sizes (Poline et al., 1996). Thus, there are different methods of data analysis, and the results and conclusions of a given blood flow study will in part depend on the statistical procedures used to analyze it.

Despite these challenges, activation paradigms have been fruitfully applied to the study of PD. Some studies have attempted to use behavioral activation of the sensorimotor system with motor control tasks (e.g., moving a joystick [Jenkins et al., 1992]) or sequential finger movements (Catalan et al., 1999) or certain cognitive systems (e.g., working memory; [Owen et al., 1998]). Although studies such as these have found specific functional abnormalities in patients with PD, one difficulty in interpreting such differences is the fact that performance on the motor or cognitive task can differ between PD and controls. Therefore, it is difficult to determine if the brain activation patterns are different because of the disease state or to the performance state. In other words, if PD patients are slower than controls on a motor activation task and their pattern of brain activity is different from controls during the task, it may reflect the fact that the brain operates differently when movement is slowed, rather than a fundamental difference caused by the disease. Alternative approaches include using tasks that PD patients perform normally or making tasks equivalently difficult for the two groups (e.g., speed up or degrade the stimuli for controls), neither of which are ideal.

Other studies have examined pharmacological activation as a way to detect neurophysiological abnormalities in PD without the confound of differing levels of performance. The promise of such studies includes potentially providing an in vivo assessment of the regional effects of drugs, thereby facilitating evaluation of new pharmacotherapies, initial selection of proper drug dose, and identification of potential unwanted effects.

Development of a new PET assay of drug function requires several steps. First, it is necessary to determine whether there is a drug-induced global shift. As stated earlier, such a shift could cloud interpretation of regional changes, as an apparent increase in globally normalized regional activity could indicate either an absolute increase in regional activity or an absolute regional decrease if there were a larger decrease in the remainder of the brain. Quantified measurements will be necessary until it is clear that a drug does not produce a global shift, which at this time complicates the use of functional magnetic resonance (MR) imaging because it does not yet provide absolute measurements of flow or metabolism. The next step in development of the PET assay of drug function is an operational validation of the method. Determination of specificity, dose-response sensitivity, and reproducibility are standards applied to any new assay and also are relevant to these PET measurements (Perlmutter et al., 1987; Black, Perlmutter, 1997).

The use of PET to assess the neurophysiological responses to dopaminergic challenges in vivo has been shown to meet these basic criteria (Black et al., 1997; Hershey et al., 1998; Black, Perlmutter, 1996). First, the effect of dopaminergic challenges on brain metabolism and blood flow have been performed in normal animals and in rat and monkey models of parkinsonism (Trugman et al., 1989; McCulloch et al., 1982; Mitchell et al., 1992). In particular, ex vivo autoradiography has produced valuable information in rat models of parkinsonism about the functional status of dopamine D_{1-} and D_2-influenced basal ganglia pathways (Trugman, Wooten, 1987). PET using $H_2{}^{15}O$ to measure regional cerebral blood flow permits an analogous in vivo assessment of the functional response of dopaminergic neuronal circuits to levodopa or specific dopamine agonists (Black et al., 1997; Black et al., 2000). Second, blood flow responses are pharmacologically specific and dose dependent. We have found that the selective D_2 dopamine agonist U91356a causes pallidal flow to decrease in sedated baboons in a dose-related fashion, and a D_2 antagonist blocked this decrease, whereas a D_1 antagonist enhanced the U91356a reduction in pallidal flow (Black et al., 1997). Antagonists of serotonin S_2 or peripheral D_2 receptors did not prevent this decrease. Additionally, the responses to a D_1 agonist are distinct from those produced by a D_2 agonist (Black et al., 1997; Black et al., 2000). Thus, PET measurements of dopaminergic challenges appear to be a well-validated and powerful method for providing important clues to the function of specific dopamine receptor-mediated pathways.

One example of the application of this method to understanding the potential mechanism underlying clinical phenomenon in PD is a recent study of the effects of levodopa activation on blood flow in PD patients with dopa-induced dyskinesias (DID [Hershey et al., 1998]). It is well known that treatment with levodopa, a precursor of dopamine, initially ameliorates the clinical symptoms of PD, including stiffness, slowness and resting tremor. However, chronic levodopa treatment can produce severe involuntary movements (called dopa-induced dyskinesias), limiting treatment. Pallidotomy, placement of a surgical lesion in the internal segment of the globus pallidus (GPi), reduces DID. Because this result is inconsistent with current theories of both basal ganglia function and DID, it appeared logical to investigate the brain's response to levodopa in these patients. A dose of levodopa that produced clinical benefit without inducing DID was used, permitting examination of the brain response to levodopa without the confounding effect of differences in motor behavior. The results of the PET study demonstrated that patients with DID had a significantly greater response to levodopa in ventrolateral thalamus than PD patients chronically treated with levodopa but who had not developed DID. Further, this abnormal response in the thalamus was associated with decreased activity in primary motor cortex. These findings provide a testable physiological explanation for the clinical efficacy of pallidotomy and demonstrate a fundamental change in how the brain responds to levodopa after chronic treatment with levodopa and the development of DID.

Finally, although agonist-mediated pharmacological activation studies may be more sensitive to detection of changes in a pathway than direct measurement of specific receptors, it is possible that responses are significantly influenced by comodulators, tone of other transmitter systems, and so forth. Pharmacologic activation may be a sensitive test, but it may also have limited specificity. Subsequent investigation of receptors or transmitter function may be necessary to identify the specific causes of the altered response to a drug.

8.4. RADIOLIGAND BINDING

Noninvasive PET measurement of radioligand binding is also a powerful tool for in vivo study of neuropharmacology. PET permits the in vivo visualization of regional uptake of an appropriate positron-emitting radioligand, and tracer kinetic models allow quantification of receptor binding.

A variety of radioligands have been developed to study different brain receptors by PET (Frost, Wagner, 1990). These radioligands are generally labeled with carbon-11 ($t_{1/2}$ = 20 min) or fluorine-18 ($t_{1/2}$ = 110 min). The shorter half-life of ^{11}C limits its utility to radioligands that require relatively short imaging times after injection into the subject, but has the potential advantage of allowing repeat studies within the same imaging session, as well as a lower absorbed radiation dose. Fluorine-18 is useful as a label for radioligands that require longer imaging sequences and has the potential advantage of greater laboratory convenience because of its longer half-life. Several ^{11}C- and ^{18}F-labeled radioligands have been developed as PET radiopharmaceuticals for investigation of brain receptor-binding activity. These include [^{11}C]raclopride and various ^{18}F- and ^{11}C-labeled butyrophenones (dopamine D2 receptor), [^{18}F]altanserin and [^{18}F]setoperone (serotonin S2 receptor), [^{11}C]flumazenil (benzodiazepine receptor), and [^{11}C]carfentanil and [^{11}C]diprenorphine (opiate receptor).

The dopamine D2 receptor system was the first to be examined by PET (Wagner et al., 1983). The ability of PET to quantify D2 receptor binding and occupancy in vivo offers a unique and powerful research tool for investigation of the role of dopamine receptors in the pathophysiology and therapy of PD.

Many radioligands have been successfully developed for PET measurements of dopamine receptor binding (Perlmutter, Moerlein, 1999). Once these methods have been validated in animals, they can then be applied to humans for examination of brain pharmacology in vivo.

Several studies have demonstrated the utility of in vivo PET measurements of radioligand binding. Application to PD revealed that D2 radioligand binding was either elevated, unchanged, or reduced, again raising the issue of methodological differences across studies until it was clarified that there is a time-dependent change in D2 receptors in nonhuman primates after induction of a nigrostriatal lesion (Perlmutter et al., 1997; Todd et al., 1996). This latter finding provided new insights into the relationship between striatal dopamine deficiency and the clinical manifestations of either parkinsonism or another condition with involuntary muscle spasms known as dystonia. In fact, a subsequent PET study in patients with primary dystonia revealed a reduction in radioligand binding in the putamen (Perlmutter et al., 1997) matching the previous animal study. Subsequent genetic studies have now confirmed that childhood-onset primary dystonia may be associated with a specific genetic defect in the *DYT1* gene that codes for a protein torsin A, which is expressed in the pars compacta of the substantia nigra, a key source of dopaminergic neurons projecting to the putamen (Augood et al., 1998).

Chronic drug treatment may change receptors and this can be measured with PET, given a judicious choice of radioligand. [^{11}C]Raclopride is well suited for such PET pharmacological studies, because the tracer kinetics have been carefully evaluated for both reproducibility and age-related effects. However, one must consider the effects of endogenous dopamine on [^{11}C]raclopride uptake because endogenous dopamine competes for binding sites with [^{11}C]raclopride. For example, there is a significant age-dependent decrease in [^{11}C]raclopride binding in the caudate nucleus and putamen (Antonini et al., 1993; Rinne et al., 1993). After age 30, the binding of the radioligand in the putamen decreases at approx 0.6% per year, which parallels the age-related decline in the presynaptic nigrostriatal dopaminergic neuronal system (Antonini et al., 1993). The age-related decrease in D2 binding by raclopride is probably the result of a decrease in receptor density (B_{max}) rather than to a change in dissociation constant (Ki) (Rinne et al., 1993). However, a age-related change in endogenous dopamine could produce the same results. There is a high test–retest reliability of PET measurements with [^{11}C]raclopride. For both short-term (24 h or less) or long-term (11 mo to 5 yr) reproducibility studies, the variability was only about 10% (Nordstroem et al., 1992; Volkow et al., 1993; Schlosser et al., 1998). This provides excellent reproducibility for identification of more robust changes induced by acute or chronic pharmacological treatment

8.5. NEUROTRANSMITTERS

Labeling transmitters or analogs with positron-emitting isotopes has the potential to provide additional insights into pathophysiology of PD. Garnett et al. (Garnett et al., 1983) first reported the preferential accumulation of radioactivity in normal human striata after administration of the dopa analog, [^{18}F]fluorodopa (FD), which crosses the blood-brain barrier (BBB) and is decarboxylated to [^{18}F]fluorodopamine, a charged molecule which is trapped extravascularly, in a manner similar to levodopa. Normally, presynaptic terminals of nigrostriatal dopaminergic neurons contain most of the striatal decarboxylase activity. Thus, FD PET presumably reflects dopaminergic innervation. However, as presynaptic neurons degenerate, a greater portion of residual decarboxylase activity resides in other compartments (Martin, Perlmutter, 1994). Nevertheless, fluorodopamine accumulation likely reflects decarboxylase activity, which may indirectly reflect, in part, residual nigrostriatal neurons (Pate et al., 1993; Snow et al., 1993).

The potential clinical utility of FD was promising as PET images clearly revealed decreased striatal accumulation in people with PD (Calne et al., 1985). There have been basically three competing methods for analysis of these types of studies. The simplest approach is the ratio approach (Martin et al., 1986). Even this can be tricky, as ratio values are time dependent (Hoshi et al., 1993). Thus, the time after FD administration must be constant to permit appropriate comparisons across subjects. The graphical approach is based on comparing time-dependent radioactivity changes in the striatum (i.e., the ratio of striatum to a reference tissue on the vertical axis) with a time-dependent measure of the input of FD to the striatum. Usually, the ratio of the striatal counts to a reference tissue is plotted on the vertical axis and the sum of all the radioactivity counts in a reference region divided by the actual counts in that region at a given time is plotted on the horizontal axis (Patlak, Blasberg, 1985; Martin et al., 1989). The slope of the line after steady state is reached represents an uptake constant, commonly called *Ki*. There are multiple variations, including the use of different reference tissues to derive the input of tracer. The original method used measurements from blood (as the reference region) and required a correction for the accumulation of radiolabeled metabolites (Garnett et al., 1983; Cummings et al., 1987; Firnau et al., 1987; Melega et al., 1991; Chan et al., 1992). Alternatively, one can use a brain region such as the occipital lobe or cerebellum assuming that this part of the brain acts as a filter and only accumulates those radiolabeled moieties that cross into the striatum. In particular, it is assumed that the reference region does not contain enzymes such as decarboxylase that would trap radioactivity within the tissue by converting FD to the charged molecule [^{18}F]fluorodopamine. The occipital lobe, as opposed to the cerebellum, has the advantage that it can be identified on the same axial PET slices as the striatum.

Compartmental modeling is another approach to the analysis of FD PET data. There have been several recent studies published using such models (Melega et al., 1991; Huang et al., 1991; Kuwabara et al., 1993). These models are complex and attempt to account for various radiolabeled metabolites of FD that not only accumulate in plasma but also enter the brain (Doudet et al., 1992; Doudet et al., 1991). Multiple assumptions are made to simplify the calculations such as fixing the relationship between the brain transport rate of FD and the O-methylated derivative. Despite these constraints, these techniques have been used to estimate decarboxylase activity in normal humans.

Other factors affect FD uptake and impact PET findings with any of the methods of data analysis. For example, administration of decarboxylase or cathechol-*O*-methyltransferase inhibitors prior to FD (Rinne et al., 1993; Laihinen et al., 1992; Guttman et al., 1993) increases the amount of FD entering the brain as well as affect radiolabeled metabolites in the blood or brain. Comparisons across studies must consider these potential differences.

Multiple investigators found that PD patients had marked diminution of striatal accumulation compared to normals (Calne et al., 1985; Hoshi et al., 1993; Leenders et al., 1990; Eidelberg et al., 1990). These studies are an important first step to determine whether FD PET will have a clinical role, but just distinguishing normal from abnormal has little clinical utility in symptomatic subjects.

Presymptomatic diagnosis of PD would be an important clinical tool once protective therapy has been found and can be an important test for research. Asymptomatic patients exposed to MPTP had intermediate values of striatal FD uptake compared to PD and normals (Calne et al., 1985). Monkeys treated with low doses of the selective dopaminergic neurotoxin MPTP that did not produce parkinsonian signs had decreased striatal uptake (Guttman et al., 1988). Other studies have reported that some unaffected monozygotic twins and non-twin relatives of PD patients had abnormally low striatal FD uptake (Piccini et al., 1997; Burn et al., 1992; Holthoff et al., 1994). Such techniques provide important adjuncts for diagnostic ascertainment in genetic studies of large pedigrees with familial PD.

The specificity of low striatal FD uptake for PD is unclear. Although low striatal FD uptake may distinguish PD from normals (Morrish et al., 1998) or those with dopa-responsive dystonia (Snow et al., 1993), it does not separate patients with multiple system atrophy (Antonini et al., 1997) or Machado-Joseph disease (Shinotoh et al., 1997). Additionally, patients with drug-induced parkinsonism who also have low FD uptake may be more susceptible to long-lasting or progressive parkinsonism on follow-up compared to those with normal FD uptake (Burn, Brooks, 1993). These studies suggest that PET may be helpful for diagnosis of presymptomatic subjects but the sensitivity, specificity, and positive predictive value have yet to be determined. Therefore, at this time, FD PET does not have clinical utility for presymptomatic diagnosis and should be limited to research studies for this application (Sawle, 1993).

[^{18}F] fluorodopa has been used to investigate the clinical response to levodopa. Leenders et al. (Leenders et al., 1986) compared mildly affected PD patients and severely affected with rapid clinical fluctuations in response to individual doses of levodopa. Severely affected patients had the greatest decrease in uptake ratios compared to normals, and mild patients had intermediate values. The authors suggested that ratios directly reflect striatal storage capacity of dopamine, and the marked decrease in severely affected patients causes rapid, symptomatic fluctuations to oral levodopa. Alterations in tracer delivery, differential effects from variable length of time patients stopped levodopa prior to study, and changes in metabolism of FD are potential alternate explanations for their findings.

Some patients have less benefit from oral levodopa if taken at mealtimes. Nutt et al. (Nutt et al., 1984) found that large neutral amino acids caused an increase in symptoms in patients treated with continuous intravenous (iv) infusions of levodopa. To investigate the mechanism of this observation, Leenders et al. (Leenders et al., 1986) found that iv amino acid loading caused a marked decrease in striatal uptake of [^{18}F] activity in a normal volunteer. These findings represent the first direct evidence in humans for competition between levodopa and other amino acids for striatal uptake.

There have been several recent studies using FD PET to monitor the efficacy of fetal dopamine cell implantation into striatum of patients with PD. Two patients that had unilateral implantation into caudate and putamen had little clinical improvement after 5–6 mo and no increased uptake of FD (Lindvall et al., 1989; Ingvar et al., 1983). Another patient had clinical improvement beginning 5 wk after surgery and his clinical condition stabilized about 3 mo later (Lindvall et al., 1990). FD PET in that subject demonstrated an increased uptake of FD 5 mo after transplant. Two of seven patients that had bilateral fetal tissue transplantation had FD PET before and as long as 33 mo after surgery (Freed et al., 1992). In one of these patients, uptake of FD in posterior putamen increased. The other patient had no increase in FD uptake at 9 mo posttransplantation. Bilateral fetal tissue implants into caudate and putamen in two MPTP-induced parkinsonian patients produced no change 5–6 mo postoperatively but a marked increase by 12–13 mo, paralleling clinical improvement (Widner et al., 1992). Two other PD patients had fetal mesencephalic tissue transplanted unilaterally into the putamen and had gradual clinical improvement beginning 6 and 12 wk later and reaching a maximum at 4–5 mo postoperatively (Lindvall et al., 1992). FD PET demonstrated increased uptake at the operative site that continued to increase after there was no further clinical improvement (Sawle et al., 1992).

All of these reports suggest that increased accumulation of FD reflects functioning grafted tissue with either increased decarboxylase activity or increased neuronal storage capacity. Support for this interpretation includes the following: (1) there is no change in the blood-brain barrier (BBB) to gadolinium MR imaging, (2) the delayed time-course of the increased FD uptake and the clinical improvement coincide, and (3) animal experiments demonstrate intact BBB. There are concerns about the preservation of an intact BBB. First, gadolinium is a large molecule and may not indicate relevant changes in BBB to smaller molecules like FD. Second, Guttman et al. (Guttman et al., 1989) reported increased accumulation of [^{18}Ga]EDTA at the surgical site that corresponded to increased accumulation of FD in patients 6 wk after adrenal medullary implants, suggesting that there is impairment of the BBB. Third, increased FD accumulation was found at the site of surgical cavitation in an MPTP-treated monkey that did not have tissue transplantation (Miletich et al., 1988; Miletich et al., 1994). Fourth, studies of transplantation in rats do not necessarily support the claim of an intact BBB (Dusart et al., 1989). There is progressive angiogenesis in the thalamus after transplantation of dissociated fetal cells (Dusart et al., 1989) : "The blood vessels are progressively more numerous in the graft and they demonstrate mature ultrastructural features 2 mo after grafting." Based on these data, it would be difficult to conclude that increased accumulation of FD in the brain exclusively reflects functioning graft tissue. Appropriate experiments to support this interpretation have yet to be done (Freed, 1990). There has been one postmortem study of a single PD patient that had a fetal transplant with postoperative increased FD uptake and viable, functioning transplanted neurons (Kordower et al., 1997). So, despite the above criticisms, there may be an important role for FD PET to play in evaluation of these procedures.

Finally, there are numerous other radiopharmaceuticals for assessment of dopaminergic pathways including those that mark presynaptic neurons. These others tracers can be divided into those that bind to presynaptic dopamine uptake sites or vesicular transport sites, or identify the activity of other enzymes in the synthesis of dopamine, like tyrosine hydroxylase. Each of these may provide additional new insights about the functions of the dopamine pathways in the brain.

ACKNOWLEDGMENTS

This work was supported by NIH grants NS31001 and NS10787-01, the Charles A. Dana Foundation (The Dana Clinical Hypotheses Research Program); the Greater St. Louis Chapter of the American Parkinson

Disease Association, the Sam & Barbara Murphy Fund, the Alzheimer's Disease and Related Disorders Program and the McDonnell Center for Higher Brain Function.

REFERENCES

Antonini, A., Leenders, K. L., Vontobel, P., Maguire, R. P., Missimer, J., Psylla, M., et al. (1997) Complementary PET studies of striatal neuronal function in the differential diagnosis between multiple system atrophy and Parkinson's disease. *Brain* **120,** 2187–2195.

Augood, S. J., Penney, J. B., Friberg, I. K., Breakefield, X. O., Young, A. B., and Ozelius, L. J (1998) Expression of the early-onset torsion dystonia gene (DYT1) in human brain. *Ann. Neurol.* **43,** 669–673.

Black, K. J., Gado, M. H., and Perlmutter, J. S. (1997) PET measurement of dopamine D2 receptor-mediated changes in striatopallidal function. *J. Neurosci.* **17,** 3168–3177.

Black, K. J. and Perlmutter, J. S. (1997) Can ^{18}F-fluorodopa positron emission tomography help determine disease status in familial Parkinson's disease? *Neurol. Network Comment.* **1,** 308–314.

Black, K. J. and Perlmutter, J. S. (1996) An in vivo test of striatopallidal sensitivity to a specific D2 agonist. *Mov. Disord.* **11,** 108.

Black, K. J., Hershey, T., Gado, M., and Perlmutter, J. S. (2000) A dopamine D1 antagonist activates temporal lobe structures in primates. *J. Neurophysiol.* (in press).

Burn, D. J. and Brooks, D. J. (1993) Nigral dysfunction in drug-induced parkinsonism: an 18F-dopa PET study. *Neurology* **43**, 552–556.

Burn, D. J., Mark, M. H., Playford, E. D., Maraganore, D. M., Zimmerman, T. R., Jr., Duvoisin, R. C., et al. (1992) Parkinson's disease in twins studied with 18F-dopa and positron emission tomography. *Neurology* **42,** 1894–1900.

Burton, H., MacLeod, A. M., Videen, T. O., and Raichle, M. E. (1997) Multiple foci in parietal and frontal cortex activated by rubbing embossed grating patterns across fingerpads: a positron emission tomography study in humans. *Cerebr. Cortex* **7,** 3–17.

Calne, D. B., Langston, J. W., Martin, W. R. W., Stoessl, A. J., Ruth, T. J., Adam, M. J., et al. (1985) Positron emission tomography after MPTP: observations relating to the cause of Parkinson's disease. *Nature* **317**, 246–248.

Catalan, M. J., Ishii, K., Honda, M., Samii, A., and Hallett, M. (1999) A PET study of sequential finger movements of varying length in patients with Parkinson's disease. *Brain* **122,** 483–495.

Chan, G., Morrison, S., Holden, J. E., and Ruth, T. J. (1992) Plasma L-[18F]fluorodopa input function: a simplified method. *J. Cereb. Blood Flow Metab.* **12,** 881–884.

Chan, P., Jiang, X., Forno, L. S., Di Monte, D. A., Tanner, C. M., and Langston, J. W. (1998) Absence of mutations in the coding region of the alpha-synuclein gene in pathologically proven Parkinson's disease. *Neurology* **50,** 1136–1137.

Doudet, D. J., Aigner, T. G., McLellan, C. A., and Cohen, R. M. (1992) PET with 18F-DOPA: Interpretation and biological correlates in nonhuman primates. *Psychiatry Res. Neuroimaging* **45,** 153–168.

Doudet, D. J., McLellan, C. A., Carson, R., Adams, H. R., Miyake, H., and Aigner, T. G. (1991) Distribution and kinetics of 3-*O*-methyl-6-[18F]fluoro-L-DOPA in the rhesus monkey brain. *J. Cereb. Blood Flow Metab.* **11,** 726–734.

Drevets, W. C., Videen, T. O., Price, J. L., Preskorn, S. H., Carmichael, S. T., and Raichle, M. E. (1992) A functional anatomical study of unipolar depression. *J. Neurosci.* **12,** 3628–3641.

Dusart, I., Nothias, F., Roudier, F., Besson, J. M., and Peschanski, M. (1989) Vascularization of fetal cell suspension grafts in the excitotoxically lesioned adult rat thalamus. *Brain Res. Dev. Brain Res.* **48,** 215–228.

Eidelberg, D., Moeller, J. R., Dhawan, V., Sidtis, J. J., Ginos, J. Z., Strother, S. C., et al. (1990) The metabolic anatomy of Parkinson's disease: complementary [^{18}F]fluorodeoxyglucose and [^{18}F]fluorodopa positron emission tomographic studies. *Mov. Disord.* **5,** 203-213.

Eidelberg, D., Moeller, J. R., Dhawan, V., Spetsieris, P., Takikawa, S., Ishikawa, T., et al. (1994) The metabolic topography of Parkinsonism. *J. Cereb. Blood Flow Metab.* **14,** 783-801.

Firnau, G., Sood, S., Chirakal, R., Nahmias, C., and Garnett, E. S. (1987) Cerebral metabolism of 6-[18F]fluoro-L-3,4-dihydroxyphenylalanine in the primate. *J. Neurochem.* **48,** 1082.

Fox, P. T., Mintun, M. A., Raichle, M. E., and Herscovitch, P. (1984) A noninvasive approach to quantitative functional brain mapping with H2(15)O and positron emission tomography. *J. Cereb. Blood Flow Metab.* **4,** 329–333.

Freed, C. R., Breeze, R. E., Rosenberg, N. L., Schneck, S. A., Kriek, E., and Qi, J. X. (1992) Survival of implanted fetal dopamine cells and neurologic improvement 12 to 46 months after transplantation for Parkinson's disease. *N. Engl. J. Med.* **327,** 1549–1555.

Freed, W. J. (1990) Fetal brain grafts and Parkinson's disease. *Science* **250,** 1434.

Friston, K. J., Holmes, A. P., Worsley, K. J., Poline, J. P., Frith, C. D., and Frackowiak, R. S. J. (1995) Statistical parametric maps in functional imaging: A general linear approach. *Hum. Brain Mapp.* **2,** 189–210.

Frost, J. J. and Wagner, H. N. (1990) *Quantitative Imaging: Neuroreceptors, Neurotransmitters and Enzymes.* Raven, New York.

Garnett, E. S., Firnau, G., and Nahmias, C. (1983) Dopamine visualised in the basal ganglia of living man. *Nature* **305,** 137–138.

Guttman, M., Burns, R. S., Martin, W. R., Peppard, R. F., Adam, M. J., and Ruth, T. J. (1989) PET studies of parkinsonian patients treated with autologous adrenal implants. *Can. J. Neurol. Sci.* **16,** 305–309.

Guttman, M., Leger, G., Reches, A., Evans, A., Kuwabara, H., and Cederbaum, J. (1993) Administration of the new COMT inhibitor OR-611 increases striatal uptake of fluorodopa. *Mov. Disord.* **8,** 298–304.

Guttman, M., Yong, V. W., Kim, S. U., Calne, D. B., Martin, W. R., Adam, M. J., et al. (1988) Asymptomatic striatal dopamine depletion: PET scans in unilateral MPTP monkeys. *Synapse* **2,** 469–473.

Hershey, T., Black, K. J., Stambuk, M. K., Carl, J. L., McGee-Minnich, L. A., and Perlmutter, J. S. (1998) Altered thalamic response to levodopa in Parkinson's patients with dopa-induced dyskinesias. *Proc. Natl. Acad. Sci.USA* **95,** 12,016–12,021.

Hoehn, M. M. and Yahr, M. D. (1967) Parkinsonism: onset, progression and mortality. *Neurology* **17,** 427–442.

Holmes, A. P. (1994) Statistical issues in functional brain mapping, Thesis, University of Glasgow.

Holmes, A. P., Blair, R. C., Watson, J. D. G., and Ford, I. (1996) Non-parametric analysis of statistic images from functional mapping experiments. *J. Cereb. Blood Flow Metab.* **16,** 7–22.

Holthoff, V. A., Vieregge, P., Kessler, J., Pietrzyk, U., Herholz, K., Bönner, J., et al. (1994) Discordant twins with Parkinson's disease: positron emission tomography and early signs of impaired cognitive circuits. *Ann. Neurol.* **36,** 176–182.

Hoshi, H., Kuwabara, H., Léger, G., Cumming, P., Guttman, M., and Gjedde, A. (1993) 6-[18F] fluoro-L-dopa metabolism in living human brain: a comparison of six analytical methods. *J. Cereb. Blood Flow Metab.* **13,** 57–69.

Huang, S. C., Yu, D. C., Barrio, J. R., Grafton, S. T., Melega, W. P., and Hoffman, J. M. (1991) Kinetics and modeling of L-6-[18F]fluoroDOPA in human positron emission tomographic studies. *J. Cereb. Blood Flow Metab.* **11,** 898–913.

Ingvar, M., Lindvall, O., and Stenevi, U. (1983) Apomorphine-induced changes in local cerebral blood flow in normal rats and after lesions of the dopaminergic nigrostriatal bundle. *Brain Res.* **262,** 259–265.

Jenkins, I. H., Fernandex, W., Playford, E. D., Lees, A. J., Frackowiak, R. S., Passingham, R. E., and Brooks, D. J. (1992) Impaired activation of the supplementary motor area in Parkinson's disease is reversed when akinesia is treated with apomorphine. *Ann. Neurol.* **32,** 749–757.

Jueptner, M. and Weiller, C. (1995) Review: does measurement of regional cerebral blood flow reflect synaptic activity? Implications for PET and fMRI. *Neuroimage* **2,** 148–156.

Kordower, J. H., Goetz, C. G., Freeman, T. B., and Olanow, C. W. (1997) Dopaminergic transplants in patients with parkinson's disease: neuroanatomical correlatates of clinical recovery. *Exp. Neurol.* **144,** 41–46.

Kuwabara, H., Cumming, P., Reith, J., Leger, G., Diksic, M., and Evans, A. C. (1993) Decarboxylase activity estimated in vivo using 6-[18F] Fluoro-DOPA and positron emission tomography: Error analysis and application to normal subjects. *J. Cereb. Blood Flow Metab.* **13,** 43–56.

Laihinen, A., Rinne, J. O., Rinne, U. K., Haaparanta, M., Ruotsalainen, U., and Bergman, J. (1992) [18F]fluorodopa PET scanning in Parkinson's disease after selective COMT inhibition with nitecapone (OR-462). *Neurology* **42,** 199–203.

Leenders, K. L., Palmer, A. J., Quinn, N., Clark, J. C., Firnau, G., and Garnett, E. S. (1986) Brain dopamine metabolism in patients with Parkinson's disease measured with positron emission tomography. *J. Neurol. Neurosurg. Psychiatry* **49,** 853–860.

Leenders, K. L., Poewe, W., Palmer, A. J., Brenton, D. P., and Frackowiak, R. S. J. (1986) Inhibition of 1-[18F]fluorodopa uptake into human brain by amino acids demonstrated by positron emission tomography. *Ann. Neurol.* **20,** 258–262.

Leenders, K. L., Salmon, E. P., Tyrrell, P., Perani, D., Brooks, D. J., and Sagar, H. J. (1990) The nigrostriatal dopaminergic system assessed in vivo by positron emission tomography in healthy volunteer subjects and patients with Parkinson's disease. *Arch. Neurol.* **47,** 1290–1298.

Lindvall, O., Brundin, P., Widner, H., Rehncrona, S., Gustavii, B., and Frackowiak, R. S. (1990) Grafts of fetal dopamine neurons survive and improve motor function in Parkinson's disease. *Science* **247,** 574–577.

Lindvall, O., Rehncrona, S., Brundin, P., Gustavii, B., Astedt, B., and Widner, H. (1989) Human fetal dopamine neurons grafed into the striaum in two patients with severe Parkinson's disease. *Arch. Neurol.* **46,** 615–631.

Lindvall, O., Widner, H., Rehncrona, S., Brundin, P., Odin, P., and Gustavii, B. (1992) Transplantation of fetal dopamine neurons in Parkinson's disease: one-year clinical and neurophysiological observations in two patients with putaminal implants. *Ann. Neurol.* **31,** 155–165.

Martin, W. R., Palmer, M. R., Patlak, C. S., and Calne, D. B. (1989) Nigrostriatal function in humans studies with positron emission tomography. *Ann. Neurol.* **26,** 535–542.

Martin, W. R., Stoessl, A. J., Adam, M. J., Ammann, W., Bergstrom, M., and Harrop, R. (1986) Positron emission tomography in Parkinson's disease: glucose and DOPA metabolism. *Adv. Neurol.* **45,** 95–98.

Martin, W. R. W. and Perlmutter, J. S. (1994) Assessment of fetal tissue transplantation in Parkinson's disease: does PET play a role? *Neurology* **44,** 1777–1780.

McCulloch, J., Kelly, P. A., and Ford, I. (1982) Effect of apomorphine on the relationship between local cerebral glucose utilization and local cerebral blood flow (with an appendix on its statistical analysis). *J. Cereb. Blood Flow Metab.* **2,** 487–499.

Melega, W. P., Grafton, S. T., Huang, S.-C., Satyamurthy, N., Phelps, M. E., and Barrio, J. R. (1991) L-6-[18F] Fluoro-DOPA metabolism in monkeys and humans: Biochemical parameters for the formulation of tracer kinetic models with positron emission tomography. *J. Cereb. Blood Flow Metab.* **11,** 890–897.

Miletich, R. S., Bankiewicz, K. S., and Plunkett, R. (1988) L-[18F]6-Fluorodopa PET images of catecholaminergic tissue implants in hemi-parkinsonian monkeys. *Neurology* **38,** S145.

Miletich, R. S., Quarantelli, M., and Di Chiro, G. (1994) Regional cerebral blood flow imaging with ^{99m}Tc-Bicisate SPECT in asymmetric Parkinson's disease: studies with and without chronic drug therapy. *J. Cereb.Blood Flow Metab.* **14,** S106–S114

Mitchell, I. J., Boyce, S., Sambrook, M., and Crossman, A. R. (1992) A 2-deoxyglucose study of the effects of dopamine agonists on the parkinsonian primate brain. Implications for the neural mechanisms that mediate dopamine agonist-induced dyskinesia. *Brain* **115,** 809–824.

Morrish, P. K., Rakshi, J. S., Bailey, D. L., Sawle, G. V., and Brooks, D. J. (1998) Measuring the rate of progression and estimating the preclinical period of Parkinson's disease with [^{18}F]dopa PET. *J. Neurol. Neurosurg. Psychiatry* **64,** 314–319.

Nordstroem, A. L., Farde, L., Pauli, S., Litton, J.-E., and Halldin, C. (1992) PET analysis of central [11C]raclopride binding in healthy young adults and schizophrenic patients—reliability and age effects. *Hum. Psychopharmacol.* **2,** 157–165.

Nutt, J. G., Woodward, W. R., Hammerstad, J. P., Carter, J. H., and Anderson, J. L. (1984) The "on-off" phenomenon in Parkinson's disease: relations to levodopa absorption and transport. *N. Engl. J. Med.* **310,** 483–487.

Owen, A. M., Doyon, J., Dagher, A., Sadikot, A., and Evans, A. C. (1998) Abnormal basal ganglia outflow in Parkinson's disease identified with PET. Implications for higher cortical functions. *Brain* **121,** 949–965.

Parsian, A., Racette, B. A., Zhang, Z. H., Chakraverty, S., Rundle, M., Goate, A. M., et al. (1998) Mutation, sequence analysis, and association studies of alpha-synuclein in Parkinson's disease. *Neurology* **51,** 1757–1759.

Pate, B. D., Kawamata, T., Yamada, T., McGeer, E. G., Hewitt, K. A., Snow, B. J., et al. (1993) Correlation of striatal fluorodopa uptake in the MPTP monkey with dopaminergic indices. *Ann. Neurol.* **34,** 331–338.

Patlak, C. S. and Blasberg, R. G. (1985) Graphical evaluation of blood-to-brain transfer constants from multiple-time uptake data: Generalizations. *J. Cereb. Blood Flow Metab.* **5,** 584–590.

Perlmutter, J. S., Kilbourn, M. R., Raichle, M. E., and Welch, M. J. (1987) MPTP-induced up-regulation of in vivo dopaminergic radioligand-receptor binding in humans. *Neurology* **37,** 1575–1579.

Perlmutter, J. S. and Moerlein, S. M. (1999) PET measurements of dopaminergic pathways in the brain. *Quart. J. Nucl. Med.* **43,** 140–154.

Perlmutter, J. S. and Raichle, M. E. (1984) Pure hemidystonia with basal ganglion abnormalities on positron emission tomography. *Ann. Neurol.* **15,** 228–233.

Perlmutter, J. S. and Raichle, M. E. (1985) Regional blood flow in hemiparkinsonism. *Neurology* **35,** 1127–1134.

Perlmutter, J. S., Stambuk, M. K., Markham, J., Black, K. J., McGee-Minnich, L., Jankovic, J., et al. (1997) Decreased [18F]spiperone binding in putamen in idiopathic focal dystonia. *J. Neurosci.* **17,** 843–850.

Perlmutter, J. S., Tempel, L. W., Black, K. J., Parkinson, D., and Todd, R. D. (1997) MPTP induces dystonia and parkinsonism: clues to the pathophysiology of dystonia. *Neurology* **49,** 1432–1438.

Piccini, P., Morrish, P. K., Turjanski, N., Sawle, G. V., Burn, D. J., Weeks, R. A., et al. (1997) Dopaminergic function in familial Parkinson's disease: a clinical and 18F-dopa positron emission tomography study. *Ann. Neurol.* **41,** 222–229.

Poline, J.-B., Holmes, A. P., Worsley, K. J., and Friston, K. J. (1996) Statistical inference and the theory of random fields, in *SPM96 Course Notes* (http://www.fil.ion.ucl.ac.uk/spm/course/notes97/Ch4.pdf).

Polymeropoulos, M., Lavedan, C., Leroy, E., Ide, S. E., Dehejia, A., Dutra, A., et al. (1997) Mutation in the alpha-synuclein gene identified in families with Parkinson's disease. *Science* **27,** 2045–2047.

Raichle, M. E. (1987) Circulatory and metabolic correlates of brain function in normal humans, in *Handbook of Physiology: The Nervous System V*, (Plum, F., ed.), American Physiological Society, Bethesda, MD, Part 2, pp 643–674.

Rinne, J. O., Heitala, J., Ruotsalainen, U., Sako, E., Laihinen, A., and Nagren, K. (1993) Decrease in human striatal dopamine D2 receptor density with age: a PET study with [11c]raclopride. *J. Cereb. Blood Flow Metab.* **13,** 310–314.

Sawle, G. V. (1993) The detection of preclinical Parkinson's disease: what is the role of positron emission tomography? *Mov. Disord.* **8,** 271–277.

Sawle, G. V., Bloomfield, P. M., Bjorklund, A., Brooks, D. J., Brundin, P., and Leenders, K. L. (1992) Transplantation of fetal dopamine neurons in Parkinson's disease: PET [18F]6-L-fluorodopa studies in two patients with putaminal implants. *Ann. Neurol.* **31,** 166–173.

Schlosser, R., Brodie, J. D., Dewey, S. L., Alexoff, D., Wang, G. J., and Fowler, J. S. (1998) Long-term stability of neurotransmitter activity investigated with 11C-raclopride PET. *Synapse* **28,** 66–70.

Shinotoh, H., Thiessen, B., Snow, B. J., Hashimoto, S., MacLeod, P., Silveira, I., et al. (1997) Fluorodopa and raclopride PET analysis of patients with Machado-Joseph disease. *Neurology* **49,** 1133–1136.

Shulman, G. L., Fiez, J.A. , Corbetta, M., Buckner, R. L., Miezin, F. M., Raichle, M. E., et al. (1997) Common blood flow changes across visual tasks: II. Decreases in cerebral cortex. *J. Cogn. Neurosci.* **9,** 648–663.

Snow, B. J., Nygaard, T. G., Takahashi, H., and Calne, D. B. (1993) Positron emission tomographic studies of dopa-responsive dystonia a*nd early-onset idiopathic parkinsonism.* Ann. Neurol. **34,** 733–738.

Snow, B. J., Tooyama, I., McGeer, E. G., Yamada, T., Calne, D. B., Takahashi, H., et al. (1993) Human positron emission tomographic [18F]fluorodopa studies correlate with dopamine cell counts and levels. *Ann. Neurol.* **34,** 324–330.

Todd, R. D., Carl, J., Harmon, S., O'Malley, K. L., and Perlmutter, J. S. (1996) Dynamic changes in striatal dopamine D2 and D3 receptor protein and mRNA in response to 1-methyl-4-phenyl-1,2,3,6-tetrahydropyridine (MPTP) denervation in baboons. *J. Neurosci.* **16,** 7776–7782.

Trugman, J. M., Arnold, W. S., Touchet, N., and Wooten, G. F. (1989) D1 dopamine agonist effects assessed in vivo with [14C]-2-deoxyglucose autoradiograhpy. *J. Pharmacol. Exp. Ther.* **250,** 1156–1160.

Trugman, J. M. and Wooten, G. F. (1987) Selective D1 and D2 dopamine agonists differentially alter basal ganglia glucose utilization in rats with unilateral 6-hydroxydopamine substantia nigra lesions. *J. Neurosci.* **7,** 2927–2935.

Volkow, N. D., Fowler, J. S., Wang, G. J., Dewey, S. L., Schyler, D. J., and MacGregor, R. R. (1993) Reproducibility of repeated measures of carbon-11-raclopride binding in the human brain. *J. Nucl. Med.* **34,** 609–613.

Wagner, H. N., Burns, H. D., Dannals, F. R., Wong, D. F., Langstrom, B., and Duffy, J. D. (1983) Imaging dopamine receptors in the human brain by positron emission tomography. *Science* **221,** 1264–1266.

Widner, H., Tetrud, J., Rehncrona, S., Brundin, P., and Gustavii, B. (1992) Bilateral fetal mesencephalic grafting in two patients with parkinsonism induced by 1-methyl-4-phenyl-1,2,3,6-tetrahydropyridine. *N. Engl. J. Med.* **327,** 1556–1563.

9

Dopamine Neurotoxicity and Neurodegeneration

BethAnn McLaughlin

9.1. INTRODUCTION

Dopamine is the major catecholamine neurotransmitter in the central nervous system (CNS) and is critically involved in motor control, motivation, emotion, and affect. Dopamine is also an inherently unstable compound and can be easily oxidized under physiological conditions leading to production of a host of compounds that are potentially neurotoxic. Alterations in dopaminergic function have been observed in a variety of motor and psychiatric disorders (Stevens, 1981; Klawans, 1973; Carlsson, 1988). Moreover, dopamine-rich regions are vulnerable to several neurodegenerative conditions, including Parkinson's disease, Huntington's disease, and ischemia (Jenner, 1996; Lin et al., 1997). This chapter will focus on the cellular and molecular mechanisms of dopamine toxicity, the effects of other neurotransmitters and cellular stressors on dopamine-induced cell death, and the relevance of dopamine toxicity to neurodegenerative events.

9.2. MECHANISMS OF DOPAMINE TOXICITY

It has been known for some time that in vivo and in vitro exposure to dopamine and its metabolites can cause neuronal death (Cheng et al., 1996; Filloux and Townsend, 1993; Michel and Hefti, 1990; Rosenberg, 1988). Dopamine toxicity can involve either the compound itself, products of its metabolism, or a combination thereof (Graham et al., 1978; Seiden and Vosmer, 1984; Yamamoto et al., 1994). Dopamine toxicity has been attributed to the following: (1) direct inhibition of the respiratory chain by the amine (Ben-Shachar et al., 1995; but *see* Morikawa et al., 1996 and McLaughlin et al., 1998a to the contrary); (2) free radicals and hydrogen peroxide production by auto-oxidation and metabolism (Chiueh et al., 1993); (3) generation of quinones and semiquinones (Graham et al., 1978; Hastings

From: *Contemporary Clinical Neuroscience: Molecular Mechanisms of Neurodegenerative Diseases*
Edited by: M.-F. Chesselet © Humana Press Inc., Totowa, NJ

and Zigmond, 1994); and (4) depletion of endogenous antioxidants (Hastings et al., 1996). Dopamine toxicity may also occur through receptor-dependent pathways by altering ion homeostasis and activating endonucleases, proteases, and members of the mitogen-activated protein kinase family that are associated with apoptotic cell death (Ziv et al., 1994; Luo et al.,1998; McLaughlin et al., 1998b).

Inhibition of Oxidative Phosphorylation

The hypothesis that dopamine could have direct inhibitory effects on oxidative phosphorylation is appealing in that insufficient ATP production has been linked to cell death in dopamine-producing regions and in dopaminergic target areas (Schapira and Cooper 1992; Bowling and Beal, 1995). For example, although Huntington's disease is characterized by striatal degeneration, the disease protein, huntingtin, is widely expressed throughout the CNS (Fusco et al., 1999; Bhide et al., 1996). If expression of huntingtin impairs cellular respiration, further inhibition of ATP production by dopamine could produce the selective vulnerability observed in the dopamine-rich striatum.

Although Ben-Shachar and co-workers reported that dopamine is a rather potent inhibitor of mitochondrial complex I (NADH dehydrogenase) in isolated organelles with an inhibitory concentration at which 50% of activity is blocked (IC_{50}) of 8 μ*M* (Ben-Shachar et al., 1995), other groups have failed to reproduce this finding. Fu and colleagues reported that 10 μ*M* dopamine had no effect on 3-(4,5-dimethylthiazol-2-yl)-2,5-diphenyltetrazolium bromide (MTT) reduction and that only much higher doses (200 μ*M*) were able to induce modest (approx 30%) decreases in MTT conversion and rhodamine 123 fluorescence (Fu et al., 1998). The MTT assay is based on the cleavage of the yellow tetrazolium salt MTT to purple formazan crystals, a process that requires NADH and NADPH and has been used as a measure of "metabolically active cells." MTT is, at best, an indirect measure of cellular respiration and is based on the assumption that the only process that depletes NADH and NADPH is oxidative phosphorylation. This tenant is problematic given that other cellular processes consume nicotinamide, the precursor for both of NADH and NADPH. One example of this is poly-ADP ribose polymerase, a caspase-activated DNA repair enzyme, which uses both nicotinamide and ATP when activated. It is therefore possible that MTT reduction can be caused by processes that are not solely related to mitochondrial inhibition. Indeed, Berridge and co-workers (1993) reported that most MTT reduction did not occur within the mitochondria but rather was caused by NADH- and NADPH- dependent mechanisms that were not reflective of changes in the respiratory chain. Perturbations in rhodamine 123, which is a voltage-sensitive fluorescent dye, were also used to assess the effects of dopamine on oxidative phosphorylation by Fu and colleagues (1998). The limitation of this technique is that it does not directly measure ATP production but rather is a semiquantitative means to assess mitochondrial membrane potential that is dependent on oxidative phosphorylation. The finding that rhodamine 123 fluorescence decreases with 200 μ*M* dopamine remains equivocal however, as studies by Hoyt and colleagues (1997) found that exposure to slightly higher concentrations of dopamine (250 μ*M*) did not cause any disruption in mitochondrial membrane potential measured with the same fluorescent dye.

Direct measurement of the effects of dopamine on respiration and adenine nucleotide levels, which are the best means to assess oxidative phosphorylation, were performed by Morikawa and co-workers who reported very little effect on either NADH dehydrogenase activity (Complex I) or mitochondrial respiration at concentrations of the amine as high as 10 m*M* (Morikawa et al., 1996). Similarly, we have observed no appreciable alteration in the [ATP]/[ADP] in striatal cultures after 6 h exposure to 250 μ*M* dopamine (McLaughlin et al., 1998). It is noteworthy that we measured nucleotide ratios very late in the cell death process, as determined by time-lapse video microscopy recordings. These direct measurements of oxidative phosphorylation strongly suggest that dopamine is not, in fact, a direct inhibitor of the respiratory chain. Dopamine may, however, inhibit enzymes that are not a part of oxidative phosphorylation, but which can, under some highly energetically stressful circumstances, contribute to ATP production. Maker and co-workers (1986) found that dopamine decreased the activity of creatine kinase and adenylate kinase. Although these enzymes do not generate ATP, they do help maintain its level under adverse conditions (reviewed by Erecinska and Silver, 1989). Hence, the inhibition of creatine kinase and adenylate kinase would not be expected to decrease ATP content in intact cells.

Generation of Reactive Species

Another mechanism that has been implicated in dopamine neurotoxicity is the generation of reactive oxygen species. Dopamine metabolism occurs either through spontaneous auto-oxidation or by enzymatic degradation, either intracellularly by monoamine oxidase B (MAO-B) or extracellularly by catechol-*o*-methyl-transferase (COMT). At physiological pH, the catechol moiety of dopamine is relatively easily auto-oxidized leading to the production of superoxide radicals and hydrogen peroxide (which can form hydroxyl radicals in the presence of transition metals such as iron) and quinones (Graham, 1984). These species readily remove electrons from cellular nucleophiles causing lipid peroxidation, DNA strand breakage, and protein damage (Halliwell, 1991). The enzymatic oxidative removal of the amino group from dopamine by the mitochondrial enzyme MAO-B leads to the formation of hydrogen peroxide in addition to DOPAC (Graham et al., 1978; Graham, 1984; Spina and Cohen, 1989). Dopamine can also be hydroxylated to form 6-hydroxydopamine, a potent neurotoxin (Slivka et al., 1988). Further, auto-oxidation of dopamine may actually inhibit normal enzymatic degradation of the amine, as the quinone species 6-amino dopamine-*p*-quinone can inactivate COMT (Cheng et al., 1987). Normally, COMT promotes the methylation of the 3-hydroxy group of the catechol ring of dopamine, leading to the formation of 3-methyltyramine which can subsequently be metabolized by MAO-B to homovanillic acid (HVA). Reactive quinones can interfere with the intracellular degradative pathways for dopamine metabolism as well. For example, dopamine transporters are cysteine-rich proteins that can be oxidized by dopamine-derived quinones (Berman et al., 1996), a process that could hinder dopamine reuptake and MAO-B-linked degradation of the amine.

We and others have observed that the cell death induced by concentrations of dopamine as high as 250 μM can be attenuated with antioxidants and free-radical scavengers (McLaughlin et al., 1998; Hoyt et al., 1997; Gabby et al., 1996; Offen et al., 1995, 1996; Ziv et al., 1994). This is perhaps not surprising given the host of reactive oxygen species formed by dopamine metabolism. As there is very little degradation of the amine at times when we observe gross cellular changes and neurotoxicity, auto-oxidation products of dopamine such as quinones, which are toxic only at relatively high concentrations, are not likely to contribute to cell death in this system. However, it is likely that the types of free radicals formed from dopamine, as well as their sites of action, may depend on the particular model used. This postulate is supported by results that show that not all antioxidants are equally effective in decreasing dopamine-induced cell death. For example, ascorbate, which is cell impermeant, is not neuroprotective in mesencephalic cultures (Michel and Hefti, 1990), neuroblastoma cells (Gabby et al., 1996) or PC12 cells (Offen et al., 1996), but is effective against high doses of dopamine in striatal neurons (Cheng et al., 1996). Even cell-permeant antioxidants such as α-tocopherol are not protective in some systems (Michel and Hefti, 1990). In general, thiol-containing agents have produced the most consistent neuroprotective results (Gabby et al., 1996; Offen et al., 1995; Offen et al., 1996; Zilkha-Falb et al., 1997).

Role of Dopamine Transport

In order to understand the mode of action of reactive oxygen species that contribute to dopamine toxicity, one must know the site of dopamine toxicity; that is, if dopamine transport is required for cell death, this would suggest that the by-products of MAO-B metabolism or direct mitochondrial effects of the amine are more likely to contribute to the cell death process. If, however, dopamine toxicity does not require reuptake, this would suggest that receptor-dependent mechanisms or auto-oxidation products are responsible for cell death.

Filloux and Townsend (1993) demonstrated that intrastriatal injection of dopamine results in both presynaptic and postsynaptic damage. Given that there are no postsynaptic dopamine transporters in the striatum, this suggests that striatal cell death in this system is the result of the loss of innervation from the substantia nigra, the presence of a toxic diffusible factor, or a dopamine receptor-dependent mechanism. These investigators also reported that removal of dopaminergic terminals from the striatum enhanced striatal cell death induced by dopamine, presumably by increasing the extracellular concentration of the amine. Although it has been reported that expression of antisense to the dopamine transporters attenuated dopamine-induced cell death in a human neuroblastoma cell line (Simantov et al., 1996), cultures that lack dopamine transporters are still highly vulnerable to dopamine (Offen et al., 1995; McLaughlin et al., 1998). Taken with the aforementioned studies which suggest that cell-impermanent antioxidants can, in some instances, protect cells from dopamine toxicity, these data suggest that reuptake is not required for dopamine-induced cell death to occur. Therefore, dopamine can elicit its toxic effects by both intracellular and extracellular mechanisms depending on the presence or absence of uptake sites in the target cells.

Receptor-Dependent Mechanisms of Dopamine Toxicity

Because dopamine is easily degraded and converted to highly reactive species, its toxic effects are thought to be largely receptor independent. However, dopamine receptors are coupled to a variety of intracellular signaling pathways, which, in turn, are critically linked to ion homeostasis, transcriptional regulation, and cellular repair processes. Dopamine receptors are divided into two broad classes: D1-like and D2-like.

D1-Like Receptors

The D1-like receptors are positively coupled to adenylyl cyclase through G-protein stimulatory subunit (Gs). In this way, ligand binding increases protein kinase A activity which phosphorylates L-type calcium channels thereby increasing whole-cell calcium currents (Trautwein and Hescheler, 1990; Hartzell et al., 1991; Surmeier et al., 1995; Hernandez-Lopez et al., 1997). Further, activation of D1-like receptors can also cause release of calcium from intracellular sites (Liu et al., 1992; Seabrook et al., 1994a; 1994b).

Augmentation of intracellular calcium concentration caused by activation of these receptors could initiate a number of reactions, including activation of proteases and endonucleases and the generation of free radicals which could induce cell death (reviewed by McConkey and Orrenius, 1996). High micromolar concentrations of dopamine cause gradual increases in intracellular calcium from extracellular pools in a subpopulation of forebrain neurons in culture (Hoyt et al., 1997). However, the fact that dopamine toxicity in this system is not attenuated by the removal of extracellular calcium suggests that perturbations in calcium ion homeostasis may not underlie dopamine toxicity in all systems (Hoyt et al., 1997). This finding is consistent with the work of Alagarsamy et al. (1997) who reported that D1 receptor antagonists do not prevent dopamine toxicity in cortical cell cultures. We have found that a saturating concentration of SCH 23390, a D1 receptor competitive antagonist, reduces but does not completely inhibit dopamine toxicity in dissociated striatal cultures. Although this effect is relatively small (approximately a 25% decrease in toxicity induced by 250 μM dopamine), it nevertheless suggests that D1 receptor activation may contribute to dopamine toxicity in the striatum. Given that the striatum expresses the highest density of D1 receptors in the brain (Boyson et al., 1986), this suggests that this region may be more vulnerable to dopamine-induced calcium dysregulation.

It has also been shown that binding to D1 receptors activates members of the mitogen-activated protein kinase (MAPK) family through a PKA-dependent mechanism (Zhen et al., 1998). MAPKs integrate and communicate extracellular signals to intracellular targets, generally resulting in alterations in gene expression. Activation of these kinases is also a critical link to induction or suppression of apoptotic cell death cascades (*see* Subheading 9.3.). MAPKs are activated by upstream kinases (MAPKK or MEKs) which, in turn, are activated by MAPKK kinases (MAPKKK; [Treisman, 1996]). MAPKKK receive information from cell surface receptors via interactions with GTP-binding proteins of the ras superfamily, from cytoplasmic tyrosine kinases, or following a variety of environmental stresses (Treisman, 1996; Chakraborti and Chakraborti, 1998). Three MAPK subtypes have been identified in mammalian cells, ERK (extracellular regulated kinase), JNK (c-Jun N-terminal kinase), and p38. Both p38 and JNK, the putative "stress-activated protein kinases," are upregulated by the D1 receptor agonist SKF 38393 in neuroblastoma cells (Zhen et al., 1998). If activation of these pathways is toxic in culture, this finding could be consistent with the work of Kelley et al. (1990) who reported that intrastriatal injection of this D1 agonist induced cell death in vivo. It should be noted, however, that the authors of this work suggested that non-receptor-dependent effects of SKF 38393 contributed to the observed neurotoxicity because an inactive isomer of this compound was also neurotoxic.

D2-Like Receptors

The D2-like receptors are generally coupled to the Gi/Go family of G-proteins, and activation of these receptors can inhibit adenylyl cyclase and calcium currents (Weiss et al., 1985; Lledo et al., 1992), activate potassium channels (Einhorn et al., 1991) and potentiate arachidonic acid release (Kanterman et al., 1991; Piomelli et al., 1991). Upon dopamine binding to D2 receptors, βγ-subunits are released from associated G-proteins resulting in ras activation and recruitment of raf kinases. These kinases phosphorylate and activate MEK-1, which, in turn, upregulates the MAPKs ERK and JNK (reviewed by Karin, 1998). Coexpression of proteins that sequester βγ–subunits from ras block dopamine induced ERK and JNK activation (Faure et al., 1994), as does expression of dominant negative ras mutants or co-incubation with the MEK-1 inhibitor PD 98059 (Luo et al., 1998a). These results require confirmation in other systems as the D2 receptor agonist quinpirole does not activate JNK in neuroblastoma cells (Zhen et al., 1998). It is likely that the heterogeneity in the MAPK family activation is dependent on the subtype of D2-like receptors expressed in a given population of cells, the density of receptors, and the presence of other proteins that can also sequester βγ-subunits.

Receptor-Dependent Activation of Other Transcription Factors

The specific cellular and molecular processes that contribute to receptor-dependent neurotoxicity are just beginning to be elucidated. In addition to the changes induced in cAMP, ion homeostasis and MAPK family members, both D1-like and D2-like receptors are linked to the induction of a number of immediate early genes that are important for development, signal integration, drug addiction, plasticity, and, possibly, neurotoxicity (Drago et al., 1998; Smith et al., 1997; Jung et al., 1996; Ishida et al., 1998). Activation of D1-like receptors increases expression of the immediate early genes (IEGs) fos and zinc-finger immediate early gene (zif) 268 (Keefe and Gerfen, 1996). It has been reported that activation of D2 receptors alone decreases zif268 but increases zif268 in the presence of D1 receptor agonists (Gerfen et al., 1995; Keefe and Gerfen, 1995). A similar balance of transcriptional activities is also observed when dopamine is released in the presence of glutamate (*see* Subheading 9.4., Glutamate). These studies suggest that the cellular neurotransmitter receptor profiles (i.e., whether cells express D1-like or D2-like receptors, glutamate receptors and other receptors which interact with the ras signaling pathways) are critically involved with transcriptional activation and, ultimately, cell viability. As genes under the control of these IEGs and other DNA-binding proteins (such as Nurr 1 and the nuclear retinoid receptor RXR) are elucidated, the importance of these compounds in mediating dopamine toxicity can be determined.

9.3. INDUCTION OF APOPTOSIS BY DOPAMINE

We have found that striatal cells exposed to 250 μ*M* dopamine express hallmark features of apoptosis, including DNA laddering, chromatin condensation, and membrane blebbing (McLaughlin et al., 1998a). Using time-lapse video microscopy we observed that approximately 80% of cells that died underwent apoptosis. This is in general agreement with studies in the literature that reported apoptotic death in other systems at comparable concentrations of the amine (Hoyt et al., 1997; Offen et al., 1995; Offen et al., 1996; Simantov et al., 1996). Indeed, dopamine has been shown to be a potent apoptogen in a variety of cells, including postmitotic sympathetic chick neurons, rat pheochromocytoma (PC12) cells, striatal, forebrain and cerebellar neurons, human neuroblastoma cells, and a clonal catecholaminergic cell line (Ziv et al., 1994; Walkinshaw et al., 1995; Simantov et al., 1996; Hoyt et al., 1997; McLaughlin et al., 1998b; Takai et al., 1998). However, exact comparisons of the extent of apoptotic versus necrotic cell death are difficult to make because previous investigators have not used dynamic assessment of this phenomenon.

In the past 5 yr, proteases known as caspases (cysteine proteases cleaving after an aspartate residue) have been shown to be critical components of apoptotic cell death (reviewed by Nunez et al., 1998). These proteins are present constitutively throughout the cell as inactive zymogens comprised of a prodomain and two subunits. Caspases are activated by cleavage by other activated caspases, by self-association with other caspase proenzymes or apoptosis protease activating factor 1 (Apaf-1), by granzyme B or by recognition of at least four amino acids upstream of a requisite aspartate residue (reviewed by Thornberry and Lazebnik, 1998; and Buckley et al., 1999). The ability of caspases to cleave other cellular components such as lamin A, fodrin, Rb, poly-ADP ribose polymerase, gelsolin, and an inhibitor of caspase-activated DNase (ICAD) result in many of the hallmark changes in cell morphology and DNA integrity indicative of apoptosis (reviewed by Stroh et al., 1998). However, the mechanism of caspase activation, the specific caspases that contribute to dopamine-induced apoptosis, the sequence in which they are activated, and the final steps that are needed for the commitment to die after dopamine exposure remain largely unknown.

Caspases and MAPK family members can mutually influence one anothers' activation and are important in the induction of apoptosis. The MAPK family members JNK and p38 can be potently activated by dopamine receptor stimulation (as previously described) as well as by cellular stressors, including Ca^{2+} influx and the presence of reactive oxygen species (Kawasaki et al., 1997; Schwarzschild et al., 1997). Although JNK activation has been linked with both apoptosis and cell survival, overexpression of c-Jun clearly induces apoptotic cell death (Park et al., 1997; Bossy-Wetzel et al., 1997). Moreover, dominant negative mutants deficient in JNK signaling are markedly less sensitive to otherwise apoptotic stimuli (Ham et al., 1995; Yujiri et al. 1998). Emerging evidence suggests that dopamine induction of JNK (either by a receptor-dependent mechanism or by oxidative stress induced by dopamine metabolism) contributes to the apoptotic process induced by the catecholamine. Induction of c-Jun occurs in a time- and concentration-dependent manner after exposure to 500 μM dopamine with maximal JNK induction at 3 h and c-Jun protein expression at 18 h (Luo et al., 1998b). Coincubation with agents that decrease the generation of reactive oxygen species such as *n*-acetyl cysteine and catalase blocks both apoptosis and JNK activation (Luo et al., 1998b) suggesting that this induction is less dependent on receptor activation and more on oxidative stress in this system. Similar studies with shorter incubations and lower doses of dopamine in younger striatal cultures failed to find JNK-activated transcription (Schwartzchild et al., 1997). In the older cultures in which dopamine did upregulate JNK, transfection of primary striatal cultures with a dominant negative kinase upstream of JNK blocks apoptosis induced by the amine (Luo et al., 1998b).

Given that caspases are constitutively expressed in an inactive form and that MAPK family members can both activate and be activated by caspases, new protein synthesis may not be required for dopamine-induced apoptosis. However, transcription of "prodeath" proteins clearly is important for some forms of apoptosis with the most notable example being the upregulation of Bax by p53 (Miyashita and Reed, 1995). Yet, the role of new protein synthesis in dopamine toxicity is unclear. Addition of cyclohexamide or actinomycin D does not attenuate death induced by 300 μM dopamine in chick sympathetic neurons or cerebellar granular cells (Zilka-Falb et al., 1997). On the other hand, protein synthesis inhibitors block changes in cell cycle indicative of apoptosis in neuroblastoma cells treated with 500 μM dopamine (Simantov et al, 1996; Luo et al., 1998b).

Apoptosis can be induced not only by upregulating death-promoting pathways, such as caspase dependent pathways, but also by downregulating cell survival mechanisms. Indeed, overexpression of prosurvival members of the Bcl-2 family can block apoptosis in many systems, including in cultures exposed to dopamine and mice subject to 6-OHDA lesions (Offen et al., 1997, 1998). Although the mechanism of action of Bcl-2 family members is still being elucidated, these proteins can act prior to caspase and JNK activation to halt cell death in some systems (Park et al., 1997; Yujiri et al., 1998).

Members of the Bcl-2 family can have prosurvival or prodeath properties which are thought to be heavily dependent on the relative abundance of prodeath and antideath proteins, as there is extensive heterodimerization and inactivation that occurs within this family. Consistent with the idea that Bcl-2 family members exert influence over the dopamine toxicity pathways, cultures derived from mice deficient in Bcl-2 are more vulnerable to dopamine toxicity (Hochman et al., 1998). PC12 cells exposed to 100 μM auto-oxidized dopamine may also increase Bax expression, a pro-death member of the Bcl-2 family, prior to undergoing apoptosis (Kang et al., 1998). Other factors such as protein phosphorylation can also alter the activity of Bcl-2 family members. For instance, phosphorylation of BAD by mitochondrially anchored cAMP-dependent PKA leads to suppression of its normal proapoptotic activities (Harada et al., 1999). It will be of particular interest to resolve the effects of D1 receptor activation with this work given that D1 receptor activation also stimulates PKA activity and thus to determine the contribution of Bax phosphorylation to physiological and pathophysiological events involving dopamine.

9.4. DOPAMINE NEUROTOXICITY CAN BE INFLUENCED BY MULTIPLE FACTORS

Mitochondrial Toxins

Mitochondrial toxins have proved to be invaluable tools in the study of neurological diseases and in understanding basic cellular and molecular mechanisms that contribute to neurodegeneration. For instance, the combination of mitochondrial inhibitors with glutamate has provided support for the excitotoxic model of cell death, which predicts that minor mitochondrial deficiencies can lead to membrane depolarization, massive calcium influx through NMDA receptors, and cell death (Albin and Greenamyre, 1992). In vitro and in vivo studies have supported this theory as intrastriatal injection of malonate, a potent blocker of succinate dehydrogenase (or complex II; Alston et al., 1977; Dutra et al., 1993), produce age-dependent lesions of the striatum and decreases in tissue [ATP] (Beal et al., 1993a). These injections increase the extracellular concentration of glutamate (Messam et al., 1995), and NMDA receptor antagonists attenuate malonate-induced cell death (Greene et al., 1993; 1995b). However, malonate injections also produce long-term decreases in striatal dopamine content and tyrosine hydroxylase activity, suggesting that dopaminergic neurotransmission is altered by mitochondrial inhibitors.

We have found that mitochondrial toxins also potentiate the cell death caused by dopamine. Subthreshold doses of dopamine and methylmalonate cause massive cell death when coapplied to striatal cultures (McLaughlin et al., 1998a). Further, removal of the dopaminergic projections to the striatum attenuates cell death induced by mitochondrial toxins such as malonate and 3-nitropropionic acid (Reynolds et al., 1998; Maragos et al., 1998). In order to understand the mechanism by which mitochondrial toxins potentiate dopamine toxicity, one must understand the potential sites of convergence of these two toxins. One mechanism by which subtoxic levels of dopamine and mitochondrial toxins may cause cell death is via lethal increases in intracellular calcium levels. Activation of D1 receptors has been linked to increases in the cytosolic concentration of calcium caused by release from intracellular stores as well as influx from outside the cell via phosphorylation and activation of calcium channels (Liu et al., 1992; Seabrook et al., 1994a,b; Hoyt et al., 1997). We found that treatment with a D1 receptor antagonist partially attenuates the cell death caused by dopamine suggesting that calcium accumulation may contribute to dopamine toxicity. As methylmalonate alone induces a massive increase in intracellular calcium levels (McLaughlin et al., 1998a), accumulation of intracellular calcium caused by a lower dose of methylmalonate together with calcium influx through a D1-mediated pathway may be sufficient to induce cell death in a manner similar to that observed in the excitotoxic model of neurodegeneration.

We have also found that the potentiating cell death observed by coincubation of dopamine and methylmalonate can be attenuated by antioxidants (McLaughlin et al., 1998b). This suggests that reactive oxygen species are generated by both compounds and can work in a concerted manner to elicit cell death. As previously discussed, dopamine auto-oxidation and metabolism produces free radicals, semiquinones, quinones and depletion of endogenous antioxidants (reviewed by Hastings et al., 1996). Further, interruption of the electron transport chain by mitochondrial toxins such as methylmalonate can also produce reactive oxygen species (reviewed by Halliwell, 1991), and combined production of free radicals by dopamine and methylmalonate may overwhelm endogenous antioxidant defenses and result in potentiated toxicity. The intracellular targets that are altered by reactive oxygen species produced by both agents remain to be elucidated.

Taken together, this work predicts that even a minor inhibition of mitochondrial function combined with dopamine release could produce massive neuronal damage in vivo. Such a situation is created not only during hypoxia/ischemia but also in situations such as mild hypoglycemia in which a relatively small metabolic insult results in a large amount of damage to the CNS.

Serotonergic and Noradrenergic Influences

The dopaminergic systems is influenced by both serotonin and noradreneline under physiological and pathophysiological circumstances. There is extensive serotonergic innervation of dopaminergic cell bodies and terminal regions, and these amines mutually influence one another's release. One of the most well-understood neurotoxic compounds that affects the serotonin and dopaminergic systems is the amphetamine analog, 3,4-methylenedioxymethamphetamine (MDMA or ''ectasy''). In both rodents and monkeys, MDMA is cytotoxic to serotonergic neurons, and although this effect is less clear in humans, substantially diminished serotonergic activity is observed in patients who abuse MDMA (Ricaurte et al., 1992; reviewed by Steele et al., 1994). Like methamphetamine (METH), MDMA is a psychostimulant that can elicit excess dopamine outflow and dopamine-derived oxidation products that have been implicated in the pathological changes observed in serotonergic systems. The observation that there is a proportional depletion of serotonin and dopamine in several brain regions has led to speculation that the degeneration in these two pathways is related (Wrona et al., 1997). Sprague et al. (1998) have suggested that the acute release of serotonin and dopamine caused by MDMA activates postsynaptic $5\text{-HT}_{2A}/5\text{-HT}_{2C}$ receptors on GABAergic interneurons, thus decreasing GABAergic transmission and further increasing MDMA-induced dopamine release. Through metabolic and auto-oxidation pathways, reactive oxygen species are generated by dopamine, and this process is ultimately toxic to serotonergic axons and terminals. The reactive oxygen species that underlie MDMA toxicity are presently unknown. However, free radicals produced by dopamine can drive generation of the neurotoxic oxidation products 6-hydroxydopamine and 5,6-dihydroxytryptamine from dopamine and serotonin, respectively. Alternatively, dopamine can alter serotoninergic functioning by interfering with serotonin metabolic pathways. For example, dopamine can form a redox cycling agent that is an endogenous toxin to serotonergic neurons (Kuhn and Arthur, 1998), and dopamine concentrations as low as 10 μM can significantly impair the activity of tryptophan hydroxylase (Schmidt et al., 1986; Stone et al., 1986), the rate-limiting step in the production of serotonin (Jequier et al., 1967).

Serotonin and dopamine can also influence one another's signal transduction pathways. The serotonin-1A (5-HT_{1A}) receptor subtype is an inhibitory protein that can provide neuroprotection against excitotoxic damage by decreasing the activity of N-type Ca^{2+} channels and NMDA receptors (Strosznajder et al., 1996; Oosternik et al., 1998). Stimulation of 5-HT_{1A} receptors may also prove neuroprotective against a variety of otherwise apoptotic stimuli by activation of the ERK pathway. Indeed, 5-HT_{1A} receptor agonists decrease anoxia-induced DNA damage and caspase activation in a hippocampal neuron-derived cell line, which does not possess Ca^{2+} channels (Adayev et al., 1999). This protection appears to be mediated by activation of ERK and can be blocked by the MEK1 inhibitor PD98059. As dopamine can induce activation of prodeath members of the MAPK family (JNK and p38), it will be of particular interest to determine if serotonin receptor agonist treatment can protect against dopamine toxicity.

Norepinephrine is synthesized from dopamine by dopamine β-hydroxylase and, like dopamine, norepinephrine causes dose-dependent cell death in striatal, cortical, and cerebellar cultures (Rosenberg, 1988; Zilka-Falb et al., 1997; McLaughlin et al., 1998b). However, norepinepherine produces appreciably less cytotoxicity than comparable doses of dopamine (McLaughlin et al., 1998a). More chronic exposure paradigms suggest that this disparity may be lessened with longer, lower-dose exposures (Rosenberg, 1988). Like dopamine, norepinephrine can also produce toxic oxidation products, including adrenochrome and noradrenochrome, which are derived from adrenaline and noradrenaline, respectively. These compounds can be conjugated with glutathione in the presence of glutathione transferase in a detoxification reaction that may prevent redox cycling in adverse conditions (Baez et al., 1997).

There is intriguing evidence that suggests that norepinephrine may provide protection against dopamine toxicity both in vitro and in vivo. Indeed, norepinephrine (1 m*M*) blocks the toxicity of 50 μ*M* 6-OHDA in neuroblastoma cells and attenuates the cell death induced by dopamine itself (Graham et al., 1978). The authors speculate that this process may be the result of norepinephrine's ability to quench superoxide anions which are normally present and can be produced by dopamine and its derivatives. This effect may also be applicable to other neurotransmitter systems that involve oxidative stress, such as glutamate. Indeed, there is evidence to suggest that much lower doses of norepinephrine (0.1 and 1 μ*M*) can also provide protection against NMDA toxicity in vitro (Gepdiremen et al., 1998). Further, limited in vivo evidence suggests that prior lesioning of the locus coeruleus, which contains the majority of the norepinephrine cell bodies in the CNS, increases METH-induced dopaminergic loss. These effects can also be reproduced by pretreating animals with clonidine, which decreases noradrenergic cell firing (Fornai et al., 1998, 1999).

Glutamate

It has been known for some time that glutamate and dopamine can mutually influence one another's release in physiologically important ways (Giorguieff et al., 1977; Roberts and Anderson, 1979). For instance, dopamine can inhibit the release of glutamate through activation of presynaptic D2 receptors (Godukhin et al., 1984; Kerkerian et al., 1987; Maura et al., 1988). In addition, oxidation by products of the amine can block glutamate uptake from striatal synaptosomes (Maura et al., 1988; Yamamoto et al., 1992; Berman and Hastings, 1997). D1 receptor stimulation can also influence postsynaptic glutamatergic signaling by increasing NMDA evoked excitation (Cepeda et al., 1993). Activation of D1 receptors can also lead to phosphorylation of DARPP-32 (dopamine and cAMP-regulated phosphoprotein) and increase AMPA receptor currents within the striatum — a process that has been implicated in neuronal plasticity and local regulation of glutamatergic neurotransmission (Yan et al., 1999). It remains to be determined what role DARPP-32 signaling has in integrating excitotoxic signaling through glutamate receptors, although it has been suggested that regulation of this phosphoprotein may dampen calcium signaling elicited by NMDA receptor stimulation (Hemmings et al., 1987).

Glutamate can also enhance or inhibit dopamine outflow, and these interactions are of particular physiological relevance in the striatum where projection neurons receive dopaminergic input from the substania nigra and glutamatergic input from the cortex. Dysregulation of basal ganglia activity has been implicated in the motor manifestations of several neurological diseases. In vivo studies support the functional significance of the interaction between these two neurotransmitter systems, as prior lesioning of the nigrostriatal dopaminergic projections decreases striatal cell death associated with local infusion of NMDA and NMDA agonists, as well as kainate (Chapman et al., 1989; Buisson et al., 1991). This effect may be the result, in part, of activation of D1 receptors, as D1 receptor agonists increase swelling induced by NMDA in striatal cultures (Cepeda et al., 1998) suggesting that D1 receptor induced increases in intracellular calcium may be particularly deleterious when combine with that which is elicited by NMDA. Given the complex interdependent relationship of dopamine and glutamate that occurs within the striatum, it is likely that application of glutamate receptor antagonists has both short- and long- term effects on dopaminergic transmission within this region. That is to say that the use of glutamate receptor antagonists such as MK-801, which have been shown to decrease cell death caused by mitochondrial toxins (Greene and Greenamyre, 1995 a,b; Greene et al., 1993; Greene et al., 1996; Beal et al., 1993b), likely also decreases dopamine release within the striatum, which may also contribute to the observed neuroprotection.

Dopamine and glutamate interactions are not simply limited to mutual influences on each others neurotransmitter pools, but extend to transcriptional and translational modification of neuronal receptor expression patterns as well as metabolic activity (Porter et al., 1994). The cellular and molecular integration of signals from dopaminergic and glutamatergic systems within regions such as the striatum are just beginning to be understood. For instance, dopamine and glutamate are both important for the activation of the cyclic AMP responsive element binding protein (CREB) and expression of IEGs such as cfos, c-Jun, and zif 268 (Konradi et al., 1996). This convergence of synaptic inputs at the transcriptional level has been implicated in synaptic plasticity, but the role of these elements in neurotoxicity and disease is unclear.

9.5. INVOLVEMENT OF DOPAMINE TOXICITY IN ACUTE AND CHRONIC NEURODEGENERATION

Ischemia and Hypoxia

Loss of blood flow (ischemia) and insufficient oxygenation of blood (hypoxia) occurs in a number of acute conditions such as cardiac arrest, respiratory arrest, and stroke. The brain receives over 20% of the body's oxygen supply, contains a high concentration of polyunsaturated fatty acids which are highly vulnerable to lipid peroxidation, and is relatively deficient in antioxidants. This makes the brain a prime target for ischemic damage (Olanow, 1990). Certain anatomical sites and subsets of cells within these sites are particularly vulnerable to hypoxia and ischemia. These regions include the CA1 layer of the hippocampus, neocortical layers 3, 5, and 6 and the medium-sized spiny neurons of the striatum (Petito and Pulsinelli, 1984; Pulsinelli et al., 1982; Chesselet et al., 1990). These sites are thought to be targets of ischemic damage based on the relative abundance of glutamatergic projections and cell bodies, but a role for dopamine-induced damage in ischemia, especially in the striatum and hippocampus, has been suggested by several studies.

The striatum receives the densest dopaminergic innervation of any brain region (Carlsson et al., 1965; Carlsson, 1959; Dahlstrom, 1971; Anden et al., 1965), and projections arise predominately from the substantia nigra pars compacta but also emanate from the ventral tegmentum and the retrorubral area (Fallon and Moore, 1978; Graybiel and Ragsdale, 1978). Large increases in striatal extracellular dopamine have been observed in experimental models of ischemia (Phebus et al., 1986; Slivka et al., 1988; Clemens and Phebus, 1988; Yao et al., 1988; Baker et al., 1991). Indeed, Slivka et al. (1988) reported extracellular dopamine levels as high as 200 μ*M* in the striatum 40 min after unilateral ligation of the common carotid artery in gerbils. The magnitude of the increase in striatal dopamine compared to other neurotransmitters that are released during ischemia is particularly noteworthy. Buisson et al. (1992) reported that striatal dopamine levels increase 1100 times after middle cerebral artery occlusion in rats, whereas glutamate increased only sixfold. The impact of this dopamine efflux in contributing to ischemia-induced neurodegeneration is evident from studies in which it was found that prior lesioning of the nigrostriatal dopaminergic pathway decreases the amount of striatal cell death following some, although not all, forms of ischemia (Clemens and Phebus, 1988; Globus et al., 1987; Wieloch et al, 1990; Lin, 1997). Similar neuroprotective effects were observed when catecholamine stores were depleted by pretreatment with α-methyl-*para*-tyrosine (Weinberger et al., 1985).

Given that dopamine toxicity appears to be mediated in large part by the production of reactive oxygen species, the observed ability of antioxidants to attenuate ischemic injury is consistent with a role for dopamine in ischemic injury (Clemens et al., 1994; Cao et al., 1994). Indeed, hydrogen peroxide produced by MAO-B may contribute to cell death during ischemia, as inhibitors of this pathway enhance the survival of rats subject to carotid artery occlusion (Damsma et al., 1990). Although the combined influence of energetic stress and glutamate release have been the primary focus of much of the ischemia literature, the complex interplay between dopamine and glutamate release may account for some part of the protective effects of nigrostriatal lesions.

Nigrostriatal lesions block ischemia-induced striatal dopamine release, but they also partially attenuate striatal glutamate efflux (Globus et al., 1988). The ability of dopamine to decrease glutamate efflux may, however, be limited to pathophysiological events such as ischemia, where there is large neurotransmitter efflux, as electrophysiological evidence suggests that dopamine depletion increases glutamate release in the striatum under normal circumstances (Calabresi et al., 1993). The role of D1 receptors in mediating striatal ischemic injury is not known, but it has been suggested that D2 receptor antagonists may improve the functional recovery of striatal neurons exposed to ischemic conditions (Benefenati et al., 1989) and the cell death induced by forebrain ischemia (Hashimoto et al., 1994).

Other regions, such as the hippocampus, also receive dopaminergic input and are exquisitely sensitive to ischemia. The dorsal hippocampus receives dopaminergic innervation from the ventral tegmentum and the substania nigra (Scatton et al., 1980; Swanson, 1982), and dopamine levels in this region also increase during ischemic injury (Bhardwaj et al., 1990). The density of D2 receptors in the hippocampus is, however, substantially less than that found in the striatum, and D1-like receptors appear to play a more important role in mediating ischemic injury in this region. Indeed, D1 antagonists, but not D2 antagonists, attenuate ischemia-induced decreases in presynaptic potentials in hippocampal slices. Further, coadministration of the D1 receptor antagonist SCH 23390 with an NMDA antagonist provides better protection against some forms of hippocampal ischemic injury than is achieved by either compound alone (Globus et al., 1989).

In spite of these compelling findings, many of the strategies to rescue cells following ATP loss are focused largely on glutamate and its receptors. This is not to suggest that glutamate does not contribute to the death of striatal cells under many pathological conditions, but should serve as an example that multiple factors influence neuronal vulnerability following ischemia and hypoxia.

Methamphetamine Toxicity

METH is a CNS stimulant that is neurotoxic to monoaminergic systems and can cause decreases in dopamine and the loss of both presynaptic dopamine transporters and of tyrosine hydroxylase activity in regions including the striatum, cortex and the core of the nucleus accumbens (Bowyer et al., 1995; Hotchkiss and Gibb, 1980; Pu et al., 1993; Broening et al., 1997). It has been hypothesized that excessive dopamine release and subsequent generation of reactive oxygen species is responsible for METH toxicity. This theory is supported by work demonstrating that multiple injections of METH cause 5- to 36-fold increases in dopamine release in the striatum (O'Dell et al., 1991). Pharmacological blockade of the dopamine transporters blocks METH-induced degeneration suggesting that dopamine release and reuptake are critically involved in this neurotoxic cascade (O'Dell et. al., 1991; Stephans and Yamamoto, 1994). Indeed, mice lacking the dopamine transporter are protected against METH toxicity (Fumagalli et al., 1998). Further, inhibition of dopamine synthesis with α-methyl-*p*-tyrosine prevents METH neurotoxicity and enhancement of monoamine release with reserpine increases METH toxicity (Wagner et al., 1983; Axt et al., 1990). Once dopamine is released, the toxic process propagates by stimulation of dopamine receptors and it has been shown that coadministration of D2 receptor antagonists attenuates METH toxicity (Albers and Sosalla, 1995; Sonsalla et al., 1986). The ability of dopamine receptor antagonists to block toxicity is, however, complicated by the finding that many of these agents also prevent METH-induced hyperthermia (Albers and Sonsalla, 1995).

Another aspect of METH toxicity that may be attributable to increased dopamine release is the ability of METH to perturb oxidative phosphorylation and produce oxidative stress. The generation of reactive oxygen species has been observed following METH administration (Cubells et al., 1994; Hirata et al., 1995), and dopamine quinone formation is enhanced by METH, suggesting that modification of existing proteins by dopamine may also contribute to METH toxicity (LaVoie and Hastings, 1999). Cellular targets that are damaged by reactive oxygen generated by METH administration include the transporters for dopamine and serotonin (Kokoshka et al., 1998). The finding that the blockade of reuptake and metabolic systems for dopamine attenuates METH toxicity suggests that these effects may be related to excessive dopamine outflow or aberrant dopamine metabolism and sequestration rather than direct properties of METH. It is therefore not surprising that administration of antioxidants and upregulation of antioxidant proteins have both been efficacious in decreasing METH toxicity (DeVito and Wagner, 1989; Cadet et al., 1994). Further, coadministration of agents that can both increase oxidative phosphorylation and scavenge free radicals also attenuate METH toxicity (Stephans et al., 1998). The importance of impaired oxidative phosphorylation as a pathological mechanism for METH toxicity is supported by the finding that there are increased extracellular concentrations of lactate in both the striatum and prefrontal cortex following METH administration, suggesting that alterations in energy metabolism may also contribute to the observed toxicity (Stephans et al., 1998). Indeed, inhibition of glucose metabolism potentiates dopamine depletion in the striatum and the transient loss of ATP observed after repeated administration of METH (Chan et al., 1994), and combination of METH with the complex II inhibitors malonate or 3-nitropropionic acid potentiates neurochemical and pathological changes observed in the striatum (Albers et al., 1996; Bowyer et al., 1996; Reynolds et al., 1998).

It has, however, been shown that mice heterozygous for the vesicular monoamine oxidase transporter 2, which sequesters dopamine into synaptic vesicles, have less extracellular dopamine (DA) overflow and hydroxyl radical formation but greater DA and metabolite depletion and dopamine transporter activity following METH treatment compared to wild-type littermates (Fon et al., 1997; Wang et al., 1997; Fumagalli et al., 1999). This suggests that METH toxicity is more closely linked with failed dopamine compartmentalization and sequestration rather than the actual amount of dopamine that is released following METH administration. These mice will indubitably prove useful for determining the cellular and subcellular sites of dopamine neurotransmission that are important for neurotoxicity in other pathological events such as hypoxia and ischemia.

Like dopamine, methamphetamine toxicity in vitro can also be blocked by overexpression of the anti-apoptotic protein Bcl-2 (Cadet et al., 1997). METH induces appreciable DNA damage and cleavage of the DNA binding protein poly(ADP-ribose) polymerase (PARP). PARP is a DNA repair enzyme that is activated by caspases during apoptosis. This protein uses NAD^+ as a substrate and can deplete cellular ATP stores when activated (reviewed by Pieper et al., 1999). Indeed, inhibition of this protein with benzamine or nicotinamide decreases METH neurotoxicity in vivo and in vitro, suggesting that methamphetamine-induced DNA damage may produce an unrecoverable energetic stress (Cosi et al., 1996; Stephans et al., 1998; Sheng et al., 1996). Further evidence that DNA damage is an important signal for methamphetamine-induced degeneration can be found with studies using p53 knockout mice. p53 is a tumor-suppresser gene that is activated by DNA damage and can cause cell cycle arrest or apoptosis depending on the extent of the injury. In p53 null mice, many of the long-term anatomical and neurochemical changes commonly observed following METH treatment were not found (Hirata and Cadet, 1997). This suggests that these nuclear signals are critical for the neurotoxic cascade induced by METH and, possibly, dopamine itself.

Parkinson's Disease

Parkinson's disease (PD) is a chronic, progressive neurodegenerative disease characterized by loss of the dopaminergic projection neurons in the substantia nigra pars compacta. Like Alzheimer's disease (AD), PD can occur in either a familial or sporadic form. The mechanism of dopaminergic cell death in PD is unclear, but there is mounting evidence of significant oxidative stress in the substania nigra pars compacta of PD patients (reviewed by Cassarino and Bennett, 1999). The relative abundance of iron in the substania nigra of PD patients may facilitate reactive oxygen species (ROS) generation and has been implicated in the degenerative process (Sofic et al., 1988; Jenner, 1989). Indeed, dopamine neurotoxicity can be attenuated in vitro with iron chelators (Tanaka et al., 1991).

Energetic dysfunction has been implicated in contributing to the pathogenesis of PD and loss of mitochondrial complex I activity has been reported in PD patients (Schapira, 1994). Further, administration of the mitochondrial complex I inhibitor MPTP causes akinesia, bradykinesia, and loss of dopaminergic cell bodies in the substantia nigra in primates (reviewed by Bloem et al. 1990). The relatively selective degeneration induced by this compound has been attributed to the uptake of this toxin by the dopamine transporter and its intramitochondrial conversion to MPP^+. Another commonly used model of PD is intranigral injection of 6-OHDA, a hydroxylated derivative of dopamine. 6-OHDA is also taken up by high-affinity dopamine transporters, where it is auto-oxidized to produce hydrogen peroxide, superoxide, and hydroxyl radicals (Heikkila and Cohen, 1971). 6-OHDA is formed in patients with Parkinson's disease (reviewed by Glinka et al., 1997) and has been hypothesized to contribute to PD pathology (Jellinger et al., 1997). In cultures, 6-OHDA induces apoptosis that can be delayed, but not prevented, by the broad-spectrum caspase inhibitor zVAD or the more caspase 3 specific inhibitor Acetyl-L-Asp-Glu-Val-Asp Chloromethylketone (Ac-DEVD-CHO) (Dodel et al., 1999). DEVD is also neuroprotective in microglial cultures and PC12 cells exposed to 6-OHDA (Takai et al., 1998) suggesting that 6-OHDA induces apoptotic cell death in a variety of systems. This is supported by the finding that overexpression of Bcl-2 attenuates 6-OHDA toxicity in PC12 cells (Takai et al., 1998) although similar studies could not reproduce this effect in a dopaminergic cell line (Oh et al., 1995).

The commonly used strategy of dopamine replacement with levodopa (L-Dopa) in PD patients is problematic given that L-Dopa has been shown to be even more toxic than dopamine in vitro (Michel and Hefti, 1990; Nako et al., 1997). L-Dopa can form many deleterious oxidizing agents, quinones, semiquinones, and radicals (Wilkinshaw et al., 1995). In vivo studies suggest that L-Dopa is not toxic to normal animals at physiologically relevant doses. It is, however, toxic in animals subject to 6-OHDA lesions of the substantia nigra (Blunt et al., 1993; Ogawa et al., 1994a). Dopamine receptor agonists, particularly D2 family agonists, have proved to be a clinically useful means of decreasing L-Dopa requirements and relieving parkinsonian symptoms. The relative success of this strategy has led to the suggestion that these compounds may actually be providing neuroprotection against further disease-related degeneration. Indeed, the D2-like receptor agonists bromocriptine and pergolide are effective in blocking 6-OHDA-induced lipid peroxidation and nigral degeneration (Ogawa et al., 1994b; Yoshikawa et al., 1994). As dopamine release and metabolism generates a variety of toxic ROS, depressing dopaminergic neuronal firing through D2 autoreceptor activation may also decrease generation of these toxins. Alternatively, the neuroprotective action of these compounds may be the result of their endogenous antioxidant effects (Yoshikawa wet al., 1994; Ogawa et al., 1994b; Kondo et al., 1998), or by increasing the expression of antioxidant defenses such as Cu/Zn^{2+} superoxide dismutase (Clow et al., 1993). Carefully controlled studies of the efficacy of these agents as neuroprotective agents to block substania nigra cell death are ongoing, but initial studies sited by Olanow and co-workers (1998) revealed no change in PD progression as a result of dopamine agonist treatment.

Alzheimer's Disease

Alzheimer's disease (AD) is the most common neurodegenerative disease affecting 5–10% of the population over 65. AD can result from either a familial genetic mutation inherited in an autosomal-dominant manner or from an unidentified mutation(s) that results in the sporatic acquisition of the disease. The inherited forms of AD account for 5% of all AD cases and are caused by point mutations in either the amyloid precursor protein or the presenilin proteins. The pathological hallmarks of AD are the presence of intraneuronal fibrillary tangles and the secretion of the amyloidogenic peptides that form extracellular plaques. The neurotoxicity observed in AD can be modeled in vitro by treating cultures with Aβ peptide fragments (Fu et al., 1998 *see also*, Chapter 1). Coincubation of subtoxic concentrations of Aβ with catecholamines, including dopamine, potentiates the cell death observed in hippocampal cultures. This effect has been attributed to increased intracellular calcium through non-receptor-dependent mechanisms and generation of reactive oxygen species (Fu et al., 1998). Further, the observed neurotoxicity can be attenuated with antioxidants and is not present in cultures treated with other neurotransmitters such as serotonin and acetylcholine (Fu et al., 1998). It remains to be determined if physiological, rather than pathophysiological, concentrations of dopamine can also elicit the same accelerated Aβ toxicity and if this is relevant to neurotoxicity observed in AD.

Huntington's Disease

Huntington's disease (HD) is an autosomal dominantly inherited neurodegenerative disorder first described by George Huntington (Huntington, 1872). The most prominent neuropathological change in postmortem brains from patients affected with HD is atrophy of the striatum with accompanying ventricular enlargement (Vonsattal et al., 1985). The dense dopaminergic and glutamatergic innervation of the striatum converge on the dendrites of medium-sized spiny neurons (Smith and Bolam, 1990). These cells are GABAergic and comprise 95% of the neurons within the striatum, and their axons are the efferent projections of the striatum. These neurons are in a unique position to integrate information from cortical and subcortical structures and are also exquisitely sensitive to HD pathology.

Until the gene responsible for HD was characterized recently (*see* Chapter 14), the best available models for HD were based on energetic dysfunction and overactivation of the NMDA receptors. The finding that there is decreased glucose utilization and increased lactate generation in the brains of presymptomatic HD patients (Young et al., 1986; Jenkins et al., 1993; Browne et al., 1997) supported the role of mitochondrial dysfunction in HD pathology. Further, it was found that otherwise normal individuals who accidentally ingested the mitochondrial poison 3-nitropropionic acid presented with a strikingly similar pathological and motor phenotype to that observed in HD patients (Ludolph et al., 1991). These observations suggested that a primary defect in energy metabolism could result in an inability to maintain ion homeostasis and a secondary excitotoxic cell death when striatal NMDA receptors are stimulated by glutamate released from cortical projection neurons (Albin and Greenamyre, 1992; Beal, 1992). This excitotoxic model would, however, be more compelling if striatal cells were particularly vulnerable to mitochondrial toxins, which is not the case (McLaughlin et al., 1998b) or if the striatum received the densest glutamatergic innervation in the brain, which is also not true.

Although glutamatergic innervation or inherent deficiencies in oxidative phosphorylation of striatal cells do not appear to provide unique vulnerability to the striatum, the striatum is unique in that it receives the densest dopaminergic input of any brain region. Indeed, dopamine and agents that increase dopamine release potentiate the toxicity of mitochondrial toxins such as 3-NPA both in vitro and in vivo (McLaughlin et al., 1998a; Bowyer et al., 1996; Bazzett et al., 1996; Reynolds et al., 1998). This effect is, in part, mediated by stimulation of D1 receptors in cultures (McLaughlin et al., 1998a) and can be prevented by prior lesioning of dopaminergic projections to the striatum in vivo (Maragos et al., 1998; Reynolds et al., 1998). These studies suggest that dopamine release may enhance striatal cell death in HD, although D2 receptor antagonists have not, however, been successful in halting the long-term progression of the disease.

Animal models of HD with mutations in the huntingtin gene can partially recapitulate the behavioral and pathological hallmarks of this disease (*see* Chapter 14). In cell cultures, full-length or partial constructs of genes from the trinucleotide disease family (which includes HD) induce protein aggregation similar to that observed in HD and apoptotic cell death (Ikeda et al., 1996) (*see* Chapter 14). It remains to be determined what subcellular distribution of the huntingtin protein aggregates is required for induction of cell death, why this ubiquitously expressed protein causes selectively cell degeneration, and how dopamine, glutamate, and energetic dysfunction contribute to HD pathology.

9.6. SUMMARY

In recent years, it has become increasingly clear that dopamine is critically involved in a number of physiologically important activities such as normal synaptic function, development, synaptic plasticity, cognition, emotion, affect, motivation and movement. The complexity of dopaminergic transmission is not simply related to subtypes of receptors but is also dependent on individual cell energetic status, receptor profile, signal transduction pathways, and a plethora of potentially toxic oxidative and enzymatic by-products derived from dopamine. This complexity is particularly relevant to a number of neurological conditions in which dopamine can promote cell death when released in abundance or when improperly trafficked. A greater understanding of the transcriptional, translational, and signal transduction pathways activated by cell stressors such as dopamine will indubitably allow researchers to develop more sophisticated strategies to assess and prevent neurodegeneration.

ACKNOWLEDGMENTS

The author wishes to express her gratitude to Drs. Gregg D. Stanwood and Elias Aizenman for their helpful comments and suggestions while assembling this chapter. I am also appreciative of the outstanding secretarial services provided by Ms. Lucy Tran.

REFERENCES

Adayev, T., El-Sherif, Y., Barua, M., Penington, N. J., and Banerjee, P. (1999) Agonist stimulation of the serotonin1A receptor causes suppression of anoxia-induced apoptosis via mitogen-activated protein kinase in neuronal HN2-5 cells. *J. Neurochem.* **72,** 1489–1496.

Alagarsamy, S., Phillips, M., Pappas, T., and Johnson, K. M. (1997) Dopamine neurotoxicity in cortical neurons. *Drug Alcohol Depend.* **48,** 105–111.

Albers, D. S. and Sonsalla, P. K. (1995) Methamphetamine-induced hyperthermia and dopaminergic neurotoxicity in mice: pharmacological profile of protective and nonprotective agents. *J. Pharmacol. Exp. Ther.* **275,** 1104–1114.

Albers, D. S., Zeevalk, G. D., and Sonsalla, P. K. (1996) Damage to dopaminergic nerve terminals in mice by combined treatment of intrastriatal malonate with systemic methamphetamine or MPTP. *Brain Res.* **718,** 217–220.

Albin, R. L. and Greenamyre, J. T. (1992) Alternative excitotoxic hypotheses. *Neurology* **42,** 733–738.

Alston T. A., Mela, L., and Bright, H. J. (1977) Nitropropionate, the toxic substance of *Indigofera,* is a suicide inactivator of succinate dehydrogenase. *Proc. Natl. Acad. Sci. USA* **74,** 3767–3771.

Anden, N. E., Dahlstrom, A., Fuxe, K., and Larsson, K. (1965) Mapping out of catecholamine and 5-hydroxytryptamine neurons innervating the telencephalon and diencephalon. *Life Sci.* **4,** 1275–1279.

Axt, K. J., Commins, D. L., Vosmer, G., and Seiden, L. S. (1990) α-Methyl-*p*-tyrosine pretreatment partially prevents methamphetamine-induced endogenous neurotoxin formation. *Brain Res.* **515,** 269–276.

Baez, S., Segura-Aguilar, J., Widersten, M., Johansson, A. S., and Mannervik, B. (1997) Glutathione transferases catalyse the detoxication of oxidized metabolites (*o*-quinones) of catecholamines and may serve as an antioxidant system preventing degenerative cellular processes. *Biochem. J.* **324,** 25–28.

Baker, A. J., Zornow, M. H., Scheller, M. S., Yaksh, T. L., Skilling, S. R., Smullin, D. H., et al. (1991) Changes in extracellular concentrations of glutamate, aspartate, glycine, dopamine, serotonin, and dopamine metabolites after transient global ischemia in the rabbit brain. *J. Neurochem.* **57,** 1370–1379.

Bazzett, T. J., Falik, R. C., Becker, J. B., and Albin, R. L. (1996) Synergistic effects of chronic exposure to subthreshold concentrations of quinolinic acid and malonate in the rat striatum. *Brain Res.* **718,** 228–232.

Beal, M. F. (1992) Does impairment of energy metabolism result in excitotoxic neuronal death in neurodegenerative illnesses? *Ann. Neurol.* **31,** 119–130.

Beal, M. F., Brouillet, E., Jenkins, B., Henshaw, R., Rosen, B., and Hyman, B. T. (1993a) Age-dependent striatal excitotoxic lesions produced by the endogenous mitochondrial inhibitor malonate. *J. Neurochem.* **61,** 1147–1150.

Beal, M. F., Brouillet, E., Jenkins, B. G., Ferrante, R. J., Kowall, N. W., Miller, J. M., et al. (1993b) Neurochemical and histologic characterization of striatal excitotoxic lesions produced by the mitochondrial toxin 3-nitropropionic acid. *J. Neurosci.* **13,** 4181–4192.

Ben-Shachar, D., Zuk, R., and Glinka, Y. (1995) Dopamine neurotoxicity: inhibition of mitochondrial respiration. *J. Neurochem.* **64,** 718–723.

Benfenati, F., Pich, E. M., Grimaldi, R., Zoli, M., Fuxe, K., Toffano, G., et al. (1989) Transient forebrain ischemia produces multiple deficits in dopamine D_1 transmission in the lateral neostriatum of the rat. *Brain Res.* **498,** 376–380.

Berman, S. B., Zigmond, M. J., and Hastings, T. G. (1996) Modification of dopamine transporter function: effect of reactive oxygen species and dopamine. *J. Neurochem.* **67,** 593–600.

Berman, S. B. and Hastings, T. G. (1997) Inhibition of glutamate transport in synaptosomes by dopamine oxidation and reactive oxygen species. *J. Neurochem.* **69,** 1185–1195.

Berridge, M. V. and Tan, A. S. (1993) Chartacterization of the cellular reduction of 3-(4,5-dimethylthiazol-2-yl)-2,5-diphenyltetrazolium bromide (MTT): subcellular localization, substrate dependence, and involvement of mitochondrial electron transport in MTT reduction. *Arch. Biochem. Biophys.* **303,** 474–482.

Bhardwaj, A., Brannan, T., Martinez-Tica, J., and Weinberger, J. (1990) Ischemia in the dorsal hippocampus is associated with acute extracellular release of dopamine and norepinephrine. *J. Neural Transm.* **80,** 195–201.

Bhide, P. G., Day, M., Sapp, E., Schwarz, C., Sheth, A., Kim, J., et al. (1996) Expression of normal and mutant huntingtin in the developing brain. *J. Neurosci.* **16,** 5523–5535.

Bloem, B. R., Irwin, I., Buruma, O. J., Haan, J., Roos, R. A., Tetrud, J. W. et al. (1990) The MPTP model: versatile contributions to the treatment of idiopathic Parkinson's disease. *J. Neurol. Sci.* **97,** 273–293.

Blunt, S. B., Jenner, P., and Marsden, C. D. (1993) Suppressive effect of λ-DOPA on dopamine cells remaining in the ventral tegmental area of rats previously exposed to the neurotoxin 6-hydroxydopamine. *Mov. Dis.* **8,** 129–133.

Bossy-Wetzel, E., Bakiri, L., and Yaniv, M. (1997) Induction of apoptosis by the transcription factor c-Jun. *EMBO J.* **16,** 1695–1709.

Bowling, A. C. and Beal, M. F. (1995) Bioenergetic and oxidative stress in neurodegenerative diseases. *Life Sci.* **56,** 1151–1171.

Bowyer, J. F., Clausing, P., Gough, B., Slikker, W., Jr., and Holson, R. R. (1995) Nitric oxide regulation of methamphetamine-induced dopamine release in caudate/putamen. *Brain Res.* **699,** 62–70.

Bowyer, J. F., Clausing, P., Schmued, L., Davies, D. L., Binienda, Z., Newport, G. D., et al. (1996) Parenterally administered 3-nitropropionic acid and amphetamine can combine to produce damage to terminals and cell bodies in the striatum. *Brain Research.* **712,** 221–229.

Boyson, S. J., McGonigle, P., and Molinoff, P. B. (1986) Quantitative autoradiographic localization of the D_1 and D_2 subtypes of dopamine receptors in rat brain. *J. Neurosci.* **6,** 3177–3188.

Broening, H. W., Cunfeng, P., and Vorhees, C. V. (1997) Methamphetamine selectively damages dopaminergic innervation to the nucleus accumbens core while sparing the shell. *Synapse.* **27,** 153–160.

Browne, S. E., Bowling, A. C., MacGarvey, U., Baik, M. J., Berger, S. C., Muqit, M. M., et al. (1997) Oxidative damage and metabolic dysfunction in Huntington's disease: selective vulnerability of the basal ganglia. *Ann. Neurol.* **41,** 646–653.

Buckley, C. D., Pilling, D., Henriquez, N. V., Parsonage, G., Threlfall, K., Scheel-Toellner, D., et al. (1999) RGD peptides induce apoptosis by direct caspase-3 activation. *Nature* **397,** 534–539.

Buisson, A., Pateau, V., Plotkine, M., and Boulu, R. G. (1991) Nigrostriatal pathway modulates striatum vulnerability to quinolinic acid. *Neurosci. Lett.* **131,** 257–259.

Buisson, A., Callebert, J., Mathieu, E., Plotkine, M., and Boulu, R. G. (1992) Striatal protection induced by lesioning the substantia nigra of rats subjected to focal ischemia. *J. Neurochem.* **59,** 1153–1157.

Cadet, J. L., Ordonez, S. V., and Ordonez, J. V. (1997) Methamphetamine induces apoptosis in immortalized neural cells: protection by the proto-oncogene, bcl-2. *Synapse.* **25,** 176–184.

Calabresi, P., Mercuri, N. B., Sancesario, G., and Bernardi, G. (1993) Electrophysiology of dopamine-denervated striatal neurons: implications for Parkinson's disease. *Brain* **116,** 433–452.

Cao, X. and Phillis, J. W. (1994) Alpha-phenyl-tert-butyl-nitrone reduces cortical infarct and edema in rats subjected to focal ischemia. *Brain Res.* **644,** 267–272.

Carlsson, A. (1959) The occurrence, distribution and physiological role of catecholamines in the nervous system. *Pharmacol. Rev.* **11,** 490–493.

Carlsson, A., Lindqvist, M., Dahlstrom, A., Fuxe, K., and Masuoka, D. (1965) Effects of the amphetamine group on intraneuronal brain amines in vivo and in vitro. *J. Pharm. Pharmacol.* **17,** 521–523.

Carlsson, A. (1988) The current status of the dopamine hypothesis of schizophrenia. *Neuropsychopharmacology* **1,** 179–186.

Cassarino, D. S. and Bennett, J. P., Jr. (1999) An evaluation of the role of mitochondria in neurodegenerative diseases: mitochondrial mutations and oxidative pathology, protective nuclear responses, and cell death in neurodegeneration. *Brain Res. Brain Res. Rev.* **29,** 1–25.

Cepeda, C., Colwell, C. S., Itri, J. N., Gruen, E., and Levine, M. S. (1998) Dopaminergic modulation of early signs of excitotoxicity in visualized rat neostriatal neurons. *Eur. J. Neurosci.* **10,** 3491–3497.

Cepeda, C., Buchwald, N. A., and Levine, M. S. (1993) Neuromodulatory actions of dopamine in the neostriatum are dependent upon the excitatory amino acid receptor subtypes activated. *Proce. Natl. Acad. Sci. USA* **90,** 9576–9580.

Chakraborti, S. and Chakraborti, T. (1998) Oxidant-mediated activation of mitogen-activated protein kinases and nuclear transcription factors in the cardiovascular system: a brief overview. *Cell. Signal.* **10,** 675–683.

Chan, P., DiMonte, D. A., Luo, J.-J., DeLanney, L. E., Irwin, I., and Langston, J. W. (1994) Rapid ATP loss caused by methamphetamine in mouse striatum: relationship between energy impairment and dopaminergic neurotoxicity. *J. Neurochem.* **62,** 2484–2487.

Chapman, A. G., Durmuller, N., Lees, G. J., and Meldrum, B. S. (1989) Excitotoxicity of NMDA and kainic acid is modulated by nigra-striatal dopamine fibers. *Neurosci. Lett.* **107,** 256–260.

Cheng, B. Y., Origitano, T. C., and Collins, M. A. (1987) Inhibition of catechol-*O*-methyltransferase by 6,7-dihydroxy-3,4-dihydroisoquinolines related to dopamine: demonstration using liquid chromatography and a novel substrate for O-methylation. *J. Neurochem.* **48,** 779–786.

Cheng, N., Maeda, T., Kume, T., Kaneko, S., Kochiyama, H., Akaike, A., et al. (1996) Differential neurotoxicity induced by L-DOPA and dopamine in cultured striatal neurons. *Brain Res.* **743,** 278–283.

Chesselet, M. F., Gonzales, C., Lin, C. S., Polsky, K., and Jin, B. K. (1990) Ischemic damage in the striatum of adult gerbils: relative sparing of somatostatin and cholinergic interneurons contrasts with loss of efferent neurons. *Exp. Neurol.* **110,** 209–218.

Chiueh, C. C., Miyake, J., and Peng, M. T. (1993) Role of dopamine autooxidation, hydroxyl radical generation and calcium overload in underlying mechanisms involved in MPTP induced Parkinsonism. *Adv. Neurol.* **60,** 251–258.

Clemens, J. A. and Phebus, L. A. (1988) Dopamine depletion protects striatal neurons from ischemia-induced cell death. *Life Sci.* **42,** 707–713.

Clemens, J. A. and Panetta, J. A. (1994) Neuroprotection by antioxidants in models of global and focal ischemia. *Ann. NY Acad. Sci.* **738,** 250–256.

Clow, A., Freestone, C., Lewis, E., Dexter, D., Sandler, M., and Glover, V. (1993) The effect of pergolide and MDL 72974 on rat brain CuZn superoxide dismutase. *Neurosci. Lett.* **164,** 41–43.

Cosi, C., Chopin, P., and Marien, M. (1996) Benzamide, an inhibitor of poly(ADP-ribose) polymerase, attenuates methamphetamine-induced dopamine neurotoxicity in the C57B1/6N mouse. *Brain Res.* **735,** 343–348.

Cubells, J. F., Rayport, S., Rajendran, G., and Sulzer, D. (1994) Methamphetamine neurotoxicity involves vacuolation of endocytic organelles and dopamine-dependent intracellular oxidative stress. *J. Neurosci.* **14,** 2260–2271.

Dahlstrom, A. (1971) Regional distribution of brain catecholamines and serotonin. *Neurosci. Res. Prog. Bull.* **9,** 197–205.

Damsma, G., Boisvert, D. P., Mudrick, L. A., Wenkstern, D., and Fibiger, H. C. (1990) Effects of transient forebrain ischemia and pargyline on extracellular concentrations of dopamine, serotonin, and their metabolites in the rat striatum as determined by in vitro microdialysis. *J. Neurochem.* **54,** 801–808.

DeVito, M. J. and Wagner, G. C. (1989) Methamphetamine-induced neuronal damage: a possible role for free radicals. *Neuropharmocology.* **28,** 1145–1150.

Dodel, R. C., Du, Y., Bales, K. R., Ling, Z., Carvey, P. M., and Paul, S. M. (1999) Caspase-3-like proteases and 6-hydroxydopamine induced neuronal cell death. *Brain Res. Mol. Brain Res.* **64,** 141–148.

Drago, J., Padungchaichot, P., Accili, D., and Fuchs, S. (1998) Dopamine receptors and dopamine transporter in brain function and addictive behaviors: insights from targeted mouse mutants. *Dev. Neurosci.* **20,** 188–203.

Dutra, J. C., Dutra-Filho, C. S., Cardozo, S. E. C., Wannmacheer, C. M. D., Sarkis, J. J. F., and Wajner, M. (1993) Inhibition of succinate dehydrogenase and β-hydroxybutyrate dehydrogenase activities by methylmalonate in brain and liver of developing rats. *J. Inherited Metab. Dis.* **16**, 147-153.

Einhorn, L. C., Gregerson, K.A., and Oxford, G. S. (1991) D2 dopamine receptor activation of potassium channels in identified rat lactotrophs: whole-cell and single-channel recording. *J. Neurosci.* **11,** 3727–3737.

Erecinska, M. and Silver, I. A. (1989) ATP and brain function. *J. Cerebr. Blood Flow Metab.* **9,** 2–19.

Fallon, J. H. and Moore, R. Y. (1978) Catecholamine innervation of the basal forebrain. IV. Topography of the dopamine projection of the basal forebrain and neostriatum. *J. Comp. Neurol.* **180,** 545–580.

Faure, M., Voyno-Yasenetskaya, T. A., and Bourne, H. R. (1994) cAMP and beta gamma subunits of heterotrimeric G proteins stimulate the mitogen-activated protein kinase pathway in COS-7 cells. *J. Biol. Chem.* **269,** 7851–7854.

Filloux, F. and Townsend, J. J. (1993) Pre- and postsynaptic neurotoxic effects of dopamine demonstrated by intrastriatal injection. *Exp. Neurol.* **119,** 79–88.

Fink, S. L., Ho, D. Y., and Sapolsky, R. M. (1996) Energy and glutamate dependency of 3-nitropropionic acid neurotoxicity in culture. *Exp. Neurol.* **138,** 298–304.

Fon, E. A., Pothos, E. N., Sun, B. C., Killeen, N., Sulzer, D., and Edwards, R. H. (1997) Vesicular transport regulates monoamine storage and release but is not essential for amphetamine action. *Neuron* **19,** 1271–1283.

Fornai, F., Giorgi, F. S., Alessandri, M. G., Giusiani, M., and Corsini, G. U. (1999) Effects of pretreatment with *N*-(2-chloroethyl)-*N*-ethyl-2-bromobenzylamine (DSP-4) on methamphetamine pharmacokinetics and striatal dopamine losses. *J. Neurochem.* **72,** 777–784.

Fu, W., Luo, H., Parthasarathy, S., and Mattson, M. P. (1998) Catecholamines potentiate amyloid beta-peptide neurotoxicity: involvement of oxidative stress, mitochondrial dysfunction, and perturbed calcium homeostasis. *Neurobiol. Dis.* **5,** 229–243.

Fumagalli, F., Gainetdinov, R. R., Valenzano, K. J., and Caron, M. G. (1998) Role of dopamine transporter in methamphetamine-induced neurotoxicity: evidence from mice lacking the transporter. *J. Neurosci.* **18,** 4861–4869.

Fumagalli, F., Gainetdinov, R. R., Wang, Y.-M., Valenzano, K. J., Miller, G. W., and Caron, M. G. (1999) Increased methamphetamine neurotoxicity in heterozygous vesicular monoamine transporter 2 knockout mice. *J. Neurosci.* **19,** 2424–2431.

Fusco, F. R., Chen, Q., Lamoreaux, W. J., Figueredo-Cardenas, G., Jiao, Y., Coffman, J. A., et al. (1999) Cellular localization of huntingtin in striatal and cortical neurons in rats: lack of correlation with neuronal vulnerability in Huntington's disease. *J. Neurosci.* **19,** 1189–1202.

Gabby, M., Tauber, M., Porat, S., and Simantov, R. (1996) Selective role of glutathione in protecting human neuronal cells from dopamine-induced apoptosis. *Neuropharmacology* **35,** 571–578.

Giorguieff, M. F., Kemel, M. L., and Glowinski, J. (1977) Presynaptic effects of L-glutamic acid on dopamine release in striatal slices. *Neurosci. Lett.* **6,** 77–78.

Gepdiremen, A., Sonmez, S., Kiziltunc, A., Ikbal, M., Erman, F., and Duzenli, S. (1998) Effects of norepinephrine on NMDA-induced neurotoxicity in cerebellar granular cell culture of rat pups. *Fundam. Clin. Pharm.* **12,** 517–520.

Gerfen, C. R., Keefe, K. A., and Gauda, E. B. (1995) D1 and D2 dopamine receptor function in the striatum: coactivation of D1- and D2-dopamine receptors on separate populations of neurons results in potentiated immediate early gene response in D1-containing neurons. *J. Neurosci.* **15,** 8167–8176.

Glinka, Y., Gassen, M., and Youdim, M. B. (1997) Mechanism of 6-hydroxydopamine neurotoxicity. *J. Neural Transm.* **50(Suppl.),** 55–66.

Globus, M. Y.-T., Ginsberg, M. D., Dietrich, W. D., Busto, R., and Scheinberg, P. (1987) Substantia nigra lesion protects agains ischemic damage in the striatum. *Neurosci. Lett.* **80,** 251–256.

Globus, M. Y.-T., Busto, R., Dietrich, W. D., Martinez, E., Valdes, I., and Ginsberg, M. D. (1988) Effect of ischemia on the in vivo release of striatal dopamine, glutamate, and γ-aminobutyric acid studied by intracerebral microdialysis. *J. Neurochem.* **51,** 1455–1464.

Globus, M. Y.-T., Dietrich, W. D., Busto, R., Valdes, I., and Ginsberg, M. D. (1989) The combined treatment with a dopamine D_1 antagonist (SCH-23390) and NMDA receptor blocker (MK-801) dramatically protects against ischemia-induced hippocampal damage. *J. Cereb. Blood Flow Metab.* **9(Suppl. 1),** S5.

Godukhin, O. V., Zharikova, A. D., and Yu, A. B. (1984) Role of presynaptic dopamine receptors in regulation of the gulatmatergic neurotransmission in rat neostriatum. *Neuroscience* **12,** 377–383.

Graham, D. G., Tiffany, S. M., Bell, W. R., and Gutnecht, W. F. (1978) Autooxidation versus covalent binding of quinones as the mechanism of toxic-

ity of dopamine, 6-hydroxydopamine, and related compounds toward c1300 neuroblastoma cells in vitro. *Mol. Pharmacol.* **14,** 644–653.
Graham, D. G. (1984) Catecholamine toxicity: a proposal for the molecular pathogenesis of manganese neurotoxicity and Parkinson's disease. *Neurotoxicology* **5,** 83–96.
Graybiel, A. M. and Ragsdale, C. W. (1978) Histochemically distinct compartments in the striatum of human, monkey and cat demostrated by acetylcholinesterase staining. *Proc. Natl. Acad. Sci. USA* **75,** 5723–5726.
Greene, J. G., Porter, R. H., Eller, R. V., and Greenamyre, J. T. (1993) Inhibition of succinate dehydrogenase by malonic acid produces an "excitotoxic" lesion in rat striatum. *J. Neurochem.* **61,** 1151–1154.
Greene, J. G. and Greenamyre, J. T. (1995a) Characterization of the excitotoxic potential of the reversible succinate dehydrogenase inhibitor malonate. *J. Neurochem.* **64,** 430–436.
Greene, J. G. and Greenamyre, J. T. (1995b) Exacerbation of NMDA, AMPA, and L-glutamate excitotoxicity by the succinate dehydrogenase inhibitor malonate. *J. Neurochem.* **64,** 2332–2338.
Greene, J. G., Porter, R. H., and Greenamyre, J. T. (1996) ARL-15896, a novel *N*-methyl-D-aspartate receptor ion channel antagonist: neuroprotection against mitochondrial metabolic toxicity and regional pharmacology. *Exp. Neurol.* **137,** 66–72.
Halliwell, B. (1991) Reactive oxygen species in living systems: source, biochemistry, and role in human disease. *Am. J. Med.* **91,** 14s–22s.
Ham, J., Babij, C., Whitfield, J., Pfarr, C. M., Lallemand, D., Yaniv, M., et al. (1995) A c-Jun dominant negative mutant protects sympathetic neurons against programmed cell death. *Neuron* **14,** 927–939.
Harada, H., Becknell, B., Wilm, M., Mann, M., Huang, L. J., Taylor, S. S., et al. (1999) Phosphorylation and inactivation of BAD by mitochondria-anchored protein kinase A. *Mol. Cell.* **3,** 413–422.
Hartzell, H. C., Mery, P. F., Fischmeister, R., and Szabo, G. (1991) Sympathetic regulation of cardiac calcium current is due exclusively to cAMP-dependent phosphorylation. *Nature.* **351,** 573–576.
Heikkila, H. and Cohen, G. (1971) Inhibition of biogenic amine uptake by hydrogen peroxide: a mechanism for toxic effects of 6-hydroxydopamine. *Science* **172,** 1257–1258.
Hashimoto, N., Matsumoto, T., Mabe, H., Hashitani, T., and Nishino, H. (1994) Dopamine has inhibitory and accelerating effects on ischemia-induced neuronal cell damage in the rat striatum. *Brain Res. Bull.* **33,** 281–288.
Hastings, T. G., Lewis, D. A., and Zigmond, M. J. (1996) Role of oxidation in the neurotoxic effects of intrastriatal dopamine injections. *Proc. Natl. Acad. Sci. USA.* **93,** 1956–1961.
Hastings, T. G. and Zigmond, M. J. (1994) Identification of catechol-protein conjugates in neostriatal slices incubated with [3H] dopamine: impact of ascorbic acid and glutathione. *J. Neurochem.* **63,** 1126–1132.
Hemmings, H. C., Jr., Nairn, A. C., Elliott, J. I., and Greengard, P. (1990) Synthetic peptide analogs of DARPP-32 (Mr 32,000 dopamine- and cAMP-regulated

phosphoprotein), an inhibitor of protein phosphatase-1. Phosphorylation, dephosphorylation, and inhibitory activity. *J. Biol. Chem.* **265,** 20369–20376.

Hernandez-Lopez, S., Bargas, J., Surmeier, D. J., Reyes, A., and Galarraga, E. (1997) D1 receptor activation enhances evoked discharge in neostriatal medium spiny neurons by modulating an L-type Ca2+ conductance. *J. Neurosci.* **17,** 3334–3342.

Hirata, H. and Cadet, J. L. (1997) p53-knockout mice are protected against the long-term effects of methamphetamine on dopaminergic terminals and cell bodies. *J. Neurochem.* **69,** 780–790.

Hirata, H., Ladenheim, B., Rothman, R. B., Epstein, C., and Cadet, J. L. (1995) Methamphetamine-induced serotonin neurotoxicity is mediated by superoxide radicals. *Brain Res.* **677,** 345–347.

Hochman, A., Sternin, H., Gorodin, S., Korsmeyer, S., Ziv, I., Melamed, E., et al. (1998) Enhanced oxidative stress and altered antioxidants in brains of Bcl-2-deficient mice. *J. Neurochem.* **71,** 741–748.

Hotchkiss, A. and Gibb, J. W. (1980) Long-term effects of multiple doses of methamphetamine on tryptophan hydroxyalse and tyrosine hydroxylase activity in rat brain. *J. Pharmacol. Exp. Ther.* **214,** 257–262.

Hoyt, K. R., Reynolds, I. J., and Hastings, T. G. (1997) Mechanisms of dopamine-induced cell death in cultured rat forebrain neurons: interactions with and differences from glutamate-induced cell death. *Exp. Neurol.* **143,** 269–281.

Huntington, G. (1872) On chorea, *Med. Surg. Reporter.* **26,** 320–321.

Ikeda, H., Yamaguchi, M., Sugai, S., Aze, Y., Narumiya, S., and Kakizuka, A. (1996) Expanded polyglutamine in the Machado-Joseph disease protein induces cell death in vitro and in vivo. *Nat. Genet.* **13,** 196–202.

Ishida, Y., Todaka, K., Kuwahara, I., Ishizuka, Y., Hashiguchi H., Nishimori, T., et al. (1998) Methamphetamine induces Fos expression in the striatum and the substantia nigra pars reticulata in a rat model of Parkison's disease. *Brain Res.* **809,** 107–114.

Jellinger, K. A. (1997) Morphological substrates of dementia in parkinsonism. A critical update. *J. Neural Transm.* **51(Suppl.),** 57–82.

Jenkins, B., Koroshetz, W., Beal, M. F., and Rosen, B. (1993) Evidence for an energy metabolism defect in Huntington's disease using localized proton spectroscopy. *Neurology* **43,** 2689–2695.

Jenner, P. (1989) Clues to the mechanism underlying dopamine cell death in Parkinson's disease. *J. Neurol. Neurosurg. Psychiatry* **Suppl l**, 22–28.

Jenner, P. (1996) Oxidative stress in Parkinson's disease and other neurodegenerative disorders. *Pathol. Biol.* **44,** 57–64.

Jequier, E., Lovenberg, W., and Sjoerdsma, A. (1967) Tryptophan hydroxylase inhibition: the mechanism by which *p*-chlorophenylalanine depletes rat brain serotonin. *Mol. Pharmacol.* **3,** 274–278.

Jung, A. B. and Bennett, J. P., Jr. (1996) Development of striatal dopaminergic function. II: Dopaminergic regulation of transcription of the immediate early gene zif268 and of D1 (D1a) and D2 (D2a) receptors during pre- and postnatal development. *Brain Res. Deve. Brain Res.* **94,** 121–132.

Kang, C. D., Jang, J. H., Kim, K. W., Lee, H. J., Jeong, C. S., Kim, C. M., et al. (1998) Activation of c-jun N-terminal kinase/stress-activated protein kinase and the decreased ratio of Bcl-2 to Bax are associated with the auto-oxidized dopamine-induced apoptosis in PC12 cells. *Neurosci. Lett.* **256,** 37–40.

Kanterman, R. Y., Mahan, L. C., Briley, E. M., Monsma, F. J., Sibley, D. R., Axelrod, J., et al. (1991) Transfected D2 dopamine receptors mediate the potentiation of arachidonic acid release in Chinese hamster ovary cells. *Mol. Pharm.* **39,** 364–369.

Karin, M. (1998) Mitogen-activated protein kinase cascades as regulators of stress responses. *Ann. NY Acad. Sci.* **851,** 139–146.

Kawasaki, H., Morooka, T., Shimohama, S., Kimura, J., Hirano, T., Gotoh, Y., et al. (1997) Activation and involvement of p38 mitogen-activated protein kinase in glutamate-induced apoptosis in rat cerebellar granule cells. *J. Biol. Chem.* **272,** 18,518–18,521.

Keefe, K. A. and Gerfen, C. R. (1995) D1-D2 dopamine receptor synergy in striatum: effects of intrastriatal infusions of dopamine agonists and antagonists on immediate early gene expression. *Neuroscience* **66,** 903–913.

Keefe, K. A. and Gerfen, C. R. (1996) D1 dopamine receptor-mediated induction of zif268 and c-fos in the dopamine-depleted striatum: differential regulation and independence from NMDA receptors. *J. Comp. Neurol.* **367,** 165–176.

Kelley, A. E., Delfs, J. M., and Chu, B. (1990) Neurotoxicity induced by the D-1 agonist SKF 38393 following microinjection into rat brain. *Brain Res.* **532,** 342–346.

Kerkerian, L., Dusticier, N., and Nieoullon, A. (1987) Modulatory effect of dopamine on high-affinity glutamate uptake in the rat striatum. *J. Neurochem.* **48,** 1301–1306.

Klawans, H. L., Jr. (1973) The pharmacology of extrapyramidal movement disorders. *Mono. Neural Sci.* **2,** 1–136.

Kokoshka, J. M., Metzger, R. R., Wilkins, D. G., Gibb, J. W., Hanson, G. R., and Fleckenstein, A. E. (1998) Methamphetamine treatment rapidly inhibits serotonin, but not glutamate, transporters in rat brain. *Brain Res.* **799,** 78–83.

Kondo, T., Shimada, H., Hatori, K., Sugita, Y., and Mizuno, Y. (1998) Talipexole protects dopaminergic neurons from methamphetamine toxicity in C57BL/6N mouse. *Neurosci. Lett.* **247,** 143–146.

Konradi, C., Leveque, J. C., and Hyman, S. E. (1996) Amphetamine and dopamine-induced immediate early gene expression in striatal neurons depends on postsynaptic NMDA receptors and calcium. *J. Neurosci.* **16,** 4231–4239.

Kuhn, D. M. and Arthur, R., Jr. (1998) Dopamine inactivates tryptophan hydroxylase and forms a redox-cycling quinoprotein: possible endogenous toxin to serotonin neurons. *J. Neurosci.* **18,** 7111–7117.

LaVoie, M. J. and Hastings, T. G. (1999) Dopamine quinone formation and protein modification associated with the striatal neurotoxicity of methamphetamine: evidence against a role for extracellular dopamine. *J. Neurosci.* **19,** 1484–1491.

Lin, M. T. (1997) Heatstroke-induced cerebral ischemia and neuronal damage. Involvement of cytokines and monoamines. *Ann. NY Acad. Sci.* **813,** 572–580.

Liu, Y. F., Civelli, O., Zhou, Q. Y., and Albert, P. R. (1992) Cholera toxin-sensitive 3',5'-cyclic adenosine monophosphate and calcium signals of the human dopamine-D1 receptor: selective potentiation by protein kinase A. *Mol. Endocrinol.* **6,** 1815–1824.

Lledo, P. M., Homburger, V., Bockaert, J., and Vincent, J. D. (1992) Differential G protein-mediated coupling of D2 dopamine receptors to K^+ and $Ca2^+$ currents in rat anterior pituitary cells. *Neuron* **8,** 455–463.

Ludolph, A. C., He, F., Spencer, P. S., Hammerstad, J., and Sabri, M. 3-Nitropropionic acid-exogenous animal neurotoxin and possible human striatal toxin (1991) *Canadian J. Neurological Sci.* **18,** 492–498.

Ludolph, A. C., Seelig, M., Ludolph, A., Novitt, P., Allen, C. N., Spencer, P. S., et al. (1992) 3-Nitropropionic acid decreases cellular energy levels and causes neuronal degeneration in cortical explants. *Neurodegeneration* **1,** 155–161.

Luo, Y., Kokkonen, G. C., Wang, X., Neve, K. A., and Roth, G. S. (1998) D2 dopamine receptors stimulate mitogenesis through pertussis toxin-sensitive G proteins and Ras-involved ERK and SAP/JNK pathways in rat C6-D2L glioma cells. *J. Neurochem.* **71,** 980–990.

Luo, Y., Umegaki, H., Wang, X., Abe, R., and Roth, G. (1998) Dopamine induces apoptosis through an oxidation-involved SAPK/JNK activation pathway. *J. Biol. Chem.* **273,** 3756.

Maker, H. S., Weiss, C., and Brannan, T. S. (1986) Amine-mediated toxicity. The effects of dopamine, norepinephrine, 5-hydroxytryptamine, 6-hydroxydopamine, ascorbate, glutathione and peroxide on the in vitro activities of creatine and adenylate kinases in the brain of the rat. *Neuropharmacology* **25,** 25–32.

Maragos, W. F., Jakel, R. J., Pang, Z., and Geddes, J. W. (1998) 6-Hydroxydopamine injections into the nigrostriatal pathway attentuate striatal malonate and 3-nitropropionic acid lesions. *Exp. Neurol.* **154,** 637–644.

Maura, G., Giardi, A., and Raiteri, M. (1988) Release-regulating D-2 dopamine receptors are located on striatal glutamatergic nerve terminals. *J. Pharmacol. Exp. Ther.* **247,** 680–684.

McConkey, D. J. and Orrenius, S. (1996) The role of calcium in the regulation of apoptosis. *J. Leukocyte Biol.* **59,** 774–783.

McLaughlin, B. A., Nelson, D., Erecinska, M., and Chesselet, M.-F. (1998) Toxicity of dopamine to striatal neurons in vitro and potentiation of cell death by a mitochondrial inhibitor. *J. Neurochem.* **70,** 2406–2415.

McLaughlin, B. A., Nelson, D., Silver, I. A., Erecinska, M., and Chesselet, M.-F. (1998) Methylmalonate toxicity in primary neuronal cultures. *Neuroscience* **86,** 279–290.

Messam, C. A., Greene, J. G., Greenamyre, J. T., and Robinson, M. B. (1995) Intrastriatal injections of the succinate dehydrogenase inhibitor, malonate, cause a rise in extracellular amino acids that is blocked by MK-801. *Brain Res.* **684,** 221–224.

Michel, P. P. and Hefti, F. (1990) Toxicity of 6-hydroxydopamine and dopamine for dopaminergic neurons in culture. *J. Neurosci. Res.* **26,** 428–435.

Miyashita, T. and Reed, J. C. (1995) Tumor suppressor p53 is a direct transcriptional activator of the human bax gene. *Cell* **80,** 293–299.

Morikawa, N., Nakagawa-Hattori, Y., and Mizuno, Y. (1996) Effect of dopamine, dimethoxyphenylethylamine, papaverine, and related compounds on mitochondrial respiration and complex I activity. *J. Neurochem.* **66,** 1174–1181.

Nakao, N., Nakai, K., and Itakura, T. (1997) Metabolic inhibition enhances selective toxicity of L-DOPA toward mesencephalic dopamine neurons in vitro. *Brain Res.* **777,** 202–209.

Nunez, G., Benedict, M. A., Hu, Y., and Inohara, N. (1998) Caspases: the proteases of the apoptotic pathway. *Oncogene* **17,** 3237–3245.

O'Dell, S. J., Weihmuller, F. B., and Marshall, J. F. (1991) Multiple methamphetamine injections induce marked increases in extracellular striatal dopamine which correlate with subsequent neurotoxicity. *Brain Res.* **564,** 256–260.

Offen, D., Ziv, I., Gorodin, S., Barzilai, A., Malik, Z., and Melamed, E. (1995) Dopamine-induced programmed cell death in mouse thymocytes. *Biochim. Biophys. Acta* **1268,** 171–177.

Offen, D., Ziv, I., Sternin, H., Melamed, E., and Hochman, A. (1996) Prevention of dopamine-induced cell death by thiol antioxidants: possible implications for treatment of Parkinson's disease. *Exp. Neurol.* **141,** 32–39.

Offen, D., Ziv, I., Panet, H., Wasserman, L., Stein, R., Melamed, E., et al. (1997) Dopamine-induced apoptosis is inhibited in PC12 cells expressing Bcl-2. *Cell. Mol. Neurobiol.* **17,** 289–304.

Offen, D., Beart, P. M., Cheung, N. S., Pascoe, C. J., Hochman, A., Gorodin, S., et al. (1998) Transgenic mice expressing human Bcl-2 in their neurons are resistant to 6-hydroxydopamine and 1-methyl-4-phenyl-1,2,3,6-tetrahydropyridine neurotoxicity. *Proc. Natl. Acad. Sci. USA* **95,** 5789–5794.

Ogawa, N., Asanuma, M., Kondo, Y., Kawada, Y., and Yamamoto, M., and Mori, A. (1994a) Differential effects of chronic L-DOPA treatment on lipid peroxidation in the mouse brain with or without pretreatment with 6-hydroxydopamine. *Neurosci. Lett.* **171,** 55–58.

Ogawa, N., Tanaka, K., Asanuma, M., Kawai, M., Masumizu, T., Kohno, M., and Mori, A. (1994b) Bromocriptine protects mice against 6-hydroxydopamine and scavenges hydroxyl free radicals in vitro. *Brain Res.* **657,** 207–213.

Oh, Y. J., Wong, S. C., Moffat, M., and O'Malley, K. L. (1995) Overexpression of Bcl-2 attenuates MPP+, but not 6-ODHA, induced cell death in a dopaminergic neuronal cell line. *Neurobiol. Dis.* **2,** 157–167.

Olanow, C. W. (1990) Oxidation reactions in Parkinson's disease. *TINS* **40,** 32–37.

Olanow, C. W., Jenner, P., and Brooks, D. (1998) Dopamine agonists and neuroprotection in Parkinson's disease. *Ann. Neurol.* **44,** S167–S174.

Oosternik, B. J., Korte, S. M., Nyakas, C., Korf, J., and Luiten, P. G. (1998) Neuroprotection against *N*-methyl-D-aspartate-induced excitotoxicity in rat magnocellular nucleus basalis by the 5-HT1A receptor agonist 8-OH-DPAT. *Eur. J. Pharmacol.* **358,** 147–152.

Park, J., Kim, I., Oh, Y.J., Lee, K., Han, P. L., and Choi, E. J. (1997) Activation of c-Jun N-terminal kinase antagonizes an anti-apoptotic action of Bcl-2. *J. Biol. Chem.* **272,** 16,725–16,728.

Petito, C. K. and Pulsinelli, W. A. (1984) Delayed neuronal recovery and neuronal death in rat hippocampus following severe cerebral ischemia: possible relationship to abnormalities in neuronal processes. *J. Cereb. Blood Flow Metab.* **4,** 194–205.

Phebus, L. A., Perry, K. W., Clemens, J. A., and Fuller, R.W. (1986) Brain anoxia releases striatal dopamine in rats. *Life Sci.* **38,** 2447–2453.

Pieper, A. A., Verma, A., Zhang, J., and Snyder, S. (1999) Poly (ADP-ribose) polymerase, nitric oxide and cell death. *Trends Pharmacol. Sci.* **20,** 171–181.

Piomelli, D. and Di Marzo, V. (1991) Dopamine D2 receptor signaling via the arachidonic acid cascade: modulation by cAMP-dependent protein kinase A and prostaglandin E2. *J. Lipid Mediators* **6,** 433–443.

Porter, R. H. P., Greene, J. G., Higgins, D. S., and Greenamyre, J. T. (1994) Polysynaptic regulation of glutamate receptors and mitochondrial enzyme activities in the basal ganglia of rats with unilateral dopamine depletion. *J. Neurosci.* **14,** 7192–7199.

Pu, C. and Vorhees, C. V. (1993) Developmental dissociation of methamphetamine-induced depletion of dopaminergic terminals and astrocyte reaction in rat striatum. *Dev. Brain Res.* **72,** 325–328.

Pulsinelli, W. A., Brierley, J. B., and Plum, F. (1982) Temporal profile of neuronal damage in a model of transient forebrain ischemia. *Ann. Neurol.* **11,** 491–498.

Reynolds, D. S., Carter, R. J., and Morton, A. J. (1998) Dopamine modulates the susceptibility of striatal neurons to 3-nitropropionic acid in the rat model of Huntington's disease. *J. Neurosci.* **18,** 10,116–10,127.

Ricaurte, G. A. and McCann, U. D. (1992) Neurotoxic amphetamine analogues: effects in monkeys and implications for humans. *Ann. NY Acad. Sci.* **648,** 371–382.

Roberts, P. J. and Anderson, S. D. (1979) Stimulatory effect of L-glutamate and related amino acids on [3H]dopamine release from rat striatum: an in vitro model of glutamate actions. *J. Neurochem.* **32,** 1539–1545.

Rosenberg, P. A. (1988) Catachloamine toxicity in cerebral cortex of dissociated cell culture. *J. Neurosci.* **8,** 2887–2894.

Scatton, B., Simon, H., Le Moal, M., and Bishoff, S. (1980) Origin of dopaminergic innervation of the rat hippocampal formation. *Neurosci. Lett.* **18,** 125–131.

Schapira, A. H. and Cooper, J. M. (1994) Inborn and induced defects of the mitochondrial respiratory chain. *Biochem. Soc. Trans.* **22,** 996–1001.

Schmidt, C. J., Wu, L., and Lovenberg, W. (1986) Methylenedioxymethamphetamine: a potentially neurotoxic amphetamine analog. *Eur. J. Pharmocol.* **124,** 175–178.

Schwarzschild, M. A., Cole, R. L., and Hyman, S. E. (1997) Glutamate, but not dopamine, stimulates stress-activated protein kinase and AP-1-mediated transcription in striatal neurons. *J. Neurosci.* **17,** 3455–3466.

Seabrook, G. R., Kemp, J. A., Freedman, S. B., Patel, S., Sinclair, H. A., and McAllister, G. (1994a) Functional expression of human D3 dopamine receptors in differentiated neuroblastoma x glioma NG108-15 cells. *Br. J. Pharmacol.* **111,** 391–393.

Seabrook, G. R., Knowles, M., Brown, N., Myers, J.. Sinclair, H., Patel, S., et al. (1994b) Pharmacology of high-threshold calcium currents in GH4C1 pituitary cells and their regulation by activation of human D2 and D4 dopamine receptors. *Br. J. Pharmacol.* **112,** 728–734.

Seiden, L. S. and Vosmer, G. (1984) Formation of 6-hydroxydopamine in caudate nucleus of the rat brain after a single large dose of methylamphetamine. *Pharmacol. Biochem. Behav.* **21,** 29–31.

Schapira, A. H. and Cooper, J. M. (1992) Mitochondrial function in neurodegeneration and aging. *Mutat. Res.* **275,** 133–143.

Sheng, P., Cerruti, C., Ali, S., and Cadet, J. L. (1996) Nitric oxide is a mediator of methamphetamine (METH)-induced neurotoxicity. In vitro evidence from primary cultures of mesencephalic cells. *Ann. NY Acad. Sci.* **801,** 174–186.

Simantov, R., Blinder, E., Ratovitski, T., Tauber, M., Gabbay, M., and Porat, S. (1996) Dopamine-induced apoptosis in human neuronal cells: inhibition by nucleic acids antisense to the dopamine transporter. *Neuroscience* **74,** 39–50.

Slivka, A., Brannan, T. S., Weinberger, J., Knott, P. J., and Cohen, G. (1988) Increase in extracellular dopamine in the striatum during cerebral ischemia: a study utilizing cerebral microdialysis. *J. Neurochem.* **50,** 1714–1718.

Smith, A. D. and Bolam, J. P. (1990) The neural network of the basal ganglia as revealed by the study of synaptic connections of identified neurones. *Trends Neurosci.* **13,** 259–265.

Smith, T. S., Trimmer, P. A., Khan, S. M., Tinklepaugh, D. L., and Bennett, J. P., Jr. (1997) Mitochondrial toxins in models of neurodegenerative diseases. II: Elevated zif268 transcription and independent temporal regulation of striatal D1 and D2 receptor mRNAs and D1 and D2 receptor-binding sites in C57BL/6 mice during MPTP treatment. *Brain Res.* **765,** 189–197.

Sofic, E., Riederer, P., Heinsen, H., Beckmann, H., Reynolds, G. P., Hebenstreit, G., et al. (1988) Increased iron (III) and total iron content in post mortem substantia nigra of parkinsonian brain. *J. Neural Transm.* **74,** 199–205.

Sonsalla, P., Gibb, J., and Hanson, G., (1986) Roles D_1 and D_2 dopamine receptor subtypes in mediating the methamphetamine-induced changes in monoamine systems. *J. Pharmacol. Exp. Ther.* **238,** 932–937.

Spina, M. B. and Cohen, G. (1989) Dopamine turnover and glutathione oxidation: implications for Parkinson disease. *Proc Natl. Acad. Sci. USA* **86,** 1398–1400.

Sprague, J. E., Everman, S. L., and Nichols, D. E. (1998) An integrated hypothesis for the serotonergic axonal loss induced by 3,4-methylenedioxymethamphetamine. *Neurotoxicology* **19,** 427–441.

Steele, T. D., McCann, U. D., and Ricaurte, G. A. (1994) 3,4-Methylenedioxymethamphetamine (MDMA, "Ecstasy"): pharmacology and toxicology in animals and humans. *Addiction* **89,** 539–551.

Stephans, S. E. and Yamamoto, B. K. (1994) Methamphetamine-induced neurotoxicity: roles for glutamate and dopamine efflux. *Synapse* **17,** 203–209.

Stephans, S. E., Whittingham, T. S., Douglas, A. J., Lust, W. D., and Yamamoto, B. K. (1998) Substrates of energy metabolism attenuate methamphetamine-induced neurotoxicity in striatum. *J. Neurochem.* **71,** 613–621.

Stevens, J. R. (1981) Receptor supersensitivity: relationships to cerebral anatomy and histopathology of schizophrenia. *Biol. Psychiatry* **16,** 1119--1122.

Stone, D. M., Stahl, D. C., Hanson, G. R., and Gibb, J. W. (1986) The effects of 3,4-methylenedioxymethamphetamine (MDMA) and 3,4-methylenedioxyamphetamine (MDA) on monoamine systems in the rat brain. *Eur. J. Pharmacol.* **128,** 41–48.

Stroh, C. and Schulze-Osthoff, K. (1998) Death by a thousand cuts: an ever increasing list of caspase substrates. *Cell Death Differ.* **5,** 997–1000.

Strosznajder, J., Chalimoniuk, M., and Samochocki, M. (1996) Activation of serotonergic 5-HT1A receptor reduces Ca(2+)- and glutamatergic receptor-evoked arachidonic acid and No/cGMP release in adult hippocampus. *Neurochem. Int.* **28,** 439–444.

Surmeier, D. J., Bargas, J., Hemmings, H. C., Nairn, A. C., and Greengard, P. (1995) Modulation of calcium currents by a D1 dopaminergic protein kinase/phosphatase cascade in rat neostriatal neurons. *Neuron* **14,** 385–397.

Swanson, L. W. (1982) The projections of the ventral tegmental area and adjacent regions: a combined flurorescent retrograde tracer and immunofluorescence study in the rat brain. *Brain Res. Bull.* **9,** 321–353.

Takai, N., Nadanishi, H., Tanabe, K., Nishioku, T., Sugiyama, T., Fujiwara, M., et al. (1998) Involvement of caspase-like proteinases in apoptosis of neuronal PC12 cells and primary cultured microglia induced by 6-hydroxydopamine. *J. Neurosci. Res.* **54,** 214–222.

Tanaka, M., Sotomatsu, A., Kanai, H., and Hirai, S. (1991) Dopa and dopamine cause cultured neuronal death in the presence of iron. *J. Neurol. Sci.* **101,** 198–203.

Thornberry, N. A. and Lazebnik, Y. (1998) Caspases: enemies within. *Science* **281,** 1312–1316.

Trautwein, W. and Hescheler, J. (1990) Regulation of cardiac L-type calcium current by phosphorylation and G proteins. *Ann. Rev. Physiol.* **52,** 257–274.

Treisman, R. (1996) Regulation of transcription by MAP kinase cascades. *Curr. Opin. Cell Biol.* **8,** 205–215.

Vonsattel, J. P., Ferrante, R. J., Stevens, T. J., et al. (1985) Neuropathologic classification of Huntington's disease. *J. Neuropathol. Exp. Neurol.* **44,** 559–577.

Wagner, G., Lucot, J. B., Schuster, C. R., and Seiden, L. S. (1983) Alpha-methyltyrosine attentuates and reserpine increases methamphetamine-induced neuronal changes. *Brain Res.* **270,** 285–288.

Walkinshaw, G. and Waters, C. M. (1995) Induction of apoptosis in catecholaminergic PC12 cells by L-DOPA. Implications for the treatment of Parkinson's disease. *J. Clin. Invest.* **95,** 2458–2464.

Wang, Y. M., Gainetdinov, R. R., Fumagelli, F., Xu, F., Jones, S. R., Bock, C. B., et al. (1997) Knockout of the vesicular monoamine transporter-2 gene results in neonatal death and supersensitivity to cocaine and amphetamine. *Neuron* **19,** 1285–1296.

Weinberger, J., Nieves-Rosa, J., and Cohen, G. (1985) Nerve terminal damage in cerebral ischemia: protective effect of alpha-methyl-para-tyrosine. *Stroke* **16,** 864–870.

Weiss, S., Sebben, M., Garcia-Sainz, J. A., and Bockaert, J. (1985) D2-dopamine receptor mediated inhibition of cyclic AMP formation in striatal neurons in primary culture. *Mol. Pharm.* **27,** 595–599.

Wieloch, T., Miyauchi, Y., and Lindvall, O. (1990) Neuronal damage in the striatum following forebrain ischemia: lack of effect of selective lesions of mesostriatal dopamine neurons. *Exp. Brain Res.* **83,** 159–163.

Wrona, M. Z., Yang, Z., Zhang, F., and Dryhurst, G. (1997) Potential new insights into the molecular mechanisms of methamphetamine-induced neurodegeneration. *NIDA Res, Monogr.* **173,** 146–174.

Yamamoto, B. K. and Davy, S. (1992) Dopaminergic modulation of glutamate release in striatum as measured by microdialysis. *J. Neurochem.* **58,** 1736–1742.

Yamamoto, Y., Tanaka, T., Shibata, S., and Watanabe, S. (1994) Involvement of D_1 dopamine receptor mechanism in ischemia-induced impairment of CA1 presynaptic fiber spikes in rat hippocampal slices. *Brain Res.* **665,** 151–154.

Yan, Z., Hsieh-Wilson, L., Feng, J., Tomizawa, K., Allen, P. B., Fienberg, A. A., et al. (1999) Protein phosphatase 1 modulation of neostriatal AMPA channels: regulation by DARPP-32 and spinophilin. *Nature Neurosci.* **2,** 13–17.

Yao, H., Sadoshima, S., Ishitsuka, T., Nagao, T., Fujishima, M., Tsutsumi, T., et al. (1988) Massive striatal dopamine release in acute cerebral ischemia in rats. *Experientia* **44,** 506–508.

Yoshikawa, T., Minamiyama, Y., Naito, Y., and Kondo, M. (1994) Antioxidant properties of bromocriptine, a dopamine agonist. *J. Neurochem.* **62,** 1034–1038.

Young, A. B., Penney, J. B., and Starosta-Rubinstein, S. (1986) PET scan investigations of Huntington's disease: cerebral metabolic correlates of neurological features and functional decline. *Ann. Neurol.* **20,** 296–303.

Yujiri, T., Sather, S., Fanger, G. R., and Johnson, G. L. (1998) Role of MEKK1 in cell survival and activation of JNK and ERK pathways defined by targeted gene disruption. *Science* **282,** 1911–1914

Zhen, X., Uryu, K., Wang, H. Y., and Friedman, E. (1998) D1 dopamine receptor agonists mediate activation of p38 mitogen-activated protein kinase and c-Jun amino-terminal kinase by a protein kinase A-dependent mechanism in SK-N-MC human neuroblastoma cells. *Mol. Pharmacol.* **54,** 453–458.

Zilkha-Falb, R., Ziv, I., Nardi, N., Offen, D., Melamed, E., and Barzilai, A. (1997) Monoamine-induced apoptotic neuronal cell death. *Cell. Mol. Neurobiol.* **17,** 101–118.

Ziv, I., Melamed, E., Nardi, N., Luria, D., Achiron, A., Offen, D., et al. (1994) Dopamine induces apoptosis-like cell death in cultured chick sympathetic neurons—a possible novel pathogenetic mechanism in Parkinson's disease. *Neurosci. Lett.* **170,** 136–140.

10
Mitochondria and Parkinson's Disease

Russell H. Swerdlow

10.1. INTRODUCTION

Parkinson's disease (PD) affects 1–2% of the United States population, and after Alzheimer's disease it is the second most common of the neurodegenerative diseases (Lilienfeld, 1994). Although the cause or causes of PD are not entirely understood, ongoing research continues to refine our understanding of its underlying pathogenesis. Much of this research suggests that mitochondrial dysfunction plays a central role in PD neurodegeneration. This chapter will recount the chain of events that implicated mitochondria as players in this disease and will review past and current controversies regarding this subject. It will also review data addressing a potential mitochondrial molecular basis for the sporadic form of idiopathic PD.

10.2. CLINICAL AND HISTOPATHOLOGIC ASPECTS

In 1817, the English physician James Parkinson described the clinical syndrome that would bear his name (Parkinson, 1817). What Parkinson actually detailed were persons presenting with tremors of resting limbs and an unusually hunched gait. Others also confirmed the presence of this stereotyped syndrome and advanced various names for it such as "paralysis agitans." Refined observations of the signs and symptoms of this entity would follow, and include the presence of slow, often difficult-to-initiate movements (bradykinesia) and muscle rigidity. Currently, Parkinson's disease (PD) is diagnosed when particular combinations of rest tremor, bradykinesia, rigidity, or postural instability are noted by patients or observed by physicians (Gibb and Lees, 1988; Ward and Gibb, 1990; Calne et al., 1992). Thus, the exact definition of Parkinson's disease depends on the operational criteria one chooses to use.

Because the clinical diagnosis always was (and remains) somewhat arbitrary, the 1900s saw the linking of the syndrome to neuropathologic

From: *Contemporary Clinical Neuroscience: Molecular Mechanisms of Neurodegenerative Diseases*
Edited by: M.-F. Chesselet © Humana Press Inc., Totowa, NJ

phenomena. In 1912, Frederich Lewy observed the presence of intracytoplasmic inclusions in the vagal dorsal motor nucleus and substantia innominata of persons diagnosed with Parkinson's disease (Lewy, 1912). In 1919, Tretiakoff described the presence of similar inclusions in the substantia nigra of Parkinson's patients and designated them "Lewy bodies" (Tretiakoff, 1919). Although Lewy bodies are neither specific to Parkinson's disease nor encompass all those who present with the stereotyped clinical syndrome (Mark et al., 1996), for lack of a better marker the "gold standard" diagnosis of Parkinson's disease is now defined by the demonstration of nigral Lewy bodies in the patient meeting any of the accepted clinical diagnostic criteria.

Although classic histopathologic studies failed to explain why neurodegeneration occurs in PD, they facilitated the development of treatments useful (at least in the initial stages) for the management of PD symptoms ("parkinsonism"). It became apparent in 1938 that dopaminergic neurons of the substantia nigra pars compacta are lost in PD patients over and above the typical dropout of these neurons that occurs with aging (Hassler, 1938). This observation permitted the prediction of dopamine deficiency within certain nigral projection nuclei called the striatum (Ehringer and Hornykiewicz, 1960). The components of the striatum, the caudate and putamen, are important relay centers for the production of planned movement (Evarts and Thach, 1969). The dopamine precursor levodopa, which crosses the blood-brain barrier and elevates striatal dopaminergic tone by increasing dopamine production in remaining nigral neurons, constitutes the most effective available symptomatic treatment (Cotzias et al., 1967). As was the case with neuropathology surveys, though, development of this drug in the 1960s would contribute only minimally to our understanding of why neurons degenerate in PD. Certain events occurring a decade later would provide greater insight into this mystery.

10.3. INITIAL CLUES THAT MITOCHONDRIAL DYSFUNCTION OCCURS IN PD: THE MPTP STORY

In 1979, Davis and co-workers reported the case of a young drug abuser with severe and chronic parkinsonism (Davis et al., 1979). The patient first presented in 1976 as a 23-yr-old male with 3 mo of progressive rigidity, tremor, bradykinesia, and masked facies. He experienced symptomatic improvement on relatively high doses of levodopa. After dying from a subsequent drug overdose in 1978, autopsy of the brain revealed degeneration of the substantia nigra and at least one example of an eosinophilic, intracytoplasmic inclusion that resembled a classic Lewy body.

Just prior to the onset of symptoms, the patient began synthesizing and consuming meperidine analogs. Synthesis was initially based on a previously published procedure (Ziering et al., 1947) which was subsequently simplified. His deterioration started several days after these procedural modifications were adopted. In addition to the presence of the intended compound, 1-methyl-4-phenyl-4-propionoxy-piperidine (MPPP), analysis of the subject's laboratory glassware revealed the presence of several unsolicited pyridine molecules. Davis et al. concluded that these by-products were likely responsible for the patient's syndrome.

This interpretation was supported and extended by Langston and colleagues in 1983 (Langston et al., 1983). Over the course of the previous year, recreational abusers of meperidine analogs began presenting to California physicians with chronic parkinsonism. All four of the affected persons described in Langston et al.'s original report developed symptoms within 1 wk of using MPPP contaminated with substantial amounts of 1-methyl-4-phenyl-1,2,3,6-tetrahydropyridine (MPTP). This latter molecule was reported by Davis et al. as one of the pyridine molecules inadvertently synthesized and consumed by their original patient.

The mechanisms by which MPTP induces specific neurodegeneration of the substantia nigra and therefore causes parkinsonism began to unravel in 1984. In April of that year, Chiba et al. demonstrated that MPTP is eventually metabolized to its derivative 1-methyl-4-phenylpyridine (MPP+) (*see* Fig. 1) by the mitochondrially localized enzyme monoamine oxidase (MAO) (Chiba et al., 1984). In October, Markey et al. and Heikkila et al. showed that specific inhibition of the MAO B form blocked conversion of MPTP to MPP+, with subsequent prevention of toxic consequences in experimental animals (Markey et al., 1984; Heikkila et al., 1984). These experiments established that MPP+ rather than MPTP was the actual neurotoxic molecule. By April of 1985, animal studies revealed that conversion of MPTP to MPP+ likely occurred (or was as least initiated) extraneuronally in glial cells (Uhl et al., 1985). Also in April, the same group reported that MPP+ diffusing within the extracellular compartment underwent selective neuronal internalization via dopamine reuptake sites located on the striatally projecting synaptic endings of nigral neurons (Javitch et al., 1985). This phenomenon provided a potential explanation for the anatomic specificity of MPTP-induced neurodegeneration.

Nicklas and co-workers published in July 1985 that MPP+ inhibits activity of the mitochondrial electron transport chain (ETC) enzyme NADH : ubiquinone oxidoreductase (complex I) (Nicklas et al., 1985). The relevance of this discovery to MPTP-induced neurodegeneration was underscored by the subsequent

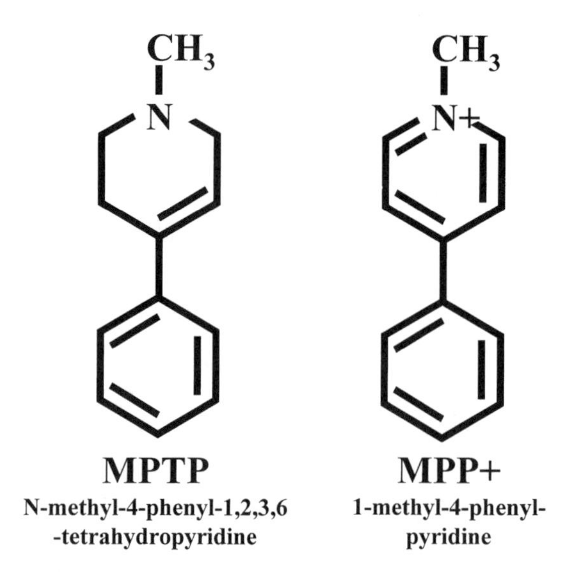

Fig. 1. MPTP and MPP+.

report of Ramsay et al. that mitochondria possess an energy-dependent uptake system that facilitates sequestration of MPP+ within the organelle (Ramsey et al., 1986; Ramsay and Singer, 1986). Meanwhile, the relevance of all this work to idiopathic PD was reinforced by the development of MPTP-generated animal models, which appeared to recapitulate key clinical and neuropathologic features of the human disease (Burns et al., 1983). Of particular importance was the demonstration in 1986 that structures resembling Lewy bodies appeared within the substantia nigra of aged primates exposed to the toxin (Forno et al., 1986).

10.4. THE DISCOVERY OF COMPLEX I DYSFUNCTION IN IDIOPATHIC PD

By the mid 1980s, it was established that MPP+-induced complex I dysfunction caused a parkinsonian syndrome that clinically and histopathologically resembled idiopathic PD. The question of whether similar mitochondrial dysfunction occurred in idiopathic PD patients arose. In 1988, W. Davis Parker began studying PD patients for the potential presence of ETC dysfunction. Parker chose to assay platelets because they were procurable from living patients, thus bypassing the pitfalls of working with autopsy tissue. Parker additionally believed assay of a “nontarget” rather than degenerating tissue would decrease the likelihood of detecting epiphenomenal ETC impairment.

Parker and colleagues used plateletpheresis technology to collect and concentrate platelets from 10 PD and 10 control subjects. Highly enriched mitochondrial fractions were prepared by centrifuging already concentrated mitochondria through a Percoll density gradient. Activities for the ETC enzymes complex I, II/III (measured as succinate : cytochrome-*c* oxidoreductase activity), and IV were spectrophotometrically determined. Compared to that of control subjects, complex I activity was decreased by 55%, a statistically significant difference. Complex II/III and IV activities were comparable between groups (Parker et al., 1989a). This work was submitted in May 1989 to the *Annals of Neurology* after undergoing rejection by *Science*, *The New England Journal of Medicine*, and *Lancet*. It was accepted in June.

While the paper of Parker and colleagues was undergoing peer review, an editorial letter appeared in the June 3, 1989 issue of *Lancet* that reported that relative to controls, PD substantia nigra complex I activity was reduced 28% (Schapira et al., 1989). Assays in this study were performed on autopsy homogenates from nine PD and nine age-matched subjects. In July, *Lancet* published another letter to the editor describing ETC analysis of PD muscle (Bindoff et al.,1989). Five PD subject and four control-derived samples were studied. In addition to a 40% complex I activity decrement in the PD group, a 49% decrease in complex II activity and a 40% decrease in complex IV activity were observed. In September 1989, Mizuno et al. reported the results of ETC immunoblotting studies comparing PD (n = 5) and control (n = 4) subject striata. Several small complex I subunits (the 30-, 25-, and 24-kDa subunits) were generally decreased in PD striata, whereas complex III and IV immunoblotting appeared similar between groups (Mizuno et al., 1989).

10.5. IS MITOCHONDRIAL DYSFUNCTION IN PD FOCAL OR SYSTEMIC?

Although three of the four initial reports detailing PD ETC impairment derived from studies of non-nigral tissue, controversy soon developed over whether complex I dysfunction in PD patients was systemic or limited to the substantia nigra. In 1990, Schapira and co-workers reported they were unable to detect complex I dysfunction in any part of the brain other than substantia nigra, where a 42% decrease was found relative to controls (Schapira et al., 1990a). Similar results were again reported from this same laboratory 2 and 5 yr later (Mann et al., 1992a; Cooper et al., 1995). Other studies also entered the literature that failed to corroborate the presence of ETC dysfunction in either platelet or muscle mitochondria from PD subjects. Confident resolution of this issue is critical, because the presence of mitochondrial dysfunction in nondegenerating tissues would support the contention that ETC perturbation constitutes an early, perhaps even primary, lesion in this disease.

To better understand the discrepancies between positive and negative studies of ETC dysfunction in PD, it is worthwhile to review mitochondrial assay techniques. Procured tissue is first homogenized to disrupt cell membranes. The assay of mitochondria suspended in the resultant homogenate is possible, and perhaps mandatory if only very limited amounts of tissue are available. The determination of ETC activities in homogenates is insensitive, however, because the organelles are diluted in the solution and much nonmitochondrial protein is present. Because of this, some investigators also measure the activity of citrate synthase, a soluble matrix enzyme that at this stage can provide an estimate of mitochondrial mass, and reference ETC activites to citrate synthase activity.

Assay sensitivities are greatly increased by then preparing enriched mitochondrial fractions. To accomplish this, a relatively slow centrifugation step is performed to pellet out nuclei. The remaining postnuclear supernatant contains the mitochondria. Fast centrifugation of this supernatant then yields a mitochondrial pellet, which is resuspended in a buffer following its separation from the lipid-containing supernatant. This resuspension constitutes the "crude" mitochondrial fraction, the purity of which depends on the nature of the starting tissue. Crude mitochondrial fractions from lipid-rich tissues, including the brain, are generally less pure than those produced from other tissues, such as muscle. Measured ETC activities are higher in crude mitochondrial fractions because mitochondria are concentrated. Because a large proportion of protein in crude mitochondrial fractions is mitochondrial, determination of protein concentration in the solution provides an appropriate reference for the amount of mitochondria in the assay. Citrate synthase activity is occasionally used as an additional surrogate of mitochondrial concentration. The value of this step is dubious, because citrate synthase is a soluble enzyme that may preferentially leak from fragile mitochondria during preparation of the crude fraction. The resulting erroneously depressed denominator then leads to overestimation of the final ETC rate determination.

Assay sensitivity is further improved by subsequent gradient centrifugation of the crude mitochondrial fraction, which allows for density separation of mitochondria from other similarly sized organelles. Enzyme activities are best referenced to total protein in these "pure mitochondrial fractions."(*See* Fig. 2.)

Regardless of whether mitochondria in homogenates, crude mitochondrial fractions, or pure mitochondrial fractions are assayed, complex I

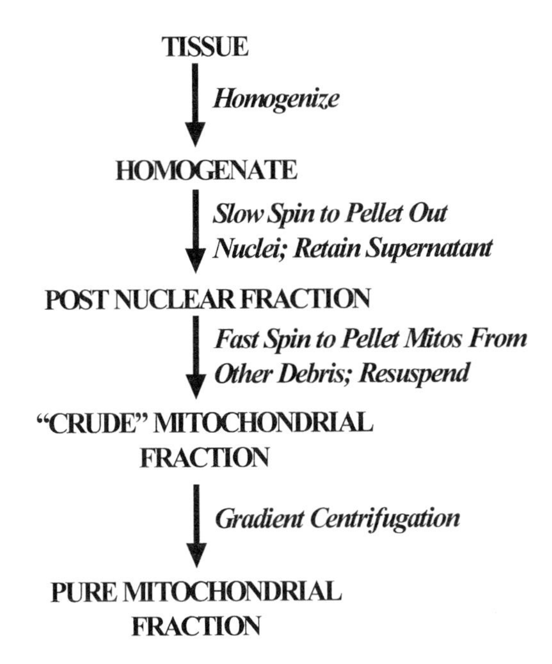

Fig. 2. Preparation of mitochondria for electron transport chain assays.

activity is determined by spectrophotometrically measuring its ability to oxidize NADH and reduce a downstream electron acceptor. The optimal strategy is to follow the enzyme's direct reduction of an ubiquinone analog, such as coenzyme Q1 or decylubiquinone (DB). Ultimate rate determinations will vary depending on the ubiquinone analog used. Alternatively, measuring electron transfer from NADH to cytochrome-*c* (which is actually reduced by cytochrome-*c* oxidoreductase, or complex III, the next step along the ETC) and subtracting out the rotenone-sensitive portion of the reaction provides an estimate of complex I activity but is less sensitive than analyzing ubiquinone reduction. Finally, the specific environment within the reaction cuvet, such as the presence or absence of bovine serum albumin, influences activity determinations. (*See* Fig. 3.)

The relevance of such technical issues in this field is illustrated by a review of the PD platelet ETC literature. In 1992, one group published two studies of PD platelet ETC measurements. In the first, published in April, ETC assays were performed on homogenates and no biochemical defects were found (Mann et al., 1992a). In December, assays were performed with crude mitochondrial fractions, and a significant 16% complex I defect was detected (Krige et al., 1992). Even using crude mitochondrial fractions probably underestimates the magnitude of the defect in platelets, as pure

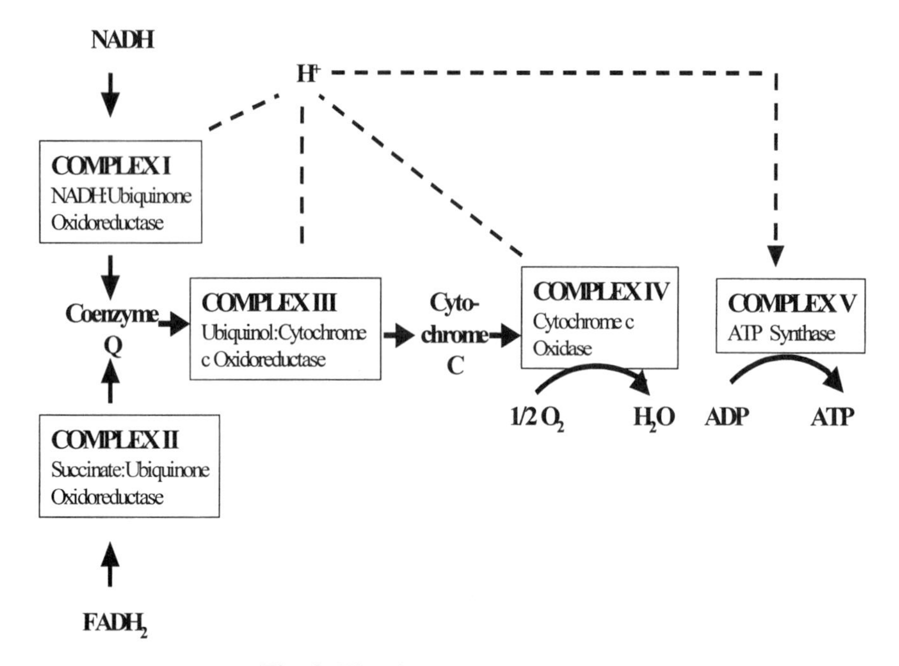

Fig. 3. The electron transport chain.

mitochondrial fraction studies reveal a 25–54% complex I defect (Parker et al., 1989a; Benecke et al., 1993; Haas et al., 1995).

Regarding ETC catalytic analysis of PD brain tissue, the majority of studies prepared only homogenates. Mostly, an isolated depression of complex I (Schapira et al.,1989, 1990a, 1990b; Mann et al., 1992; Janetzky et al., 1994) or rarely complex III (Reichmann et al., 1990) was detected within substantia nigra. Only one study assayed crude mitochondrial fractions. Interestingly, this study detected a 36% decline in striatal complex III activity (Mizuno et al., 1990). Given the experience of PD platelet mitochondrial assays, however, any conclusion that brain complex I deficiency in this disease is limited to substantia nigra is premature. Indeed, a recent preliminary study of pure mitochondrial fractions prepared from PD brain frontal lobes revealed a significant decrease in complex I activity (Parks et al., 1999). A summary of PD brain ETC studies is provided in Table 1.

Multiple studies concur with Parker et al.'s initial discovery of complex I deficiency in PD subject platelets (Parker et al., 1989a). The first of these, the study of Yoshino and colleagues, also reported decreased complex II activity in this tissue (Yoshino et al., 1992). As discussed earlier, Krige et al. found a 16% complex I defect in PD platelet crude mitochondrial fractions (Krige et al., 1992). In 1993, Benecke et al. performed assays on pure PD

Table 1
PD ETC Brain Data

Study	Journal	Defect(s)	Comments
Schapira et al., 1989	Lancet	SN (Comp. I,↓30%)	Homogenates
Mizuno et al., 1989	BBRC	Striatum (Comp. I↓)	Immunoblotting
Reichmann et al., 1990	Mov. Disord.	SN (Comp. III, ↓35-40%)	Homogenates(?)
Schapira et al., 1990	J. Neurochem.	SN (Comp. I,↓30%)	Homogenates
Schapira et al., 1990	J. Neurochem.	Only SN (Comp. I, ↓42%)	Homogenates
Mizuno et al., 1990	J. Neurol. Sci.	Striatum (Comp. III, ↓36%)	Crude mito fractions
Hattori et al., 1991	Ann. Neurol.	SN (Comp. I↓)	Immuno-histochemistry
Mann et al., 1992	Brain	SN (Comp. I,↓37%)	Homogenates
Janetzky et al., 1994	Neurosci. Lett.	SN (Comp. I,↓34%)	Homogenates
Parks et al., 1999	Soc. Neurosci. Abstr.	Cortex (Comp. I↓45%, Comp. IV↓28%)	Pure mitochondria

SN = substantia nigra

platelet mitochondria and detected a 52% decrease in complex I activity, a 30% decrease in complex IV activity, and a nonsignificant "trend" toward decreased complex III activity (Benecke et al., 1993). These authors also reassayed five of their more recently diagnosed PD subjects 1 yr after initial ETC determinations were performed, and over this time period the already reduced complex I and IV activities of these subjects showed significant further decline. Two years later, Haas and colleagues studied ETC activities in platelets from untreated PD subjects using methods similar to those of Parker et al. (Parker et al., 1989a), and reported both complex I and II/III activity decrements (Haas et al., 1995). These results addressed the question of whether mitochondrial ETC defects in PD were occurring secondary to medication effects and, indeed, suggested that drug therapy is not responsible for observed ETC deficiencies. The magnitude of the complex I defect in this study of newly diagnosed PD patients was 25%, about half as profound as the defect found by Parker et al. This is consistent with the hypothesis that complex I activity might decline with disease progression, as parkinsonism was less severe in the subjects of Haas et al.

In addition to the negative platelet study of Mann et al., which not surprisingly failed to detect complex I deficiency in PD subject platelet homogenates (Mann et al., 1992a), two other published studies report normal complex I activity in this tissue. Bravi et al. did not detect diminished ETC activity in PD samples using crude mitochondrial fractions and the less sensitive NADH to cytochrome-*c* assay (Bravi et al., 1992). Blake et al. (Blake et al., 1997) also found no defect in pure mitochondrial fractions, although in this study, plateletpheresis, mitochondrial enrichment, and complex I assay methodology differed from that of Parker et al. (Parker et al., 1989a). The mean complex I activity of the 13 PD samples analyzed was 20% less than that of the 9 control samples, but this discrepancy did not reach significance (p=0.095). A summary of both positive and negative platelet ETC studies is provided in Table 2.

Bindoff et al.'s finding that complex I activity was reduced in PD muscle (in addition to complex II and IV) (Bindoff et al., 1989,1991) was corroborated by several other studies, which are listed in the Table 3 (Shoffner et al., 1991; Nakagawa-Hattori et al., 1992; Cardellach et al., 1993; Blin et al., 1993). In PD muscle crude mitochondrial fractions, ETC dysfunction appears to extend beyond complex I. One study also found evidence for PD muscle mitochondrial dysfunction using ^{31}P magnetic resonance spectroscopy (Penn et al., 1995).

Reminiscent of the PD platelet mitochondria literature, several groups studying muscle ETC functioning failed to detect significant complex I decrements. The first negative report was that of Mann et al., which also failed to detect the defect in platelets (Mann et al., 1992a). The negative 1993 study of Anderson et al. used the insensitive NADH to cytochrome-*c* assay to calculate complex I activities (Anderson et al., 1993). DiDonato et al. reported citrate synthase-corrected ETC assay values that were comparable between PD and control subject homogenates and crude mitochondrial fractions (DiDonato et al., 1993), although citrate synthase activity was about 40% less in the PD crude mitochondrial fraction group. Potential "leakage" of this enzyme from defective, fragile mitochondria during the enrichment procedure raises the possibility of a confounding negative bias; a similar phenomenon was observed to a lesser extent in an earlier PD muscle study (Bindoff et al., 1991). In 1994, Reichmann et al. were unable to detect any ETC defects from muscle in six PD subjects versus five control subjects (Reichmann et al., 1994).

Information on PD lymphocyte ETC functioning is far less extensive than that for brain, platelets, and muscle. Yoshino et al. reported a complex II defect in PD lymphocyte crude mitochondrial preparations (Yoshino et al., 1992). In 1993, Barroso et al. found complex I and IV defects in PD lymphocyte homogenates (Barroso et al., 1993). Interestingly, 3 yr later this group was unable to corroborate this result using lymphocyte crude mitochondrial preparations (Martin et al., 1996).

Table 2
PD ETC Platelet Data

Study	Journal	Defect(s)	Comments
(A. Positive Studies)			
Parker et al., 1989	Ann. Neurol.	Comp. I,↓54%	Pure mitochondria
Yoshino et al., 1992	J. Neurol. Transm.	Comp. I↓27%, II↓20%	Crude mito fraction
Krige et al., 1992	Ann. Neurol,	Comp. III↓16%	Crude mito fraction
Benecke et al., 1993	Brain	Comp. I↓52%, IV↓30%	Pure mitochondria
Haas et al., 1995	Ann. Neurol.	Comp. I,↓25%, II/III↓20%	Pure mitochondria
(B. Negative Studies)			
Mann et al., 1992	Brain	None	Homogenate
Bravi et al., 1992	Mov. Disord.	None	Crude mito fraction; NADH–CCR
Blake et al., 1997	Mov. Disord.	None	Pure mitochondria; DB and albumin

In 1994, Mytelineou et al. found that oxidative decarboxylation of radiolabeled pyruvate was reduced in cultured PD fibroblasts (Mytelineou, 1994). Beyond simply demonstrating the presence of complex I impairment in another PD tissue, this observation is important because it indirectly addresses the question of why such impairment occurs (or, at least, does not occur) in this disease. Perpetuation of ETC dysfunction over time in cultured cells suggests environmental toxins are not a common cause of ETC dysfunction in PD. More recently, a reduction in $\Delta\psi m$ (the mitochondrial membrane potential) was demonstrated in PD fibroblasts (Tatton and Chalmers-Redman, 1998) (*See* Table 4.)

Clearly, complex I is defective in PD patients, although ETC aberration in this disease is not necessarily restricted to complex I. Complex I dysfunction in PD is present in multiple tissues and likely represents a systemic defect. Observation of ETC dysfunction in nondegenerating tissues indicates this phenomenon is not simply a consequence of neuronal demise. The magnitude of ETC impairment in affected patients is unclear and may increase as disease progression occurs. Because such defects are potentially subtle and/or their clear demonstration is technically difficult, use of appropriate methodology is absolutely essential. Negative studies using insufficiently enriched mitochondrial fractions require cautious interpretation.

Table 3
PD ETC Muscle Data

Study	Journal	Defect(s)	Comments
A. Positive Studies			
Bindoff et al., 1989	Lancet	Comp. I, II, IV↓40, 49, 40%	Crude mitos
Bindoff et al., 1991	J. Neurol. Sci.	Same as above	Same as above
Shoffner et al., 1991	Ann. Neurol.	Variably↓'d Comp. I, III, IV	Crude mitos
Nakagawa-Hattori et al., 1992	J. Neurol. Sci.	Comp. I↓49%	Crude mitos
Blin et al., 1993	J. Neurol. Sci.	Comp. I, III↓71, 35%	Crude mitos
Cardellach et al., 1993	Neurology	Comp. I, IV↓26, 68%	Crude mitos
Penn et al., 1995	Neurology	Nonspecific mito defect	In vivo ^{31}P MRS
(B. Negative Studies)			
Mann et al., 1992	Brain	None	Crude mitos
Anderson et al., 1993	J. Neurol. Neurosurg. Psychiatry	None	Crude mitos/NADH–CCR
DiDonato et al., 1993	Neurology	None	Homogenates/crude mitos
Reichmann et al., 1992	Eur. Neurol.	None	Crude mitos

Table 4
PD ETC Lymphocyte and Fibroblast Data

Study	Journal	Defect(s)	Comments
Yoshino et al., 1992	J. Neural Transm.	Comp II↓	Lymphocyte crude mitos
Barroso et al., 1993	Clin. Chem.	Comp. I, III, IV↓	Lymphocyte homogenates
Martin et al., 1996	Neurology	None	Lymphocyte crude mitos
Mytelineou et al., 1994	J. Neural Transm.	Comp I↓	Fibroblast pyruvate decarbox

Spectrophotometric assays of complex I should measure electron transfer from NADH to ubiquinone rather than to cytochrome-*c*. Finally, until data are presented to suggest otherwise, the use of citrate synthase ratioing for the "correction" of enriched mitochondrial fraction ETC determinations is contraindicated (Bindoff et al., 1991; DiDonato et al., 1993; Haas et al., 1995).

10.6. ORIGIN OF PD ETC DYSFUNCTION

Toxic Inhibition as a Potential Cause

Because ETC dysfunction in PD does not arise as a consequence of the diseased nigra's biochemical milieu, potential explanations for its presence are limited: either systemic inhibition of a normal ETC chain occurs or else the ETC chain itself is intrinsically abnormal.

Enthusiasm for an environmental etiology was fueled by the discovery of MPTP parkinsonism. Some epidemiologic studies suggested increased PD risks for those exposed to pesticides, herbicides, and well water (Rajput 1987; Hubble et al., 1993), but others did not corroborate this (Wong et al 1991; Semchuck et al., 1991). Recognition that multiple parkinsonism-inducing compounds (MPP+, cyanide, carbon monoxide) were ETC inhibitors (Ginsberg, 1980; Nicklas et al., 1985; Rosenberg et al., 1989) prompted additional screening for similar compounds. The most scrutinized group of environmental toxins belong to the isoquinolone family, which in 1986 were shown to weakly inhibit complex I (Hirata et al., 1986). More specifically, administration of tetrahydroisoquinolone (TIQ) to laboratory animals causes parkinsonism (Nagatsu and Yoshida, 1988; Yoshida et al., 1990), and TIQ compounds are found in certain common foods (Makino et al., 1988). Other candidate environmental toxins include the β-carboline compounds, which also may occur in the food chain (Melchior and Collins,

1982). Several β-carboline derivatives are weak inhibitors of complex I, although related β-carbolinium compounds provide more potent inhibition (Hoppel et al., 1987; Drucker et al., 1990).

Still, direct evidence for an exogenous toxin in idiopathic PD is lacking. Experimental data inconsistent with this idea come from the study of Mytelineou and colleagues, who observed that fibroblast complex I dysfunction perpetuates even within a controlled environment (Mytelineou et al., 1994). Theoretical arguments against the exogenous toxin hypothesis include the fact that the idiopathic form presents insidiously, unlike what is observed with MPTP exposure, and is always progressive, which is not the case with MPTP parkinsonism. Disease progression due to exogenous toxins suggests a need for serial re-exposure in all patients, an improbable scenario. Other lessons learned from the MPTP experience of the 1980s may apply. Not all of those exposed to MPTP developed parkinsonism. Therefore, even if exogenous toxins do contribute to PD etiology by causing ETC inhibition, other factors must play a role in determining why some exposed to a particular toxin develop disease, whereas others exposed to similar quantities over similar periods of time do not.

Systemic complex I dysfunction in PD patients could also potentially arise from the endogenous production of ETC inhibitory molecules. Isoquinolone and β-carboline derivatives, for example, are physiologic byproducts of human metabolism (Melchior and Collins, 1982). Matsubara et al. found that β-carbolinium cations were increased in the cerebrospinal fluid of PD patients (Matsubara et al., 1995). Some studies suggest that TIQ is increased in PD brain (Niwa et al., 1987), whereas others suggest the opposite (Ohta et al., 1987). The literature addressing this issue is inconclusive. Certainly, differential production of such molecules between PD and control populations would suggest that PD patients possess a unique nuclear genetic background, which remains neither proven nor disproven. Somewhat relevant to this issue are data from cybrid studies, which are described in a subsequent section of this chapter. These studies indicate that complex I dysfunction in PD mitochondria perpetuates in culture even after these mitochondria are separated from their original host nuclear background (Swerdlow et al., 1996; Gu et al., 1998; Shults and Miller, 1998).

Genetic Factors as a Potential Cause and Introduction to Mitochondrial Genetics

Genetic factors can potentially account for the pervasive depression of complex I activity seen in even nondegenerating PD tissues. Complex I is a multimeric enzyme that consists of 41 known protein subunits. Seven of these subunits are coded for by individual "NADH dehydrogenase" (ND) genes on mitochondrial DNA (mtDNA), and the rest are encoded by nuclear DNA (nDNA). There is little data to date addressing whether particular nuclear complex I gene mutations or polymorphisms associate with PD. One survey of the NDUFV2 complex I subunit-encoding gene on chromosome 18 found that persons with PD were more likely homozygous for a particular amino acid changing polymorphism than were controls (23.8% of PD patients, compared to 11.5% of controls) (Hattori et al., 1998).

More information regarding this issue comes from studies of mtDNA. Mitochondrial DNA consists of 16,569 nucleotides that form a plasmidlike circle. This DNA contains 37 genes, which encode 13 ETC proteins, 2 synthetic rRNA molecules, and 22 synthetic tRNA molecules. Although it is highly polymorphic, its entire "normal" sequence is known (Anderson et al., 1981; Howell et al., 1992) and is referred to as the "Cambridge sequence." The copy number of mtDNA molecules per mitochondrion also varies between tissues. Platelet mitochondria each contain only one copy, whereas brain mitochondria may carry more than five (Nass, 1969; Shmookler-Reis and Goldstein, 1983; Shuster et al., 1988). Cells containing hundreds of mitochondria therefore can possess thousands of copies of this genome. Variation can occur between individual mtDNA molecules within a cell, a condition known as heteroplasmy. The proportion of a heteroplasmic non-Cambridge mtDNA species is often expressed by what percentage of the cell's total mtDNA it constitutes. Wide ranges are possible, and deviant sequences may not necessarily confer phenotypic consequences if their percent composition is below a particular "threshold." Heteroplasmic ratios within a cell can vary over time (Shoubridge et al., 1990; Hayashi et al., 1991), so that initially silent subthreshold mtDNA variants may not remain so.

Mitochondrial DNA is maternally inherited. Although sperm contain mtDNA, it appears that male gametes make no functionally meaningful contribution to the ova's mtDNA content. As ontogeny proceeds through mitosis of the fertilized egg, daughter cells randomly acquire mitochondria and, hence, mtDNA through cytoplasmic partition, a process called "replicative segregation." When a developing female is herself generating an egg pool within her own forming ovaries, this random cytoplasmic partitioning can lead to differences in mtDNA between individual ova. Replicative segregation may even account for mitochondrial genetic differences between identical twins.

Because of interplay among heteroplasmic variability, threshold effects, and replicative segregation, even though mtDNA is maternally inherited expression of mtDNA-dependent phenotypes is often sporadic. For example, most cases of the classic mtDNA-derived disease Leber's Hereditary Optic

Neuropathy (LHON) present sporadically rather than within the confines of a recognizable, matrilineally dependent pedigree (Johns et al., 1991; Newman, 1993).

Parkinson's disease and LHON actually share several epidemiologic and biochemical features. Male predominance is seen in both diseases (Oostra et al., 1996; Tanner and Goldman, 1996). Most affected individuals in either case present sporadically or pseudosporadically (without a family history suggestive of autosomal dominant or recessive inheritance). Complex I impairment occurs in apparently unaffected tissues, but causes only anatomically specific neurologic dysfunction (substantia nigra projections to the striatum in PD, optic nerve in LHON). The magnitude of the complex I defect in pure mitochondrial fractions derived from platelets is perhaps upward of 50% (Parker et al., 1989a, 1989b; Benecke et al., 1993).

Data from mitochondrial DNA deletion, restriction fragment-length polymorphism (RFLP), and sequencing surveys do not conclusively rule in or out mtDNA aberration as either a primary or secondary event in PD. Such surveys for mtDNA mutation are complicated by issues of heteroplasmy and high polymorphism frequencies. Potential mutations may appear "invisible" to current screening techniques. Furthermore, it is often unclear whether or not detected mutations or polymorphic Cambridge sequence deviations are phenotypically relevant.

As discussed earlier, in 1989 Mizuno et al. performed immunoblotting studies of complex I subunits in PD striatum and found quantitative decreases in several small, likely mtDNA-encoded subunits (Mizuno et al., 1989). This result was potentially consistent with the presence of mtDNA deletion in the PD subjects, and in 1990, Ikebe et al. reported the presence of the approx 5-kb "common deletion" in PD striatum (Ikebe et al., 1990). Subsequent studies, however, failed to demonstrate increased heteroplasmic percentages of this deletion in PD brain or muscle as compared to controls (Schapira et al., 1990c; Lestienne et al., 1990, 1991; Mann et al., 1992b; Sandy et al., 1993; DiDonato et al., 1993). By themselves, large deletions of mtDNA are probably insufficient to cause PD, although some contribution to mitochondrial pathophysiology from such mutations is difficult to rule out.

Shoffner et al. used RFLP analysis to screen PD subjects for mtDNA polymorphisms (Shoffner et al., 1993). An A to G transition at position 4336 of the tRNA (Gln) was found in 5.3% of PD subjects but only in 0.7% of control subjects, an almost eightfold increase. Another study failed to replicate this result (Mayr-Wohlfart et al., 1997). Subsequent RFLP studies suggest other mtDNA polymorphisms may occur more commonly in those with PD. The study of Kosel et al. reported that G to A transition at position 5460 of the mtDNA

ND2 gene was present in 4 of 21 PD brains (19%), but only in 5 of 77 control brains (6%) (Kosel et al., 1996). An obvious limitation of RFLP surveys is that they cannot accurately screen an entire genome for mutations.

Sequencing surveys generally extend RFLP studies by revealing that Cambridge sequence mtDNA deviations are more common in PD subjects than in control subjects. Ikebe et al. performed polymerase chain reaction (PCR)-based sequencing of the mitochondrial genomes from five PD patients, and they found multiple nucleotide substitutions that were uncommon in a control cohort (Ikebe et al., 1995). Kosel et al. also used PCR-based sequencing to screen PD mtDNA ND genes for changes from the Cambridge sequence, and they found nucleotide substitutions in 19 of 22 subjects (Kosel et al., 1998). The functional consequences of the detected sequence changes is unclear. As mtDNA is highly polymorphic, it is not unusual to find Cambridge sequence deviation in healthy persons as well. A conservative generalization suggested by these studies is that no single, high-abundance mtDNA point mutation/polymorphism accounts for the majority of PD. This observation does not imply that seemingly random mtDNA polymorphisms are unimportant in this disease. For example, it is worth considering whether mtDNA polymorphisms influence PD risk, much like apolipoprotein E allele polymorphisms influence Alzheimer's disease risk.

It is important to note that PCR-based mtDNA sequencing is unable to detect low-percentage heteroplasmic substitutions. Even under optimal circumstances, heteroplasmic species whose abundance is under 30% lie in a sequencing "blind spot." This limited resolution is improved by ligating individual PCR copies of the gene under study into plasmids, transfecting these plasmids into *Escherichia coli*, and sequencing the bacteria-amplified clonal plasmids following their recovery. The probability of detecting a low-abundance mutation increases as more clonal *E. coli* plasmids are studied. This technique is laborious and expensive, and no such study of PD mtDNA exists.

In summary, definitive studies of PD mtDNA remain unperformed. This will likely remain the case until sequencing technologies advance to allow for the routine screening of low-abundance heteroplasmic mutations. Also intriguing but of unclear significance is the observation that mtDNA Cambridge sequence deviations are more common in PD than in non-PD subjects. Finally, if mtDNA aberration is important in PD, then the pathophysiologic relevance of point mutations or polymorphisms is likely greater than that of large deletions.

Cybrid studies provide powerful support for the contention that mtDNA aberration is important in PD. Cybrid experimental data suggest that pursuance of mtDNA mutational analysis is worthwhile in PD. Cybrid theory and methodology are described in Subheading 10.7., and data from PD cybrid studies are discussed.

10.7. CYBRID STUDIES OF PARKINSON'S DISEASE MITOCHONDRIAL DNA

Cytoplasmic nucleic acid was described (Ephrussi et al., 1949) years before it was further characterized as residing within mitochondria (Nass and Nass, 1963). Originally, it was termed "ρ" DNA. In the 1960s, it was observed that yeast cells could adapt themselves to anaerobic environments by depleting their mtDNA (Goldring et al., 1970; Nagley and Linnane, 1970). Several groups explored methods for the experimental inducement of this condition (Wiseman and Attardi, 1978; Desjardins et al., 1985, 1986; Morais et al., 1988). The DNA intercalator ethidium bromide (ETBR) proved useful in this regard. ETBR carries a net positive charge, so that it is concentrated within negatively charged mitochondrial matrices. Through trial and error, it is possible to determine a concentration that will interrupt mtDNA replication without overwhelming nDNA replication. Chronic exposure to this compound can then lead to varying degrees of mtDNA depletion. With long enough exposure total mtDNA depletion occurs, and cells manipulated in this way are traditionally called "ρ^0" cells. Maintenance of ρ^0 cells requires specific metabolic support, as these cells lack a functional ETC. High glucose is necessary to enhance anaerobic metabolism. Pyruvate is essential for the recycling of reducing equivalents. Because the function of the pyrimidine pathway enzyme dihydroorotate dehydrogenase is normally coupled to complex I, uridine supplementation is also essential if ongoing nucleotide synthesis is to occur (Gregoire et al., 1984; King and Attardi, 1989; Martinus et al., 1993).

In 1989, King and Attardi succeeded in repopulating an osteosarcoma ρ^0 cell line with exogenous mtDNA derived from enucleated fibroblasts (King and Attardi, 1989). In 1994, Chomyn et al. demonstrated that platelets could serve as an exogenous source of mtDNA for ρ^0 cell transfection (Chomyn et al., 1994). Platelets were chosen because they contain mitochondria and hence mtDNA, but do not contain nDNA. In experiments such as this, platelets and ρ^0 cells are coincubated in polyethylene glycol (PEG), which causes cell membrane disruption. Upon dilution of PEG membrane, integrity is restored, and occasionally ρ^0 cells incorporate platelet cytoplasmic contents to form cytoplasmic hybrids, or "cybrids." Cybrid cells are then selected for by maintaining post-PEG exposed cells in a pyruvate- and uridine- deficient media, in which unfused ρ^0 cells become nonviable and die. As this selection takes place, transferred mitochondria populate their host cells, which requires expression and replication of their mtDNA. After

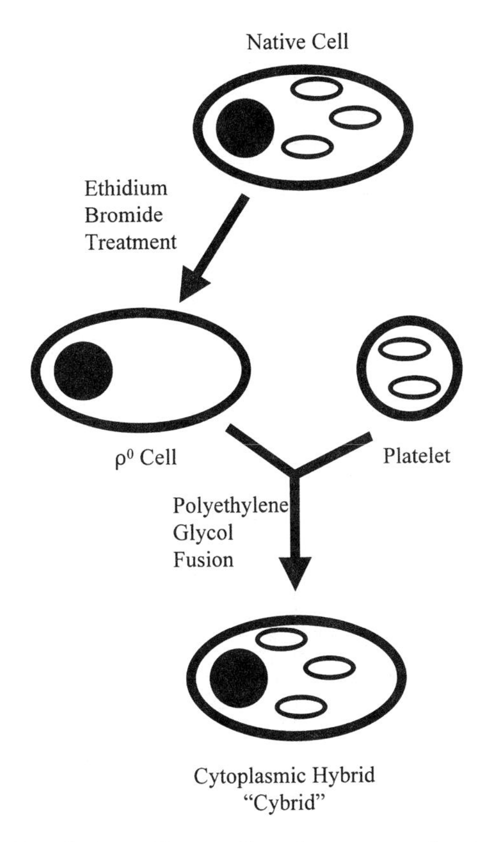

Fig. 4. The cybrid technique. The small ovals represent functional mitochondria that contain mtDNA, and the dark circles represent cell nuclei. Endogenous mitochondrial genes are removed from a culturable cell line. These ρ^0 cells are then fused with platelets to form a cytoplasmic hybrid ("cybrid"). Mitochondria from these platelets populate the cybrid cell, and as that cell propagates, its transferred mitochondria/mtDNA propagate as well. The cybrid cell expresses the mtDNA of the platelet donor individual. The functional ETC in these cells therefore contains subunits encoded by the native cell line's nDNA and subunits encoded by the platelet donor's mtDNA.

repeated cycles of replication, a functional ETC is created that contains nuclear encoded subunits from the original ρ^0 cell line nDNA and mitochondrially encoded subunits from the platelet donor's mtDNA. (*See* Fig. 4.)

Each time a donor individual's platelet mitochondria/mtDNA is transferred, unique cell lines are created that differ only in the origin of their mtDNA. The clonal nDNA of the original ρ^0 stock is maintained, so nDNA is identical between resultant cybrid cell lines. Culture conditions are also standardized, so environmental factors are taken into account. Theoretically, any differences in ETC functioning between cybrid cell lines must arise from differences in mtDNA. Evidence supporting this comes from studies in which mitochondria from patients with known mtDNA mutations were used to generate cybrid lines. Compared to cybrid lines generated from control subjects, cybrids containing mtDNA from persons with LHON (Vergani et al., 1995; Jun et al., 1996; Hofhaus et al., 1996) and mitochondrial encephalomyopathies (King et al., 1992; Chomyn et al., 1994; Massucci et al., 1995) are ETC impaired.

As conceived by Parker, cybrid technology was subsequently adapted for and applied to the screening of mtDNA aberration in PD. In order to optimize the utility of the intended PD cybrid model for anticipated studies of neurodegenerative pathophysiology, a "neuronallike" SH-SY5Y neuroblastoma ρ^0 cell line was prepared (MitoKor, San Diego, CA) (Miller et al., 1996). Platelets were obtained from 24 PD and 28 control subjects to generate 52 individual cybrid cell lines. ETC assays were performed on enriched but unpurified mitochondrial fractions. Complex I activity in the PD cell lines was 20% less than that of the controls, a significant difference. Complex IV activities were overall not significantly different, although some PD lines with low complex IV activites were observed (Swerdlow et al., 1996). These data indicate that mtDNA is defective (or suboptimally effective) in those with PD, to the extent that complex I dysfunction results. (*See* Fig. 5.)

Two other PD cybrid studies obtained similar results and reached similar conclusions. The study of Gu et al. found that mean complex I activity in seven PD cybrid lines was 25% less than that of seven control cybrid lines (Gu et al., 1998). Complex I activities in platelets used to generate cybrids was also determined in assays of crude mitochondrial fractions and revealed a comparable 24% reduction. This suggests that at least in PD plalelets, mtDNA can entirely account for and is solely responsible for ETC dysfunction. Also, Shults and Miller generated PD cybrid lines and reported that mean complex I activity in four disease lines was 31% less than that of six control lines (Shults and Miller, 1998).

Therefore, cybrid data strongly suggest complex I dysfunction in PD arises from mtDNA. The magnitude of the complex I defect detected in these studies ranges from 20% to 31% (Swerdlow et al., 1996: Gu et al., 1998; Shults and Miller, 1998), similar to what is found in assays of PD platelet crude mitochondrial fractions (Krige et al., 1992; Yoshino et al., 1992). This is not surprising because mitochondrial enrichment strategies in these studies approximates what is obtained by preparing crude mitochondrial fractions. Generalizing past experience from direct platelet ETC assays, failure to

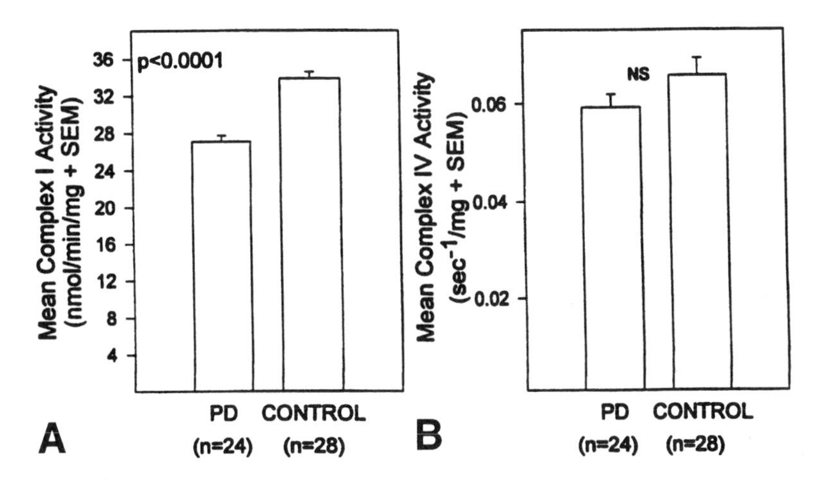

Fig. 5. Complex I and complex IV activities in Parkinson's disease and control cybrid cell lines. Complex I activity is significantly reduced in PD cybrids **(A)**. Activity of complex IV is not significantly reduced **(B)**. Data are from Swerdlow et al., 1996, with permission from Lippincott Wilkins and Wilkins, and are shown as means ± standard error of the mean.

prepare gradient-purified mitochondria probably leads to underestimation of the actual defect in the system under study. Other factors may also contribute to diminish the size of the observed cybrid complex I defect. Cybrid cultures represent a dynamic system in which cells are constantly replicating. If the putative mtDNA defect of PD is heteroplasmic, then cells within a newly created cybrid line may not all carry a similar mutational burden. High mutation levels could confer a replicative disadvantage, and so underrepresentation of cells with the greatest ETC impairment might occur (Wiseman and Attardi, 1978). Also when considering this issue, it is important to recognize that mtDNA in each of these cybrid studies was obtained from platelets. The complex I defect in this tissue may not equal that of substantia nigra.

Moreover, the magnitude of complex I dysfunction observed in these three published PD cybrid studies is quantitatively comparable to that seen in a LHON cybrid study (Jun et al., 1996). LHON arises from mtDNA ND gene mutation (Wallace et al., 1988), such mutation causes complex I dysfunction (Parker et al., 1989b), and anatomically specific neurodegeneration results (Howell, 1997). Therefore, the magnitude of the complex I defect in PD cybrid studies does not reflect upon the pathophysiologic relevance of mtDNA-encoded complex I dysfunction in this disease. This issue is further addressed by the fact that PD cybrid ETC impairment independently drives other neurodegenerative-related events, as is reviewed in Subheading 10.8.

10.8. FUNCTIONAL CONSEQUENCES OF mtDNA-DETERMINED COMPLEX I DYSFUNCTION IN PD CYBRIDS

Cybrid cell lines appear useful for the evaluation of how mtDNA-encoded ETC defects impact upon various aspects of cell physiology. Because mitochondria are major generators of reactive oxygen species (ROS), PD cybrids were surveyed for evidence of oxidative stress. In the study of Swerdlow et al., PD and control cybrid lines were incubated with the dye 2'7'-dichlorodihydrofluorescein diacetate (DCFDA). Exposure to ROS causes DCFDA to fluoresce, and as oxidative stress increases so does fluorescence of the dye (Royall and Ischiropoulos, 1993). PD cybrids displayed higher DCFDA fluorescence per cell than did control cybrids, consistent with the presence of either increased ROS generation resulting from ETC dysfunction or decreased free radical scavenging (Swerdlow et al., 1996). To address this, Cassarino et al. measured activities of the free-radical scavenging enzymes glutathione peroxidase, glutathione reductase, total superoxide dismutase (SOD), manganese SOD, copper-zinc SOD, and catalase. In PD cybrids, activities for all these enzymes were increased over those of the control cybrid cell lines, indicating that oxidative stress in PD cybrids is the result of increased ROS production (Cassarino et al., 1997). Upregulation of radical scavenging enzyme activities also occurred in native SH-SY5Y neuroblastoma cells placed in MPP+-supplemented media over several days. These data are consistent with several studies showing increased oxidative stress (Dexter et al., 1994; Sanchez-Ramos et al., 1994) and free-radical scavenging enzyme activities in PD brain (Martilla et al., 1988; Saggu et al., 1989; Kalra et al., 1992; Damier et al., 1993). (*See* Fig. 6.)

Mitochondria are also important in the regulation of neuronal calcium homeostasis (Werth and Thayer, 1994). Calcium handling in PD cybrids was quantitatively evaluated by fluorescence of the dye fura-2. At baseline, PD and control cybrid cytosolic calcium concentrations were equivalent. Cybrids were then exposed to the ETC uncoupler carbonyl cyanide *m*-chlorophenylhydrazone (CCCP), which dissipates the mitochondrial membrane potential and causes subsequent efflux of calcium from the organelle to the cytosol. Following CCCP, cytosolic calcium was higher in the control cell lines, suggesting that the ability of PD cybrid mitochondria to sequester calcium was diminished. In a related experiment, carbachol was used to induce inositol triphosphate-mediated cytosolic calcium transients. Recovery back to basal cytosolic calcium concentrations was retarded in PD

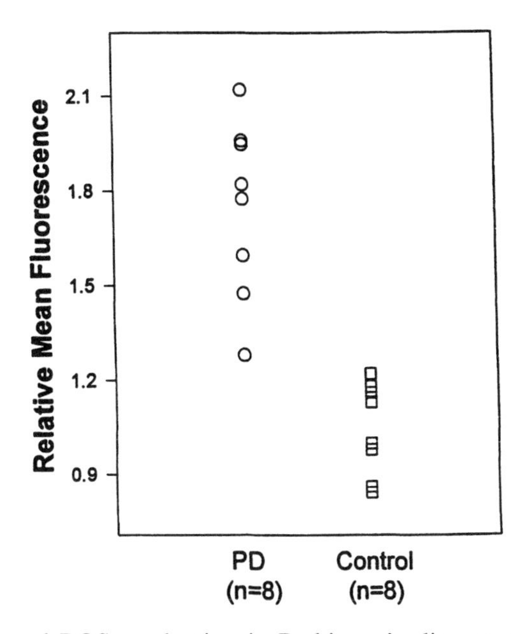

Fig. 6. Increased ROS production in Parkinson's disease cybrid lines. ROS generation was determined by fluorescence of the dye DCFDA, and higher values are indicative of increased ROS. The fluorescence of each individual cell line is plotted as its fluorescence relative to that of native SH-SY5Y cells, which was arbitrarily set as 1.0. (Data are from Swerdlow et al., 1996, with permission from Lippincott Wilkins and Wilkins.)

cybrid lines, most likely the result of this ETC-determined decrease in mitochondrial buffering capacity (Sheehan et al., 1997).

Cybrids may also provide a system for studying the interaction of mitochondrial toxins with ETC variants. In one experiment, PD and control cybrids underwent identical exposures to MPP+. After 24 and 48 h, more cell death was evident in the PD cell lines. Terminal deoxynucleotidyl transferase-mediated dUTP nick-end-labeling analysis further determined that cell demise occurred via apoptotic pathways. This observation illustrates that a specific, genetically determined biochemical lesion can increase a particular cell's vulnerability to an environmental toxin, potentially explaining why some individuals develop parkinsonism following a particular toxic exposure, whereas others similarly subjected to the same toxin at the same dose do not (Swerdlow et al., 1996).

Parkinson's disease cybrids thus support the contention that complex I impairment in this disease arises at the mtDNA genomic level. PD cybrids model several pathophysiologic phenomena that are either observed in PD subjects or are suspected of contributing to neurodegenerative cell death in general. Still, direct correlation of cybrid physiology with actual mtDNA mutation(s) or polymorphisms remains undone, and until this is accomplished, discourse regarding the nature of mtDNA aberration in PD will remain speculative.

10.9. mtDNA ABERRATION IN SPORADIC PD: INHERITED OR ACQUIRED, PRIMARY OR SECONDARY?

Cybrid models will not express ETC or other related pathophysiology unless these defects ultimately arise from defective mtDNA (Swerdlow et al., 1999). When cybrid studies do infer the presence of mtDNA aberration, as they do in PD, it is difficult to tell whether the implied mitochondrial genetic lesion is inherited or acquired. If mtDNA aberration in PD is inherited, then an etiologic role for mtDNA mutation in this disease is likely. On the other hand, the incidence of somatic mutation for mtDNA is fairly high (Linnane et al., 1990; Mecocci et al., 1994), and so mtDNA aberration may simply develop as a downstream event (albeit an important one) of a more basic underlying pathogenesis.

Relevant to this issue are data from a preliminary study in which cybrids expressing Contursi kindred mtDNA were created and characterized. The Contursi kindred is a large Italian family with autosomal dominant PD (Golbe et al., 1990). A highly penetrant mutation of the α-synuclein gene on chromosome 4 accounts for PD in this cohort (Polymeropoulos et al., 1997). Unlike cybrids that express mtDNA from sporadic PD subjects, cybrids expressing mtDNA from Contursi kindred PD patients do not exhibit decreased complex I activity (Parker et al., 1999). Thus, systemic degradation of mtDNA does not occur in α-synuclein PD. Mitochondrial DNA mutation in PD, therefore, cannot arise simply as part of a neurodegenerative "final common pathway." Absence of systemic mtDNA aberration in autosomal dominant PD suggests mtDNA-encoded complex I dysfunction is specific to sporadic PD. Such specificity argues against a somatic origin for mtDNA mutation in this disease.

Consistent with the inheritance hypothesis is a report of maternal inheritance bias in a large PD clinical database (Wooten et al., 1997a). Out of 265 PD probands, 5 instances of an affected parent with 2 affected children were observed. In each case, the affected parent was a mother ($p < 0.05$). Overall, however, maternal inheritance bias is not readily obvious in most PD epidemiologic studies (Zweig, 1992). This is possibly the result of the fact that many more men than women are affected, so that in this common, sporadic disease when a chance parent–child association occurs, the affected parent is more likely a father and maternal inheritance trends are obscured. When reviewing PD epidemiologic data that address inheritance in PD, therefore, correction for male predominance is required.

Because of male predominance, if no gender-specific genetic factor contributes to this disease, the expected gender ratio for affected parents of probands is not 1.0 father to 1.0 mother. Rather, the expected parent gender ratio should equal the gender ratio of the probands themselves. Meta-analysis of the PD epidemiologic literature yields more male than female probands (about 1.5 men to 1.0 women). Ascertainment of these probands for parents with PD, however, reveals more affected mothers than fathers (about 1.0 fathers to 1.2 mothers). These two intergenerational gender ratios are in statistical dysequilibrium, which can result only from an underrepresentation of affected fathers or else an overrepresentation of affected mothers (Swerdlow et al 1998b). The latter interpretation is certainly consistent with possible mitochondrial inheritance. Other factors (such as increased female longevity), however, complicate this type of analysis.

In 1997, Wooten et al. reported a kindred in which multiple members over at least three generations were affected with PD (Wooten et al., 1997b). Interestingly, in this family, transmission of the disease respects maternal inheritance lines. Bayesian analysis of this kindred suggests Mendelian inheritance of PD is possible but unlikely. Platelets were obtained from15 family members over 2 generations, including PD affected and unaffected individuals, and used to generate cybrid cell lines (Swerdlow et al., 1998c). These cell lines were assayed for ETC activities, oxidative stress, and the presence of morphologically abnormal mitochondria. For statistical analysis, cybrid lines were classified as belonging to one of two groups. If the platelet donor for a particular cybrid cell line was connected to the family through an uninterrupted maternal lineage, that cell line was placed in a matrilineal descendent group. Likewise, platelet donors descended from males in this family (who would not possess the hypothesized pathogenic mitochondrial genome) were placed in a patrilineal descendent group. Mean complex I activity in the matrilineal descendent group cybrids was less than that of the paternal descendent cell lines. Oxidative stress was increased in the matrilineal descendent cybrids, as was the quantitative presence of abnormal mitochondrial morphologies. These analyses were then carried out just for the eight cybrid lines representing the youngest generation, which contained mtDNA from individuals who were mostly in their third decade and who were all phenotypically normal. Similar results were obtained. Thus, an inherited mtDNA species seems to cause complex I dysfunction in this family, and this ETC defect may precede onset of a symptomatic PD phenotype. (*See* Fig. 7.)

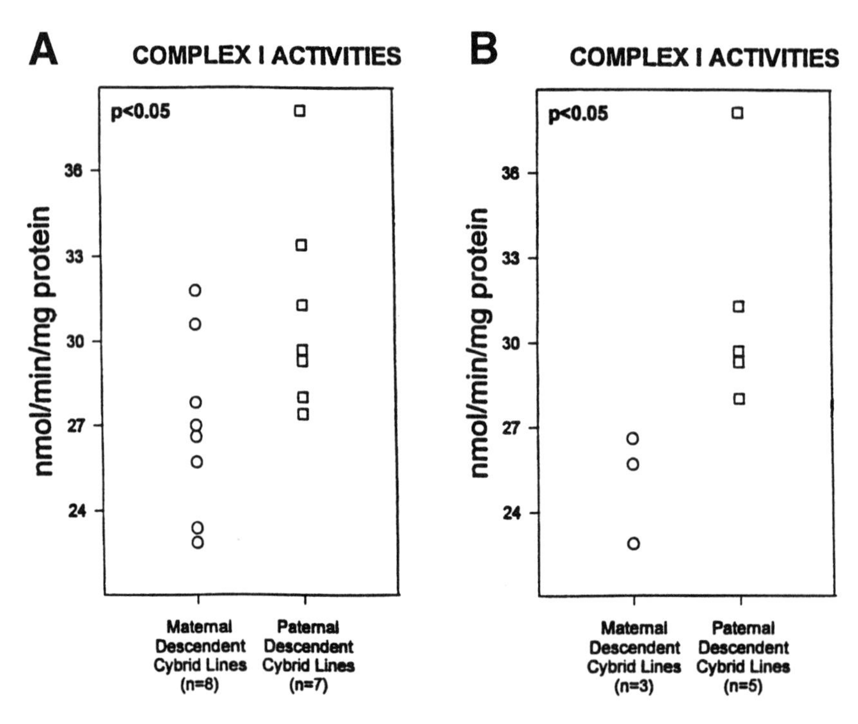

Fig. 7. Complex I activities in a family with maternally inherited Parkinson's disease over several generations. Individual cybrid lines are grouped depending on whether the platelet donor is a paternally or maternally descended member of the family. See text for details. **(A)** All cybrid lines; **(B)** data from youngest generation; no one in this generation is yet affected with Parkinson's disease. (Data are from Swerdlow et al., 1998c., with permission from Lippincott Wilkins and Wilkins.)

10.10. OTHER EVIDENCE OF MITOCHONDRIAL PATHOLOGY IN PD

Krebs cycle dysfunction may also occur in PD. The Krebs cycle takes place within mitochondrial matrices, and electrons from this pathway are transferrable to the ETC via succinate dehydrogenase. Immunostaining of the Krebs enzyme α-ketoglutarate dehydrogenase (α-KGDH) appears reduced in PD brain (Mizuno et al., 1994). Activity of α-KGDH is also decreased in mitochondrial preparations exposed to MPP+ (Mizuno et al., 1987), although this may simply arise as a consequence of ETC-dependent free radical generation (Joffe et al., 1998). A particular allelic polymorphism of the α-KGDH dihydrolipoyl transsuccinylase gene on chromosome 14 is possibly more common in PD subjects than controls, even though this nucleotide variation does not alter the amino acid sequence of the protein

(Kobayashi et al., 1998). Regardless of its origin, impairment of this enzyme could augment adverse consequences of complex I dysfunction by impeding compensatory electron flux through complex II.

Mitochondria may also play a role in Lewy body formation. In 1997, Gai et al. reported that Lewy bodies in part represent degenerating mitochondria (Gai et al., 1997). However, to date, this work is described only in abstract form.

10.11. MITOCHONDRIA AND CELL DEATH MECHANISMS IN PD

Neuronal demise in PD substantia nigra appears to proceed through apoptotic pathways (Mochizuki et al., 1996; Anglade et al., 1997; Tatton et al., 1998). This observation highlights the potential importance of mitochondrial impairment in this disease, because mitochondria essentially "hold the keys" to apoptosis (Yang et al., 1997; Kluck et al., 1997; Petit et al., 1997; Zamzami et al., 1997). Mitochondrial failure may also enable excitotoxicity, and in this capacity further facilitate cell death (Novelli et al., 1988; Beal, 1998).

Cybrid studies reveal that in sporadic PD, mtDNA-encoded complex I dysfunction drives a diverse set of pathophysiologic processes that either alone or in combination can trigger apoptotic or excitotoxic cell death. These include diminished oxidative phosphorylation, decreased $\Delta\psi m$, elevated free radical generation with increased oxidative stress, and impaired calcium handling (Swerdlow et al., 1996; Cassarino et al., 1997; Sheehan et al., 1997; Gu et al., 1998). Therefore, although complex I dysfunction may represent the primary biochemical pathology in PD, cell death in this disease probably does not occur simply because of ATP depletion. In fact, it is not clear whether ATP depletion substantially contributes at all to PD neuronal death (Sheehan et al., 1997). Thus, even subtle ETC dysfunction in this disease could indirectly account for insidious and progressive late-onset neurodegeneration through the initiation of other pathophysiologic events.

10.12. CONCLUSIONS

Although PD is currently defined as a stereotyped clinical and histopathologic syndrome, at the molecular level it actually consists of several different diseases that share a common phenotype. This is not surprising, as PD can present within Mendelian, maternal, or sporadic contexts. Autosomal gene mutations that account for some Mendelian cases are now known (Polymeropoulos et al., 1997; Kruger et al., 1998; Kitada et al., 1998). For these Mendelian subsets, pathogenic insights will result from the study of defined nonmitochondrial genes and gene products. Meanwhile, in most persons with PD, mitochondrial dysfunction occurs in the form of a

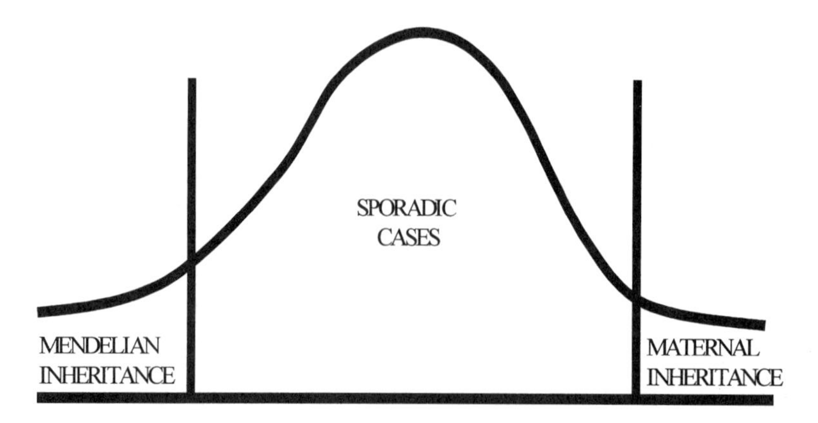

Fig. 8. The epidemiology of Parkinson's disease. Most patients present "sporadically," although Mendelian and maternal inheritance patterns are occasionally seen. Parkinson's disease is a clinical syndrome that arises from multiple molecular defects.

systemic complex I defect. This defect, which appears to arise from mutation of mtDNA, can lead to oxidative stress and alter calcium homeostasis. Such pathophysiology may underly and explain apoptotic neurodegeneration in PD. In some and perhaps many, mtDNA may represent a primary or determinant cause of PD. Multifactorial causes of mitochondrial impairment could also apply, and interplay among a person's mitochondrial genome, nuclear background, and environmental experience is not ruled out. (*See* Fig. 8.)

REFERENCES

Anderson, J. J., Bravi, D., Ferrari, R., Davis, T. L., Baronti, F., Chase, T. N., et al. (1993) No evidence for altered muscle mitochondrial function in Parkinson's disease. *J. Neurol. Neursurg. Psychiatry* **56,** 477–480.

Anderson, S., Bankier, A. T., Barell, B. G., de Bruijn, M. H., Coulson, A. R., Drouin, J., et al. (1981) Sequence and organization of the human mitochondrial genome. *Nature* **290,** 457–465.

Anglade, P., Vyas, S., Javoy-Agid, F., Herrero, M. T., Michel, P. P., Marquez, J., et al. (1997) Apoptosis and autophagy in nigral neurons of patients with Parkinson's disease. *Histol. Histopathol.* **12,** 25–31.

Barroso, N., Campos, Y., Huertas, R., Esteban, J., Molina, J. A., Alonso, A., et al. (1993) Respiratory chain enzyme activities in lymphocytes from untreated patients with Parkinson disease. *Clin. Chem.* **39,** 667–669.

Beal, M. F. (1998) Excitotoxicity and nitric oxide in Parkinson's disease pathogenesis. *Ann. Neurol.* **44(Suppl.1),** S110–S114.

Benecke, R., Strumper, P., and Weiss, H. (1993) Electron transfer complexes I and IV of platelets are abnormal in Parkinson's disease but normal in Parkinson-plus syndromes. *Brain* **116,** 1451–1455.

Bindoff, L. A., Birch-Machin, M., Cartlidge, N. E. F., Parker, W. D. Jr., and Turnbull, D. M. (1989) Mitochondrial function in Parkinson's disease. *Lancet* **2,** 49.

Bindoff, L. A., Birch-Machin, M. A., Cartlidge, N. E., Parker, W. D. Jr., and Turnbull, D. M. (1991) Respiratory chain abnormalities in skeletal muscle from patients with Parkinson's disease. *J. Neurol. Sci.* **104,** 203–208.

Blake, C. I., Spitz, E., Leehey, M., Hoffer, B. J., and Boyson, S. J. (1997) Platelet mitochondrial respiratory chain function in Parkinson's disease. *Mov. Disord.* **12,** 3–8.

Blin, O., Desnuelle, C., Rascol, O., Borg, M., Peyro Saint Paul, H., Azulay, J. P., et al. (1994) Mitochondrial respiratory failure in skeletal muscle from patients with Parkinson's disease and multiple system atrophy. *J. Neurol. Sci.* **125,** 95–101.

Bravi, D., Anderson, J. J., Dagani, F., Davis, T. L., Ferrari, R., Gillespie, M., et al. (1992) Effect of aging and dopaminomimetic therapy on mitochondrial respiratory function in Parkinson's disease. *Mov. Disord.* **7,** 228–231.

Burns, R. S., Chiueh, C. C., Markey, S. P., Ebert, M. H., Jacobowitz, D. M., and Kopin, I. J. (1983) A primate model of parkinsonism: selective destruction of dopaminergic neurons in the pars compacta of the substantia nigra by *N*-methyl-4-phenyl-1,2,3,6-tetrahdropyridine. *Proc. Natl. Acad. Sci. USA* **80,** 4546–14550.

Calne, D. B., Snow, B. J., and Lee, C. (1992) Criteria for diagnosing Parkinson's disease. *Ann. Neurol.* **32(Suppl.),** S125–S127.

Cardellach, F., Marti, M. J., Fernandez-Sola, J., Marin, C., Hoek, J. B., Tolosa, E., et al. (1993) Mitochondrial respiratory chain activity in skeletal muscle from patients with Parkinson's disease. *Neurology* **43,** 2258–2262.

Cassarino, D. S., Fall, C. P., Swerdlow, R. H., Smith, T. S., Halvorsen, E. M., Miller, S. W., et al. (1997) Elevated reactive oxygen species and antioxidant enzyme activities in animal and cellular models of Parkinson's disease. *Biochim. Biophys. Acta* **1362,** 77–86.

Chiba, K., Trevor, A., and Castagnoli, N., Jr. (1984) Metabolism of the neurotoxic tertiary amine, MPTP by brain monoamine oxidase. *Biochem. Biophys. Res. Commun.* **120,** 574–578.

Chomyn, A., Lai, S. T., Shakeley, R., Bresolin, N., Scarlatto, G., and Attardi, G. (1994) Platelet-mediated transformation of mtDNA-less human cells: analysis of phenotypic variability among clones from normal individuals — and complementation behavior of $tRNA^{lys}$ mutation casusing myoclonic epilepsy and ragged red fibers. *J. Hum. Genet.* **54,** 966–974.

Cooper, J. M., Saniel, S. E., Marsden, C. D., and Schapira, A. H. V. (1995) L-Dihydroxyphenylalanine and complex I deficiency in Parkinson's disease brain. *Mov. Disord.* **10,** 295–297.

Cotzias, G. C., Van Woert, M. H., and Schiffer, L. M. (1967) Aromatic amino acids and modification of parkinsonism. *N. Engl. J. Med.* **276,** 374–379.

Damier, P., Hirsch, E. C., Zhang, P., Agid, Y., and Javoy-Agid, F. (1993) Glutathione peroxidase, glial cells and Parkinson's disease. *Neuroscience* **52,** 1–6.

Davis, G. C., Williams, A. C., Markey, S. P., Ebert, M. H., Caine, E. D., Reichert, C. M., et al. (1979) Chronic Parkinsonism secondary to intravenous injection of meperidine analogues. *Psychol. Res.* **1,** 249–254.

Desjardins, P., Frost, E., and Morais, R. (1985) Ethidium bromide-induced loss of mitochondrial DNA from primary chicken embryo fibroblasts. *Mol. Cell. Biol.* **5,** 1163–1169.

Desjardins, P., de Muys, J. M., and Morais, R. (1986) An established avian fibroblast cell line without mitochondrial DNA. *Somat. Cell. Mol. Genet.* **12,** 133–139.

Dexter, D. T., Holley, A. E., Flitter, W. D., Slater, T. F., Wells, F. R., Daniel, S. E., et al. (1994) Increased levels of lipid hydroperoxides in the Parkinsonian substantia nigra, an HPLC and ESR study. *Mov. Disord.* **9,** 92–97.

DiDonato, S., Zeviani, M., Giovannini, P., Savarese, N., Rimoldi, M., Mariotti, C., et al. (1993) Respiratory chain and mitochondrial DNA in muscle and brain in Parkinson's disease patients. *Neurology* **43,** 2262–2268.

Drucker, G., Raikoff, K., Neafey, E. J., and Collins, M. A. (1990) Dopamine uptake inhibitory capacities of β-carboline and 3,4-dihydro- β-carboline analogs of *N*-methyl-4-phenyl-1,2,3,6-tetrahydropyridine (MPTP) oxidation products. *Brain Res.* **509,** 125–133.

Ehringer, H. and Hornykiewicz, O. (1960) Verteilung von noradrenalin und dopamin im Gehirn des Menschen und ihr Verhalten bei Erkrankungen des extrapyramidalen systems. *Wien Kin. Wochenschr.* **72,** 1236–1239.

Ephrussi, B., Hottinger, H., and Chimenes, A. M. (1949) Action de l'acriflavine sur les levures. I. La mutation "petite clonie" *Ann. Instit. Pasteur* **76,** 531.

Evarts, E. V. and Thach, W. T. (1969) Motor mechanisms of the CNS: cerebrocerebellar interrelations. *Ann. Rev. Physiol.* **31,** 451–498.

Forno, L. S., Langston, J. W., DeLanney, L. E., Irwin, I., and Ricuarte, G. A. (1986) Locus ceruleus lesions and eosinophilic inclusions in MPTP-treated monkeys. *Ann. Neurol.* **20,** 449–455.

Gai, W. P., Blumbergs, P. C., and Blessing, W. W. (1997) The ultrastructure of Lewy neurites. *Mov. Disord.* 1**2(Suppl. 1),** 5.

Gibb, W. R. and Lees, A. J. (1988) The relevance of the Lewy body to the pathogenesis of idiopathic Parkinson's disease. *J. Neurol. Neurosurg. Psychiatry* **51,** 745–752.

Ginsberg, M. D. (1980) Carbon monoxide. in *Experimental and Clinical Neurotoxicology*. (Spencer, P. S. and Schaumberg, H. H., eds.) Williams & Wilkins, Baltimore, pp. 374–394.

Golbe. L. I., Di Iorio, G., Bonavita, V., Miller, D. C., Duvoisin, R. C. (1990) A large kindred with autosomal dominant Parkinson's disease. *Ann. Neurol.* **27,** 276–282.

Goldring, E. S., Grossman, L. I., Krupnick, D., Cryer, D. R., and Marmur, J. (1970) The petite mutation in yeast. Loss of mitochondrial deoxyribonucleic acid during induction of petites with ethidium bromide. *J. Mol. Biol.* **52,** 323–335.

Gregoire, M., Morais, R., Quilliam, M. A., and Gravel, D. (1984) On auxotrophy for pyrimidines of respiration-deficient chick embryo cells. *Eur. J. Biochem.* **142,** 49–55.

Gu, M., Cooper, J. M., Taanman, J. W., and Schapira, A. H. V. (1998) Mitochondrial DNA transmission of the mitochondrial defect in Parkinson's disease. *Ann. Neurol.* **44,** 177–186.

Haas, R. H., Nasirian, F., Nakano, K., Ward, D., Pay, M., Hill, R., et al. (1995) Low platelet mitochondrial complex I and complex II/III activity in early untreated Parkinson's disease. *Ann. Neurol.* **37,** 714–722.

Hassler, R. (1938) Zur pathologie der paralysis agitans und des postenzephalitisschen parkinsonismus. *J. Psychol. Neurol.* **48,** 387–476.

Hattori, N., Tanaka, M., Ozawa, T., and Mizuno, Y. (1994) Immunohistochemical studies on complexes I, II, III, and IV of mitochondria in Parkinson's disease. *Ann. Neurol.* **30,** 563–571.

Hattori, N., Yoshino, H., Tanaka, M., Suzuki, H., and Mizuno, Y. (1998) Allele in the 24-kDa subunit gene (NDUFV2) of mitochondrial complex I and susceptibility to Parkinson's disease. *Genomics* **49,** 52–58.

Hayashi, J. I., Ohta, S., Kikuchi, A., Takemitsu, M., Goto, Y., and Nonaka, I. (1991) Introduction of disease-related mitochondrial DNA deletions into HeLa cells lacking mitochondrial DNA results in mitochondrial dysfunction. *Proc. Natl. Acad. Sci. USA* **88,** 10614–10618.

Heikkila, R. E., Manzino, L., Cabbat, F. S., and Duvoisin, R. C. (1984) Protection against the dopaminergic neurotoxicity of 1-methyl-4-phenyl-1,2,5,6-tetrahydropyridine by monoamine oxidase inhibitors. *Nature* **311,** 467–469.

Hirata, Y., Sugimura, H., Takei, H., and Magatsu, T. (1986) The effects of pyridinium salts, structurally related compounds of 1-methyl-4-phenylpyridinium ion (MPP+), on tyrosine hydroxylation in rat striatal tissue slices. *Brain Res.* **397,** 341–344.

Hofhaus, G., Johns, D. R., Hurko, O., Attardi, G., and Chomyn, A. (1996) Respiration and growth defects in transmitochondrial cell lines carrying the 11778 mutation associated with Leber's hereditary optic neuropathy. *J. Biol. Chem.* **271,** 13,155–13,161.

Hoppel, C. L., Grinblatt, D., Kwok, H. C., Arora, P. K., Singh, M. P., Sayre, L. M., et al. (1987) Inhibition of mitochondrial respiration by analogs of 4-phenylpyridine and 1-methyl-4-phenylpyridinium cation (MPP+), the neurotoxic metabolite of MPTP. *Biochem. Biophys. Res. Commun.* **148,** 684–693.

Howell, N., McCullough, D. A., Kubacka, I., Halvorson, S., and Mackey, D. (1992) The sequence of human mtDNA, the question of errors versus polymorphisms. *Am. J. Hum. Genet.* **50,** 1333–1340.

Howell, N. (1997) Leber hereditary optic neuropathy, mitochondrial mutations and degeneration of the optic nerve. *Vision Res.* **37,** 3495–3507.

Hubble, J. P., Cao, T., Hassanein, R. E., Neuberger, J. S., and Koller, W. C. (1993) Risk factors for Parkinson's disease. *Neurology* **43,** 1693–1697.

Ikebe, S. I., Tanaka, M., Ohno, K., Sato, W., Hattori, K., Kondo, T., et al. (1990) Increase of deleted mitochondrial DNA in the striatum in Parkinson's disease and sensescence. *Biochem. Biophys. Res. Commun.* **170,** 1044–1048.

Ikebe, S. I., Tanaka, M., and Ozawa, T. (1995) Point mutations of mitochondrial genome in Parkinson's disease. *Mol. Brain Res.* **28,** 281–295.

Janetzky, B., Hauck, S., Youdim, M. B., Reiderer, P., Jellinger, K., Pantucek, F., et al. (1994) Unaltered aconitase activity, but decreased complex I activity in substantia nigra pars compacta of patients with Parkinson's disease. *Neurosci. Lett.* **169,** 126–128.

Javitch, J. A., D'Amato, R. J., Strittmatter, S. M., and Snyder, S. H. (1985) Parkinsonism-inducing neurotoxin, methyl-4-phenyl-1,2,3,6-tetrahydropyridine: uptake of the metabolite *N*-methyl-4-phenylpyridine by dopamine neurons explains selective toxicity. *Proc. Natl. Acad. Sci. USA* **82,** 2173–2177.

Joffe, G. T., Parks, J. K., and Parker, W. D., Jr. (1998) Secondary inhibition of α-ketoglutarate dehydrogenase complex by MPTP. *NeuroReport* **9,** 2781–2783.

Johns, D. R., Lessell, S., and Miller, N. R. (1991) Molecularly confirmed Leber's hereditary optic neuropathy. *Neurology* **41(Suppl. 1),** 347.

Jun, A. S., Trounce, I. A., Brown, M. D., Shoffner, J. M., and Wallace, D. C. (1996) Use of transmitochondrial cybrids to assign a complex I defect to the mitochondrial DNA-encoded NADH dehydrogenase subunit 6 gene mutation at nucleotide pair 14459 that causes Leber Hereditary Optic Neuropathy and Dystonia. *Mol. Cell. Biol.* **16,** 771–777.

Kalra, J., Rajput, A. H., Mantha, S. V., and Prasad, K. (1992) Serum antioxidant enzyme activity in Parkinson's disease. *Mol. Cell. Biochem.* **110,** 165–168.

King, M. P. and Attardi, G. (1989) Human cells lacking mtDNA**,** repopulation with exogenous mitochondria by complementation. *Science* **246,** 500–503.

King, M. P., Koga, Y., Davidson, M., and Schon, E. A. (1992) Defects in mitochondrial protein synthesis and respiratory chain activity segregate with the tRNA$^{Leu\ (UUR)}$ mutation associated with mitochondrial myopathy, encephalopathy, lactic acidosis, and strokelike episodes. *Mol. Cell. Biol.* **12,** 480–490.

Kitada, T., Asakawa, S., Hattori, N., Matsumine, H., Yamamura, Y., Minoshima, S., et al. (1998) Mutations in the parkin gene cause autosomal recessive juvenile parkinsonism. *Nature* **392,** 605–608.

Kluck, R. M., Bossy-Wetzel, E., Green, D. R., and Newmeyer, D. D. (1997) The release of cytochrome c from mitochondria**,** A primary site for Bcl-2 regulation of apoptosis. *Science* **275,** 1132–1136.

Kobayashi, T., Matsumine, H., Matsubayashi, S., Matuda, S., and Mizuno, Y. (1998) Polymorphism of the gene encoding dihydrolipoamide succinyltransferase, a subunit of α-ketoglutarate dehydrogenase complex, is associated with the susceptibility to Parkinson disease: a population based study. *Ann. Neurol.* **43,** 120–123.

Kosel, S., Lucking, C. B., Egensperger, R., Mehraein, P., and Graeber, M. B. (1996) Mitochondrial NADH Dehydrogenase and CYP2D6 genotypes in Lewy-Body parkinsonism. *J. Neurosci. Res.* **44,** 174–183.

Kosel, S., Grasbon–Frodl, E. M., Mautsch, U., Egensperger, R., von Eitzen, U., Frishman, D., et al. (1998). Novel mutations of mitochondrial complex I in pathologically proven Parkinson disease. *Neurogenetics* **1,** 197–204.

Krige, D., Carrol, M. T., Cooper, J. M., Marsden, C. D., and Schapira, A. H. V. (1992) Platelet mitochondrial function in Parkinson's disease. The Royal Kings and Queens Parkinson Disease Research Group. *Ann. Neurol.* **32,** 782–788.

Kruger, R., Kuhn, W., Muller, T., Woitalla, D., Graeber, M., Kosel, S., et al. (1998) Ala30Pro mutation in the gene encoding α-synuclein in Parkinson's disease. *Nature Genet.* **18,** 106–108.

Langston, J. W., Ballard, P. A., Tetrud, J. W., and Irwin, I. (1983) Chronic parkinsonism in humans due to a product of meperidine-analog synthesis. *Science* **219,** 979–980.

Lestienne, P., Nelson, J., Riederer, P., Jellinger, K., and Reichmann, H. (1990)Normal mitochondrial genome in brain from patients with Parkinson's disease and complex I defect. *J. Neurochem.* **55,** 1810–1812.

Lestienne, P., Nelson, I., Reiderer, P., Reichmann, H., and Jellinger, K. (1991) Mitochondrial DNA in postmortem brain from patients with Parkinson's disease. *J. Neurochem.* **56,** 1819.

Lewy, F. H. (1912) Paralysis agitans, I. Pathologische anotomie, in *Handbuch der Neurologie III.* Springer-Verlag, New York, pp. 920–933.

Lilienfeld, D. E. (1994) An epidemiologic overview of amyotrophic lateral sclerosis, Parkinson's disease, and dementia of the Alzheimer type, in *Neurodegenerative Diseases* (Calne, D. B., ed.) W. B. Saunders, Philadelphia, pp. 399–425.

Linnane, A. W., Marzuki, S., Ozawa, T., and Tanaka, M. (1990) Mitochondrial DNA mutations as an important contributor to aging and degenerative diseases. *Lancet* **1(8639),** 642–645.

Makino, Y., Ohta, S., Tachikawa, O., and Hirobe, M. (1988) Presence of tetrahydroisoquinoline and 1-methyl-tetrahydroisoquinoline in foods: compounds related to Parkinson's disease. *Life Sci.* **43,** 373–378.

Mann, V. M., Cooper, J. M., Krige, D., Daniel, S. E., Schapira, A. H. V., and Marsden, C. D. (1992) Brain, skeletal muscle and platelet homogenate mitochondrial function in Parkinson's disease. *Brain* **115,** 333–342.

Mann, V. M., Cooper, J. M., and Schapira, A. H. V. (1992) Quantitation of a mitochondrial DNA deletion in Parkinson's disease. *FEBS* **299,** 218–222.

Mark, M. H., Dickson, D. W., Sage, J. I., and Duvoisin, R. C. (1996) The clinicopathologic spectrum of Lewy body disease, in *Parkinson's Disease,* Advances in Neurol Vol. 69, (Battistin, L., Scarlato, G., Caraceni, T., and Ruggieri, S. eds.), Lippincott-Raven, Philadelphia,pp. 315–318.

Markey, S. P., Johannessen, J. H., Chiueh, C. C., Burns, R. S., and Herkenham, M. A. (1984) Intraneuronal generation of a pyridinium metabolite may cause drug-induced parkinsonism. *Nature* **311,** 464–467.

Martilla, R. J., Lorentz, H., and Rinne, U. K. (1988) Oxygen toxicity protecting enzymes in Parkinson's disease. Increase of superoxide dismutase-like activity in the substantia nigra and basal nucleus. *J. Neurol. Sci.* **86,** 321–331.

Martin, M. A., Molina, J. A., Jimenenz-Jimenez, F. J., Benito-Leon, J., Orti-Pareja, M., Campos, Y., et al. (1996) Respiratory-chain enzyme activities in isolated mitochondria of lymphocytes from untreated Parkinson's disease patients. *Neurology* **46,** 1343–1346.

Martinus, R. D., Linnane, A. W., and Nagley, P. (1993) Growth of ρ^0 human namalwa cells lacking oxidative phosphorylation can be sustained by redox compounds potassium ferricyanide or coenzyme Q_{10} putatively acting through the plasma membrane oxidase. *Biochem. Mol. Biol. Int.* **31,** 997–1005.

Masucci, J. P., Davidson, M., Koga, Y., Schon, E. A., and King, M. P. In vitro analysis of mutations causing myoclonus epilepsy with ragged–red fibers in the mitochondrial $tRNA^{Lys}$ gene, two genotypes produce similar phenotypes. Mol Cell Biol 1995;15, 2872–2881.

Matsubara, K., Kobayashi, S., Kobayashi, Y., Yamashita, K., Koide, H., Hatta, M., et al. (1995) β-Carbolinium cations, endogenous MPP+ analogs, in the lumbar cerebrospinal fluid of patients with Parkinson's disease. *Neurology* **45,** 2240–2245.

Mayr-Wohlfart, U., Rodel, G., and Hennesber, A. (1997) Mitochondrial tRNA (Gln) and tRNA (Thr) gene variants in Parkinson's disease. *Eur. J. Med. Res.* **2,** 111–113.

Mecocci, P., MacGarvey, U., and Beal, M. F. (1994) Oxidative damage to mitochondrial DNA is increased in Alzheimer's disease. *Ann. Neurol.* **36,** 747–751.

Melchior, C. and Collins, M. A. (1982) The route and significance of endogenous synthesis of alkaloids in animals. *Crit. Rev. Toxicol.* **9,** 313–356.

Miller, S. W., Trimmer, P. A., Parker, W. D. Jr., and Davis, R. E. (1996) Creation and characterization of mitochondrial DNA depleted cell lines with "neuronal-like" properties. *J. Neurochem.* **67,** 1897–1907.

Mizuno, Y., Saitoh, T., and Sone, N. (1987) Inhibition of mitochondrial alpha-ketoglutarate dehydrogenase by 1-methyl-4–phenyl-pyridinium ion. *Biochem. Biophys. Res. Commun.* **143,** 971–976.

Mizuno, Y., Ohta, S., Tanaka, M., Takamiya, S., Suzuki, K., Sato, T., et al. (1989). Deficiencies in complex I subunits of the respiratory chain in Parkinson's disease. *Biochem. Biophys. Res. Commun.* **163,** 1450–1455.

Mizuno, Y., Suzuki, K., and Ohta, S. (1990) Postmortem changes in mitochondrial respiratory enzymes in brain and a preliminary observation in Parkinson's disease. *J. Neurol. Sci.* **96,** 49–57.

Mizuno, Y., Matuda, S., Yoshino, H., Mori, H., Hattori, N., Ikebe, S. (1994) An immunohistochemical study on α-ketoglutarate dehydrogenase complex in Parkinson's disease. *Ann. Neurol.* **35,** 204–210.

Mochizuki, H., Goto, K., Mori, H., and Mizuno, Y. (1996) Histochemical detection of apoptosis in Parkinson's diease. *J. Neurol. Sci.* **131,** 120–123.

Morais, R., Desjardins, P., Turmel, C., and Zinkewich-Peotti, K. (1988) Development of continuous avian cell lines depleted of mitochondrial DNA. *In Vitro Cell Dev. Biol.* **24,** 649–658.

Mytilineou, C., Werner, P., Molinari, S., Di Roco, A., Cohen, G., and Yahr, M. D. (1994) Impaired oxidative decarboxylation of pyruvate in fibroblasts from patients with Parkinson's disease. *J. Neural Transm.* **8,** 223–228.

Nagatsu, T. and Yoshida, M. (1988) An endogenous substance of the brain, tetrahydroisoquinolone, produces parkinsonism in primates with decreased dopamine, tyrosine hydroxylase and biopterin in the nigrostriatal regions. *Neurosci. Lett.* **87,** 178–182.

Nagley, P. and Linnane, A. W. (1970) Mitochondrial DNA deficient petite mutants of yeast. *Biochem. Biophys. Res. Commun.* **39,** 989–996.

Nakagawa-Hattori, Y., Yoshino, H., Kondo, T., Mizuno, Y., and Horai, S. (1992) Is Parkinson's disease a mitochondrial disorder? *J. Neurol. Sci.* 1**107,** 22–33.

Nass, S. and Nass, M. M. (1963) Intramitochoncrial fibers and DNA Characteristics. II. Enzymatic and other hydrolytic treatments. *J. Cell Biol.* **19,** 613–629.

Nass, M. M. K. (1969) Mitochondrial DNA: advances, problems, and goals. *Science* **165,** 25–35.

Newman, N. J. (1993) Leber's hereditary optic neuropathy. *Arch. Neurol.* **50,** 540–548.

Nicklas, W. J., Vyas, I., and Heikkila, R. E. (1985) Inhibition of NADH-linked oxidation in brain mitochondria by 1-methyl-4-phenylpyridine, a metabolite of the neurotoxin, 1-methyl-4-phenyl-1,2,3,6-tetrahydropyridine. *Life Sci.* **36,** 2503–2508.

Niwa, T., Takeda, N., Kaneda, N., Hashizume, Y., and Nagatsu, T. (1987) Presence of tetrahydroisoquinoline and 2-methyl-tetrahydroquinoline in parkinsonian and normal human brains. *Biochem. Biophys. Res. Commun.* **144,** 1084–1089.

Novelli, A., Reilly, J. A., Lysko, P. G., and Henneberry, R. C. (1988) Glutamate becomes neurotoxic via the methyl–D–aspartate receptor when intracellular energy levels are reduced. *Brain Res.* **451,** 205–212.

Ohta, S., Kohno, M., Makino, Y., et al. (1987) Tetrahydroisoquinoline and 1-methyl-tetrahydroisoquinoline are present in the human brain: relation to Parkinson's disease. *Biomed. Res.* **8,** 453–456.

Oostra, R. J., Kemp, S., Bolhuis, P. A., and Bleeker-Wagemakers, E. M. (1996) No evidence for 'skewed' inactivation of the X–chromosome as cause of Leber's hereditary optic neuropathy in female carriers. *Hum. Genet.* **97,** 500–505.

Parkinson, J. (1817) *An Essay on the Shaking Palsy.* Sherwood, Neely and Jones, London.

Parker, W. D., Boyson, S. J., and Parks, J. K. (1989) Electron transport chain abnormalities in idiopathic Parkinson's disease. *Ann. Neurol.* **26,** 719–723.

Parker, W. D. Jr., Oley, C. A., and Parks, J. K. (1989) A defect in mitochondrial electron-transport activity (NADH-coenzyme Q oxidoreductase) in Leber's hereditary optic neuropathy. *N. Engl. J. Med.* **320,** 1331–1333.

Parker, W. D., Golbe, L. I., Parks, J. K., Di Iorio, G., Wooten, G. F., and Swerdlow, R. H. (1999) Cybrid analysis of Contursi kindred mtDNA. *Soc. Neurosci. Abstr.* **25,** 1336.

Parks, J. K., Swerdlow, R. H., and Parker, W. D. (1999) Decreased NADH ubiquinone oxidoreductase in frontal cortex in Parkinson's disease. *Soc. Neurosci. Abstr.* **25,** 1337.

Penn, A. M. W., Roberts, T., Hodder, J., Allen, P. S., Zhu, G., and Martin, W. R. W. (1995) Generalized mitochondrial dysfunction in Parkinson's disease detected by magnetic resonance spectroscopy of muscle. *Neurology* **45,** 2097–2099.

Petit, P. X., Zamzami, N., Vayssiere, J. L., Mignotte, B., Kroemer, G., and Castedo, M. (1997) Implication of mitochondria in apoptosis. *Mol. Cell. Biochem.* **174,** 185–188.

Polymeropoulos, M. H., Lavedan, C., Leroy, E., Ide, S. E., Dehajia, A., Dutra, A., et al. (1997) Mutation in the α-synuclein gene identified in families with Parkinson's disease. *Science* **276,** 2045–2047.

Rajput, A. H., Uitti, R. J., Stern, W., Laverty, W., O'Donnell, K., O'Donnell, D., et al. (1987) Geography, drinking water chemistry, pesticides and heribicides and the etiology of Parkinson's disease. *Can. J. Neurol. Sci.* **14,** 414–418.

Ramsay, R. R., Salach, J. I., and Singer, T. P. (1986) Uptake of the neurotoxin 1-methyl-4-phenylpyridine (MPP+) by mitochondria and its relation to the inhibition of the mitochondrial oxidation of NAD+ linked substrates by MPP+. *Biochem. Biophys. Res. Commun.* **134,** 743–748.

Ramsay, R. R. and Singer, T. P. (1986) Energy-dependent uptake of *N*-methyl-4-phenylpyridinium, the neurotoxic metabolite of 1-methyl-4-phenyl-1,2,3,6-tetrahydropyridine, by mitochondria. *J. Biol. Chem.* **261,** 7585–7587.

Reichmann, H., Riederer, P., Seufert, S., and Jellinger, K. (1990) Disturbances of the respiratory chain in brain from patients with Parkinson's disease. *Mov. Disord.* **5(Suppl. 1),** 28.

Reichmann, H., Janetzky, B., Bischof, F., Seibel, P., Schols, L., Kuhn, W., et al. (1994) Unaltered respiratory chain enzyme activity and mitochondrial DNA in skeletal muscle from patients with idiopathic Parkinson's syndrome. *Eur. Neurol.* **34,** 263–267.

Rosenberg, N. L., Myers, J. A., and Martin, W. R. W. (1989) Cyanide-induced parkinsonism: clinical, MRI, and 6-fluorodopa PET studies. *Neurology* **39,** 142–144.

Royall, J. A. and Ischiropoulos, H. (1993) Evaluation of 2'7'-dichlorofluorescin and dihydrorhodamine 123 as fluorescent probes for intracellular H_2O_2 in cultured endothelial cells. *Arch. Biochem. Biophys.* **302,** 348–355.

Saggu, H., Cooksey, J., Dexter, D., Wells, F. R., Lees, A., Jenner, P., et al. (1989) A selective increase in particulate superoxide dismutase activity in parkinsonian substantia nigra. *J. Neurochem.* **53,** 692–697.

Sanchez-Ramos, J. R., Overvik, E., and Ames, B. N. (1994) A marker of oxyradical-mediated DNA damage (8-hydroxy-2'deoxyguanosine) is increased in nigro-striatum of Parkinson's disease brain. *Neurodegeneration* **3,** 197–204.

Sandy, M. S., Langston, J. W., Smith, M. T., and Di Monte, D. A. (1993) PCR analysis of platelet mtDNA: lack of specific changes in Parkinson's disease. *Mov. Disord.* 8, 74–82.

Schapira, A. H. V., Cooper, J. M., Dexter, D., Jenner, P., Clark, J. B., and Marsden, C. D. (1989) Mitochondrial complex I deficiency in Parkinson's disease. *Lancet* **1,** 1289.

Schapira, A. H. V., Mann, V. M., Cooper, J. M., Dexter, D., Daniel, S. E., Jenner, P., et al. (1990) Anatomic and disease specificity of NADH CoQ1 reductase (Complex I) deficiency in Parkinson's disease. *J. Neurochem.* **55,** 2142–2145.

Schapira, A. H. V., Cooper, J. M., Dexter, D., Clark, J. B., Jenner, P., and Marsden, C. D. (1990) Mitochondrial complex I deficiency in Parkinson's disease. *J. Neurochem.* **54,** 823–827.

Schapira, A. H. V., Holt, I. J., Sweeney, M., Harding, A. E., Jenner, P., and Marsden, C. D. (1990) Mitochondrial DNA analysis in Parkinson's disease. *Mov. Disord.* **5,** 294–297.

Semchuk, K. M., Love, E. J., and Lee, R. G. (1991) Parkinson's disease and exposure to rural environmental factors: a population based case-control study. *Can. J. Neurol. Sci.* **18,** 279–286.

Sheehan, J. P., Swerdlow, R. H., Parker, W. D., Miller, S. W., Davis, R. E., and Tuttle, J. B. (1997) Altered calcium homeostasis in cells transformed by mitochondria from individuals with Parkinson's disease. *J. Neurochem.* **68,** 1221–1233.

Shmookler-Reis, R. J. and Goldstein, S. (1983) Mitochondrial DNA in mortal and immortal human cells. *J. Biol. Chem.* **258,** 9078–9085.

Shoffner, J. M., Watts, R. L., Juncos, J. L., Torroni, A., and Wallace, D. C. (1991) Mitochondrial oxidative phosphorylation defects in Parkinson's disease. *Ann. Neurol.* **30,** 332–339.

Shoffner, J. M., Brown, M. D., Torroni, A., Lott, M. T., Cabell, M. F., Mirra, S. S., et al. (1993) Mitochondrial DNA variants observed in Alzheimer disease and Parkinson disease patients. *Genomics* **17,** 171–184.

Shoubridge, E. A., Karpati, G., and Hastings, E. M. (1990) Deletion mutants are functionally dominant over wild-type mitochondrial genomes in skeletal muscle fiber segments in mitochondrial disease. *Cell* **62,** 43–49.

Shults, C. W. and Miller, S. W. (1998) Reduced complex I activity in parkinsonian cybrids. *Mov. Disord.* **13(Suppl. 2),** 217.

Shuster, R. C., Rubenstein, A. J., and Wallace, D. C. (1988) Mitochondrial DNA in anucleate human blood cells. *Biochem. Biophys. Res. Commun.* **155,** 1360–1365.

Swerdlow, R. H., Parks, J. K., Miller, S. W., Tuttle, J. B., Trimmer, P. A., Sheehan, J. P., et al. (1996) Origin and functional consequences of the complex I defect in Parkinson's disease. *Ann. Neurol.* **40,** 663–671.

Swerdlow, R. H., Wooten, G. F., and Parker, W. D. (1998) Epidemiologic support for mitochondrial genetics in Parkinson's disease. *Neurology* **50,** A373.

Swerdlow, R. H., Parks, J. K., Davis, J. N., Cassarino, D. S., Trimmer, P. A., Currie, L. J., et al. (1998) Matrilineal inheritance of complex I dysfunction in a multigenerational Parkinson's disease family. *Ann. Neurol.* **44,** 873–881.

Swerdlow, R. H., Parks, J. K., Cassarino, D. C., Schilling, A. T., Bennett, J. P., Jr., Harrison, M. B., and Parker, W. D., Jr. (1999) Characterization of cybrid lines containing mtDNA from Huntington's disease patients. *Biochem. Biophys. Res. Comm.* **261,** 701–704.

Tanner, C. M. and Goldman, S. M. (1996) Epidemiology of Parkinson's disease. *Neurol. Clin.* **14,** 317–335.

Tatton, W. G. and Chalmers-Redman, R. M. E. (1998) Mitochondria in neurodegenerative apoptosis: an opportunity for therapy? *Ann. Neurol.* **44(Suppl. 1),** S134–S141.

Tatton, N. A., Maclean-Fraser, A., Tatton, W. G., Perl, D. P., and Olanow, C. W. (1998) A fluorescent double-labeling method to detet and confirm apoptotic nuclei in Parkinson's disease. *Ann. Neurol.* **44(Suppl. 1),** S142–S148.

Tretiakoff, M. C. (1919) Contribution a l'etude d'anatomie pathologique de Locus Niger de Soemmerling. Thesis, University of Paris.

Uhl, J. R., Javitch, J. A., and Snyder, S. H. (1985) Normal MPTP binding in parkinsonian substantia nigra: evidence for extraneuronal toxin conversion in human brain. *Lancet* **1,** 956–957.

Vergani, L., Martinuzzi, A., Carelli, V., Cortelli, P., Montagna, P., Schievano, G., et al. (1995) mtDNA mutations associated with Leber's hereditary optic neuropathy, studies on cytoplasmic hybrid (cybrid) cells. *Biochem. Biophys. Res. Commun.* **210,** 880–888.

Wallace, D. C., Singh, G., Lott, M. T., Hodge, J. A., Schurr, T. G., Lezza, A. M. S., et al. (1988) Mitochondrial DNA mutation associated with Leber's hereditary optic neuropathy. *Science* **242,** 1427–1430.

Ward, C. D. and Gibb, W. R. (1990) Research diagnostic criteria for Parkinson's disease, in *Parkinson's Disease: Anatomy, Pathology, and Therapy*, Advances in Neurology, Vol. 53, (Streifler, M. B., Korczyn, A. D., Melamed, E., and Youdim, M. B. H., eds.), Raven, New York, pp. 245–249.

Werth, J. L. and Thayer, S. A. (1994) Mitochondria buffer physiological calcium loads in cultured rat dorsal ganglion neurons. *J. Neurosci.* **14,** 348–356.

Wiseman, A. and Attardi, G. (1978) Reversible tenfold reduction in mitochondrial DNA content of human cells treated with ethidium bromide. *Mol. Gen. Genet.* **167,** 51–63.

Wong, G. F., Gray, C. S., Hassanein, R. S., and Koller, W. C. (1991) Environmental risk factors in siblings with Parkinson's disease. *Arch. Neurol.* **48,** 287–289.

Wooten, G. F., Currie, L. J., Bennett, J. P., Harrison, M. B., Trugman, J. M., and Parker, W. D., Jr. (1997) Maternal inheritance in Parkinson's disease. *Ann. Neurol.* **41,** 265–268.

Wooten, G. F., Currie, L. J., Bennett, J. P., Trugman, J. M., and Harrison, M. B. (1997) Maternal inheritance in two large kindreds with Parkinson's disease. *Neurology* **48,** A333.

Yang, J., Liu, X., Bhalla, K., Kim, C. N., Ibrado, A. M., Cai, J., et al. (1997) Prevention of apoptosis by Bcl-2: release of cytochrome c from mitochondria blocked. *Science* **275,** 1129–1132.

Yoshida, M., Niwa, T., and Nagatsu, T. (1990) Parkinsonism in monkeys produced by chronic administratin of an endogenous substance of the brain, tetrahydroisoquinoline: the behavioral and biochemical changes. *Neurosci. Lett.* **119,** 109–113.

Yoshino, H., Nakagawa-Hattori, Y., Kondo, T., and Mizuno, Y. (1992) Mitochondrial complex I and II activities of lymphocytes and platelets in Parkinson's disease. *J. Neural Transm.* **41,** 27–34.

Zamzami, N., Hirsch, T., Dallaporta, B., Petit, P. X., and Kroemer, G. (1997) Mitochondrial implication in accidental and programmed cell death: apoptosis and necrosis. *J. Bioenerg. Biomembr.* **29,** 185–193.

Ziering, A., Berger, L., Heineman, S. D., and Lee, J. (1947) Piperidine derivatives: Part III. 4–aryl piperidines. *J. Org. Chem.* **12,** 894–903.

Zweig, R. M., Singh, A., Cardillo, J. E., and Langston, J. W. (1992) The familial occurrence of Parkinson's disease, Lack of evidence for maternal inheritance. *Arch. Neurol.* **49,** 1205–1207.

11

Pathophysiology of SCA1

Harry T. Orr and Huda Y. Zoghbi

11.1. INTRODUCTION

In 1991, a novel mutational mechanism in human genetics was discovered: the expansion of an unstable trinucleotide repeat (Fu et al., 1991; La Spada et al., 1991). To date, trinucleotide repeat expansions have been found to be associated with 16 neurological disorders. Although the sequence of the unstable repeat and its location within the affected gene varies among these disorders, by far the largest category of disorders are those in which the neurodegenerative disease results from the expansion of a CAG repeat. Because the CAG tract is located in the coding region of each gene and encodes a polyglutamine stretch in each respective protein, these disorders are often designated as polyglutamine diseases (Ross, 1997). The eight polyglutamine repeat diseases currently include Kennedy disease or spinobulbar muscular atrophy (SBMA), Huntington disease (HD), and the spinocerebellar ataxias (SCA1, SCA2, SCA3, Machado–Joseph disease [MJD], SCA6, and SCA7), including dentatorubropallidoluysian atrophy (DRPLA). Except for Kennedy disease (SBMA), these neurodegenerative disorders are dominantly inherited. All eight polyglutamine disorders are progressive, often with an onset in mid-life with an increase in neuronal dysfunction and eventual neuronal loss 10–20 yr after onset. Other features that characterize this group of diseases are (1) an inverse relationship between the number of CAG repeats on expanded alleles and age of onset and severity of disease and (2) an intergenerational instability that leads to repeat expansions and earlier age of onset and more rapid disease progression in affected offspring of affected parents. Most interesting, despite the widespread expression of the relevant protein throughout the brain and other tissues, only a subset of neurons that is unique to each disease appears to be vulnerable to the mutation in each of these diseases. This review focuses on one of these polyglutamine disorders, spinocerebellar

From: *Contemporary Clinical Neuroscience: Molecular Mechanisms of Neurodegenerative Diseases*
Edited by: M.-F. Chesselet © Humana Press Inc., Totowa, NJ

ataxia type 1 (SCA1). The reader is referred to other chapters for reviews on some of the other polyglutamine disorders: Huntington disease (Chapters 9–11 and 13), Kennedy disease (Chapter 14), and SCA3/MJD (Chapter 15).

11.2. SCA1: THE HUMAN DISEASE

Spinocerebellar ataxia type 1 is a member of a large group of dominantly inherited spinocerebellar ataxias. This group of disorders is characterized by neural degeneration in the cerebellum, spinal tracts, and brain stem (Greenfield, 1954; Koeppen 1998). A considerable effort to develop a standardized clinical classification of the dominant SCAs was blocked by their interfamilial and intrafamilial variability in neurological and pathological features (Harding, 1982). For example, patients from the same kindred were occasionally diagnosed with two or more forms of SCA. The first *SCA* gene cloned was *SCA1* from the short arm of chromosome 6 (Orr et al., 1993). With the cloning of each subsequent *SCA* gene, it has become possible to reliably distinguish among them using a DNA test.

The clinical features of SCA1 vary depending on the stage of the disease (Zoghbi & Ballabio, 1995). Typically, an SCA1 patient will have ataxia, dysarthria, and bulbar dysfunction. The features present in addition to these often vary between patients. Early in the course of SCA1, patients complain of a mild loss of limb and gait coordination, slurred speech, and poor handwriting skills. Some can have hyperreflexia, hypermetric saccades, and nystagmus. With disease progression, the ataxia worsens, dysmetria, dysdiadochokinesis, and hypotonia develop, and patients often experience vibration and prorioceptive loss. Some patients develop ophthalmoparesis, and a mild optic atrophy and deep tendon reflexes may be decreased or absent. As the disease reaches an advanced stage, usually around 10 yr after the onset of symptoms, the ataxia becomes very severe and brainstem dysfunction results in facial weakness and swallowing and breathing problems. Patients typically die 10–15 yr after onset from the loss of the ability to cough effectively, food aspiration, and respiratory failure.

The predominant neuropathologic findings in SCA1 are cerebellar atrophy with severe loss of Purkinje cells, dentate nucleus neurons, and neurons in the inferior olive and cranial nerve nuclei III, IV, IX, X, and XII (Schut & Haymaker, 1951; Robitaille et al., 1995; Koeppen 1998). Eosinophilic spheres, also known as torpedoes, are present in the internal granule cell layer and some are related to Purkinje cell bodies. The dorsal and ventral spinocerebellar tracts and dorsal columns are demyelinated; gliosis of the molecular layer of the cerebellum is marked, whereas gliosis of the anterior horn of the spinal cord is milder.

11.3. SCA1: MOLECULAR AND CELL BIOLOGY

The highly polymorphic CAG repeat lies in the coding region of the *SCA1* gene and encodes a glutamine tract. Wild-type, unaffected alleles contain 6–44 repeats. A normal allele with over 20 repeats is interrupted with 1–4 CAT repeat units encoding histidine, which most likely maintains CAG-repeat tract stability (Chung et al, 1993). Disease alleles, in contrast, not only contain a longer repeat tract length (39–82) but are uninterrupted by CAT sequences (Goldfarb et al., 1996; Jodice et al., 1994; Quan et al., 1995). Paternal transmissions tend to produce expansions, whereas maternal transmissions often result in contractions (Chung et al., 1993; Jodice et al., 1994).

The fact that mice either heterozygous or homozygous for the null mutation did not develop ataxia (Matilla et al., 1998) supports the idea that SCA1 is not caused by loss of normal ataxin-1 function. Furthermore, large deletions on human chromosome 6p22–23 spanning *SCA1* do not result in SCA1, but cause mental retardation and seizures (Davies et al., 1999). Although this phenotype could result from the loss of genes in addition to *SCA1*, it confirms that haploinsufficiency does not lead to ataxia.

The *SCA1*-encoded protein, ataxin-1, is a novel protein with no apparent extensive homology to other proteins. Wild-type ataxin-1 consists of 792–830 amino acids, depending on the length of the polyglutamine tract (Banfi et al., 1994). The murine *SCA1* gene product is 89% identical to human ataxin-1 (Banfi et al., 1996). Interestingly, the mouse protein contains two glutamines and three prolines at the location of the polyglutamine tract in human ataxin-1. The molecular mass of wild-type ataxin-1 is predicted to be 87 kDa, but has an altered electrophoretic mobility, likely the result of the glutamine tract. Importantly, the mobility of the mutant protein varies directly according to the number of CAG repeats confirming that the CAG repeat is, in fact, within the coding region of ataxin-1 (Servadio et al., 1995). Ataxin-1 is widely expressed in neurons throughout the central nervous system and in cells in peripheral tissues. In lymphoblasts, heart, skeletal muscle, and liver, the protein is localized to the cytoplasm. In neurons, ataxin-1 is predominantly nuclear, with some cytoplasmic staining in Purkinje cells and brainstem nuclei (Servadio et al., 1995). In SCA1 patients, mutant ataxin-1 localized to a single large nuclear inclusion in brainstem neurons (Skinner et al., 1997). These inclusions stain positively for ubiquitin, the proteasome, and the molecular chaperone HDJ–2/HSDJ (Cummings et al., 1998).

The leucine-rich acidic nuclear protein (LANP) associates with ataxin-1 in a glutamine repeat-length-dependent manner (Matilla et al., 1997). The highest level of murine LANP expression occurs in Purkinje cells around postnatal day 14 (Matsuoka et al., 1994), similar to the transient burst of ataxin-1 expression in the mouse (Banfi et al., 1996). Like ataxin-1, LANP localizes to the nucleus, and in transfected cells, it is redistributed into nuclear inclusions formed by mutant ataxin-1. That LANP seems to interact more strongly with ataxin-1 containing an expanded polyglutamine coupled with its more restricted pattern of cellular expression in the brain, a pattern that more closely correlates with the cellular pathology of SCA1, strongly suggest that LANP's interaction with ataxin-1 may be the molecular basis for the cell specificity of pathology seen in SCA1 brains.

Another nuclear protein recently found to be capable of interacting with ataxin-1 is A1Up (Davidson et al., 2000). The A1Up protein consist of 601 amino acids and showed substantial homology to a cDNA clone reported by Ueki et al. (1998). Analysis of the A1Up amino acid sequence identified a region in its N-terminal with substantial similarity to an ubiquitinlike domain. Thus, A1Up is a member of the family of ubiquitinlike domain proteins. All of these proteins contain the ubiquitin domain in their amino-termini. For example, one member of the ubiquitinlike family is Rad23, a protein important for nucleotide excision repair (Guzder et al., 1995). Rad23 and interacts with the 26S proteasome through its N-terminal ubiquitinlike domain (Schauber et al., 1998) and the C-terminus of Rad23 interacts with the DNA repair complex. Thus, it has been suggested that by virtue of its dual interactions, Rad23 provides a molecular link between the DNA repair and proteasome system (Schauber et al., 1998).

11.4. SCA1 PATHOGENESIS

Mouse Models

The first transgenic mouse model of a polyglutamine disease utilized a strong Purkinje cell-specific promoter from the *Pcp2/L7* gene to direct expression of the human *SCA1* cDNA encoding full-length ataxin-1 (Burright et al., 1995). These lines expressed high levels of either a wild-type *SCA1* allele, ataxin-1[30Q], or an expanded allele, ataxin1[82Q]. The ataxin-1[82Q] transgenic mice developed severe ataxia and progressive Purkinje cell pathology. In contrast, mice expressing ataxin-1[30Q] failed to develop any signs of neurologic or pathological abnormalities and were indistinguishable from nontransgenic littermates (Clark et al., 1997). These studies demonstrated that Purkinje cell pathological changes are induced specific to the expression of ataxin-1 with an expanded polyglutamine tract.

In transgenic mice from an ataxin-1[30Q] line, ataxin-1 localized to several approx 0.5-μm nuclear inclusions. In contrast, in ataxin-1[82Q] mice ataxin-1 localized to a single approx 2-μm ubiquitinated nuclear aggregate, as in patient tissue (Skinner et al., 1997). The appearance of these aggregates, which stained positive for the 20S proteasome and the HDJ-2/HSDJ (Hsp40) chaperone protein, preceded the onset of ataxia by approximately 6 wk (Cummings et al., 1998). The only notable difference between the pathology observed in *SCA1* transgenic mice and that of SCA1 patients is that the mice lack axonal dilatations (torpedoes). *SCA1* transgenic animals from the ataxin-1[82Q] line had mild cerebellar impairment at 5 wk of age; there was no evidence of gait abnormalities or balance problems at that age (Clark et al., 1997). By 12 wk, the motor-skill impairment progressed to overt ataxia, which worsened over time.

The first histologic change detected was the development of cytoplasmic vacuoles within Purkinje cell bodies at postnatal day 25; by 5 wk, loss of proximal dendritic branches and a decrease in the number of dendritic spines became apparent, indicating that ataxin-1[82Q] impairs the maintenance of dendritic arborization (Clark et al., 1997). By 12–15 wk, the complexity of the dendritic arborization of Purkinje cells was markedly reduced, the molecular layer atrophied, and there were several heterotopic Purkinje cells within the molecular layer (Fig. 1). The Purkinje cells heterotopia was not detected in young animals and, thus, is not a developmental abnormality. Most likely it reflects an attempt to preserve synaptic function in the presence of severely reduced dendritic arborization. Importantly, Purkinje cell loss was minimal at the time of progressive gait abnormality. Thus, although it had long been assumed that the neurological phenotype in SCA1 patients results from neuronal death, the *SCA1* transgenic mice indicated that the neurological impairment is the result of neuronal dysfunction and not the result of neuronal loss.

To ascertain whether ataxin-1 must be in the nucleus to cause disease, Klement et al. (1998) generated and characterized transgenic mice that express expanded ataxin-1[82Q] with a mutated nuclear localization sequence, ataxin-1^{K772T}. Although these mice expressed high levels of ataxin-1 in Purkinje cells, similar to those observed in the original *SCA1*, ataxin-1[82Q] transgenic mice, they never developed Purkinje cell pathology or motor dysfunction. Ataxin-1 was diffusely distributed throughout the cytoplasm and formed no aggregates, even when the mice were a year old. Nuclear localization is clearly critical for pathogenesis and ataxin-1 aggregation. Furthermore, the ataxin-1^{K772T} mice expressed levels of *SCA1* mRNA as high as the ataxin-1[82] mice and, yet, they failed to develop disease. Thus, these studies provided direct evidence that it is the ataxin-1 protein with an expanded number of glutamine residues that is pathogenic and not *SCA1* RNA with an expanded number of CAG triplet repeats.

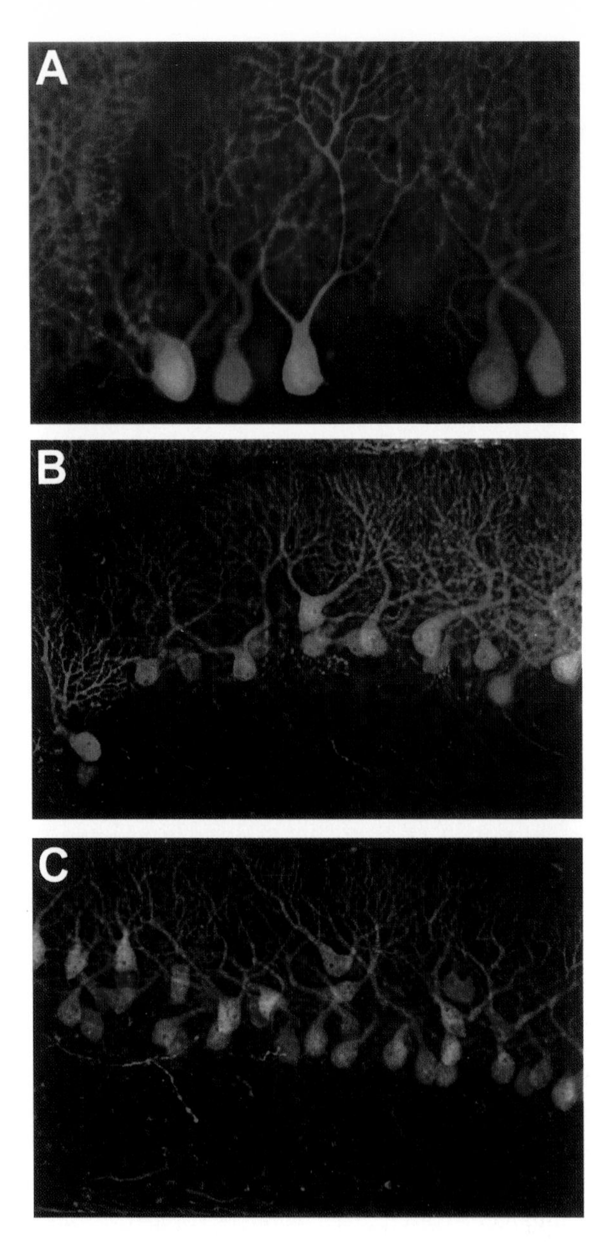

Fig. 1. Progressive Purkinje cell pathology in transgenic mice expressing ataxin-1[82Q]. Cerebellar sections were examined for Purkinje cell morphological alterations using calbindin immunofluroesence. **(A)** A cerebellar section from a 6-wk-old transgenic mouse demonstrating the loss of proximal dendritic processes in some Purkinje cells. **(B)** A cerebellar section from a 15-wk-old transgenic mouse showing that Purkinje cell dendritic atrophy has progressed. **(C)** A cerebellar section from a 28-wk-old transgenic mouse demonstrating the severe atrophy of the Purkinje cell dendrites and the presence of Purkinje cell bodies within the molecular layer. (Reproduced from Burright et al., 1997b, with permission.)

To directly assess the role of nuclear aggregates in causing disease, Klement et al. (1998) also generated transgenic mice using ataxin-1[77Q] with amino acids deleted from the self-association region found to be essential for ataxin-1 dimerization. These mice developed ataxia and Purkinje cell pathology similar to the original ataxin-1[82Q] transgenic mice, but without detectable nuclear ataxin-1 aggregates at either the light or electron microscopic levels. Thus, although nuclear localization of ataxin-1 is necessary, nuclear aggregation of ataxin-1 appears not to be required for initiation of Purkinje cell pathogenesis in transgenic mice. It is important to note that the deletion of 122 amino acids might have compromised ataxin-1 in various ways (e.g., its folding, turnover rate, or ability to interact with other cellular factors). This seems unlikely because this truncated ataxin-1 retained its ability to produce all of the neurobehavioral and unique pathologic features observed in the ataxin-1[82Q] mice.

Recently, Cummings et al. (1999) crossed ataxin-1[82Q] transgenic mice with mice lacking expression of the *Ube3a* gene (Jiang et al., 1998). *Ube3a* encodes the E6-associated protein, E6-AP, which is a member of the E3 class of ubiquitin ligases (Huibregtse et al., 1995). Previous work had demonstrated that the nuclear aggregates of ataxin-1 seen in SCA1 patient brain material and *SCA1* transgenic mouse Purkinje cells were positive for ubiquitin (Skinner et al., 1997). Thus, these studies were undertaken to examine the role of the ubiquitin–proteasome pathway in SCA1 pathogenesis. Two important observations were made in the double mutant mice (i.e., mice expressing the full-length ataxin-1[82Q] protein in the absence of *Ube3a* expression). The presence of nuclear aggregates of ataxin-1[82Q] were reduced significantly both in terms of their frequency and their size. Yet, the Purkinje cell pathology was markedly worse compared to that seen in the ataxin-1[82Q] mice. These studies demonstrated the importance of ataxin-1 ubiquitination for the formation of the nuclear aggregates. Furthermore, they showed that pathology is not dependent on the formation of nuclear aggregates. Taken together, the *SCA1* transgenic mice studies provide two distinct examples in which nuclear aggregates are shown not to be necessary for the development of polyglutamine-induced neuronal disease.

Insight into the molecular basis of SCA1 pathogenesis was recently obtained by the observation that an expanded allele of SCA1 has the ability to alter Purkinje cell gene expression early on in the *SCA1* transgenic mice (Lin et al., 2000). Using the *SCA1* transgenic mice and a polymerase chain reaction (PCR)-based cDNA subtractive hybridization strategy, several genes, all expressed by Purkinje cells, were found to be downregulated at an early stage of disease, prior to any detectable pathological or neurological alteration. Interestingly, a number of the genes found to be downregulated encoded proteins involved in neuronal calcium signaling. These included IP3R1, SERCA2, TRP3, and inositol polyphosphate 5-phosphatase type 1. Intriguingly, all of the genes whose expression was found to be altered early on in the *SCA1* transgenic mice were downregulated. Although the mechanism of this downregulation could involve either transcription and/or posttranscription events, given that ataxin-1 is an RNA-binding protein in vitro (Yue and Orr, unpublished data) and may bind RNA in vivo, it is possible to speculate that the downregulation in gene expression seen in SCA1 involves a mechanism at the RNA level. It is also worth noting that the same downregulation in gene expression was found in the mice expressing ataxin-1 [77Q] with amino acids deleted from the self-association region, lacking aggregates, as well as the ataxin-1[82] expressing mice, containing aggregates. This provides further evidence that the Purkinje cell disease process is the same with or without nuclear aggregates of ataxin-1.

SCA1: In Vitro and Cell Culture Models

Overexpression of full-length ataxin-1 in COS cells caused the protein to behave similar to how it does in vivo (i.e., it localized to the nucleus and formed multiple aggregates whose size correlated with the number of glutamine repeats [Skinner et al., 1997]). Deletion of the self-association domain (Burright et al., 1997) prevented aggregation. Within the nucleus, ataxin-1 associated with the nuclear matrix, and the mutant form caused redistribution of the promyelocytic oncogenic domain.

An important aspect of SCA1 pathogenesis elucidated by cell culture studies is that protein misfolding may play a role in pathogenesis. Cummings et al. (1998) demonstrated that ataxin-1 aggregates induce Hsp70 expression and redistribute it to the site of aggregation, along with HDJ-2/HSDJ and components of the proteasome. Overexpression of the HDJ-2/HSDJ chaperone decreased both the size and frequency of mutant ataxin-1 nuclear aggregates. Furthermore, the importance of the ubiquitin–proteasomal pathway in the degradation of mutant ataxin-1 has also been demonstrated using a cell culture system (Cummings et al., 1999). Although both ataxin-1[2Q] and ataxin-1[82Q] were ubiquitinated to a similar level in transfected cells, ataxin-1[82Q] was less susceptible to degradation than ataxin-1[2Q]. In addition, inhibition of the proteasomal pathway enhanced the aggregation of ataxin-1 in transfected cells.

11.5. SCA1: A PATHOGENIC MODEL

All pathogenesis models of polyglutamine disease are based on the expansion of CAG repeats conferring a toxic gain of function (i.e., disease develops because the mutant form of the protein gains a new function, not because the protein loses its normal function). Data in support of this model is substantial and comes from both patient studies and mice with targeted deletions.

The best available model of SCA1 pathogenesis, at least within Purkinje cells, maintains that sequences within ataxin-1 along with the polyglutamine tract are critical in specifying the cellular sites and course of disease. The neuronal dysfunction that causes the SCA1 symptoms begins with the localization of ataxin-1 to the nucleus (Fig. 2). Once there, mutant ataxin-1 can follow either of two pathways, a pathogenic pathway or a sequestration pathway. Both pathways are very likely triggered by the same feature of ataxin-1 that is triggered the expansion of the polyglutamine tract. Data from cell culture studies suggest that one such feature might be protein misfolding and/or conformational alterations. Although it seems that the cytoplasm of Purkinje cells is able to handle mutant ataxin-1, perhaps because of appropriate chaperone function, at this time it is unclear whether mutant ataxin-1 misfolds upon entering the nucleus or whether nuclear chaperone function is simply unable to handle the levels of mutant ataxin-1 presented to it. In any case, our data are most consistent with pathogenesis being dependent on high levels of free mutant ataxin-1 in the nucleus. Therefore, the sequestration pathway, at least initially, is proposed to function to clear mutant ataxin-1 within the nucleus. This process is dependent on mutant ataxin-1 being ubiquitinated and can lead either to proteasomal degradation or the formation of aggregates as the proteasomal system becomes unable to handle the mutant ataxin-1. Both animal and cell culture studies indicate strongly that aggregation is neither necessary nor sufficient for neuronal dysfunction (Klement et al., 1998; Saudou et al., 1998; Cummings et al., 1999). Thus, we propose that with continued expression of mutant ataxin-1, its levels eventually exceed the capability of the sequestration pathway and enough mutant ataxin-1 becomes available to initiate pathogenesis. It is likely that changes in nuclear architecture and the interaction of mutant ataxin-1 with other nuclear factors alters gene expression, all of which may well contribute to pathogenesis and eventually lead to neuronal dysfunction, symptomatology, and, finally, cell loss. The cell specificity of SCA1 is likely the result of a number of factors such as relative levels of the polyglutamine protein, protein interactors like LANP, and cell specificity of certain modifying proteins. Notwithstanding evidence that aggregates are likely not the first step in pathogenesis, it is possible that, over time, the Purkinje cell, no longer able to cope with accumulating mutant polyglutamine proteins, eventually succumbs to the large inclusions that disrupt cellular functions at the latter stages of pathogenesis. Future research investigating the effects of

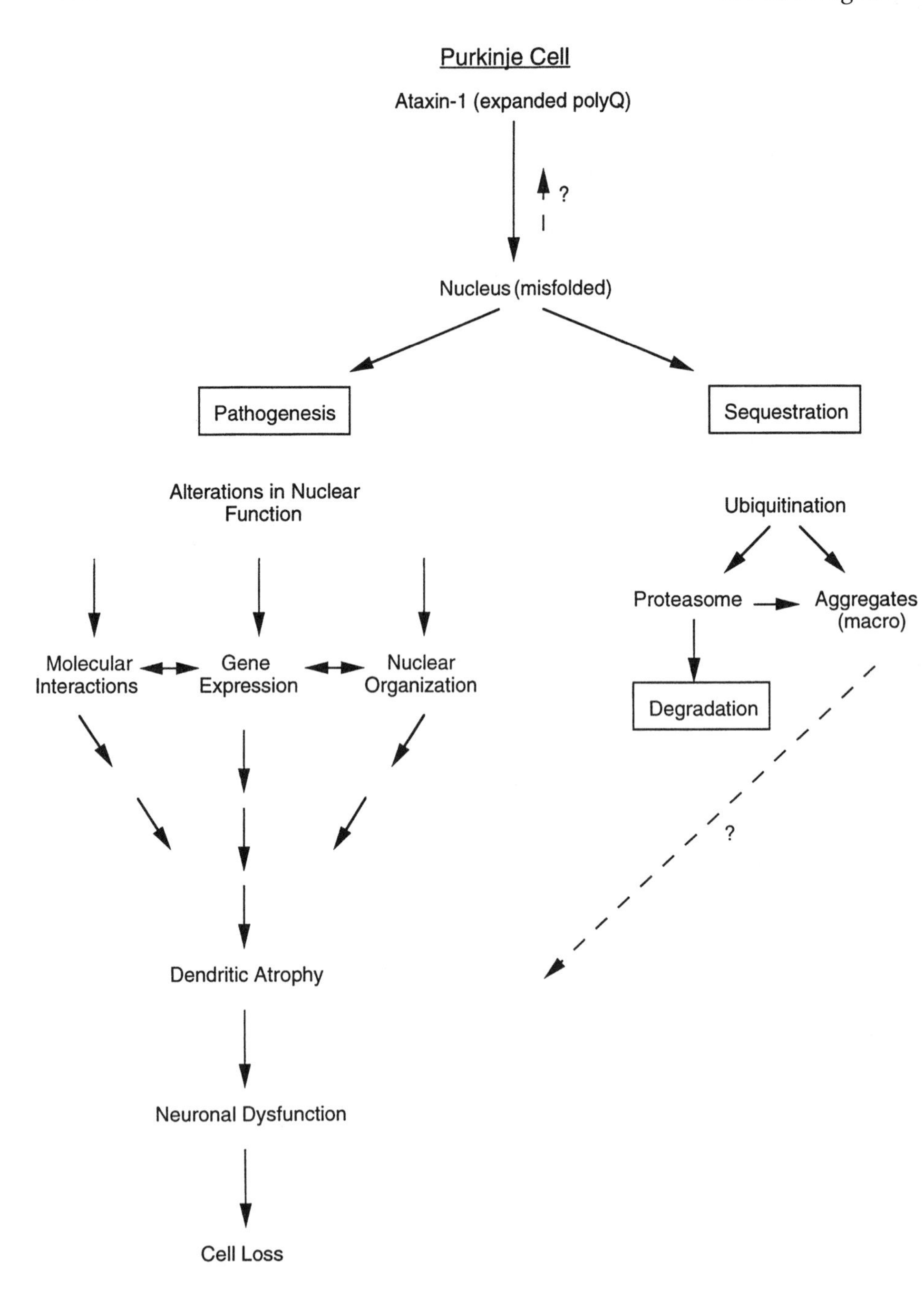

Fig. 2. A model for the pathogenesis of SCA1. This model depicts the presence of two pathways, one pathogenic and one leading to the protein sequestration, triggered by the movement of mutant ataxin-1 into the nucleus of a Purkinje cell.

the expanded polyglutamine tract on ataxin-1 molecular interactions and its turnover along with studies directed at mitigating the disease process by enhancing proper folding and enhanced degradation of mutant ataxin-1 will establish the accuracy of this model.

ACKNOWLEDGMENTS

The authors gratefully acknowledge the support of the NIH/HINDS and the Howard Hughes Medical Institute.

REFERENCES

Banfi, S., Servadio, A., Chung, M.-Y., Capozzoli, F., Duvick, L. A., Elde, R., et al. (1996) Cloning and developmental expression analysis of the murine homolog of the spinocerebellar ataxia type 1 gene (SCA1). *Hum. Mol. Genet.* **5,** 33–40.

Banfi, S., Servadio, A., Chung, M.-Y,. Kwiatkowski, T. J., Jr., McCall, A. E., Duvick, L. A., et al. (1994) Identification and characterization of the gene causing type 1 spinocerebellar ataxia. *Nature Genet.* **7,** 513–519.

Burright, E. N., Clark, H. B., Servadio, A., Matilla, T., Feddersen, R. M., Yunis, W. S., et al. (1995) SCA1 transgenic mice: a model for neurodegeneration caused by an expanded CAG trinucleotide repeat. *Cell* **82,** 937–948.

Burright, E. N., Davidson, J. D., Duvick, L. A., Koshy, B., Zoghbi, H. Y., and Orr, H. T. (1997) Identification of a self-association region within the SCA1 gene product, ataxin-1. *Hum. Mol. Genet.* **6,** 513–518.

Burright, E. N., Orr, H. T., and Clark, H. B. (1997b) Mouse models of human CAG repeat disorders. *Brain Pathol.* **7,** 965–977.

Chung, M.-Y., Ranum, L. P. W., Duvick, L., Servadio, A., Zoghbi, H. Y., and Orr, H. T. (1993) Analysis of the CAG repeat expansion in spinocerebellar ataxia type I: evidence for a possible mechanism predisposing to instability. *Nature Genet.* **5,** 254–258.

Clark, H. B., Burright, E. N., Yunis, W. S., Larson, S., Wilcox, C., Hartman, B. et al. (1997) Purkinje cell expression of a mutant allele of SCA1 in transgenic mice leads to disparate effects on motor behaviors, followed by a progressive cerebellar dysfunction and histological alterations. *J. Neurosci.* **17**, 7385–7395.

Cummings, C. J., Mancini, M. A., Antalffy, B., DeFranco, D. B., Orr, H. T., and Zoghbi, H. Y. (1998) Chaperone suppression of aggregation and altered subcellular proteasome localization imply protein misfolding in SCA1. *Nature Genet.* **19,** 148–154.

Cummings, C. J., Reinstein, E., Sun, Y., Antalffy, B., Jiang, Y.-H., Ciechanover, A., et al. (1999) Mutation of the E6-AP ubiquitin ligase reduces nuclear inclusion frequency while accelerating polyglutamine-induced pathology in SCA1 transgenic mice. *Neuron*, **24,** 879–892..

Davidson, J. D., Riley, B., Burright, E. N., Duvick, L. A., Zoghbi, H. Y., and Orr, H. T. Identification and characterization of an ataxin-1 interacting proteins: A1Up a ubiquitin-like nuclear protein, submitted.

Davies, A. F., Mirza, G., Sekhon, G., Turnpenny, P., Leroy, F., Speleman, F., et al. (1999) Delineation of two distinct 6p deletion syndromes. *Hum. Genet.* **104,** 64–72.

Fu, Y.-H., Kuhl, D. P. A., Pizutti, A., Pieretti, M., Sutcliffe, J. S., Richards, S., et al. (1991) Variation of the CGG repeat at the fragile X site results in genetic instability: resolution of the Sherman paradox. *Cell* **67,** 1047–1058.

Goldfarb, L. G., Vasconcelos, O., Platonov, F. A., Lunkes, A., Kipnis, V., Kononova, S., et al. (1996) Unstable triplet repeat and phenotypic variability of spinocerebellar ataxia type 1. *Ann. Neurol.* **39,** 500–506.

Greenfield, J. G. (1954) *The Spino-cerebellar Degenerations.* Charles C Thomas, Springfield, IL.

Guzder, S. M., Bailly, V., Sung, P., Prakash, L., and Prakash, S. (1995) Yeast DNA repair protein RAD23 promotes complex formation between transcription factor TFIIH and DNA damage recognition factor RAD14. *J. Biol. Chem.* **270,** 8385–8388.

Harding, A. E. (1982) The clinical features and classification of the late onset autosomal dominant cerebellar ataxias. *Brain* **105,** 1–28.

Huibregtse, J. M., Scheffner, M., Beaudenon, S., and Howley, P. M. (1995) A family of proteins structurally and functionally related tot he E6-AP ubiquitin-protein ligase. *Proc. Natl. Acad. Sci. USA* **92,** 2563–2567.

Jiang, Y. H., Armstrong, D., Albrecht, U., Atkins, C. M., Noebels, J. L., Eichele, G., et al. (1998) Mutation of the Angelman ubiquitin ligase in mice causes increased cytoplasmic p53 and deficits of contextual learning and longterm potentiation. *Neuron* **21,** 799–811.

Jodice, C., Malaspina, P., Persichetti, F., Novelletto, A., Spadaro, M., Giuinti, P., et al. (1994) Effect of trinucleotide repeat length and parental sex on phenotypic variation in spinocerebellar ataxia 1. *Am. J. Hum. Genet.* **54,** 959–965.

Klement, I. A., Skinner, P. J., Kaytor, M. D., Yi, H., Hersch, S. M., Clark, H. B., et al. (1998) Ataxin-1 nuclear localization and aggregation: Role in polyglutamine-induced disease in SCA1 transgenic mice. *Cell* **95,** 41–53.

Koeppen, A. H. (1998) The herediatary ataxias. *J. Neuropathol. Exp. Neurol.* **57,** 531–543.

La Spada, A. R., Wilson, E. M., Lubahn, D. B., Harding, A. E., and Fischbeck, H. (1991) Androgen receptor gene mutations in X-linked spinal and bulbar muscular atrophy. *Nature* **352,** 77–79.

Lin, X., Antalffy, B., Kang, D., Orr, H. T., and Zoghbi, H. Y. (2000) Polyglutamine expansion in ataxin-1 downregulates specific neuronal genes prior to any known pathogenic changes in spinocerebellar ataxia type 1. *Nature Neurosci.* **3,** 137–163.

Matilla, T., Koshy, B., Cummings, C. J., Isobe, T., Orr, H. T., and Zoghbi, H. Y. (1997) The cerebellar leucine-rich acidic nuclear protein interacts with ataxin-1. *Nature* **389,** 974–978.

Matilla, A., Roberson, E. D., Banfi, S., Morales, J., Armstrong, D. L., Burright, E. N., et al. (1998) Mice lacking ataxin-1 display learning deficits and decreased hippocampal paired-pulse facilitation. *J. Neurosci.* **18,** 5508–5516.

Matsuoka, K., Taoka, M., Satozawa, N., Nakayama, H., Ichimura, T., Takahashi, N., et al. (1994) A nuclear factor containing the leucine-rich repeats expresses in murine cerebellar neurons. *Proc. Natl. Acad. Sci. USA* **91,** 9670–9674.

Orr, H. T., Chung, M.-Y., Banfi, S., Kwiatkowski, T. J., Jr., Servadio, A., Beaudet, A. L., et al. (1993) Expansion of an unstable trinucleotide CAG repeat in spinocerebellar ataxia type 1. *Nature Genet.* **4,** 221–226.

Quan, F., Janas, J., and Popovich, B. W. (1995) A novel CAG repeat configuration in the SCA1 gene: implication for the molecular diagnosis of spinocerebellar ataxia type 1. *Hum. Mol. Genet.* **4,** 2411–2413.

Roitaille, Y., Schut, L., and Kish, S. J. (1995) Structural and immunocytochemical features of olivopontocerebellar atrophy caused by the spinocerebellar ataxia type 1 (SCA-1) mutation define a unique phenotype. *Acta Neuropathol.* **90,** 572–581.

Ross, C. A. (1997) Intranuclear neuronal inclusions: a common pathogenic mechanism for glutamine-repeat neurodegenerative diseases? *Neuron* **19,** 1147–1150.

Saudou, F., Finkbeiner, S., Devys, D., and Greenberg, M. E. (1998) Huntingtin acts in the nucleus to induce apoptosis but death does not correlate with the formation of intranuclear inclusions. *Cell* **95,** 55–66.

Schauber, C., Chen, L., Tongaonkar, P., Vega, I., Lambertson, D., Potts, W., et al. (1998) Rad23 Links DNA Repair to the ubiquitin/proteasome pathway. *Nature* **391,** 715–718.

Schut, L. and Haymaker, W. (1951) Hereditary ataxia: a pathological study of five cases of common ancestry. *J. Neuropathol. Clin. Neurol.* **1,** 183–213.

Servadio, A., Koshy, B., Armstrong, D., Antalfy, B., Orr, H. T., and Zoghbi, H. Y. (1995) Expression analysis of the ataxin-1 protein in tissues from normal and spinocerebellar ataxia type 1 individuals. *Nature Genet.* **10,** 94–98.

Skinner, P. J., Koshy, B., Cummings, C., Klement, I. A., Helin, K., Servadio, A., et al. (1997) Ataxin-1 with extra glutamines induces alterations in nuclear matrix-associated structures. *Nature* **389,** 971–974.

Ueki, N., Oda, T., Kondo, M., Yano, K., Noguchi, T., and Muramatsu. M. (1998) Selection system for genes encoding nuclear targeted proteins. *Nature Biotech.* **16,** 1338–1342.

Zoghbi, H. Y. and Ballabio, A. (1995) Spinocerebellar ataxia type 1, In *The Metabolic and Molecular Bases of Inherited Disease.* 7th ed., (Scriver, C. R., Beaudet, A. L., Sly, W. S. et al., eds.), McGraw-Hill, New York, pp. 4559–4567.

12
Pathophysiology of SCA3

Puneet Opal and Henry Paulson

Once thought to be rare, spinocerebellar ataxia type 3 (SCA3) is now believed to be the most common dominantly inherited ataxia. Of the polyglutamine disorders described to date, SCA3 has perhaps the most interesting history—one that reflects past confusion about the clinical spectrum of this remarkably pleiotropic disease.

The reason for its highly variable phenotype became clear in 1994, when Kakizuka and colleagues determined that the genetic defect in SCA3 is the dynamic expansion of a CAG trinucleotide repeat that encodes expanded polyglutamine in the disease protein (Kawaguchi et al., 1994). The last few years have seen tremendous growth in our understanding of the pathophysiology of this and other polyglutamine diseases. In this chapter, we first provide a brief history of the recognition of SCA3 as a distinct clinical and genetic entity; then we discuss possible disease mechanisms and therapeutic strategies.

12.1. ORIGIN AND HISTORY OF SCA3

In the 1970s, three varieties of a dominantly inherited ataxic syndrome were described in families of Portuguese extraction from the Islands of the Azores. The first to be described was the Machado family afflicted with a degenerative condition characterized by progressive cerebellar ataxia and a distal sensory neuropathy (Nakano et al., 1972). Soon thereafter, Woods and Schaumburg described an unusual form of nigral–spinal–dental degeneration in the Thomas family that was characterized by spastic–ataxic gait, external ophthalmoplegia, and facial and lingual fasciculations (Woods and Schaumburg, 1972). The last of the Azorean families to be described was the Joseph family, in which affected members typically developed ataxia in the second or third decade, accompanied by dystonia, parkinsonism, spasticity, and oculomotor disturbances (Rosenberg et al., 1976). As a group, these familial diseases were referred to as "Azorean neurodegeneration" to reflect their common

From: *Contemporary Clinical Neuroscience: Molecular Mechanisms of Neurodegenerative Diseases*
Edited by: M.-F. Chesselet © Humana Press Inc., Totowa, NJ

geographical provenance. Because of their different clinical manifestations, they were thought to represent distinct genetic disorders. Only when families were described with a phenotype spanning the full clinical spectrum did it become evident that Azorean degeneration might represent a single genetic entity. As this became clear and as additional families were described in locales beyond the Azores (e.g., Portugal, Spain, Brazil, Japan, and India), Azorean degeneration was renamed Machado-Joseph disease (MJD).

In the 1990s, molecular genetic advances resulted in further unexpected findings about the disease. Linkage analysis in Japanese families assigned MJD to the long arm of chromosome 14 (14q24.3–q32) (Takiyama et al., 1993). Simultaneously, a European group narrowed the genetic locus for SCA3, which was thought to be an unrelated dominantly inherited ataxia. SCA3 differed from MJD in that it displayed considerably less dystonia and bradykinesia, and thus it came as a surprise when SCA3 mapped to the same region of chromosome 14 (Stevanin et al., 1994). After the genetic defect in MJD was identified (Kawaguchi et al., 1994), it soon became clear that MJD and SCA3 were caused by the same mutation — an unstable CAG repeat expansion in the *MJD1* gene (Cancel et al., 1995; Schols et al., 1995). Thus, SCA3 and MJD are genetically identical disorders with subtly different clinical features. This illustrates the important point that a genetic disorder may not manifest the same way in every genetic background and can be influenced by additional genetic and/or environmental factors.

Some have speculated that SCA3/MJD originated in the Portuguese Azores, then spread throughout the world via a single founder mutation. However, genetic studies using flanking microsatellite markers and intragenic polymorphisms make the single-founder hypothesis untenable (Stevanin et al. 1995; Gaspar et al., 1996; Stevanin et al., 1997). There are at least four distinct haplotypes associated with SCA3, including at least two in the Azores, implicating multiple ancestral mutations in the worldwide distribution of SCA3. Nonetheless, there is evidence for local-founder effects in some geographically distinct populations in France, Brazil, and Japan.

Because of its complex history, the disease has variously been referred to as MJD, SCA3, or SCA3/MJD. We prefer calling the disorder SCA3 to conform to the currently accepted genetic classification of dominantly inherited ataxias.

12.2. THE *MJD1* GENE

By 1993, the discovery of CAG/glutamine-repeat expansions in three progressive neurodegenerative diseases led researchers to suspect that SCA3 might also be a CAG repeat disease. With this in mind, Kawaguchi and colleagues carried out a directed search for candidate CAG repeat genes and

Table 1
Polyglutamine Diseases

Disease	Protein	Repeat length in disease	Evidence for misfolding
HD	Huntingtin	38–180	NI, neuropil aggregates
DRPLA	Atrophin	49–88	NI
SBMA	Androgen receptor	38–65	NI
SCA1	Ataxin-1	39–83	NI
SCA2	Ataxin-2	34–59	Dense cytoplasmic staining
SCA3	Ataxin-3	55–84	NI
SCA6	Calcium channel	21–30	Cytoplasmic aggregates
SCA7	Ataxin-7	34–>200	NI

Note: HD, Huntington disease; DRPLA, dentatorubral–pallidoluysian atrophy; SBMA, spinobulbar muscular atrophy; SCA, spinocerebellar ataxia types 1, 2, 3, 6, and 7; MJD, Machado-Joseph disease. (References for aggregation in disease tissue are: Davies et al., 1997; DiFiglia et al., 1997; Paulson et al., 1997; Skinner et al., 1997; Becher et al., 1998; Holmberg et al., 1998; Li et al., 1998; Huynh et al., 1999; Gutekunst et al 1999; Ishikawa et al 1999).

discovered that the genetic defect in SCA3 was indeed an expanded CAG repeat (Kawaguchi et al., 1994). From a human brain cDNA library, they isolated a CAG repeat-containing cDNA that mapped to the same region of 14q as MJD (Takiyama et al., 1993). The CAG repeat in the gene (designated *MJD1*) proved to be expanded in all SCA3 patients and was predicted to encode an expanded polyglutamine tract (polyQ) in the disease protein. SCA3 thus became the fourth polyQ disease identified, of which there are now at least eight (*see* Table 1).

The initial report of the *MJD1* mutation in Japanese patients indicated a significant gap between repeats lengths of normal and expanded alleles. This has since been confirmed by researchers studying SCA3 in a variety of ethnic groups (Cancel et al., 1995; Giunti et al., 1995; Maruyama et al., 1995; Matilla et al., 1995; Durr et al., 1996; Soong et al., 1997; Schols et al., 1995). As shown in Fig. 1, normal alleles range from 12 to approx 42 repeats, whereas expanded alleles are nearly always at least 60 repeats in length. In this respect, SCA3 differs from most other polyglutamine disease, in which the ranges for normal and expanded alleles are either contiguous or only narrowly separated. Although the reason for this wide gap in SCA3 is unknown, the implications are clear. First, results of genetic testing for SCA3 are unambiguous: An individual's repeat length will fall unequivocally into either the normal or disease distribution. Second, there is no "zone of reduced penetrance" in SCA3, as there is in Huntington disease (HD) or

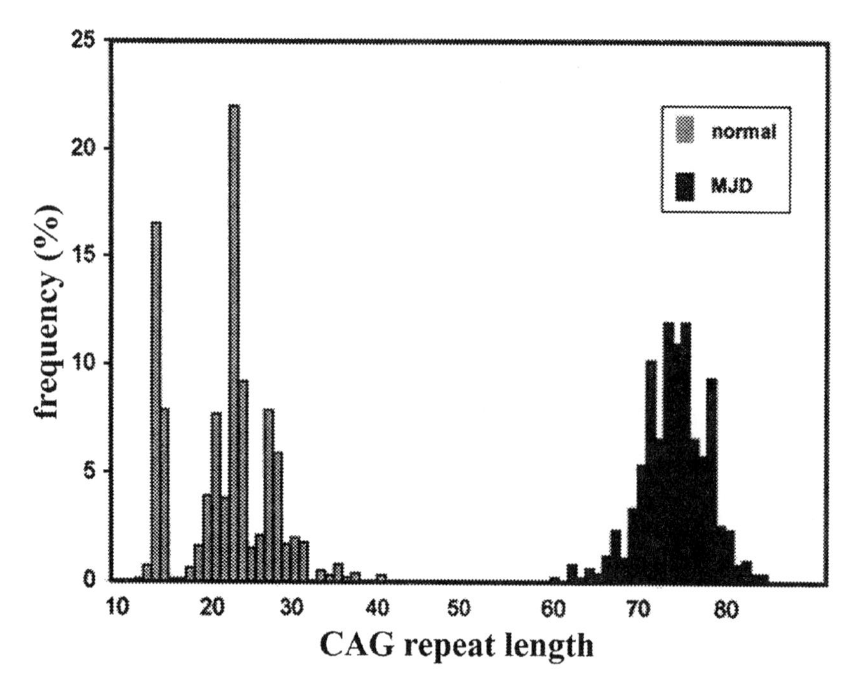

Fig. 1. CAG repeat lengths in normal and SCA3 (MJD) chromosomes. In SCA3, there is a large gap separating the range of normal and disease alleles. Nearly all disease alleles have 60 or more repeats (a single disease allele shorter than this, 55 repeats, was reported by Quan et al., 1997). Data for this graph were pooled from five representative studies of SCA3 in different populations (Maciel et al., 1995; Maruyama et al., 1995; Matilla et al., 1995; Schols et al., 1995; Durr et al. 1996).

SCA2. In other words, there is no intermediate MJD1 repeat length at which some individuals will, and others will not, develop signs of disease during a normal life span. Third, the rather large jump in repeat size that a normal allele would need to make in order to expand into the disease range suggests that *de novo* mutations occur only rarely. One would thus expect sporadic SCA3/MJD to seldom occur, and this is indeed the case.

12.3. GENOTYPE–PHENOTYPE RELATIONSHIPS

Before describing possible pathophysiologic mechanisms in SCA3, several interesting genotype–phenotype relationships warrant discussion. A few, such as anticipation, are relatively well understood, whereas others are not so well appreciated. Any successful model of SCA3 pathogenesis must take into account these features.

Table 2
Clinical Subtypes of SCA3/MJD

Onset (yr)	Repeat length	Major features	Associated findings
<25	>75[a]	Dystonia, rigidity, and ataxia	Oculomotor disturbance
20–50	>73	Ataxia, pyramidal and bulbar signs	Peripheral signs late in disease
>40	<72	Ataxia, peripheral signs	Distal wasting, hyporeflexia cramping, neuropathy

[a] The CAG repeat lengths indicated are not absolute cutoffs but represent *approximate* transition points at which the subtype is increasingly likely to develop.

1. As in all CAG triplet repeat diseases, the size of the MJD1 repeat correlates well with disease severity. Larger repeats cause earlier disease onset and are associated with faster disease progression than are smaller, disease alleles (Klockgether et al., 1996; Klockgether et al., 1998).
2. SCA3 (MJD at the time) had been classified into three clinical subtypes long before the genetic defect was known (*see* Table 2). It is now clear that CAG repeat length is the major determinant of the different clinical subtypes (Durr et al., 1996; Maciel et al., 1995; Schols et al., 1996). Patients with type I disease have the earliest onset (usually before age 20), develop prominant pyramidal and extrapyramidal signs (including dystonia, spasticity, and ataxia), and typically have greater than 75 CAG repeats. Type II disease, the most common subtype, typically begins in the third through sixth decades, is characterized by ataxia and bulbar signs, and usually has a repeat length of 73 or greater. Type III disease with the latest onset, beginning after age 40, is characterized by peripheral neuropathy and ataxia without significant bulbar signs, and it is most commonly seen with repeat lengths of less than 72. It is important to recognize that not all SCA3 falls neatly into one of the subtypes. Rather, the clinical spectrum of SCA3 represent a continuum spanning all three subtypes.
3. Anticipation commonly occurs in SCA3. At the molecular level, there is a tendency for expanded repeats to enlarge during transmission. This phenomenon, coupled with the fact that larger repeats cause more severe disease phenotype, means that SCA3 often worsens in successive generations.
4. Although SCA3 is an autosomal dominant disease, there may be a gene dosage effect, because the few individuals who are homozygous for the expanded MJD1 allele seem to have a more severe phenotype than would be predicted by the size of their CAG repeats (Lerer et al., 1996; Sobue et al., 1996).
5. The threshold repeat length for developing SCA3 is significantly larger than the roughly 35–40 threshold range observed in most other polyQ diseases. For instance, an MJD1 repeat of 42 does not cause SCA3, whereas the same repeat size causes disease in HD and SCAs 1, 2 and 7. Similarly, relatively mild and late-onset disease occurs with MJD1 repeat lengths of 60–65, but the same

size repeats in other polyQ diseases cause devastating, early-onset degeneration. Surprisingly, this higher threshold in SCA3 exists despite the fact that the SCA3 disease protein, ataxin-3, is the smallest polyQ disease protein. Given the small size of the ataxin-3 protein, one might have expected SCA3 to have the lowest repeat threshold for disease because there is less surrounding polypeptide to "buffer" the polyQ domain than in other larger polyQ disease proteins. One possible explanation for the unexpectedly high threshold in SCA3 is that the surrounding ataxin-3 polypeptide may have unique structural features that modulate the toxicity of its polyQ domain.
6. Unlike other polyQ diseases, SCA3 does not display marked paternal bias in intergenerational instability: Juvenile-onset SCA3 can occur through paternal or maternal transmission. Several studies have shown no difference in repeat instability with paternally or maternally transmitted alleles, whereas others have shown a small bias toward increased paternal instability (Cancel et al., 1995; Giunti et al., 1995; Durr et al., 1996; Soong et al., 1997; Maciel et al., 1995; Maruyama et al., 1995; Igarashi et al., 1996; Takiyama et al., 1997). The molecular mechanism underlying trinucleotide repeat instability is an area of great research interest, but beyond the scope of this review. Excellent reviews of repeat instability can be found elsewhere (Pearson and Sinden, 1998).

12.4. THE *MJD1* GENE PRODUCT, ATAXIN-3

Ataxin-3, the smallest polyQ disease protein, is an intracellular protein of approx 42 kDa. The polyQ domain is located near the C-terminus (Fig. 2). The actual protein size varies, depending on at least three factors: (1) the length of the glutamine repeat; (2) a single nucleotide polymorphism (nucleotide 1118 A to C) that converts the original published stop codon to a tyrosine residue when C replaces A, extending the polypeptide by 16 amino acids (Kawaguchi et al., 1994); and (3) differential splicing. Perhaps the most interesting splice variant occurs near the C-terminus, where alternative splicing replaces the last 17 amino acids of the originally published ataxin-3 sequence with a different C-terminus of about the same size (Schmidt et al., 1998). Which C-terminal isoform is more prevalent in disease tissue is unknown, but studies suggest that both are expressed in disease brain (Paulson et al., 1997a; Schmidt et al., 1998). Although there is evidence for additional splice variants near the amino-terminus, full-length ataxin-3 appears to be the predominant isoform expressed in brain and elsewhere (Paulson et al., 1997a).

Ataxin-3 has been found in every mammalian tissue and cell line studied so far (Paulson et al., 1997a; Wang et al., 1997; Tait et al., 1998; Schmidt et al., 1998). As with most polyQ disease proteins, the function of ataxin-3 is unknown. Ataxin-3 does not show extensive homology to known proteins, although there is a predicted ortholog in *C. elegans* (accession no. Z81071,

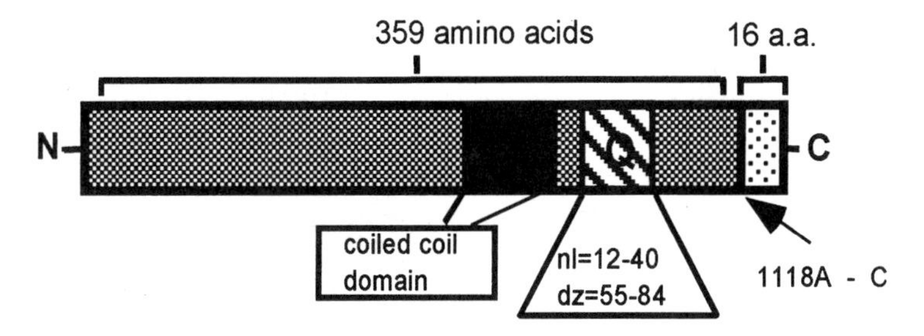

Fig. 2. The *MJD1* gene product, ataxin-3. Ataxin-3 is a small hydrophilic protein with the gln repeat (Q) near the carboxyl terminus. Ranges for normal and disease repeats are shown. Arrow indicates an intragenic polymorphism 1118 A–C, that alters the stop codon, extending the protein by 16 amino acids. A predicted coiled-coil region is indicated just before the gln repeat.

gene F28F8.6). The protein is predicted to have a high degree of helical secondary structure, including a coiled-coil domain situated just before the polyglutamine domain (Fig. 2) that is similar to structural domains in certain cytoskeletal associated proteins. Coiled-coil domains often mediate protein–protein interactions, but whether it does so in ataxin-3 is not yet known.

Although ataxin-3 is highly conserved between rat and human ataxin-3 (93% identical), the glutamine repeat itself is not: The repeat in rat ataxin-3 contains only five glutamine residues and is interrupted by a histidine residue (QQHQQQ). This difference in an otherwise highly conserved protein suggests that a homopolymeric glutamine repeat is not essential for normal ataxin-3 function.

The subcellular localization of ataxin-3 appears to be complex. Some reports indicate predominantly cytoplasmic staining for ataxin-3, whereas others suggest nuclear staining (Paulson et al., 1997a; Wang et al., 1997; Tait et al., 1998; Schmidt et al., 1998; Trottier et al., 1998). This apparent discrepancy may provide a clue to the protein's biology. Ataxin-3 is likely to be both a cytoplasmic and nuclear protein whose subcellular localization is regulated by one or more factors, including the type of cell, the state of the cell cycle, and the presence or absence of particular splice variants. The most detailed study to date suggests multiple isoforms of ataxin-3 with heterogeneous patterns of subcellular localization (Trottier et al., 1998). In many cells, a fraction of the ataxin-3 pool is intranuclear, bound to the nuclear matrix (Tait et al., 1998). This nuclear pool of ataxin-3 may be important to pathogenesis, in light of the fact that the mutant protein forms intranuclear inclusions in disease brain.

12.5. PATHOLOGICAL FEATURES

Just as the clinical features in SCA3 vary, so do the pathological features. However, some generalizations can be made based on numerous reports (Sachdev et al., 1982; Coutinho et al., 1982; Yuasa et al., 1986; Takiyama et al., 1994; Cancel et al., 1995; Durr et al., 1996; Kinoshita et al., 1995; Lopes-Cendes et al., 1996; Coutinho et al., 1986). First, the pathological changes are degenerative, involving neuronal loss and gliosis. Second, certain brain regions are typically affected, although the pattern may differ depending on the patient's ethnicity, age of onset, age of death, and repeat length. Commonly affected regions include the globus pallidus, subthalamic nucleus, substantia nigra, dentate nucleus, pontine nuclei, various cranial nerve nuclei (III, IV, VI, VII, VIII, and X), anterior horn cells, and neurons in the Clarke's column. Unaffected brain regions include the cerebral cortex, striatum, thalamus, cerebellar cortex, and inferior olives. Third, pathological changes are concentrated in certain pathways, especially the pallidothalamic, dentatorubral, pontocerebellar, and spinocerebellar tracts. Fourth, peripheral neuropathy is common, especially in later-onset cases. The neuropathy is usually symmetric, involving sensory and motor neurons and both unmyelinated and myelinated fibers.

Until recently, there were no known cellular hallmarks of SCA3 pathology. Now, however, it is clear that the disease protein forms neuronal intranuclear inclusions (NI) (Paulson et al., 1997a; Schmidt et al., 1998). NI are spherical, ubiquitinated aggregates that are preferentially found in neurons from susceptible brain regions (Fig. 3). In SCA3, they are particularly abundant in affected pontine neurons, but are absent in unaffected regions such as cerebral and cerebellar cortex. *Extranuclear* ubiquitin staining has also been noted in SCA3 brain; in fact, skeinlike ubiquitin-positive deposits in the cytoplasm of motor neurons were described several years before the discovery of NI (Suenaga et al., 1993). Our own studies of disease tissue suggest that extranuclear deposits are present, but not common in SCA3 brain (H. Paulson, unpublished observations). A systematic search for abnormal ubiquitin staining in neuronal processes is still needed to assess whether cytoplasmic protein aggregation contributes to the axonal neuropathy of SCA3/MJD.

Neuronal intranuclear inclusions are now recognized to be a common pathological hallmark of polyglutamine diseases, having been found in all but two of the polyglutamine diseases. Intriguingly, the two exceptions, SCA2 and SCA6, have the shortest expanded repeats. The discovery of NI was exciting because they seemed to represent a unifying pathological structure found in susceptible neurons. This led to the belief that NI formation

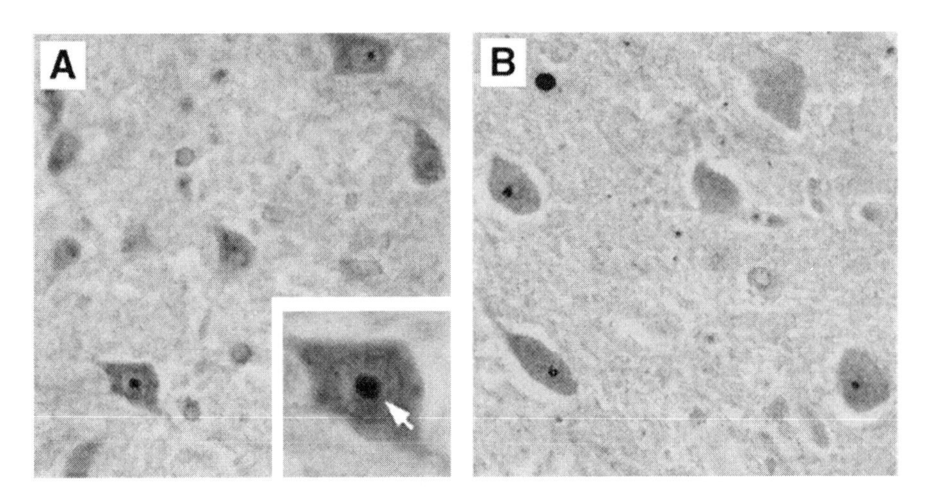

Fig. 3. Nuclear inclusions (NI) in SCA3. Ataxin-3 (**a**) and ubiquitin (**b**) immunostaining of pontine neurons in SCA3. Several neurons in both panels contain intranuclear spherical aggregates of the disease protein that are ubiquitin positive. One neuron is shown at higher power in the inset of panel a, with the NI indicated with an arrow. NI vary in size (at times spanning more than half the diameter of the nucleus) and number (typically one, but up to three per cell). The tissue sections shown were not counterstained.

might be important in pathogenesis. Recent findings, however, have dissociated NI formation from neurotoxicity. For example, certain striatal neurons in HD are among the most vulnerable neurons, yet they do not demonstrate NI (Gutekunst et al., 1999). Furthermore, several mouse models of HD suggest a dissociation between NI and degeneration (Davies et al., 1997; Reddy et al., 1998; Hodgson et al., 1999), and at least one neuronal cell culture model of HD failed to show a correlation between neuronal death and NI formation (Saudou et al., 1998). Results in SCA1 transgenic mice also suggest that visible nuclear aggregates are not required for initiating pathogenesis (Klement et al., 1998; Cummings et al., 1999).

These findings notwithstanding, it is still too early to concede that protein aggregates in general, and NI in particular, have no role in pathogenesis. The slow progression of human disease, together with results from transgenic mouse models (Burright et al., 1995; Mangiarini et al., 1997), suggest a stepwise mode of pathogenesis characterized by at least two phases: an early and prolonged period of neuronal dysfunction and a later period of neuronal death. If NI do play a role in pathogenesis it is likely to be late, near the point of neuronal demise. One way NI might compromise neuronal function is by fostering inappropriate interactions with vital housekeeping or regulatory proteins within the nucleus. Recent studies of ataxin-3 in transfected cells, transgenic flies, and human disease tissue indicate that NI can sequester certain polyglutamine-containing proteins such as the basal transcription factor TATA-binding protein (Perez et al., 1998). Once redistributed into polyglutamine aggregates, these proteins might recruit their own interacting partners, leading to deleterious downstream effects.

12.6. MECHANISM OF DISEASE

The mechanism of polyglutamine pathogenesis is still uncertain. The following discussion highlights recent findings that lead the way toward an understanding of disease mechanism.

Protein Misfolding Is Central to Pathogenesis

The basis of disease is a dominant, toxic gain of function occurring at the protein level and increasing with longer glutamine repeats. Evidence increasingly suggests that this novel toxic property is misfolding of the polyglutamine domain. Unique structural features of polyglutamine cause it to adopt an altered conformation when expanded, perhaps a β-sheet hairpin structure (Perutz, 1999). Direct evidence supporting an altered structure is the existence of antibodies that preferentially recognize and bind *expanded* polyglutamine (Trottier et al., 1995). Accumulating evidence from in vitro studies, animal models, and human disease tissue further argue for misfolding. Numerous studies in transfected cells have shown that expanded polyglutamine forms insoluble aggregates that presumably are derived from misfolded protein (Ikeda et al., 1996; Davies et al., 1997; Paulson et al., 1997a; Martindale et al., 1998; Merry et al., 1998). The presence of NI and neuropil aggregates in disease tissue argues for protein misfolding in vivo. Similar aggregates have also been observed in many transgenic animal models, both in mice and in flies (Warrick et al., 1998; Burright et al., 1995). Test tube studies of recombinant polyQ fragments indicate that polyQ aggregation can be a self-driven process that occurs in a concentration-dependent manner and results in the formation of amyloidlike insoluble fibrils (Scherzinger et al., 1997). The threshold repeat length for in vitro aggregation closely mirrors the threshold for disease, supporting the view that polyQ-induced misfolding is a central element in pathogenesis.

Do the resultant aggregates directly contribute to pathogenesis, or are they simply bystanders in the disease process? The answer is still uncertain, although recent evidence suggests that large nuclear inclusions are not necessary for pathogenesis. Aggregation likely proceeds through a series of steps: monomer misfolding, nucleation, oligomerization, and amyloidlike fibril formation (with recruitment of nondisease proteins in vivo). Which step or steps in misfolding/aggregation represent the toxic element is hotly debated. It is possible that the process of misfolding/aggregation, rather than the resultant intranuclear aggregate, is toxic. If compounds are identified that block aggregate formation, it will be possible to determine whether decreasing aggregate formation results in decreased toxicity.

Importance of Nuclear Localization

Although a direct pathogenic role for NI is uncertain, nuclear expression of polyQ protein seems important in pathogenesis. For example, in *SCA1* transgenic mice, neuronal degeneration does not occur when mutant ataxin-1 is targeted to the cytoplasm instead of its normal nuclear location (Klement et al., 1998). Also, in transfected neurons, mutant huntingtin induces apoptosis preferentially when the protein is localized to the nucleus (Saudou et al., 1998). Transfection studies with ataxin-1 and ataxin-3 further suggest that the nuclear environment favors aggregation (Klement et al., 1998; Perez et al., 1998). In transfected cells, for example, full-length mutant ataxin-3 primarily remains diffusely distributed in the cytoplasm, but forms intranuclear inclusions when it is forced into the nucleus by adding a nuclear localization signal to the protein (Chai et al., 1999b).

Why the nuclear environment might favor aggregation is unknown. Recent studies of the SCA3/MJD disease protein, ataxin-3, show that when it is in the nucleus, it may take on an altered conformation that exposes the polyglutamine domain (Perez et al., 1999). Thus, one way the nucleus may promote inclusion formation is through nuclear-specific conformational changes in the protein. Alternatively, the nucleus may be less efficient at degrading, refolding, or disaggregating misfolded protein. A third possibility is that the nucleus might concentrate mutant proteins in particular subnuclear structures that promote aggregation. It remains to be seen whether there is a causal connection between the apparent aggregation-promoting property of the nucleus and its being the site of toxicity. Yet, it is easy to imagine ways that the mutant protein, whether as misfolded monomer, oligomer, or aggregate, might alter vital nuclear functions. The existence of glutamine-rich transcription factors suggests that mutant protein might bind such factors inappropriately, altering transcription of genes critical for neuronal function. Other nuclear functions, such as mRNA splicing or the export of proteins and RNA to the cytoplasm, could be perturbed in a similar manner.

The Toxic Fragment Hypothesis: A Role for Proteases in Pathogenesis

Although a role for proteolysis in polyglutamine pathogenesis is not yet proven, mounting evidence suggests that production of a toxic fragment may be important in some, if not all, polyQ diseases. First, immunohistochemical studies in at least three diseases (including SCA3) suggest that polyQ-containing fragments, and not the full disease protein, are the constituents of NI (ataxin-1 inclusions in SCA1 may be an exception). Second, proteolytic fragments of disease protein have been reported in HD brain (Difiglia et al., 1997), transgenic mouse models of disease (Schilling et al., 1999a; Schilling et al., 1999b) and various cellular models (Goldberg et al., 1996; Merry et al., 1998; Cooper et al., 1998). Third, studies of recombinant protein and transfected cells have shown that many polyQ disease proteins, including ataxin-3, are substrates for caspases (Goldberg et al., 1996; Wellington et al., 1998). Finally, polyQ-containing fragments that have been freed from their surrounding protein context are particularly potent pathogens, more prone to aggregate and cause cell death than the full-length protein. This point was first demonstrated in cellular and transgenic models of SCA3 (Ikeda et al., 1996) and has since been confirmed in further studies of SCA3 and other polyQ diseases (Paulson et al., 1997b; Merry et al., 1998; Martindale et al., 1998; Cooper et al., 1998). Although this does not constitute evidence for proteolysis, it suggests that fragment production should accelerate the disease process. Further studies of SCA3 disease tissue are needed to determine whether a proteolytic fragment is formed during SCA3 pathogenesis.

The toxic fragment hypothesis assumes that production of a proteolytic fragment drives pathogenesis, perhaps by accelerating misfolding and aggregation. Another possibility is that proteolytic processing occurs *after* aggregation, because proteasome components are found in nuclear inclusions. The proteasome and other molecular chaperones may continue to work on the aggregated protein, partially degrading it in the process. On this view, limited proteolysis of mutant protein might be a nonessential, downstream event in some polyQ diseases.

Molecular Chaperones in Disease

Molecular chaperones such as heat shock proteins (Hsp) assist in the folding, refolding, and elimination of misfolded polypeptides that arise under conditions of cellular stress. In polyQ diseases, neurons might be expected to mount a chaperone stress response that assists in the refolding, elimination, and/or disaggregation of expanded polyglutamine protein. Studies from SCA3 and other polyglutamine diseases suggest that this may be the case. In human disease tissue, animal models, and transfected cells, certain chaperones are redistributed into polyQ aggregates (Cummings et al., 1998; Chai et al., 1999b; Stenoien et al., 1999; Warrick et al., 1999). Moreover, in cells expressing mutant polyglutamine protein, Hsp70 is upregulated (Chai et al., 1999a).

The simplest interpretation of chaperone upregulation and redistribution is that it is an appropriate cellular response to handle misfolded and aggregated polypeptide. Alternatively, it may represent a marker of polyQ-induced cellular stress that, over time, is deleterious to neurons. In either scenario, overexpression of certain chaperones might be expected to reduce polyQ aggregation and/or toxicity. Indeed, overexpression of the Hsp40 chaperones reduces aggregation of ataxin-3 and other polyQ proteins. (Cummings et al., 1998; Chai et al., 1999a; Stenoien et al., 1999). A role for chaperones has now been confirmed in a transgenic *Drosophila* model of SCA3 that recapitulates cellular features of polyQ disease, including NI formation and late-onset degeneration (Warrick et al., 1998). In the fly model, endogenous Hsp70 modulates polyglutamine toxicity and overexpression of human Hsp70 suppresses polyglutamine neurotoxicity (Warrick et al., 1999). Additional studies in the fly model are likely to yield further insights into polyQ pathogenesis.

Another major intracellular pathway implicated in disease is the ubiquitin–proteasome degradation system. The proteasome complex is responsible for the ubiquitin-dependent degradation of most cytosolic proteins, including misfolded or damaged proteins. Several studies have shown redistribution of the proteasome complex into NI (Cummings et al., 1998; Chai et al., 1999b; Stenoien et al., 1999). For example, in SCA3 brain and in cells expressing mutant ataxin-3, components of the proteasome complex redistribute into polyglutamine aggregates. This led us to test whether proteasome activity directly influences polyglutamine aggregation. When the proteasome was inactivated with specific inhibitors, polyglutamine aggregation increased in a repeat-length-dependent manner (Chai et al., 1999b). Based on this result, our working model is that the proteasome represents a first-line cellular defense that recognizes and eliminates misfolded polyglutamine protein before aggregation occurs. However, it is still unclear whether proteasome redistribution in polyQ disease is good or bad for the neuron. If proteasome redistribution into NI depletes the neuron of functioning proteasomes, the result would be deleterious to the cell: fewer active proteasomes to degrade abnormal protein, leading to a further increase in misfolded polyQ protein and further aggregation. Alternatively, proteasome recruitment into aggregates may allow for processing of the aggregated protein that renders it less toxic.

Mode of Cell Death

Cell death has classically been divided into necrosis and apoptosis on morphological grounds. However, as the cellular mechanisms underlying cell death become increasingly well understood, the distinctions between the two have blurred; not all forms of cell death fall neatly under apoptosis or necrosis. This may prove to be the case in late-onset neurodegenerative disorders like polyQ diseases. The slowly progressive nature of disease and the absence of inflammatory changes have led many to suspect that polyQ-mediated cell death is apoptotic. Yet aside from a few reports showing positive terminal deoxynucleotidyl dUTP nick end labeling (TUNEL) staining in disease tissue (Portera-Cailliau et al., 1999), there is little evidence from human disease tissue to support this view, perhaps because of the inherent limitations of postmortem analysis.

Fortunately, the recent profusion of transgenic animal models has begun to permit a more systematic analysis of polyQ-mediated degeneration. Although these animal models reveal certain degenerative cellular features that are apoptotic-like, the overall impression is that polyQ-mediated degeneration may be more complex than, for example, the programmed cell death occurring during neuronal development. On the one hand, increased TUNEL staining was found in the striatum of an HD mouse model, suggesting but not proving apoptosis (Reddy et al., 1998), and analysis of two other HD mouse models showed that neuronal loss was preceded by cellular changes such as hyperchromasia, nuclear indentation, and cell shrinkage (Hodgson et al., 1999; Davies et al., 1997). Although these changes do not fulfill classic criteria for apoptosis, they argue against necrosis. On the other hand, the potent antiapoptotic baculovirus gene *p35* showed only a modest ability to suppress degeneration in a fly model of SCA3, suggesting that polyQ-mediated cell death is not primarily a "classic" form of caspase-dependent apoptosis (Warrick et al., 1998).

Cellular models of disease have not resolved the issue. In two studies of transfected neurons, caspase inhibitors and antiapoptotic genes blocked polyQ-mediated cell death (Sanchez et al., 1999; Saudou et al., 1998), but in a third study, neuronal death was *not* blocked by caspase inhibitors(Moulder et al., 1999). Nonetheless, in this latter study, caspase activation *did* occur transiently at a sublethal level and caspase inhibitors delayed aggregate formation. These results suggest that low-level caspase activation *before* the period of cell death may itself promote further polyQ misfolding and aggregation. This raises the intriguing possibility that during the prolonged course of polyglutamine degeneration in vivo, chronic sublethal activation of caspases is one type of cellular stresses experienced by neurons that ultimately tips the balance toward cell death (other metabolic derangements might include and alterations in chaperones, loss of trophic support, and mitochondrial impairment). Consistent with this view are recent results showing that a dominant-negative form of caspase-1 delays polyQ aggregation, behavioral changes and death in an HD mouse model (Ona et al., 1999). Recent evidence demonstrating activated caspases in HD brain tissue may also reflect chronic, low-level activation of caspases rather than the final cascade of caspase activation in apoptosis (Sanchez et al., 1999; Ona et al., 1999).

A recent cellular model raised the intriguing possibility that polyQ aggregates or oligomers serve as nucleation sites for the death adaptor protein FADD, which then activates caspase 8, initiating a cascade of caspase activation that culminates in apoptosis (Sanchez et al., 1999). These studies were performed with truncated ataxin-3 constructs, but it remains to be seen whether similar caspase-8 activation occurs in SCA3 disease tissue. Another hypothesis requiring further investigation is that localization of mutant ataxin-3 to particular subnuclear structures known as PML-oncogenic domains (Chai et al., 1999b) may contribute to polyQ-mediated cell death. This possibility is suggested by recent studies implicating PML-oncogenic domains in caspase-dependent and caspase-independent cell death (Quignon et al., 1998; Wang et al., 1998).

12.7. SELECTIVE VULNERABILITY: RELATIONSHIP TO MISFOLDING

Despite widespread expression of polyQ disease proteins, only certain populations of neurons appear to be susceptible to degeneration. Factors contributing to this selective vulnerability can be grouped into two categories: (1) those that increase the level of misfolded protein or directly promote misfolding and (2) those that act downstream of misfolding. For example, the level of disease gene expression is an obvious factor in the first category. Although the various disease proteins are widely expressed, absolute levels of expression in different populations of neurons surely differ, and this would be expected to translate into corresponding differences in the intracellular concentration of misfolded monomer. Also falling under the first category are various potential posttranslational modifications that might modulate misfolding. In susceptible neurons, for example, misfolding and aggregation could be promoted by specific proteolytic events that release a polyQ fragment or by aberrant targeting of polyQ protein to the nucleus. Specific interacting proteins are likely to contribute to selective vulnerability through *both* categories. For instance, certain interacting proteins may bind to disease proteins in a way that promotes misfolding and aggregation, as was recently demonstrated for the huntingtin-interacting protein, SH3GL3 (Sittler et al., 1998). Other specific interacting proteins are likely to influence events downstream of misfolding through mechanisms that are tied to the specific functions of the disease proteins. For example, mutant protein may bind more or less avidly to specific interacting proteins, thereby altering physiologic or biochemical properties of one or both proteins. The susceptibility of a neuron to the downstream effects of the mutant protein would depend, in part, on the particular interacting proteins it expresses. Although such interacting proteins have been found for several other polyQ disease proteins, they have not yet been identified for ataxin-3.

12.8. ROUTES TO THERAPY

PolyQ diseases are fatal disorders for which there is no preventive therapy. Now that protein misfolding is thought to be central to pathogenesis, strategies to reduce the concentration of misfolded protein or block aggregation represent potential therapeutic approaches. There are many potential routes to reducing the concentration of misfolded polyQ protein: (1) Ribozyme- or antisense-mediated downregulation of disease gene expression may be useful, particularly if strategies can be developed to specifically target transcripts from the disease allele. (2) Efforts to enhance degradation of misfolded polyQ protein (e.g., by engineering ubiquitin–proteasome components to preferentially act on expanded polyglutamine) might lower the steady-state concentration of misfolded poplypeptide and thereby reduce the rate of nucleation and aggregation. (3) If specific proteases are shown to generate cytotoxic fragments, then appropriate protease inhibitors might block proteolysis and slow disease progression. (4) Compounds that block aggregation (suppressing nucleation, fibril formation, or both) may also succeed in reducing toxicity. Especially intriguing is the possibility that aggregation blockers that disrupt β-sheet formation may be relevant to other neurodegenerative proteinopathies such as Alzheimer's disease. (5) Another potential approach is overexpressing certain molecular chaperones. This approach has already been shown to be effective in cellular models and transgenic flies (Cummings et al., 1998). Now, it will be important to confirm these findings in mammalian models of disease. (6) Very likely there will be additional therapeutic targets *downstream* of polyQ misfolding. We do not yet know which regulatory pathways are perturbed in polyQ diseases, but once identified, there may be rational approaches to block or enhance these pathways. Genetically tractable animal models (e.g., *Drosophila* and *C. elegans*) should aid in identifying genes and pathways that mediate or modulate the deleterious consequences of expanded polyQ protein. (7) The final stage of disease is cell death, against which antiapoptotic agents such as caspase inhibitors or survival factors such as neurotrophins may prove useful. In slowly progressive polyQ diseases, antideath therapies may be too little, too late. However, caspase inhibitors may find a therapeutic use if, as suggested by recent results, sublethal caspase activation contributes to neurodegeneration (Ona et al., 1999). In summary, there are many avenues where one could place therapeutic barriers to pathology. Researchers now possess the knowledge and molecular tools to be ambitious in generating rational therapies for SCA3 and other polyQ diseases.

REFERENCES

Becher, M. W., Kotzuk, J. A., Sharp, A. H., Davies, S. W., Bates, G. P., Price, D. L., et al. (1998) Intranuclear neuronal inclusions in Huntington's disease and dentatorubral and pallidoluysian atrophy: correlation between the density of inclusions and IT15 CAG triplet repeat length. *Neurobiol. Dis.* **4,** 387–397.

Burright, E. N., Clark, H. B., Servadio, A., Matilla, T., Feddersen, R. M., Yunis, W. S., et al. (1995) SCA1 transgenic mice: a model for neurodegeneration caused by an expanded CAG trinucleotide repeat. *Cell* **82,** 937–948.

Cancel, G., Abbas, N., Stevanin, G., Durr, A., Chneiweiss, H., Neri, C., et al. (1995) Marked phenotypic heterogeneity associated with expansion of a CAG repeat sequence at the spinocerebellar ataxia 3/Machado-Joseph disease locus. *Am. J. Hum. Genet.* **57,** 809–816.

Chai, Y., Koppenhafer, S. L., Bonini, N. M., and Paulson, H. L. (1999) Analysis of the role of heat shock protein (Hsp) molecular chaperones in polyglutamine disease. *J. Neurosci.* **19,** 10,338–10,347.

Chai, Y., Koppenhafer, S. L., Shoesmith, S. J., Perez, M. K., and Paulson, H. L. (1999) Evidence of proteasome involvement in polyglutamine disease: localization to nuclear inclusions in SCA3/MJD and suppression of polyglutamine aggregation *in vitro*. *Hum. Mol. Genet.* **8(4),** 673–682.

Cooper, J. K., Schilling, G., Peters, M. F., Herring, W. J., Sharp, A. H., Kaminsky, Z., et al. (1998) Truncated N–terminal fragments of huntingtin with expanded glutamine repeats form nuclear and cytoplasmic aggregates in cell culture. *Hum. Mol. Genet.* **7,** 783–790.

Coutinho, P., Guimaraes, A., Pires, M. M., and Scaravilli, F. (1986) The peripheral neuropathy in Machado-Joseph disease. *Acta Neuropathol.* **71,** 119–124.

Coutinho, P., Guimaraes, A., and Scaravilli, F. (1982) The pathology of Machado-Joseph disease. Report of a possible homozygous case. Acta Neuropathol. **58,** 48–54.

Cummings, C. J., Mancini, M. A., Antalffy, B., DeFranco, D. B., Orr, H. T., and Zoghbi, H. Y. (1998) Chaperone suppression of aggregation and altered subcellular proteasome localization imply protein misfolding in SCA1. *Nature Genet.* **19,** 148–154.

Cummings, C. J., Reinstein, E., Sun, Y., Anatalffy, B., Jiang, Y., Ciechanover, A., et al. (1999) Mutation of the E6-AP ubiquitin ligase reduces nuclear inclusion frequency while accelerating polyglutamine-induced pathology in SCA1 transgenic mice. *Neuron* **24,** 879–892.

Davies, S. W., Turmaine, M., Cozens, B. A., Difiglia, M., Sharp, A. H., Ross, C. A., et al. (1997) Formation of neuronal intranuclear inclusions underlies the neurological dysfunction in mice transgenic for the HD mutation. *Cell* **90,** 537–548.

Difiglia, M., Sapp, E., Chase, K. O., Davies, S. W., Bates, G. P., Vonsattel, J. P., et al. (1997) Aggregation of huntingtin in neuronal intranuclear inclusions and dystrophic neurites in brain. Science **277,** 1990–1993.

Durr, A., Stevanin, G., Cancel, G., Duyckaerts, C., Abbas, N., Didierjean, O., et al. (1996) Spinocerebellar ataxia 3 and Machado-Joseph disease: clinical, molecular, and neuropathological features. *Ann. Neurol.* **39,** 490–499.

Gaspar, C., Lopes–Cendes, I., DeStefano, A. L., Maciel, P., Silveira, I., Coutinho, P., et al. (1996) Linkage disequilibrium analysis in Machado-Joseph disease patients of different ethnic origins. *Hum. Genet.* **98,** 620–624.

Giunti, P., Sweeney, M. G., and Harding, A. E. (1995) Detection of the Machado-Joseph disease/spinocerebellar ataxia three trinucleotide repeat expansion in families with autosomal dominant motor disorders, including the Drew family of Walworth. *Brain* **118,** 1077–1085.

Goldberg, Y. P., Nicholson, D. W., Rasper, D. M., Kalchman, M. A., Koide, HB, Graham, R. K., et al. (1996) Cleavage of huntingtin by apopain, a proapoptotic cysteine protease, is modulated by the polyglutamine tract.*Nature Genet.* **13,** 442–449.

Gutekunst, C. A., Li S–H., Hong, Y., Mulroy, J. S., Kuemmerle, S., Jones, R., et al. (1999) Nuclear and neuropil aggregates in huntington's disease: relationship to neuropathology. *J. Neurosci.* **19(7),** 2522–2534.

Hodgson, J. G., Agopyan, N., Gutekunst, C. A., Leavitt, B. R., LePiane, F., Singaraja, R., et al. A YAC mouse model for huntington's disease with full-length mutant huntingtin, cytoplasmic toxicity, and selective striatal neurodegeneration. *Neuron* **23,** 181–192.

Holmberg, M., Duyckaerts, C., Durr, A., Cancel, G., Gourfinkel-An, I., Damier, P., et al. (1998) Spinocerebellar ataxia type 7 (SCA7): a neurodegenerative disorder with neuronal intranuclear inclusions. *Hum. Mol. Genet.* **7,** 913–918.

Igarashi, S., Takiyama, Y., Cancel, G., Rogaeva, E. A., Sasaki, H., Wakisaka, A., et al. (1996) Intergenerational instability of the CAG repeat of the gene for Machado-Joseph disease (MJD1) is affected by the genotype of the normal chromosome: implications for the molecular mechanisms of the instability of the CAG repeat. *Hum. Mol. Genet.* **5,** 923–932.

Ikeda, H., Yamaguchi, M., Sugai, S., Aze, Y., Narumiya, S., and Kakizuka, A. (1996) Expanded polyglutamine in the Machado-Joseph disease protein induces cell death in vitro and in vivo. *Nature Genet.* **13,** 196–202.

Ishikawa, K., Fugigasaki, H., Saegusa, H., Ohwada, K., Fujita, T., Iwamoto, H., et al. (1999) Abundant expression and cytoplasmic aggregations of (alpha)1 A voltage-dependent calcium channel protein associated with neurodegeneration in spinocerebellar ataxia type 6. *Hum. Mol. Genet.* **8,** 1185–1193.

Kawaguchi, Y., Okamoto, T., Taniwaki, M., Aizawa, M., Inoue, M., Katayama, S., Kawakami, H., et al. (1994) CAG expansions in a novel gene for Machado-Joseph disease at chromosome 14q32. 1. Nature Genet. **8,** 221–228.

Kinoshita, A., Hayashi, M., Oda, M., and Tanabe, H. (1995) Clinicopathological study of the peripheral nervous system in Machado-Joseph disease. *J. Neurol. Sci.* **130,** 48–58.

Klement, I. A., Skinner, P. J., Kaytor, M. D., Yi, H., Hersch, S. M., Clark, H. B., et al. (1998) Ataxin-1 nuclear localization and aggregation: role in polyglutamine-induced disease in SCA1 transgenic mice. *Cell* **95,** 41–53.

Klockgether, T., Kramer, B., Ludtke, R., Schols, L., and Laccone, F. (1996) Repeat length and disease progression in spinocerebellar ataxia type 3 [letter]. *Lancet* **348,** 830

Klockgether, T., Ludtke, R., Kramer, B., Abele, M., Burk, K., Schols, L., et al. (1998) The natural history of degenerative ataxia: a retrospective study in 466 patients. *Brain* **121,** 589–600.

Lerer, I., Merims, D., Abeliovich, D., Zlotogora, J., and Gadoth, N. (1996) Machado-Joseph disease: correlation between the clinical features, the CAG repeat length and homozygosity for the mutation. *Eur. J. Hum. Genet.* **4,** 3–7.

Li, M., Miwa, S., Kobayashi, Y., Merry, D., Yamamoto, M., Tanaka, F., et al. (1998) Nuclear inclusions of the androgen receptor protein in spinal and bulbar muscular atrophy. *Ann. Neurol.* **44,** 249–254.

Lopes-Cendes, I., Silveira, I., Maciel, P., Gaspar, C., Radvany, J., Chitayat, D., et al. (1996) Limits of clinical assessment in the accurate diagnosis of Machado-Joseph disease. *Arch. Neurol.* **53,** 1168–1174.

Maciel, P., Gaspar, C., DeStefano, A. L., Silveira, I., Coutinho, P., Radvany, J., et al. (1995) Correlation between CAG repeat length and clinical features in Machado-Joseph disease. *Am. J. Hum. Genet.* **57,** 54–61.

Mangiarini, L., Sathasivam, K., Mahal, A., Mott, R., Seller, M., and Bates, G. P. (1997) Instability of highly expanded CAG repeats in mice transgenic for the Huntington's disease mutation. *Nature Genet.* **15,** 197–200.

Martindale, D., Hackam, A., Wieczorek, A., Ellerby, L., Wellington, C., McCutcheon, K., et al. (1998) Length of huntingtin and its polyglutamine tract influences localization and frequency of intracellular aggregates. *Nature Genet.* **18,** 150–154.

Maruyama, H., Nakamura, S., Matsuyama, Z., Sakai, T., Doyu, M., Sobue, G., et al. (1995) Molecular features of the CAG repeats and clinical manifestation of Machado-Joseph disease. *Hum. Mol.r Genet.* **4,** 807–812.

Matilla, T., McCall, A., Subramony, S. H., and Zoghbi, H. Y. (1995) Molecular and clinical correlations in spinocerebellar ataxia type 3 and Machado-Joseph disease. Ann. Neurol. **38,** 68–72.

Merry, D. E., Kobayashi, Y., Bailey, C. K., Taye, A. A., and Fischbeck, K. H. (1998) Cleavage, aggregation and toxicity of the expanded androgen receptor in spinal and bulbar muscular atrophy. *Hum. Mol. Genet.* **7,** 693–701.

Moulder, K. L., Onodera, O., Burke, J. R., Strittmatter, W. J., and Johnson, E. M., Jr. (1999) Generation of neuronal intranuclear inclusions by polyglutamine-GFP: analysis of inclusion clearance and toxicity as a function of polyglutamine length. *J. Neurosci.* **19,** 705–715.

Nakano, K. K., Dawson, D. M., and Spence, A. (1972) Machado disease. A hereditary ataxia in Portuguese emigrants to Massachusetts. *Neurology* **22,** 49–55.

Ona, V. O., Li, M., Vonsattel, J. P. G., Andrews, L. J., Khan, S. Q., Chung, W. M., et al. (1999) Inhibition of caspase-1 slows disease progression in a mouse model of Huntington's disease. *Nature* **399,** 263–267.

Paulson, H. L., Das, S. S., Crino, P. B., Perez, M. K., Patel, S. C., Gotsdiner, D., et al. (1997) Machado-Joseph disease gene product is a cytoplasmic protein widely expressed in brain. *Ann. Neurol.* **41,** 453–462.

Paulson, H. L., Perez, M. K., Trottier, Y., Trojanowski, J. Q., Subramony, S. H., Das, S. S., et al.(1997) Intranuclear inclusions of expanded polyglutamine protein in spinocerebellar ataxia type 3. *Neuron* **19,** 333–344.

Pearson, C. E. and Sinden, R. R. (1998) Slipped strand DNA, dynamic mutations and human disease, in *Genetic Instabilities and Hereditary Neurological Diseases* (Wells, R. D. and Warren, S. T., eds.), Academic, San Diego, pp. 585–621.

Perez, M., Paulson, H., and Pittman, R. (1999) Nuclear localization of ataxin–3 alters conformation of the polyglutamine tract. *Hum. Mol. Genet.* **8,** 2377–2385.

Perez, M. K., Paulson, H. L., Pendse, S. J., Saionz, S. J., Bonini, N. M., and Pittman, R. N. (1998) Recruitment and the role of nuclear localization in polyglutamine-mediated aggregation. *J. Cell Biol.* **143,** 1457–1470.

Perutz, M. F. (1999) Glutamine repeats and neurodegenerative diseases: molecular aspects. *TIBS* **24,** 58–63.

Portera-Cailliau, C, Hedreen, J. C., Price, D. L., and Koliatsos, V. E. (1999) Evidence for apoptotic cell death in Huntington disease and excitotoxic animal models. *J. Neurosci.* **15(5),** 3775–3787.

Quignon, F., De, B. F., Koken, M., Feunteun, J., Ameisen, J. C., and deThe, T. H. (1998) PML induces a novel caspase-independent death process. *Nature Genet.* **20,** 259–265.

Reddy, P. H., Williams, M., Charles, V., Garrett, L., Pike-Buchanan, L., Whetsell, W. O. J., et al. (1998) Behavioural abnormalities and selective neuronal loss in HD transgenic mice expressing mutated full-length HD cDNA. *Nature Genet.* **20,** 198–202.

Rosenberg, R. N., Nyhan, W. L., Bay, C., and Shore, P. (1976) Autosomal dominant striatonigral degeneration. A clinical, pathologic, and biochemical study of a new genetic disorder. *Neurology* **26,** 703–714.

Sachdev, H. S., Forno, L. S., and Kane, C. A. (1982) Joseph disease: a multisystem degenerative disorder of the nervous system. *Neurology* **32,** 192–195.

Sanchez, I., Xu, C-J., Juo, P., Kakizaka, A., Blenis, J., and Yuan, J. (1999) Caspase-8 is required for cell death induced by expanded polyglutamine repeats. *Neuron* **22,** 623–633.

Saudou, F., Finkbeiner, S., Devys, D., and Greenberg, M. E. (1998) Huntingtin acts in the nucleus to induce apoptosis but death does not correlate with the formation of intranuclear inclusions. *Cell* **95,** 55–66.

Scherzinger, E., Lurz, R., Turmaine, M., Mangiarini, L., Hollenbach, B., Hasenbank, et al. (1997) Huntingtin-encoded polyglutamine expansions form amyloid-like protein aggregates in vitro and in vivo. *Cell* **90,** 549–558.

Schilling, G., Becher, M. W., Sharp, A. H., Jinnah, H. A., Duan, K., Kotzuk, J. A., et al. (1999) Intranuclear inclusions and neuritic aggregates in transgenic mice expressing a mutant N-terminal fragment of huntingtin. *Hum. Mol. Genet.* **8(3),** 397–407.

Schilling, G., Wood, J. D., Duan, K., Slunt, H. H., Gonzales, V., Yamada, M., et al. (1999) Nuclear accumulation of truncated atrophin-1 fragments in a transgenic mouse model of DRPLA. *Neuron* **24,** 275–286.

Schmidt, T., Landwehrmeyer, G. B., Schmitt, I., Trottier, Y., Auburger, G., Laccone, F., et al. (1998) An isoform of ataxin-3 accumulates in the nucleus of neuronal cells in affected brain regions of SCA3 patients. *Brain Pathol.* **8,** 669–679.

Schols, L., Amoiridis, G., Epplen, J. T., Langkafel, M., Przuntek, H., and Riess, O. (1996) Relations between genotype and phenotype in German patients with the Machado-Joseph disease mutation. *J. Neurol. Neurosurg. Psychiatry* **61,** 466–470.

Schols, L., Vieira-Saecker, A. M., Schols, S., Przuntek, H., Epplen, J. T., and Riess, O. (1995) Trinucleotide expansion within the MJD1 gene presents clinically as spinocerebellar ataxia and occurs most frequently in German SCA patients. *Hum. Mol. Geneti.* **4,** 1001–1005.

Sittler, A., Walter, S., Wedemeyer, N., Hasenbank, R., Scherzinger, E., Eickhoff, H., et al. (1998) SH3GL3 associates with the Huntingtin exon 1 protein and promotes the formation of polygln–containing protein aggregates. *Mol. Cell* **2,** 427–436.

Skinner, P. J., Koshy, B. T., Cummings, C. J., Klement, I. A., Helin, K., Servadio, A., et al. (1997) Ataxin-1 with an expanded glutamine tract alters nuclear matrix-associated structures. *Nature* **389,** 971–974.

Sobue, G., Doyu, M., Nakao, N., Shimada, N., Mitsuma, T., Maruyama, H., et al. (1996) Homozygosity for Machado-Joseph disease gene enhances phenotypic severity [letter]. *J. Neurol. Neurosurg. Psychiatry* **60,** 354–356.

Soong, B., Cheng, C., Liu, R., and Shan, D. (1997) Machado-Joseph disease: clinical, molecular, and metabolic characterization in Chinese kindreds. *Ann. Neurol.* **41,** 446–452.

Stenoien, D. L., Cummings, C. J., Adams, H. P., Mancini, M. G., Patel, K., DeMartino, G. N., et al. (1999) Polyglutamine-expanded androgen receptors form aggregates that sequester heat shock proteins, proteasome components and SRC-1, and are suppressed by the HDJ-2 chaperone. *Hum. Mol. Genet.* **8(5),** 733–741.

Stevanin, G., Cancel, G., Didierjean, O., Durr, A., Abbas, N., Cassa, E., et al. (1995) Linkage disequilibrium at the Machado-Joseph disease/spinal cerebellar ataxia 3 locus: evidence for a common founder effect in French and Portuguese–Brazilian families as well as a second ancestral Portuguese–Azorean mutation [letter]. *Am. J. Hum. Genet.* **57,** 1247–1250.

Stevanin, G., Le, G. E., Ravise, N., Chneiweiss, H., Durr, A., Cancel, G., et al. (1994) A third locus for autosomal dominant cerebellar ataxia type I maps to chromosome 14q24. 3-qter: evidence for the existence of a fourth locus. *Am. J. Hum. Genet.* **54,** 11–20.

Stevanin, G., Lebre, A. S., Mathieux, C., Cancel, G., Abbas, N., Didierjean, O., et al. (1997) Linkage disequilibrium between the spinocerebellar ataxia 3/Machado-Joseph disease mutation and two intragenic polymorphisms, one of which, X359Y, affects the stop codon [letter]. *Am. J. Hum. Genet.* **60,** 1548–1552.

Suenaga, T., Matsushima, H., Nakamura, S., Akiguchi, I., and Kimura, J. (1993) Ubiquitin-immunoreactive inclusions in anterior horn cells and hypoglossal neurons in a case with Joseph's disease. *Acta Neuropathol.* **85,** 341–344.

Tait, D., Riccio, M., Sittler, A., Scherzinger, E., Santi, S., Ognibene, A., et al. (1998) Ataxin-3 is transported into the nucleus and associates with the nuclear matrix. *Hum. Mol. Genet.* **7,** 991–997.

Takiyama, Y., Nishizawa, M., Tanaka, H., Kawashima, S., Sakamoto, H., Karube, Y., et al. (1993) The gene for Machado-Joseph disease maps to human chromosome 14q. *Nature Genet.* **4,** 300–304.

Takiyama, Y., Oyanagi, S., Kawashima, S., Sakamoto, H., Saito, K., Yoshida, M., et al. (1994) A clinical and pathologic study of a large Japanese family with Machado-Joseph disease tightly linked to the DNA markers on chromosome 14q. *Neurology* **44,** 1302–1308.

Takiyama, Y., Sakoe, K., Soutome, M., Namekawa, M., Ogawa, T., Nakano, I., et al. (1997) Single sperm analysis of the CAG repeats in the gene for Machado-Joseph disease (MJD1): evidence for non-Mendelian transmission of the MJD1 gene and for the effect of the intragenic CGG/GGG polymorphism on the intergenerational instability. *Hum. Mol. Genet.* **6,** 1063–1068.

Trottier, Y., Cancel, G., An-Gourfinkel, I., Lutz, Y., Weber, C., Brice, A., et al. (1998) Heterogeneous intracellular localization and expression of ataxin-3. *Neurobiol. Dis.* **5,** 335–347.

Trottier, Y., Devys, D., Imbert, G., Saudou, F., An, I., Lutz, Y., et al. (1995) Cellular localization of the Huntington's disease protein and discrimination of the normal and mutated form [see comments]. *Nature Genet.* **10**, 104–110.

Wang, G., Ide, K., Nukina, N., Goto, J., Ichikawa, Y., Uchida, K., et al. (1997) Machado-Joseph disease gene product identified in lymphocytes and brain. *Biochem. Biophys. Res. Commun.* **233,** 476–479.

Wang, Z. G., Ruggero, D., Ronchetti, S., Zhong, S., Gaboli, M., Rivi, R., et al. (1998) PML is essential for multiple apoptotic pathways [see comments]. *Nature Genet.* **20,** 266–272.

Warrick, J. M., Chan, E., Gray-Board, G., Paulson, H. L., and Bonini, N. M. (1999) Suppression of polyglutamine-mediated neurodegeneration in *Drosophila* by the molecular chaperone Hsp70. *Nature Genet.* **23,** 425–428.

Warrick, J. M., Paulson, H. L., Gray-Board, G. L., Bui, Q. T., Fischbeck, KH, Pittman, R. N., et al. (1998) Expanded polyglutamine protein forms nuclear inclusions and causes neural degeneration in *Drosophila*. *Cell* **93,** 939–949.

Wellington, C. L., Ellerby, L. M., Hackam, A. S., Margolis, R. L., Trifiro, M. A., Singaraja, R., et al. (1998) Caspase cleavage of gene products associated with triplet expansion disorders generates truncated fragments containing the polyglutamine tract. *J. Biol. Chem.* **273,** 9158–9167.

Woods, B. T. and Schaumburg, H. H. (1972) Nigro-spino-dentatal degeneration with nuclear ophthalmoplegia. A unique and partially treatable clinico-pathological entity. *J. Neurol. Sci.* **17,** 149–166.

Yuasa, T., Ohama, E., Harayama, H., Yamada, M., Kawase, Y., Wakabayashi, M., et al. (1986) Joseph's disease: clinical and pathological studies in a Japanese family. *Ann. Neurol.* **19,** 152–157.

13

Pathophysiology of Spinal and Bulbar Muscular Atrophy

Diane E. Merry

13.1. INTRODUCTION: SBMA — THE DISEASE

Spinal and bulbar muscular atrophy (SBMA) is an X-linked, adult-onset motor neuronopathy that is caused by the expansion of a CAG trinucleotide repeat within the coding region of the androgen receptor gene. The disease affects males almost exclusively, although mildly affected females have been reported *(1)*. There have been several clinical reviews of this disease, beginning with the first description by Kennedy et al. *(2–5)*; a brief overview of the clinical findings will be presented here.

The symptoms of SBMA first present in men during the fourth to fifth decade of life with proximal muscle weakness and atrophy of the lower limbs and shoulder girdle; muscle cramping can precede the onset of muscle weakness, however, by as much as 20 yr. Tremor and fasciculation occur in the majority of patients; fasciculations of the chin and tongue are particularly characteristic of SBMA. Deep tendon reflexes are decreased or absent. Subclinical sensory impairment is seen with intensive testing.

Spinal and bulbar muscular atrophy patients also show signs of androgen insensitivity, including gynecomastia, testicular atrophy, and reduced fertility, despite normal testosterone levels. Sperm counts may be reduced, and impotence can occur during progression of the disease.

The X-linked pattern of inheritance for SBMA was first described by Kennedy et al. *(2)*. The disease gene was mapped to the proximal long arm of the X chromosome *(6)*; linkage to the androgen receptor gene combined with the finding of androgen insensitivity in SBMA patients suggested the androgen receptor gene as a candidate gene for this disorder *(7)*. In 1991, a CAG repeat expansion was identified as a common mutation in all SBMA patients *(8)*.

From: *Contemporary Clinical Neuroscience: Molecular Mechanisms of Neurodegenerative Diseases*
Edited by: M.-F. Chesselet © Humana Press Inc., Totowa, NJ

Pathology

Neuropathological studies of SBMA show a severe loss of motor neurons from the anterior horn of the spinal cord and from brainstem motor nuclei *(9)*. Remaining motor neurons may appear atrophied. Primary sensory neurons are also depleted, and there is a reduction of sensory nerve fibers. The neurons of Clarke's column are well preserved. The one report in which Onuf's nucleus was analyzed shows no motor neuron loss from this motor nucleus despite the expression of high androgen receptor (AR) protein levels (*see* Subheading 13.2.). Progressive muscle pathology is also seen. There is atrophy of muscle fibers, with progressive small to large group atrophy, with involvement of all fiber types. Clusters of nuclei can be seen as entire groups of muscle fibers undergo atrophy.

The most striking pathological finding in SBMA patients is the observation of ubiquitinated, neuronal nuclear inclusions (NII) in motor neurons of the anterior horn of the spinal cord *(10)*. Such inclusions have been observed in postmortem tissue of nearly every polyglutamine repeat disease to date, as well as in several model systems (*see* Subheading 13.5.). As with several other polyglutamine repeat diseases, the NII seen in SBMA tissue are detected exclusively with antibodies that recognize the amino terminus of the AR; epitopes within the middle- and carboxy-terminal regions 2/3 of the protein are either masked or absent *(10)*. In addition, whereas the appearance of NII within the nervous system appears quite selective, occurring only in anterior horn cells, inclusions are seen as well in several unaffected tissues *(11)*, including the dermis, scrotal skin, testis, heart, and kidney. Inclusions in non-neural tissues occur at a lower frequency than in neural tissue, however ($< 1\%$ vs $8.39 \pm 5.43\%$, respectively). Nonetheless, the presence of inclusions in non-neural tissues indicates that inclusion formation does not represent an aspect of the neuronal specificity of this polyglutamine repeat disease.

The acquired toxicity resulting from polyglutamine expansion and leading to abnormal protein folding and aggregation makes the lack of symptoms in carrier females curious. No assessment of AR inclusion formation in carrier females has been reported. It may be that the development of neurological symptoms and AR-induced pathology requires the presence of sufficient levels of circulating androgens. Alternatively, females may be partially protected or show a substantial delay in the onset of symptoms, resulting from random X-inactivation of the AR. It is known that the AR undergoes typical X-inactivation, despite its location in a genomic region that largely escapes inactivation *(12)*. Understanding the ligand dependence of phenotypes in model systems of SBMA should provide a molecular explanation for this finding.

13.2. THE ANDROGEN RECEPTOR GENE AND CAG EXPANSION

The androgen receptor (AR) is encoded by a 2.7-kb open reading frame, located within eight exons spanning over 100 kb at Xq12 *(12,13)*. The AR CAG repeat is polymorphic in normal individuals, ranging from 10 to 36 repeats *(14,15)*. SBMA patients show CAG tract lengths of 40–62 repeats. In other species, the androgen receptor CAG repeat is considerably shorter, although polymorphic. The mouse and rat genes contain a single glutamine at the same location, whereas they contain a polymorphic CAG repeat approximately 150 bp 3' to the location of the human repeat *(16,17)*. Primates show intraspecies and interspecies repeat-length variation *(18)*, and CAG repeat lengths increase with evolutionary distance to the lengths found in humans *(19,20)*.

As with other trinucleotide repeat diseases, the expanded CAG repeat in the androgen receptor gene is considerably more unstable than repeats in the normal size range. However, SBMA alleles do not show the extent of meiotic instability seen in other polyglutamine diseases *(21–23)*. In addition, no somatic instability of the AR CAG repeat has been observed *(23,24)*. Studies of the AR CAG repeat in single sperm from normal *(25)* and SBMA *(26)* individuals support the population studies of AR CAG repeat instability. In addition, sperm with high-normal CAG repeat lengths showed a higher mutation rate than sperm with 20–22 repeats. Contraction mutations predominated in this group of sperm, however, providing a possible mechanism for the relative paucity of new mutations in SBMA. In keeping with the lack of new mutations in SBMA, the finding of linkage disequilibrium in Japanese SBMA chromosomes suggests that a founder effect is responsible for the majority of SBMA alleles in Japan *(27)*.

A corollary to the finding of CAG repeat expansion in SBMA is the finding of an inverse correlation between the AR CAG repeat length and the incidence of prostate cancer *(28)*, as well as with the incidence of surgery for benign prostatic hyperplasia *(29)*. Both of these correlations have been found within the U.S. population, whereas a study of the French–German population revealed no such correlation *(30)*. In addition, the opposite correlation was found in a study of BRCA-1-positive women at risk for breast cancer *(31)*; *longer* AR CAG repeats were correlated with an earlier age of breast cancer onset.

13.3. THE ANDROGEN RECEPTOR PROTEIN

The androgen receptor is a 919-amino-acid phosphoprotein that is a member of the steroid/thyroid hormone receptor superfamily of transcription factors. Considerable work has been done to understand the normal domain structure and function of the androgen receptor, its regulation, and its downstream targets; much of this knowledge has come from the analysis of AR mutations that result in androgen insensitivity. Several reviews discuss structure–function relationships of the AR *(32–38)*; only the salient features are discussed here.

The androgen receptor is a phosphoprotein, whose phosphorylation status is linked to its transcriptional activation/repression function *(39)*. The C-terminal ligand-binding domain binds two natural ligands, testosterone and dihydrotestosterone, with high affinity. It binds to other steroid hormones with lower affinity. The DNA-binding domain shows high conservation among steroid hormone receptors, and there is significant overlap in the ability to bind to hormone response elements (HREs). The DNA-binding domain of the AR contains nine invariant cysteine residues; eight of these are thought to complex zinc in two zinc-finger structures formed by disulfide bonds between these residues. Specific androgen response elements (AREs) have also been identified in several androgen-responsive genes. A nuclear localization signal has been identified that borders the DNA-binding domain and extends into the hinge region. The transactivation domains of the AR appear to be more complex than the ligand-binding and DNA-binding domains, and structural motifs in this region may be of primary importance *(40)*. There is evidence for two transactivation domains (TADs) at the amino terminus of the AR *(41)*, and an additional AF-2 domain at the C-terminus of the AR. The identification of coactivators and corepressors as mediators of steroid hormone receptor function (reviewed in refs. *42* and *43*) has led to a more complete understanding of the mechanisms of steroid hormone receptor transcriptional activation. In addition, these studies have revealed a substantial number of protein interactions that may be affected by the altered protein structure of the AR.

Androgen Receptor–Protein Interactions

Understanding all aspects of the complex biological role that the AR plays in the cell, and in particular, in a neuron, will provide the identification of potential protein partners that may play a role in the pathogenesis of SBMA. In addition, the identification of protein partners should lead to an increased understanding of AR function and metabolism.

The AR interacts with many heterologous proteins, in the course of fulfilling its normal biological role. Upon translation, the AR becomes complexed with several heat shock proteins (reviewed in refs. *37* and *44*), including Hsp90, Hsp70, Hsp56, and a dnaJ protein *(45)*, and p23. This aporeceptor complex maintains the AR in a configuration "poised" to bind ligand. Upon ligand binding, the AR dissociates from the complex and translocates to the nucleus, where it binds as a dimer to specific DNA elements. Dimerization occurs through interactions between amino- and carboxy-terminal domains of the AR *(46–49)*.

In addition to these known protein interactions, additional interacting proteins have been found through two-hybrid screens, as well as the analysis of candidate proteins thought to be involved in AR function. A summary of known AR interactions can be found at The Androgen Receptor Gene Mutations Database World Wide Web Server at http://www.mcgill.ca/androgendb/.

These proteins include the following (noninclusive): D1 cyclin *(50)*, the coactivators Tip60 *(51)*, CREB-binding protein *(52,53)*, ARA24 *(54)*, ARA160 *(55)*, ARA54 *(56)*, ARA55 *(57)*, ARA70 *(58)*, Rb *(59)*, TIF2 *(60)*, Ubc9 *(61)*, and other transcription factors, including TFIIF *(62)* and c-jun *(63)*. Other interacting proteins include glyceraldehyde-3-phosphate dehydrogenase (GAPDH) *(64)*, which was also shown to interact with other polyglutamine-containing proteins *(64,65)*. Despite this interaction, no significant change in GAPDH activity was found in the brains of patients with Huntington's disease or spinocerebellar ataxias types 1, 2, or 3 *(66)*. In addition, AR interacts with another nuclear receptor, glucocorticoid receptor *(67)* and receptor accessory factor (RAF) *(68)*. A particular notable interaction is that of the coactivator ARA24 *(54)*; this interaction decreases with increasing polyglutamine repeat length, becoming, in the process, a weaker coactivator. Whether this or other of these interacting proteins play a role in the pathogenesis of SBMA remains to be determined.

13.4. A ROLE FOR THE ANDROGEN RECEPTOR IN THE NERVOUS SYSTEM

To understand the role that the normal AR function in the nervous system plays in the pathogenesis of SBMA, one must consider the fact that partial or complete loss of AR function (leading to partial or complete androgen insensitivity) does not lead to neurological disease. This indicates that the polyglutamine expansion mutation of SBMA does not cause disease via the loss of intrinsic AR function. However, this does not rule out the possibility that polyglutamine expansion reduces normal AR function. The finding of partial androgen insensitivity in SBMA patients has led to the conclusion that this mutation does, in fact, lead to a partial loss of AR function. This view likely needs re-evaluation, however, in light of the recent finding of AR inclusions in non-neural tissues, including testis and scrotal skin *(11)*. These findings suggest an alternate mechanism for the partial androgen insensitivity seen in SBMA, whereby polyglutamine expansion induces cellular dysfunction in tissues regulating testosterone production.

The AR is expressed in many regions of the nervous system, including both sexually monomorphic and dimorphic regions. These include the medial preoptic area, the lateral ventral septum, the ventromedial hypothalamus, the bed nucleus of the stria terminalis, and the medial amygdala *(69–72)*. Several other regions of the brain have been shown to accumulate androgens *(73,74)*, including midbrain, pons, medulla, cerebellum, and spinal cord. In addition, the spinal cord expresses high levels of AR protein, with the sexually dimorphic spinal nucleus of the bulbocavernosus expressing some of the highest AR protein levels in the brain *(75,76)*; these levels decrease with age *(77)*.

Androgens play a role in the structure and function of the nervous system. Relevant to SBMA is the finding of a trophic role for androgens in spinal motor neurons *(78–82)*. One particularly sensitive motor nucleus is the spinal nucleus of the bulbocavernosus (SNB), the motor neurons of which innervate the perineal levator ani and bulbocaverosus muscles. This SNB nucleus is highly androgen dependent and accumulates substantial levels of androgens *(83)*. Male rodents completely lacking the androgen receptor (Tfm mutation) show reduced motor neuron number in the SNB and atrophy of the SNB-innervated perineal muscles *(84,85)*. In these mice, androgen administration during the critical period of embryogenesis when SNB neurons normally innervate their target muscles results in motor neuron rescue *(84–91)*. This is likely not a direct androgen effect, as it is target muscle dependent *(92,93)*. Nonetheless, androgens can also regulate the structural plasticity of the SNB motor nucleus, directly increasing dendritic length of these motor neurons in adult animals *(94)*.

Recent studies of AR function in a neuronal cell culture system confirm the roles for the AR in mediating androgen as both a trophic factor and regulator of structural plasticity *(95)*. In this system, the overexpression of normal AR in the neuronal cell line MN-1 leads to increased androgen-dependent resistance to serum withdrawal, as well as increased soma and neurite size of the differentiated neurons. These collective data confirm a trophic role for the AR in some neuronal populations. They also suggest that loss of AR function in lower motor neurons may contribute to cell death in SBMA by eliminating one of the endogenous pathways to cell survival and thereby effectively lowering the threshold for polyglutamine-induced cell death.

13.5. MODEL SYSTEMS AND MOLECULAR PATHOGENESIS

Numerous laboratories have progressed in the creation of model systems that reproduce specific aspects of the pathology of SBMA. Using full-length expanded repeat AR protein, Butler et al. *(96)* demonstrated proteolytic cleavage of a 52-CAG repeat form of the AR, as well as ligand-dependent aggregation with cytoplasmic inclusions. Similarly, Stenoien et al. *(97)* found ligand-dependent aggregation of full-length expanded AR (48 CAG repeats). These studies also demonstrated sequestration of several cellular proteins that may contribute to the pathogenesis; these include SRC-1, components of the ubiquitin-dependent proteasome pathway, and the ubiquitinlike molecule NEDD-8. Electron microscopy (EM) studies revealed a fibrillar–granular appearance of aggregates. Merry et al. *(98)* used a truncated, highly expanded repeat AR and demonstrated repeat-length-dependent AR aggregation and proteolytic processing that was coupled to cellular toxicity.

Understanding the role of androgens in the pathogenesis of SBMA is important for designing effective therapies. Some conflicting results currently cloud the understanding of this issue, however. Whereas androgens had no effect on toxicity due to expanded AR overexpression in one study *(99)*, they had a trophic effect in two other studies *(100)*. Effects on aggregation were similarly disparate; although AR ligands had a striking positive effect on AR aggregation in two studies *(96,97,119)*, they had an inhibitory effect on aggregation in another *(100)*. It will be important to dissect the potential mechanistic role of androgens in polyglutamine toxicity from their well-known trophic role in motor neurons.

Intrinsic Functions of Expanded AR

The effects of polyglutamine repeat-length expansion on the intrinsic characteristics of the AR have been examined. Whereas the *Kd* for binding several physiologic ligands remains unchanged in the presence of repeat expansion, the ligand-binding capacity of the expanded repeat AR is reduced, both in patient material *(101,102)* and in several model systems *(103–105)*. Decreased ligand binding may be the result of the lower level of steady-state expanded repeat AR protein (and mRNA) that is often seen *(103,104)*.

The transcriptional competence of the expanded repeat AR has been investigated by several groups *(96,105–109)*. A decrease in the ability of the expanded repeat AR to transactivate an androgen-responsive reporter gene was found by several groups *(96,105–108)*, whereas another group *(103)* found no difference in this intrinsic property. It is possible that experimental differences, including cell lines used, expression levels achieved, and aggregation status, are responsible for this discrepancy. It is possible that the soluble monomeric AR protein with expanded repeats has a normal intrinsic transactivational capacity; this appears decreased when the protein is sequestered by aggregation. This bears clarification because the transcriptional competence (or incompetence) of AR may contribute to the pathogenesis of not only this disease but also of prostate cancer as well *(28,29)*.

Abnormal Protein Metabolism?

There is considerable evidence to indicate that an AR with a polyglutamine expansion is abnormally processed and metabolized in multiple cell types.

Aggregation

Both full-length and truncated AR proteins with expanded polyglutamine repeats have been shown to form intracellular inclusions *(96–98)*. The full-length expanded repeat AR is diffusely distributed in the cytoplasm in the absence of hormone; after hormone addition, a proportion of expanded AR-expressing cells develop cytoplasmic aggregates *(96,97)*. Using AR–GFP fusion proteins, Stenoien et al. *(97)* showed that aggregates can form within the nucleus and, during cell division, become redistributed to the cytoplasm. In addition, a small number of cells develop cytoplasmic aggregates shortly after hormone addition, indicating that the cytoplasmic milieu is competent for aggregate formation. Indeed, Becker et al. *(120)* have shown that the *cytoplasmic* milieu is important for aggregation at least in their cell culture system.

Whereas full-length AR forms aggregates in a hormone-dependent manner, a C-terminal truncated form of the expanded AR can form both nuclear and cytoplasmic aggregates. Cytoplasmic aggregates predominate in nonneuronal cells, while in neuronal cells, aggregates are predominantly nuclear; this may result from neuronal specificity or lower levels of AR protein in these cells.

Androgen receptor aggregates contain not only the androgen receptor but also several other proteins that suggest that this protein is misfolded and targeted for degradation. Aggregates observed in cultured cells *(97)* contain the ubiquitinlike protein Nedd-8, as well as several molecular chaperones (Hsp70, Hsp90, and HDJ-2/HSDJ, but not Hsp25, Hsp27 or Hsp110), and the PA700 cap structure of the 26S proteasome (but not the 20S core proteasome). Other proteins sometimes associated with the AR are also found within aggregates, namely SRC-1. Neuropathological assessment of patient material *(10)* shows that aggregates contain only the amino terminus of the AR protein and lack C-terminal epitopes; these aggregates are also ubiquitinated *(10)*. The sequestration of various proteins within aggregates suggests that these proteins might play a role in the pathogenesis of SBMA. Indeed, the overexpression of HDJ-2/HSDJ in cells expressing an expanded repeat AR decreased the frequency of AR aggregation *(97,121)*, similar to the effect of this molecular chaperone on ataxin-1 aggregation *(110)*.

The finding of molecular chaperones and proteasome components within aggregates in both cultured cells and neurons of transgenic mice suggests a cellular response to a misfolded protein. Aggregation of expanded repeat AR indicates that this mutant protein cannot be efficiently returned to its native conformation by molecular chaperones and is, instead, targeted for degradation by the 26S proteasome. The presence of aggregates reveals that this pathway of degradation is ineffective at ridding the cell of misfolded AR protein; instead, proteasome components become sequestered into AR-containing nuclear aggregates. Whether this sequestration of proteasome components contributes to the pathogenesis of SBMA by reducing the neuron's ability to degrade other proteasome-targeted proteins remains to be determined.

Proteolytic Processing

Several lines of evidence suggest that the AR is processed in a repeat-length-dependent manner. Circumstantial evidence comes from the finding by Li et al. that only amino-terminal epitopes are present in the inclusions observed in spinal cord motor neurons from SBMA patients *(10)*. This suggests that epitopes from the rest of the protein are either masked or absent. Evidence from experimental model systems supports the idea that expanded repeat AR may be aberrantly cleaved. Merry et al. *(98)* found that expanded repeat truncated forms of the AR were cleaved in both COS-7 and MN-1 cells, producing an N-terminal fragment that appears cleaved within the polyglutamine tract. Abdullah et al. *(99)* showed abnormal proteolytic processing of an expanded repeat *full-length* AR, revealed by analysis of tryptic digests of the AR. In addition, a 75-kDa fragment of the AR was identified in cells transfected with expanded repeat AR *(96,99)*. These data suggest that the polyglutamine expansion causes the AR to acquire a different conformation, altering the sites that are susceptible to enzymatic digestion. The role that proteolytic processing plays in the pathogenesis of SBMA is unclear. However, the use of these systems to identify the protease(s) involved should allow the elucidation of this role.

The AR has also been shown to be a substrate for members of the caspase family *(100,111,112)*, a group of enzymes involved in cleaving substrates during apoptotic cell death. Although the AR is cleaved inefficiently by caspase-3 *(111)*, the cleavage of a caspase-3 site 70 amino acids downstream of the polyglutamine tract is essential for AR toxicity in 293T cells, because mutation of this site (D146N) prevented AR_{exp}-induced toxicity *(100)*. In this system, aggregation of the expanded repeat AR is seen upon induction of apoptosis and in the *absence* of testosterone, in contrast to other studies *(96,97)*; addition of testosterone *prevents* caspase cleavage, aggregation and cell death.

In addition to cleavage by caspase-3, the AR is cleaved by caspases 1, 7, and 8 *(111)*. Caspase-1 and caspase-8 cleavage of AR is particularly notable, because these proteins have been implicated in the pathogenesis of polyglutamine repeat disease *(113,114)*. Inhibition of caspase-1 delayed onset of symptoms in a mouse model of Huntington's disease, as well as the appearance of intranuclear inclusions and neuronal changes in A2a, D1, and D2 receptor binding. In addition, caspase-8 was recruited and activated by an isolated, epitope-tagged Q79; inhibition of this activation blocked polyglutamine-induced cellular toxicity. Caspase-8 was also found associated with huntingtin protein in the brains from Huntington's patients *(114)*, suggesting a role for this caspase in vivo. Whether these caspase pathways contribute to the pathogenesis of SBMA can be determined with the multiple model systems now available.

Animal Models

Many attempts at creating a mouse model for SBMA have been made and, until recently, have been unsuccessful. Bingham et al. *(115)* were the first to create transgenic mice using full-length AR with either normal *(24)* or expanded *(45)* repeats. These studies used both constitutive and neuron-specific (neuron-specific enolase) and inducible and ubiquitously expressed (Mx) to drive AR expression. Mice developed in these experiments displayed neither neurologic phenotypes nor pathologic features suggestive of neuronal degeneration. No repeat instability was seen in 76 meioses.

In a subsequent experiment, Merry et al. *(116)* created transgenic mice using full-length AR with either normal *(24)* or expanded *(65)* repeats. In these experiments, expression was driven by either of two neuron-specific and constitutive promoters, the neuron-specific enolase promoter, and the neurofilament light-chain promoter. Mice showed expression from 2–5x endogenous AR expression, but also displayed no neurological phenotype or pathological features. No meiotic repeat instability was seen in 154 transmissions analyzed (unpublished observation).

In yet another experiment, La Spada et al. *(117)* created transgenic mice using large yeast artificial chromosomes (YACs) integrated into the mouse genome. The normal *(24)* or expanded *(45)* AR in these experiments was therefore driven by its cognate promoter, but showed low levels of protein expression and no neurological phenotype. In contrast to the transgene experiments using AR cDNAs, however, mice in this study displayed significant repeat instability (10%) upon meiotic transmission; unlike SBMA families, a higher rate of instability was found during female transmission. These data indicate that cis-acting sequences present in the YAC are important for the repeat instability.

Recent experiments have succeeded in creating transgenic AR-expressing mice that display neurological phenotypes *(118)*. Using a truncated AR cDNA containing either normal *(16)* or highly expanded *(112)* repeats, and expressed from either the prion protein promoter (PrP) or the neurofilament light-chain (NF-L) promoter, transgenic mice were generated that display a range of neurological phenotypes. PrP transgenic mice develop tremor, gait abnormalities, hindlimb foot clasping, weight loss, reduced fertility, handling-induced seizures, increased grooming behaviors, and early death. In contrast, NF-L promoter transgenic mice develop a phenotype that is restricted to gait abnormality (with evidence of weakness and spasticity), tremor, reduced fertility, weight loss, and early death. The most notable pathological feature in all lines of transgenic mice is the presence of large neuronal intranuclear inclusions. The development of a mouse model for AR polyglutamine expansion disease should now allow the testing of various therapies. However, there remains a need to create a mouse model using *full-length* AR in order to resolve questions of ligand dependence.

Studies of the molecular pathogenesis of SBMA, while paralleling the field of polyglutamine disease in many respects, will continue to focus on the specifics of this motor neuron disease, including ligand dependence, proteolytic processing, and determinants of motor neuron specificity. With the development of animal models for SBMA, studies can begin on the specific dysfunctional processes of lower motor neurons, in order to understand the cellular defects in this motor neuron disease. Such studies, along with the ongoing analysis of AR metabolism, will lead to the development of rational therapeutic strategies for this debilitating disease.

REFERENCES

1. Ferlini, A., Patrosso, M. C., Guidetti, D., Merlini, L., Uncini, A., Ragno, M., et al. (1995) Androgen receptor gene (CAG)n repeat analysis in the differential diagnosis between Kennedy Disease and other motoneuron disorders. *Am. J. Med. Genet.* **55,** 105–111.
2. Kennedy, W. R., Alter, M., and Sung, J. H. (1968) Progressive proximal spinal and bulbar muscular atrophy of late onset: a sex–linked recessive trait. *Neurology* **18,** 671–680.
3. Harding, A. E., Thomas, P. K., Baraitser, M., Bradbury, P. G., Morgan-Hughes, J. A., and Ponsford, J. R. (1982) X–linked recessive bulbospinal neuronopathy: report of ten cases. *J. Neurol. Neurosurg. Psychiatry* **45,** 1012–1019.

4. Arbizu, T., Santamaria, J., Gomez, J. M., Quilez, A., and Serra, J. P. (1983) A family with adult spinal and bulbar muscular atrophy, X–linked inheritance and associated testicular failure. *J. Neurol. Sci.* **59,** 371–382.
5. Zajac, J. D. and MacLean, H. E. (1998) Kennedy's disease: clinical aspects, in *Genetic Instabilities and Hereditary Neurological Diseases* (Wells, R. D. and Warren, S. T., eds), Academic , New York.
6. Fischbeck, K. H., Ionasescu, V., Ritter, A.W., Ionasescu, R., Davies, K., Ball, S., et al. (1986) Localization of the gene for X–linked spinal muscular atrophy. *Neurology* **36,** 1595–1598.
7. Fischbeck, K. H., Souders, D., and La Spada, A. R. (1991) A candidate gene for X–linked spinal muscular atrophy. *Adv. Neurol.* **56**, 209–213.
8. La Spada, A. R., Wilson, E. M., Lubahn, D. B., Harding, A. E., and Fischbeck, K. H. (1991) Androgen receptor gene mutations in X–linked spinal and bulbar muscular atrophy. *Nature* **353,** 77–79.
9. Sobue, G., Hashizume, Y., Mukai, E., Hirayama, M., Mitsuma, T., and Takahashi, A. (1989) X–linked recessive bulbospinal neronopathy: a clinicopathological study. *Brain* **112,** 209–232.
10. Li, M., Miwa, S., Kobayashi, Y., Merry, D. E., Yamamoto, M., Tanaka, F., et al. (1998) Nuclear inclusions of the androgen receptor protein in spinal and bulbar muscular atrophy. *Ann. Neurol.* **44,** 249–254.
11. Li, M., Nakagomi, Y., Kobayashi, Y., Merry, D. E., Tanaka, F., Doyu, M., et al. (1998) Nonneural nuclear inclusions of androgen receptor protein in spinal and bulbar muscular atrophy. *Am. J. Pathol.* **153,** 695–701.
12. Lubahn, D. B., Joseph, D. R., Sullivan, P. M., Willard, H. F., French, F. S., and Wilson, E. M. (1988) Cloning of human androgen receptor complementary DNA and localization to the X chromosome. *Science* **240,** 327–330.
13. Lubahn, D. B., Joseph, D. R., Sar, M., Tan, J., Higgs, H. N., Larson, R. E., et al. (1988) The human androgen receptor: complementary deoxyribonucleic acid cloning, sequence analysis and gene expression in prostate. *Mol. Endocrinol.* **2,** 1265–1275.
14. Edwards, A., Hammond, H. A., Jin, L., Caskey, C. T., and Chakraborty, R. (1992) Genetic variation at five trimeric and tetrameric tandem repeat loci in four human population groups. *Genomics* **12,** 241–253.
15. Macke, J. P., Hu, N., Hu, S., Bailey, M., King, V. L., Brown, T., et al. (1993) Sequence variation in the androgen receptor gene is not a common determinant of male sexual orientation. *Am. J. Hum. Genet.* **53,** 844–852.
16. Faber, P. W., King, A., van Rooij, H. C. J., Brinkmann, A. O., de Both, N. J., and Trapman, J. (1991) The mouse androgen receptor: functional analysis of the protein and characterization of the gene. *Biochem. J.* **278,** 269–278.
17. Tan, J.–A., Jouseph, D. R., Quarmby, V. E., Lubahn, D. B., Sar, M., French, F. S., et al. (1988) The rat androgen receptor: primary structure, autoregulation of its messenger ribonucleic acid, and immunocytochemical localization of the receptor protein. *Mol. Endocrinol.* **2,** 1276–1285.
18. Rubinsztein, D. C., Amos, W., Leggo, J., Goodburn, W. S., Jain, S., Li, S.-H., et al. (1995) Microsatellite evolution – evidence for directionality and variation in rate between species. *Nature Genet.* **10,** 337–343.

19. Rubensztein, D. C., Leggo, J., Coetzee, G. A., Irvine, R. A., Buckley, M., and Ferguson-Smith, M. A. (1995) Sequence variation and size ranges of CAG repeats in the Machado–Joseph disease, spinocerebellar ataxia type 1 and androgen receptor genes. *Hum. Mol. Genet.* **4,** 1585–1590.
20. Choong, C. S., Kemppainen, J. A., and Wilson, E. M. (1998) Evolution of the primate androgen receptor: a structural basis for disease. *J. Mol. Evol.* **47,** 334–342.
21. La Spada, A. R., Roling, D., Harding, A. E., Warner, C. L., and Speigel, R., Hausmanowa-Petrusewicz, I., et al. (1992) Meiotic stability and genotype-phenotype correlation of the expanded trinucleotide repeat in X-linked spinal and bulbar muscular atrophy. *Nature Genet.* **2,** 301–304.
22. Biancalana, V., Serville, F., Pommier, J., Julien, J., Hanauer, A., and Mandel, J. L. (1992) Moderate instability of the trinucleotide repeat in spinobulbar muscular atrophy. *Hum. Mol. Genet.* **1,** 255–258.
23. Watanabe, M., Abe, K., Aoki, M., Yasuo, K., Itoyama, Y., Shoji, M., et al. (1996) Mitotic and meiotic stability of the CAG repeat in the X–linked spinal and bulbar muscular atrophy gene. *Clin. Genet.* **50,** 133–137.
24. Spiegel, R., La Spada, A. R., Kress, W., Fischbeck, K. H., and Schmid, W. (1996) Somatic stability of the expanded CAG trinucleotide repeat in X-linked spinal and bulbar muscular atrophy. *Hum. Mutat.* **8,** 32–37.
25. Zhang, L., Leeflang, E. P., Yu, J., and Arnheim, N. (1994) Studying human mutations by sperm typing: instability of CAG trinucleotide repeats in the human androgen receptor gene. *Nature Genet.* **7,** 531–535.
26. Zhang, L., Fischbeck, K. H., and Arnheim, N. (1995) CAG repeat length variation in sperm from a patient with Kennedy's disease. *Hum. Mol. Genet.* **4,** 303–305.
27. Tanaka, F., Doyu, M., Ito, Y., Matsumoto, M., Mitsuma, T., Abe, K., et al. (1996) Founder effect in spinal and bulbar muscular atrophy (SBMA). *Hum. Mol. Genet.* **5,** 1253–1257.
28. Giovannucci, E., Stampfer, M. J., Krithivas, K., Brown, M., Dahl, D., Brufsky, A., et al. (1997) The CAG repeat within the androgen receptor gene and its relationsbip to protstate cancer. *Proc. Natl. Acad. Sci. USA* **94,** 3320–3323.
29. Giovannucci, E., Stampfer, M. J., Chan, A., Krithivas, K., Gann, P. H., Hennekens, C. H., et al. W. (1999) CAG repeat within the androgen receptor gene and incidence of surgery for benign prostatic hyperplasia in U.S. physiahs. *Prostate* **39,** 130–134.
30. Correa-Cerro, L., Wohr, G., Haussler, J., Berthon, P., Drelon, E., Mangin, P., et al. (1999) (CAG)nCAA and GGN repeats in the human androgen receptor gene are not associated with prostate cancer in a French–German population. *Eur. J. Hum. Genet.* **7,** 357–362.
31. Rebbeck, T. R., Kantoff, P. W., Krithivas, K., Neuhausen, S., Blackwood, M. A., Godwin, A. K., et al. (1999) Modification of BRCA1-associated breast cancer risk by the polymorphic androgen-receptor CAG repeat. *Am. J. Hum. Genet.* **64,** 1371–1377.
32. Brinkmann, A. O., Blok, L. J., de Ruiter, P. E., Doesburg, P., Steketee, K., Berrevoets, C. A., et al. (1999) Mechanisms of androgen receptor activation and function. *J. Steroid Biochem. Mol. Biol.* **69,** 307–313.

33. Tsai, M.–J., and O'Malley, B. W. (1994) Molecular mechanisms of action of steroid/thryoid receptor superfamily members. *Annu. Rev. Biochem.* **63,** 451–486.
34. Zhou, Z.-X., Wong, C-I., Sar, M., and Wilson, E. M. (1994) The androgen receptor: an overview. *Recent Prog. Horm. Res.* **49,** 249–274.
35. Jenster, G. (1994) Functional domains of the human androgen receptor, In *Endocrinology and Reproduction*, Erasmus University, Rotterdam.
36. Jenster, G., van der Korput, H. A. G. M., van Vroonhoven, C., van der Kwast, T. H., Trapman, J., and Brinkmann, A. O. (1991) Domains of the human androgen receptor involved in steroid binding, transcriptional activation, and subcellular localization. *Mol. Endocrinol.* **5,** 1396–1404.
37. MacLean, H. E., Warne, G. L., and Zajac, J. D. (1997) Localization of functional domains in the androgen receptor. *J. Steroid Biochem. Mol. Biol.* **62,** 233–242.
38. McPhaul, M. J. (1999) Molecular defects of the androgen receptor. *J. Steroid Biochem. Mol. Biol.* **69,** 315–322.
39. Blok, L. J., de Ruiter, P. E., and Brinkmann, A. O. (1998) Forskolin-induced dephosphorylation of the androgen receptor impairs ligand binding. *Biochemistry* **37,** 3850–3857.
40. Gast, A., Schneikert, J., and Cato, A. C. (1998) N–terminal sequences of the androgen receptor in DNA binding and transrepressing functions. *J. Steroid Biochem. Mol. Biol.* **65,** 117–123.
41. Berrevoets, C. A., Doesburg, P., Steketee, K., Trapman, J., and Brinkmann, A. O. (1998) Functional interactions of the AF–2 activation domain core region of the human androgen receptor with the amino–terminal domain and with the transcriptional coactivator TIF2 (transcriptional intermediary factor 2). *Mol. Endocrinol.* **12,** 1172–1183.
42. Jenster, G. (1998) Coactivators and corepressors as mediators of nuclear receptor function: an update. *Mol. Cell. Endocrinol.* **143,** 1–7.
43. Shibata, H., Spencer, T. E., Onate, S. A., Jenster, G., Tsai, S. Y., Tsai, M. J., et al. (1997) Role of co-activators and co–repressors in the mechanism of steroid/thyroid receptor action. *Recent Prog. Horm. Res.* **52,** 141–164.
44. Bohen, S. P., Kralli, A., and Yamamoto, K. R. (1995) Hold 'em and fold 'em: chaperones and signal transduction. *Science* **268,** 1303–1304.
45. Caplan, A. J., Langley, E., Wilson, E. M., and Vidal, J. (1995) Hormone-dependent transactivation by the human androgen receptor is regulated by a dnaJ protein. *J. Biol. Chem.* **270,** 5251–5257.
46. Langley, E., Zhou, Z.-X., and Wilson, E. M. (1995) Evidence for an anti-parallel orientation of the ligand–activated human androgen receptor dimer. *J. Biol. Chem.* **270,** 29,983–29,990.
47. Langley, E., Kemppainen, J. A., and Wilson, E. M. (1998) Intermolecular NH2-/carboxyl-terminal interactions in androgen receptor dimerization revealed by mutations that cause androgen insensitivity. *J. Biol. Chem.* **273,** 92–101.
48. Doesburg, P., Kuil, C. W., Berrevoets, C. A., Steketee, K., Faber, P. W., Mulder, E., et al. (1997) Functional in vivo interaction between the amino-terminal, transactivation domain and the ligand binding domain of the androgen receptor. *Biochemistry* **36,** 1052–1064.

49. Ikonen, T., Palvimo, J. J., and Janne, O. A. (1997) Interaction between the amino- and carboxyl–terminal regions of the rat androgen receptor modulates transcriptional activity and is influenced by nuclear receptor coactivators. *J. Biol. Chem.* **272,** 29,821–29,828.
50. Knudsen, K. E., Cavenee, W. K., and Arden, K. C. (1999) D-–type cyclins complex with the androgen receptor and inhibit its transcriptional transactivation ability. *Cancer Res.* **59,** 2297–2301.
51. Brady, M. E., Ozanne, D. M., Gaughan, L., Waite, I., Cook, S., Neal, D. E., et al. (1999) Tip60 is a nuclear hormone receptor coactivator. *J. Biol. Chem.* **274,** 17,599–17,604.
52. Aarnisalo, P., Palvimo, J. J., Janne, O. A. (1998) CREB–binding protein in androgen receptor–mediated signaling. *Proc. Natl. Acad. Sci. USA* **95,** 2122–2127.
53. Fronsdal, K., Engedal, N., Slagsvold, T., and Saatcioglu, F. (1998) CREB binding protein is a coactivator for the androgen receptor and mediates cross-talk with AP–1. *J. Biol. Chem.* **273,** 31,853–31,859.
54. Hsiao, P. W., Lin, D. L., Nakao, R., and Chang, C. (1999) The linkage of Kennedy's neuron disease to ARA24, the first identified androgen receptor polyglutamine region–associated coactivator. *J. Biol. Chem.* **274,** 20,229–20,234.
55. Hsiao, P. W. and Chang, C. (1999) Isolation and characterization of ARA160 as the first androgen receptor N-termina-associated coactivator in human prostate cells. *J. Biol. Chem.* **274,** 22,373–22,379.
56. Kang, H. Y., Yeh, S., Fujimoto, N., Chang, C. (1999) Cloning and characterization of human protstate coactivator ARA54, a novel protein that associates with the androgen receptor. *J. Biol. Chem.* **274,** 8570–8576.
57. Fujimoto, N., Yeh, S., Kang, H. Y., Inui, S., Chang, H. C., Mizokami, A., et al. (1999) Cloning and characterization of androgen receptor coactivator, ARA55, in human prostate. *J. Biol. Chem.* **274,** 8316–8321.
58. Yeh, S. and Chang, C. (1996) Cloning and characterization of a specific coactivator, ARA70, for the androgen receptor in human prostate cells. *Proc. Natl. Acad. Sci. USA* **93,** 5517–5521.
59. Yeh, S., Miyamoto, H., Nishimura, K., Kang, H., Ludlow, J., Hsiao, P., et al. (1998) Retinoblastoma, a tumor suppressor, is a coactivator for the androgen receptor in human prostate cancer DU145 cells. *Biochem. Biophys. Res. Commun.* **248,** 361–367.
60. Voegel, J. J., Heine, M. J. S., Tini, M., Vivat, V., Chambon, P., and Gronemeyer, H. (1998) The coactivator TIF2 contains three nuclear receptor-binding motifs and mediates transactivation through CBP binding-dependent and -independent pathways. *EMBO J.* **17,** 507–519.
61. Poukka, H., Aarnisalo, P., Karvonen, U., Palvimo, J. J., and Janne, O. A. (1999) Ubc9 interacts with the androgen receptor and activates receptor-dependent transcription. *J. Biol. Chem.* **274,** 19,441–19,446.
62. McEwan, I. J. and Gustafsson, J. (1997) Interaction of the human androgen receptor tranactivation function with the general transcription factor TFIIF. *Proc. Natl. Acad. Sci. USA* **94,** 8485–8490.

63. Sato, N., Sadar, M. D., Bruchovsky, N., Saatcioglu, F., Rennie, P. S., Sato, S., et al. (1997) Androgenic induction of prostate-specific antigen gene is repressed by protein–protein interaction between the androgen receptor and AP–1/c–Jun in the human prostate cell line LNCaP. *J. Biol. Chem.* **272**, 17,485–17,494.
64. Koshy, B., Matilla, T., Burright, E. N., Merry D. E., Fischbeck, K. H., Orr, H. T., et al. (1996) Spinocerebellar ataxia type-1 and spinobulbar muscular atrophy gene products interact with glyceraldehyde-3-phosphate dehydrogenase. *Hum. Mol. Genet.* **5,** 1311–1318.
65. Burke, J. R., Enghild, J. J., Martin, M. E., Jou Y.-S., Myers, R. M., Roses, A. D., et al. (1996) Huntingtin and DRPLA proteins selectively interact with the enzyme GAPDH. *Nature Med.* **2,** 347–350.
66. Kish, S. J., Lopes-Cendes, I., Guttman, M., Furukawa, Y., Pandolfo, M., Rouleau, G., et al. (1998) Brain glyceraldehyde–3–phosphate dehydrogenase activity in human trinucleotide repeat disorders. *Arch. Neurol.* **55,** 1299–1304.
67. Chen, S., Wang, J., Yu, G., Liu, W., and Pearce, D. (1997) Androgen and glucocorticoid receptor heterodimer formation. *J. Biol. Chem.* **272,** 14,087–14,092.
68. Kupfer, S. R., Wilson, E. M., and French, F. S. (1994) Androgen and glucocorticoid receptors interact with insulin degrading enzyme. *J. Biol. Chem.* **269,** 20622–20628.
69. Bingaman, E., Baeckman, L. M., Yracheta, J. M., Handa, R. J., and Gray, T. S. (1994) Localization of androgen receptor within peptidergic nerons of the rat forebrain. *Brain Res. Bull.* **35,** 379–382.
70. Huang, X. and Harlan, R.E. (1994) Androgen receptor immunoreactivity in somatostatin neruons of the periventricular nucleus but not in the bed nucleus of the stria terminalis in the male rats. *Brain Res.* **652,** 291–296.
71. Zhou, L., Blaustein, J. D., and De Vries, G. J. (1994) Distribution of androgen receptor immunoreactivity in vasopressin– and oxytocin–immunoreactive neruons in the male rat brain. *Endocrinology* **134,** 2522–2627.
72. Apostolinas S. R. G., Dobrjansky, A., and Gibson, M. J. (1999) Androgen receptor immunoreactivity in specific neural regions in normal and hypogonadal male mice: effect of androgens. *Brain Res.* **817,** 19–24.
73. Arnold, A. P., Nottebohm, F., and Pfaff, D. W. (1976) Hormone concentrating cells in vocal control and other areas of the brain of the zebra finch (Poephila guttata). *J. Comp. Neurol.* **165,** 487–512.
74. Sar, M. and Stumpf, W. E. (1977) Androgen concentration in motor neruons of crainial nerves and spinal cord. *Science* **19,** 77–79.
75. Freeman, L. M., Padgett, B. A., Prins, G. S., and Breedlove, S. M. (1995) Distribution of androgen receptor immunoreactivity in the spinal cord of wild-type, androgen–insensitive and gonadectomized male rats. *J. Neurobiol.* **27,** 51–59.
76. Jordan C. L.,Hershey, J., Prins, G., and Arnold, A. (1997) Ontogeny of androgen receptor immunoreactivity in lumbar motoneurons and in the sexually dimorphic levator ani muscle of male rats. *J. Comp. Neurol.* **379,** 88–98.
77. Matsumoto, A. and Prins, G. S. (1998) Age-dependent changes in androgen receptor immunoreactivity in motoneurons of the spinal nucleus of the bulbocavernosus of male rats. *Neurosci. Lett.* **243,** 29–32.

78. Kujawa, K. A., Emeric, E., and Jones, K. J. (1991) Testosterone differentially regulates the regenerative properties of injured hamster facial motor neurons. *J. Neurosci.* **11,** 3898–3908.
79. Kujawa, K. A., Jacob, J. M., and Jones, K. J. (1993) Testosterone regulation of the regenerative proerties of injured rat sciatic motor neurons. *J. Neurosci. Res.* **35,** 268–273.
80. Yu, W. A. (1989) Administration of testoterone attenuates neronal loss following axotomy in the brainstem motor nuclei of female rats. *J. Neurosci.* **9,** 3908–3914.
81. Jones, K. J. and Oblinger, M. M. (1994) Androgenic regulation of tubulin gene expression in axotomized hamster facial motoneurons. *J. Neurosci.* **14,** 3620–3627.
82. Perez, J. and Kelley, D. B. (1996) Trophic effects of androgen: receptor expression and the survival of laryngeal motor neurons after axotomy. *J. Neurosci.* **16,** 6625–6633.
83. Breedlove, S. M. and Arnold, A. P. (1980) Hormone accumulation in a sexually dimorphic nucleus in the rat spinal cord. *Science* **210,** 564–566.
84. Breedlove, S. M. and Arnold, A. P. (1981) Sexually dimorphic motor nucleus in the rat spinal cord: response to adult hormone manipulation, absence in androgen-insensitive rats. *Brain Res.* **225,** 297–307.
85. Breedlove, S. M. and Arnold, A. P. (1983) Hormonal control of a developing neuromuscular system. II. Sensitive periods for the androgen-induced masculinization of the rat spinal nucleus of the bulbocavernosus. *J. Neurosci.* **3,** 424–432.
86. Breedlove, S. M. and Arnold, A. P. (1983) Sex differences in the pattern of steroid accumulation by motoneurons in the rat lumbar spinal cord. *J. Comp. Neurol.* **215,** 211–216.
87. Breedlove, S. M. and Arnold, A. P. (1983) Hormonal control of a developing neuromuscular system. I. Complete demasculinization of the male rat spinal nucleus of the bulbocavernosus using the anti-androgen flutamide. *J. Neurosci.* **3,** 417–423.
88. Jordan, C. L., Breedlove, S. M., and Arnold, A. P. (1982) Sexual dimorphism and the influence of neonatal androgen in the dorsolateral motor nucleus of the rat lumbar spinal cord. *Brain Res.* **249,** 309–314.
89. Nordeen, E. J., Nordeen, K. W., Sengelaub, D. R., and Arnold, A. P. (1985) Androgens prevent normally occurring cell death in a sexually dimorphic spinal nucleus. *Science* **229,** 671–673.
90. Sengelaub, D. R., Jordan, C. L., Kurz, E. M., and Arnold, A. P. (1989) Hormonal control of neuron number in sexually dimorphic spinal nuclei of the rat. III. Differential effects of the androgen dihydrotestosterone. *J. Comp. Neurol.* **280,** 637–644.
91. Sengelaub, D. R. and Arnold, A. P. (1989) Hormonal control of neuron number in sexually dimorphic spinal nuclei of the rat. I. Testosterone-regulated death in the dorsolateral nucleus. *J. Comp. Neurol.* **280,** 622–629.
92. Al-Shamma, H. A. and Arnold, A. P. (1995) Importance of target innervation in recovery from axotomy-induced loss of androgen receptor in rat perineal motoneurons. *J. Neurobiol.* **28,** 341–353.

93. Freeman, L. M. and Breedlove, S. M. (1996) Androgen spares androgen-insensitive motoneurons from apoptosis in the spinal nucleus of the bulbocavernosus in rats. *Horm. Behav.* **30,** 424–433.
94. Kurz, E. M., Sengelaub, D. R., and Arnold, A. P. (1986) Androgens regulate the dendritic length of mammalian motoneurons in adulthood. *Science* **232,** 395–398.
95. Brooks, B. P., Merry, D. E., Paulson, H. L., Lieberman, A., Kolson, D. L., and Fischbeck, K. H. (1998) A cell culture model for androgen-inducible responses in motor neurons. *J. Neurochem.* **70,** 1054–1060.
96. Butler, R., Leigh, P. N., McPhaul, M. J., and Gallo, J. M. (1998) Truncated forms of the androgen receptor are associated with polyQ expansion in X-linked spinal and bulbar muscular atrophy. *Hum. Mol. Genet.* **7,** 121–127.
97. Stenoien, D. L., Cummings, C. J., Adams, H. P., Mancini, M. G., Patel, K., DeMartino, G. N., et al. (1999) Polyglutamine-expanded androgen receptors form aggregates that sequester heat shock proteins, proteasome components and SRC-1, and are suppressed by the HDJ-2 chaperone. *Hum. Mol. Genet.* **8,** 731–741.
98. Merry, D. E., Kobayashi, Y., Bailey, C. K., Taye, A. A., and Fischbeck, K. H. (1998) Cleavage, aggregation, and toxicity of the expanded androgen receptor in spinal and bulbar muscular atrophy. *Hum. Mol. Genet.* **7,** 693–701.
99. Abdullah, A. A. R., Trifiro, M. A., Panet-Raymond, V., Alvarado, C., de Tourreil, S., Frankel, D., et al. (1998) Spinobulbar muscular atrophy: polyglutamine-expanded androgen receptor is proteolytically resistant in vitro and processed abnormally in transfected cells. *Hum. Mol. Genet.* **7,** 379–384.
100. Ellerby, L. M., Hackam, A. S., Propp, S. S., Ellerby, H. M., Rabizadeh, S., Cashman, N. R., et al. (1999) Kennedy's Disease: caspase cleavage of the androgen receptor is a crucial event in cytotoxicity. *J. Neurochem.* **72,** 185–195.
101. Danek, A., Witt, T. N., Mann, K., Schweikert, H. U., Romalo, G., LaSpada, A. R., et al. (1994) Decrease in androgen binding and effect of androgen treatment in a case of X-linked bulbosopinal neuronopathy. *Clin. Invest.* **72,** 892–897.
102. Warner, C. L., Griffen, J. E., Wilson, J. D., Jacobs, L. D., et al. (1992) X-linked spinomuscluar atrophy; a kindred with associated abnormal androgen binding. *Neurology* **42,** 2181–2184.
103. Brooks, B. P., Paulson, H. L., Merry, D. E., Salazar-Grueso, E. F., Brinkmann, A. O., Wilson, E. M., et al. (1997) Characterization of an expanded glutamine repeat androgen receptor in a neuronal cell culture system. *Neurobiol. Dis.* **4,** 313–323.
104. Choong, C. S., Kamppainen, J. A., Zhou, Z. X., and Wilson, E. M. (1996) Reduced androgen receptor gene expression with first exon CAG repeat expansion. *Mol. Endocrinol.* **10,** 1527–1535.
105. Mhatre, A. N., Trifiro, M. A., Kaufman, M., Kazemi-Esfarjani, P., Figlewicz, D., Rouleau, G., et al. (1993) Reduced transcriptional regulatory competence of the androgen receptor in X-linked spinal and bulbar muscular atrophy. *Nature Genet.* **5,** 184–188.
106. Chamberlain, N. L., Driver, E. D., and Miesfeld, R. L. (1994) The length and location of CAG trinucleotide repeats in the androgen receptor N-terminal domain affect transactivation function. *Nucleic Acids Res.* **22,** 3181–3186.

107. Nakajima, H., Kimura, F., Nakagawa, T., Furutama, D., Shinoda, K., Shimizu, A., et al. (1996) Transcriptional activation by the androgen receptor in X-linked spinal and bulbar muscular atrophy. *J. Neurol. Sci.* **142,** 12–16.
108. Kazemi-Esfarjani, P., Trifiro, M. A., and Pinsky, L. (1995) Evidence for a repressive function of the long polyglutamine tract in the human androgen receptor: possible pathogenetic relevance for the $(CAG)_n$-expanded neuronopathies. *Hum. Mol. Genet.* **4,** 523–527.
109. Brooks, B. P. (1997) Characterization of normal- and expanded-glutamine repeat androgen receptors in a neuronal cell line, thesis in Pharmacology, University of Pennsylvania Press, Philadelphia, p. 143.
110. Cummings, C. J., Mancini, M. A., Antalffy, B., DeFranco, D. B., Orr, H. T., and Zoghbi, H. Y. (1998) Chaperone suppression of aggregation and altered subcellular proteasome localization imply protein misfolding in SCA1. *Nature Genet.* **19,** 148–154.
111. Wellington, C. L., Ellerby, L. M., Hackam, A. S., Margolis, R. L., Trifiro, M. A., Sangaraja, R., et al. (1998) Caspase cleavage of gene products associated with triplet expansion disorders generates truncated fragments containing the polyglutamine tract. *J. Biol. Chem.* **273**, 9158–9167.
112. Kobayashi, Y., Miwa, S., Merry, D. E., Kume, A., Mei, L., Doyu, M., et al. (1998) Caspase-3 cleaves the expanded androgen receptor protein of spinal and bulbar muscular atrophy in a polyglutamine repeat length-dependent manner. *Biochem. Biophys. Res. Communications* **252,** 145–150.
113. Ona, V. O., Li, M., Vonsattel, J. P. G., Andrews, L. J., Khan, S. Q., Chung, W. M., et al. (1999) Inhibition of caspase-1 slows disease progression in a mouse model of Huntington's disease. *Nature* **399,** 263–267.
114. Sanchez, I., Xu, C. J., Juo, P., Kakizaka, A., Blenis, J., and Yuan, J. (1999) Caspase-8 is required for cell death induced by expanded polyglutamine repeats. *Neuron* **22,** 623–633.
115. Bingham, P. M., Scott, M. O., Wang, S., McPhaul, M. J., Wilson, E. M., Garbern, J. Y., et al. (1995) Stability of an expanded trinucleotide repeat in the androgen receptor gene in transgenic mice. *Nature Genet.* **9,** 191–196.
116. Merry, D. E., McCampbell, A., Taye, A. A., Winston, R. L., and Fischbeck, K. H. (1996) Toward a mouse model for spinal and bulbar muscular atrophy: effect of neuronal expression of androgen receptor in transgenic mice. *Am. J. Hum. Genet.* **59(Suppl.),** A271.
117. La Spada, A. R., Peterson, K. R., Meadows, S. A., McClain, M. E., Jeng, G., Chmelar, R. S., et al. (1998) Androgen receptor YAC transgenic mice carrying CAG 45 alleles show trinucleotide repeat instability. *Hum. Mol. Genet.* **7,** 959–967.
118. Merry, D. E., Woods, J., Walcott, J., Bish, L., Fischbeck, K. H., and Abel, A. (1999) Characterization of a transgenic model for SBMA. *Am. J. Hum. Genet.* **65(Suppl.),** A30.
119. Simeoni, S., Mancini, M. A., Steusien, D. L., Marcelli, M., Weigel, N. L., Zanisi, M., Martin, L., and Poletti, A. (2000) Motoneuronal cell death is not correlated with aggregate formation of androgen receptors containing an elongated polyglutamine tract. *Hum. Molec. Genet.* **9,** 133–144.

120. Becker, M., Martin, E., Schneikert, J., Krag, H., and Cato, A. C. B. (2000) Cytoplasmic localization and the choice of ligand determine aggregate formation by androgen receptor with amplified polyglutamine stretch. *J. Cell Biol.* **149,** 255–262.
121. Kobayashi, Y., Klume, A., Li, M., Doyu, M., Hata, M., Ohtsuka, K., and Sobue, E. (2000) Chaperones Hsp70 and Hsp40 suppress aggregate formation and apoptosis in cultured neuronal cells expressing truncated androgen receptor protein with expanded polyglutamine tract. *J. Biol. Chem.* **275,** 8772–8778.

14

Mouse Models of Huntington's Disease

Marie-Françoise Chesselet and Michael S. Levine

14.1. INTRODUCTION

Molecular evidence for a common type of mutation has recently united a group of neurodegenerative diseases that were previously considered to be largely distinct. These diseases are now commonly referred to as "CAG repeat diseases" to reflect the fact that they are all caused by an expansion of the triplet encoding glutamine, albeit in different genes (The Huntington's disease Collaborative group, 1993; Perutz, 1996; Zoghbi, 1996). In addition to their common molecular cause, CAG repeat diseases share other characteristics. In their most common form, they are characterized by progressive neurodegeneration in discrete brain region and patients usually begin to show symptoms in mid-life. There are exceptions to this characteristic: The onset of symptoms occurs before age 20 in "juvenile" forms, which are now known to be the result of particularly long CAG repeats (Petersen et al., 1999). All the CAG repeat diseases are progressive and evolve inexorably toward severe motor dysfunction, with or without dementia, in the course of 15–20 yr. They are all inherited in a dominant manner, and because of their late onset, they have a profound effect on families. Indeed, affected individuals often already have children by the time they are diagnosed. Despite the availability of genetic testing, many at-risk individuals prefer to not be tested in the absence of effective treatment.

From the pathological point of view, each disease is characterized by cell death in different brain regions (Ross, 1995). The reason for this selective disease pattern is not known. Indeed, the mutated protein is expressed either ubiquitously or, at least in numerous tissues that are not primarily affected in the disease. Although the pattern of cell death in advanced cases of the diseases is usually relatively well characterized, much less is known about the pathological features of early stages of the diseases because of the paucity of corresponding postmortem material. The recent discovery of the genetic mutations responsible for these diseases has led to the generation of numerous mouse

From: *Contemporary Clinical Neuroscience: Molecular Mechanisms of Neurodegenerative Diseases*
Edited by: M.-F. Chesselet © Humana Press Inc., Totowa, NJ

models that express the mutations. These offer, for the first time, the ability to follow the effects of the CAG repeat mutations over time in vivo. Because of the experimental constraints of making mouse models and the fundamental differences between the central nervous system and life span of mice and man, it is debatable that any of these mouse models truly reproduces the diseases as they occur in humans. However, the multiplicity of approaches used to create these mice provides the opportunity to identify those characteristics that are common to different models and may be more significant for understanding the pathophysiology of each disease. A detailed account of models of SCA1 and SCA3 is given in the corresponding chapters of this book. This chapter will focus on mouse models of Huntington's disease.

14.2. SPECIFIC CHARACTERISTICS OF HUNTINGTON'S DISEASE

Huntington's disease (HD) is the result of a CAG repeat expansion in the gene *(IT15)* encoding huntingtin, a widely expressed protein of unknown function (The Huntington's disease collaborative group, 1993). Huntingtin is expressed in many brain regions and in peripheral tissues, yet cell loss occurs mostly in the striatum, where GABAergic efferent neurons die, and in the cerebral cortex where efferent glutamatergic neurons are affected (Graveland et al., 1985; Vonsattel et al., 1985; Gutekunst et al., 1995). Therefore, the mutation causing HD is well tolerated in many neurons during the life of the affected individual, yet it kills a subset of neurons that do not share either their neurotransmitter or their anatomical location. Behaviorally, the disease is characterized by involuntary movements that have a dancelike quality, hence the name of chorea, which means dance in Greek (Quarrell and Harper, 1996). Patients also show characteristic abnormal eye movements that often precede other symptoms. At later stages of the disease, and in the juvenile forms, patients become dystonic, a severe movement disorder characterized by cocontracture of opposing muscles. Cognitive and psychiatric symptoms can be present early in the disease (Morris and Scourfield, 1996) but dementia usually appears at later stages, and death is usually the result of complications of dysphagia and decubitus.

Evidently, it is unlikely that a mouse model will reproduce the type of movement disorders and cognitive deficits seen in humans. Therefore, a more realistic criteria for a successful disease model would be the reproduction of the selective pattern of cell loss induced by the mutation in humans (Vonsattel et al., 1985; Ferrante et al., 1987). This goal has proven more elusive than expected. However, the models created so far have been informative and the most recent models that are emerging appear to display increased similarities with the human illness. Instead of describing each model successively, we will review the behavioral, pathological, and cellular changes that have been observed in these various models to identify the common findings and differences that can begin to provide some insights into the pathophysiology of HD.

14.3. BEHAVIORAL ANOMALIES IN MOUSE MODELS OF HD

The occurrence of behavioral anomalies is, in a sense, a gold standard for a mouse model of neurodegenerative disease. How can we know that pathological and cellular alterations seen in these mice are meaningful for the human pathology if they do not have functional consequences at the behavioral level? The emergence of abnormal behavior is also extremely important to identify the time-course of disease progression without the need to sacrifice a large number of animals. For the same reason, behavioral measures are an ideal way to test for new therapies. Unfortunately, the behavioral equivalents in mouse of the neurological signs of adult-onset HD (abnormal eye saccade, chorea) are unknown and may be difficult to identify. Nevertheless, many of the mouse models available so far show some degree of motor impairment. A major advantage of mouse models is the ability to relate the appearance of behavioral anomalies to neuropathology, which rarely can be accomplished in humans. A more detailed account of neuropathological findings in the mice is given below. However, the time-course of the critical neuropathological features will be mentioned here as they relate to the behavior.

The first successful mouse model of HD was generated by the group of Gill Bates in London (Mangiarini et al., 1996). This success was somewhat unexpected and had a profound influence on the field. Indeed, rather than planning to make a perfect model of HD, these investigators attempted to test the effect of the CAG repeats themselves and overexpressed only exon 1 of the gene *(IT15)* encoding huntingtin, containing a 141–157 CAG repeat. One of the transgenic lines (R6/2) displayed rapid and severe motor behavior anomalies.

An overt behavioral phenotype consisting of limb clasping, stereotypical hindlimb-grooming movements, and irregular gait became evident in these mice at about 8 wk of age (Manginarini et al., 1996). However, detailed behavioral studies have shown that behavioral deficits occur as early as 5–6 wk of age (Carter et al., 1999). The first anomaly noticeable was a difficulty in swimming, manifested by a twisted posture, slower swimming, increased number of forelimbs kicks, and uncoordinated movements of both hindlimbs and forelimbs. At that age, the transgenic mice were also slower than controls in traversing the narrowest square beam. Slightly older mice (8–9 wk) also made more footslips on narrow beams and began to show a recumbent posture when attempting to traverse the beam. Transgenic mice were able to learn the rotarod test; however, as early as 5–6 wk of age, they had difficulty maintaining balance at high speed. This difficulty dramatically increased with age and made it impossible for them to maintain balance by 13–14 wk of age, even at the lowest speed. Gait anomalies, as indicated by decreased stride length in the footprint pattern test, were present by 8–9 wk. This was accompanied by an increased front base width. In contrast to these motor symptoms, the acoustic startle response of the transgenic mice did not differ from controls until 12.5 wk, when it was decreased compared to controls. Prepulse inhibition, however, was disturbed as early as 8–9 wk. At 8 wk of age, the R6/2 mice also show a decreased locomotor activity and evidence of decreased anxiety (File et al., 1998). Time-points earlier than 5–6 wk were not tested in the behavioral studies. Therefore, the earliest appearance of abnormal motor signs in these mice, when confronted to challenging situations, is not known. The first reported anomalies occur after the earliest detection of abnormal protein aggregates (by 3–4 wk). Cell death in these mice, in contrast, has only been reported much later (12.5 wk [Davies et al., 1999]).

Other mouse models have not yet been submitted to such extensive motor analysis. It appears that the type of early motor sign and the age of appearance is model dependent. Like the R6/2 mice, transgenic mice developed by Schilling et al. (1999) express a N-terminal fragment of the huntingtin gene. In this case, the transgene encodes the first 171 amino acids of huntingtin, with 82 glutamine repeats under the control of a mouse prion protein vector that drives the expression of foreign genes in every neuron of the central nervous system. These transgenic mice exhibit tremor, uncoordination, hypokinesis, abnormal gait, and frequent limb clasping, although the age on onset of these anomalies has not been reported. At 3 mo of age, the transgenic animals fail to improve their performance on the rotorod on successive days, and at 5 mo of age, they are impaired in the first trial as well. Because neuronal loss has been recently discovered in the striatum of these mice, it is unclear whether these behavioral anomalies occur before or after neuronal death. These mice show nuclear staining for huntingtin, neuropil aggregates, and numerous intranuclear inclusions, however, the relative time-course of appearance of these neuropathological features and of behavioral symptoms has not been fully explored.

Other mouse models tend to display a phase of increased rather than decreased locomotor activity. In one mouse model (Hodgson et al., 1999) yeast artificial chromosomes (YACs) containing human genomic DNA spanning the full-length gene, including all regulatory elements, were used. The transgene contained CAG repeats of 46 or 72 (i.e., comparable to adult- and juvenile-onset HD, respectively). Mice with 46 repeats did not show noticeable motor anomalies up to 20 mo of age; however, mice with 72 CAG repeats showed progressive behavioral anomalies. In one mouse that expressed high levels of the mutant protein, behavioral anomalies were observed at 6 mo of age. This mouse showed pronounced circling behavior and later developed choreoathetotic movements. It showed foot-clasping when held by the tail and was unable to complete the beam-crossing task. Unfortunately, this mouse did not breed and no line could be derived from it. Mice expressing lower levels of the mutant protein begin to show behavioral anomalies around 7 mo of age. These anomalies were limited to a mild and progressive hyperactivity during the dark phase of open field testing and only one mouse developed stereotyped turning. As indicated below, the mild hyperkinetic behavior exhibited by most mice of this line was seen before evidence of neuronal loss and in the absence of nuclear inclusions visible by light microscopy. However, translocation of huntingtin to the nucleus and electrophysiological anomalies were observed earlier in these mice, indicating neuronal dysfunction at the cellular level.

Hyperactivity also occurred in transgenic mice expressing a full-length human HD cDNA with 48 and 89 CAG repeats (Reddy et al., 1998). This behavior consisted of stereotyped rotations, backflips, and excessive grooming and was only seen in a fraction of the mice by 20 wk of age. High expressing lines with either repeat length showed feet-clasping as early as 8 wk of age, but low expressing mice with 48 repeats showed this behavior later (25 wk). Whether or not they showed a phase of hyperactive behavior, all transgenic mice with expanded repeats showed hypoactivity starting at 24 wk. This behavior worsened over 4–6 wk and led to death in a state of marked akinesia and lack of response to sensory stimuli. Massive neurodegeneration was seen in these mice, not only in the striatum but also in the cerebral cortex, hippocampus, and thalamus. In contrast to other mouse models, these transgenic animals showed evidence of DNA damage by TUNEL labeling and astrogliosis. Nuclear inclusions were seen in a very small proportion of striatal neurons, as well as in cerebral cortex, hippocampus, thalamus, and cerebellum.

Curiously, in these mice, no marked differences in age of onset or progression of the phenotypes between high expressors with 48 or 89 repeats. However, homozygotes had a much earlier onset of symptoms than heterozygotes. These features distinguish the behavioral phenotype observed in these transgenic mice from that observed in patients with HD. Indeed, humans with large expansions show an earlier onset, greater severity, and faster progression of the disease than patients with moderate expansion (Petersen et al., 1999). In contrast, heterozygotes have not been found to have marked differences in age of onset and disease progression than the rare homozygotes that have been studied (Wexler et al., 1987).

In striking contrast with the overt motor anomalies expressed by these transgenic models, mice with extended "knock-in" CAG repeats in the mouse huntingtin gene showed much more discrete symptoms, if any. Three different lines of such "knock-in mice" have been described (White et al., 1997; Shelbourne et al., 1999; Levine et al., 1999). They all share an absence of obvious motor deficits, except for an unexplained aggressive behavior in one of the mouse lines (Shelbourne et al., 1999). Preliminary data in another of the mouse lines (Levine et al., 1999; Menalled et al. 2000), revealed that a decrease in rearing and locomotor activity could be observed (Scisson, Menalled, Gotts, and Chesselet, *unpublished observation*), despite the fact that the mice do not display spontaneous clasping or obvious abnormal behavior in their home cage even when over 1 yr old. The consistency and time-course of this abnormal behavior remains to be determined. Furthermore, these different lines of knock-in mice have not been systematically tested in the same conditions (dark-phase open field) that reveled a mild hyperactivity in the YAC mice expressing low levels of the transgene (Hodgson et al., 1999).

The clinical equivalents of the behavioral anomalies observed in mouse models of HD are unclear. Spontaneous or induced feet-clasping may be a form of hyperkinetic behavior. It is tempting to equate increased locomotion and stereotyped behavior with the irrepressible movements characteristic of chorea in humans (Mangiarini et al., 1996; Reddy et al., 1998; Hodgson et al., 1999). However, these behaviors in mice are highly nonspecific and can be induced by a variety of mutations and by drugs that are not known to cause chorea in humans. Interestingly, both motor and nonmotor behaviors have been observed in mice and in humans carrying the HD mutation (Morris and Scourfield, 1996; File et al., 1998; Shelbourne et al., 1999). The interpretation of these symptoms in mice, however, remains difficult, as they have only been observed in some of the models and because strain differences can greatly affect behavior in mice.

Although comparisons of behavioral phenotypes are difficult because of these strain differences and the absence of a systematic use of similar behavioral measures, a few recurrent themes emerge at this point. Perhaps most importantly, both transgenic and knock-in models concur to show that behavioral anomalies caused by the presence of an expanded CAG repeat in huntingtin can occur in the absence of overt neuronal loss and are more likely the result of neuronal dysfunction. This has important implications for therapeutic strategies. Indeed, focusing on cell death pathways that ultimately lead to the demise of neurons after a prolonged phase of dysfunction may not lead to useful therapies for the devastating motor and psychiatric symptoms of HD. Another intriguing observation is that the severity of motor symptoms seems directly related in most cases to the level of expression of the mutant protein and not always to the length of the polyglutamine expansion. This was not predicted by previous observations in humans that homozygotes do not have a more severe from of the disease (Wexler et al., 1987). However, it suggests that decreasing the level of expression of the mutated huntingtin in gene carriers may help delay the onset of symptoms.

14.4. NEUROPATHOLOGICAL OBSERVATIONS

Neuronal Death and Abnormal Morphology

When comparing mouse models of neurodegenerative diseases for the presence or absence of neuronal loss, it is important to keep in mind that most neuropathological studies rely upon the presence of cells with morphological evidence of ongoing degeneration to establish the presence of neuronal loss. Although a marked neuronal loss, as seen in advanced cases of HD in humans, would also be noticed in these studies, a more limited decrease in neuronal number could only be detected with the use of stereological methods. To our knowledge, these have not yet been applied to mouse models of HD. Furthermore, in some cases, examination of special staining or thin plastic sections revealed neuronal degeneration that was not obvious with traditional light microscopic methods (Schilling et al., 1999).

In their original observations, Mangiarini et al (1996) and Davies et al (1997) noticed that the brains of the R6/2 transgenic mice were smaller than normal. However, no conspicuous evidence of cell death was observed at 12.5 wk. Neuronal death seems to occur in these mice but at later time points (Davies et al., 1999). When it occurred, neuronal death was restricted to the striatum and deep cortical layers. Interestingly, dying neurons did not exhibit the typical morphological features of apoptosis (Davies et al., 1999) and no evidence for DNA damage were found with *in situ* nick translation (Menalled et al. 2000). The data to date suggest that behavioral anomalies precede overt neuronal death in this model. However, no systematic stereological analyses have been done in these mice.

Although neuronal death does not appear to occur until shortly before death in these transgenics, striatal neurons show marked morphological anomalies. To date, we have examined changes in neurons from R6/2 transgenics and their wild-type littermates in mice of about 80–90 d of age. We showed first that the medium-sized neurons had reduced cross-sectional areas (Levine et al., 1999). In addition, we observed changes in biocytin-filled cells from slice preparations and in cells examined after Golgi staining. Using both of these staining paradigms, there was a marked reduction in spine density of medium-sized spiny cells, a decease in the extent of the dendritic field, and a decrease in dendritic diameter. Together, these morphological alterations suggest the occurrence of marked functional changes in these neurons.

As indicated earlier, some neuronal degeneration occurs in transgenic mice expressing a transgene encoding the first 171 amino acids of huntingtin with 82 glutamine repeats under a prion protein promoter (Schilling et al., 1999). More obvious neuronal loss has been reported in both transgenic models expressing the full-length mutated protein (Reddy et al., 1999; Hodgson et al., 1999). In the YAC mice, neurodegeneration occurred late in the progression of the disease (12 but not 8 mo of age in line YAC72 2511) and was restricted to the striatum. However, it was more pronounced in the lateral than in the medial striatum, a pattern reversed from that seen in human (Vonsattel et al., 1985). Furthermore, also in contrast to postmortem samples from patients with HD, gliosis was not observed.

Neurodegeneration was more widespread in transgenics expressing the full-length human cDNA and was accompanied by reactive gliosis. Another difference between these two types of transgenic mice is the presence of neuronal death in those mice with moderate expansions in the model generated by Reddy et al., but not in the YAC mice (Hogdson et al., 1999). A more detailed study of the time-course of neuronal loss and the use of similar pathological criteria will be necessary to further determine whether these two mouse models really differ in the extent and progression of neuronal loss.

In contrast to observations in the R6/2 mice, morphological anomalies were consistent with apoptosis in the YAC mice. Although no electron microscopic analysis has yet been reported, Reddy et al. (1998) have shown the presence of DNA damage by TUNEL staining in their mice, a feature that is associated with, but not specific for, apoptosis. Interestingly, the extent of TUNEL labeling is greater than expected based on the relatively modest neuronal loss reported in these mice, a feature that has also been observed in postmortem brains of patients with HD (Portera-Cailliau et al., 1995; Thomas et al., 1995). In contrast to the transgenic models, no evidence of neuronal loss has yet been found in the three models of knock-in mice examined so far (White et al., 1997; Shelbourne et al., 1999; Menalled et al., 2000).

In conclusion, as discussed earlier, overt neuronal loss is either undetected (knock-in mice), minor (transgenics expressing a N-terminal fragment of huntingtin), or delayed (transgenics expressing full-length mutant huntingtin). No clear correlation between the extent of neuronal loss and behavioral or cellular (*see below*) phenotypes has yet emerged. The gradient of striatal neuronal loss characteristic of adult-onset HD has not yet been reproduced in mice. No systematic study has yet determined whether striatal interneurons were spared in the mice that show striatal cell loss, a characteristic feature of HD (Ferrante et al., 1987). Similarly, it is not known whether or not neuronal loss occurs preferentially in one of the well-characterized striatal compartments, the striosomes and the matrix, that exhibit different pathology in HD (Kowall et al., 1987; Hedreen et al., 1995).

Neuronal death appears to be regionally selective in R6/2 (striatum and cortex) and YAC mice (striatum) but not in mice expressing the full-length human cDNA (Reddy et al., 1998). Although neuronal loss predominates in the striatum and to a lesser extent in the cortex of patients with adult-onset HD (Vonsattel et al., 1985; de la Monte et al., 1988; Hedreen et al., 1991), atrophy or neuronal loss in other regions has been reported in pathological studies (de la Monte et al., 1988; Kremer et al., 1990). Furthermore, neuronal loss occurs in the hippocampus and cerebellum of patients with juvenile-onset HD, which is associated with the large CAG repeat expansions used in most models of HD. Therefore, the regional selectivity usually considered a hallmark of HD is relative and may not be a critical criteria for a mouse model of the disease.

Nuclear Inclusions and Protein Aggregates

Long before neuronal death can be detected, the R6/2 transgenics display a remarkable feature: In most brain areas, prominent nuclear inclusions can be detected with immunostaining for the transgene, as well as ubiquitin and heat shock proteins (Davies et al., 1997). This suggests that at least part of the transgene product is sequestered in these inclusions in an ubiquinated form. These inclusions could not be stained with antibodies against other parts of the normal, endogenous huntingtin, suggesting that the normal protein is not recruited in the inclusions. Nuclear inclusions are not limited to brain but were also observed in skeletal muscle, heart, liver, adrenal medulla, pancreas (islets of Langherans), kidney, and myenteric and Meissner's plexus (Sathasivam et al., 1999).

Nuclear inclusions are also present in the brain of humans with the disease. In fact, prior ultrastructural studies in a rare biopsy case had evidenced such an inclusion (Roizin et al., 1979). More importantly, nuclear inclusions could also be detected with an N-terminal antibody in the brains of patients with the disease (DiFiglia et al., 1997; Beher et al., 1998; Gourfinkel-Ann et al., 1998). Numerous in some cases of juvenile HD, the inclusions are however, relatively rare in late-onset disease. Furthermore, their distribution does not clearly parallel the pattern of neurodegeneration in humans (Gutekunst et al., 1999).

Nuclear inclusions are also found in mice expressing a transgene encoding the first 171 amino acids of huntingtin with 82 CAG repeats (Schilling et al., 1999). In these mice, more than 50% of neurons in the cerebral cortex, hippocampus, cerebellum and amygdala and 10–50% of striatal neurons contained nuclear inclusions. Aparently, inclusions are much more prominent than neuronal loss in these mice. In contrast, mice expressing a full-length human cDNA encoding huntingtin with 48 or 89 glutamines (Reddy et al., 1998) showed marked cell loss in many brain regions, but nuclear inclusions were found in a much smaller number of neurons (only 1% of striatal neurons, for example). This is in agreement with the widespread distribution but relative paucity of nuclear inclusions in the brains of patients with late onset HD (Becher et al., 1998; Gutekunst et al., 1999).

Curiously, other mouse models did not show prominent nuclear inclusions despite the presence of behavioral anomalies and even cell death (Hodgson et al., 1999; Shelbourne et al., 1999; Levine et al., 1999). However, both in the high-expressing YAC mouse and in two knock-in mouse models, nuclear macroaggregates of the N-terminal portion of huntingtin were detected by immunostaining of tissue perfused with a fixative containing glutaraldehyde (Hodgson et al., 1999; Li, Menalled, and Chesselet, unpublished observation). Macroaggregates were more numerous at the ultrastructural level and were also found in low-expressing YAC mice with this technique (Hogdson et al., 1999). Although large enough to be detected with light microscopy, these aggregates were much smaller than the nuclear inclusions described in other mouse models. Furthermore, in our hands, these macroaggregates could not be detected in tissue from animals perfused with the milder fixative paraformaldehyde, suggesting that they may be more labile than nuclear inclusions. It is not yet known whether such labile aggregates also stain for chaperone proteins and ubiquitin-like nuclear inclusions.

Mutant huntingtin also forms aggregates in the cytoplasm, particularly of neuronal processes. The presence of neuropil aggregates has been reported in human brain, in R6/2 transgenic mice (Li et al., 1999), in transgenic mice expressing huntingtin 1–171 with 82 glutamines (Schilling et al., 1999) and in YAC mice (Hogdson et al., 1999). An important question that is not fully resolved is whether these precede nuclear aggregates. Therefore, the respective role of nuclear localization of huntingtin and of neuropil aggregates in the early phases of the disease process remains unclear.

In conclusion, although nuclear inclusions are a prominent feature of transgenic mice expressing a truncated mutant huntingtin, they are much less conspicuous in other mouse models. Furthermore, in postmortem human brain, the distribution of nuclear inclusions does not parallel the pattern of neurodegeneration. A debate about the significance of these inclusions has been fueled by the results of in vitro experiments and data from other mouse models of HD (*see below*). Furthermore, parallel work in other CAG repeat diseases such as SCA1 (*see* Chapter 11) and SCA3 (*see* Chapter 12) also questioned the pathological significance of the nuclear inclusions. However, abnormal location of huntingtin in the nucleus appears to be a common feature of most models. This abnormal nuclear staining is coupled to the presence of macroaggregates or microaggregates that can only be retained in tissue section with strong fixatives. A role for nuclear transport of truncated huntingtin is further suggested by evidence that preventing nuclear entry of mutated huntingtin protects tranfected primary neurons in culture, whereas preventing aggregation does not (Saudou et al., 1998). An important conclusion from these studies is that proteolytic processing of huntingtin appears to be critical for the disease process. This is further supported by the results of in vitro studies that show an increased toxicity of smaller compared to larger huntingtin fragments in transfected cells (*see below*).

14.5. METABOLIC EFFECTS

A role for oxidative stress in HD has long been suspected (Browne et al., 1999). Therefore, it is not surprising that indices of oxidative stress and susceptibility of striatal neurons to mitochondrial toxins and excitotoxic insults were among the first to be examined in mouse models of HD. No clear picture, however, has emerged so far. Transgenic mice expressing huntingtin 1–171 with 82 glutamine repeats do not show any increase in indices of oxidative stress (Schilling et al., 1999). However, the R6/2 mice show an increased vulnerability to the mitochondrial toxin 3-NP, which, in rats, produces a pattern of cell loss in striatum very similar to that seen in HD (Bogdanov et al., 1998). Surprisingly, another transgenic line expressing the same transgene but with a much milder disease course (R6/1) showed increased resistance to local injection of the NMDA agonist quinolinic acid into the striatum (Hansson et al., 1999). This is apparently at odds with evidence that striatal neurons in R6/2 mice show an increased sensitivity to the stimulation of NMDA receptors in vitro (Levine et al., 1999: *see* Subheading 14.6.). It is possible, however, that the increased resistance to excitotoxicity observed in the R6/1 mice is the result of the development of compensatory mechanisms in vivo. The analysis of excitotoxicity and sensitivity to oxidative stress in mice is complicated by the fact that different strains show marked differences in sensitivity (Alexi et al., 1998). It is important to clarify the role of oxidative stress in the mouse models because defects in mitochondrial function have been recently confirmed in cells from patients with HD (Sawa et al., 1999)

14.6. NEUROCHEMICAL EFFECTS

Neurochemical and molecular changes detected in the R6/2 mice provided the first evidence that neuronal dysfunction may be related to the expression of the huntingtin mutation. These changes clearly precede overt neuronal death and perhaps even the onset of neurological symptoms. Importantly, some of these effects are similar to those observed with in vivo brain imaging studies in patients at early stages of HD, or in the rare postmortem samples of early grade of the disease.

One of the earliest molecular changes reported in these mice (4 wk old) is a decreased level of the mRNA encoding the dopamine D1 receptor in striatum (Cha et al., 1998). This effect was followed at 8 wk by a decrease in D1-binding sites and by a decrease in both D2-binding sites and the corresponding mRNA. These effects are of particular interest because positron emission tomography studies have shown early alterations in D1- and D2- binding sites in the striatum of patients with HD (Weeks et al., 1996). Selective decreases in subtypes of glutamate receptors and their corresponding mRNA also occur in these mice (Cha et al., 1998). As early as 4 wk of age, the R6/2 mice showed a decrease in mGluR1 mRNA in the striatum, and later in the cerebral cortex. AMPA, kainate and group II (but not group I) metabotropic receptor-binding sites were significantly decreased in the striatum of 12 week old transgenic mice. AMPA, kainate, and group II metabotropic receptor-binding sites were also decreased in the cerebral cortex. Muscarinic acetylcholine-binding sites, but neither GABA A or GABA B receptors, were decreased in both striatum and cortex. Importantly, these effects are not the result of the massive loss of striatal or cortical neurons at this age, suggesting a selective neuronal dysfunction (Davies et al., 1997).

The R6/2 mice also show a deficit in dopamine and serotonin in striatum and hippocampus at 12 wk, in the absence of changes in glutamate or GABA content, except for an increase in GABA in the hippocampus (Reynolds et al., 1999). Interestingly, decreases in 5-HT are preceded by decreases in the 5-HT metabolite 5 HIAA in the striatum as early as 4 wk, suggesting that an alteration in striatal serotonin turnover precedes the earliest behavioral symptoms observed in these mice.

In addition to discrete deficits in neurotransmitter receptors, the R6/2 mice show a profound decrease in enkephalin mRNA in the striatum (Menalled et al. 2000). Enkephalin mRNA is normally present in a subset of efferent striatal neurons, and evidence suggests that these neurons may be particularly vulnerable to the disease process (Albin et al., 1989). In contrast to the early decreases in enkephalin mRNA, however, the R6/2 mice retained a high level of enkephalin immunolabeling in the globus pallidus, which contains the axons of striatal enkephalinergic neurons. Unlike the mRNA encoding enkephalin, mRNAs encoding substance P and two forms of glutamic acid decarboxylase (*Mr* 67,000 and 65,000) were not decreased in the striatum of the R6/2 mice at the same age (Menalled et al. 2000). Thus, similar to neurotransmitter receptors, mRNAs encoding neuropeptides or neurotransmitter synthesis enzymes were differentially affected in the R6/2 transgenic mice.

Few molecular studies have been performed so far in other transgenic models. However, at least some of the molecular defects noted in the R6/2 mice have also been observed in one of the knock-in lines, suggesting that they are an effect of the CAG repeat expansion. Indeed, a significant decrease in enkephalin mRNA was also observed in 3.5-mo-old knock-in mice with 94 CAG repeats, and in some of the mice with 71 CAG repeats (Menalled et al. 2000). At this age, the 94 CAG repeat mice already show weak nuclear aggregates in a subpopulation of striatal neurons and can display a decreased rearing (unpublished observation). As in the transgenic mice, immunolabeling for enkephalin, GAD, and substance P was preserved in the axon terminals of striatal neurons in these knock-in mice, consistent with data reported for another line of knock-in mice (Shelbourne et al., 1999).

The dissociation between decreased mRNA and preserved peptide labeling in striato-pallidal neurons is intriguing and could be explained by a defect in peptide release. Although this hypothesis has not yet been directly tested at the striato-pallidal synapse, it should be noted that huntingtin is associated with synaptic vesicles and interacts with proteins involved in vesicle trafficking (DiFiglia et al., 1995; Kalchman et al., 1997). Normal huntingtin is thought to influence vesicle transport in the secretory and endocytic pathway through association with clathrin-coated vesicles (Velier et al., 1998). It is not known whether the polyglutamine expansion in huntingtin alters these functions. However, electrophysiological studies in another line of knock-in mice and in YAC mice have revealed a reduced ability of synapses to sustain transmission during repetitive stimulation (*see below*).

14.7. CELLULAR EFFECTS

The mechanism by which overexpression of exon 1 of the huntingtin gene containing an expanded polyglutamine repeat causes the changes in receptor and neurotransmitter mRNAs described earlier is unknown. However, our recent data suggest that the mutation also causes marked anomalies in the functional properties of striatal and cortical neurons in these mice (Levine et al., 1999). In R6/2 mice between 70 and 90 d of age, medium-sized striatal neurons and cortical neurons were more sensitive to NMDA receptor activation than similar neurons in wild-type controls. In the striatum, this effect was accompanied by a depolarization of the resting membrane and an increase in membrane input resistance. The relationship between depolarized membrane potential and increased NMDA receptor sensitivity may be significant because it would suggest that in R6/2 mice, less excitatory input is necessary to remove the Mg^{2+} block and activate NMDA receptors. A similar increased sensitivity to NMDA receptor activation was observed in CAG94 knock-in mice of approximately the same age (Levine et al., 1999). However, there was only a small depolarization of the resting membrane potential that was not statistically significant, suggesting that in this model, membrane potential changes may not be the sole cause of the increased sensitivity to NMDA receptor stimulation.

The effect of the HD mutation on synaptic transmission in the hippocampus has been examined in the YAC mice and in one model of knock-in mice. In the YAC transgenic mice, long-term potentiation (LTP) at the Schaeffer collateral/CA1 synapses was maintained for 60 min in 4-mo-old mice of all lines examined (Hodgson et al., 1999). Slices of 6-mo-old mice with 72 repeats showed hyperexcitability and displayed a greater short-term potentiation following tetanization. Experiments with NMDA antagonists suggested that the NMDA component is predominant during fast synaptic transmission in the mice with 72 repeats. At 10 mo of age, mice with either 46 or 72 repeats demonstrated a loss of LTP in CA1 neurons. Calcium influx in response to glutamate application was abolished in hippocampal neurons of the YAC46 mice and resting levels of calcium were elevated in these neurons, suggesting an impairment of calcium-buffering capacities of the mutant neurons. Although paired-pulse facilitation was not affected in 10-mo-old mutant mice, posttetanic potentiation was reduced in these mice. This suggests an impairment of presynaptic release in response to high frequency stimulation.

Long-term potentiation was also reduced in one line of knock-in mice (Usdin et al., 1999). However, enhanced tetanic stimuli could induce LTP, suggesting that the LTP-producing mechanism is intact in these mice. Also similar to the YAC mice, the knock-in mice showed an impairment in posttetanic potentiation, further suggesting that one effect of the HD mutation is to impair the ability of nerve terminals to increase neurotransmitter release during high-frequency synaptic activation. These cellular deficits could form the basis of the neuronal dysfunction, leading to behavioral symptoms at early stages of the disease. However, the link between the

HD mutation and these cellular effects, as well as their contribution to subsequent neuronal death, remains unknown at this point.

14.8. HUNTINGTIN PROCESSING AND CASPASE ACTIVATION

It is striking that the most severe phenotype observed to date in a mouse model of HD occurs in the transgenic mice expressing only exon 1 of huntingtin with a CAG expansion. It is not possible to evaluate the level of expression of the transgene in these mice for technical reasons. However, other trangenics with a high level of the full-length mutated huntingtin have a milder phenotype. The presence of diabetes in the R6/2 mice (Hurlbert et al., 1999), perhaps because of a dysfunction of pancreatic cells that display nuclear inclusions (Sathasivam et al., 1999), does not seem to be the only explanation because behavioral symptoms and decreased gene expression appear to precede the onset of diabetes in these mice (Cha et al., 1998; Carter et al., 1999).

An explanation for this paradox may be provided by in vitro studies that have clearly demonstrated that short huntingtin fragments with expanded polyglutamine repeats are more toxic to neurons than the full-length protein with an identical mutation (Cooper et al., 1998). This observation led to the hypothesis that cleavage of huntingtin by proteases is a critical step in the pathophysiology of the disease. Huntingtin can be cleaved by caspase 3, a protease involved in the apoptotic cascade (Wellington et al., 1998), and caspase 8 has also been shown to play a role in polyglutamine-induced cell death (Sanchez et al., 1999). Interestingly, the blockade of caspase 1 in transgenic mice delays the onset of motor symptoms, of neurochemical anomalies, and the death of the mice (Ona et al., 1999). Because behavioral symptoms precede neuronal death in this model, these data suggest that caspase activation may play a role in HD, independently of its potential effect in cell death.

Noncaspase proteases, however, also seem to be important for the processing of huntingtin. Similar differences in huntingtin processing were observed in lymphoblasts and cerebral cortex of patients with HD and in mice with a knock-in HD mutation, compared to their respective controls (Wu et al., 1999). Whether this abnormal processing contributes to the pathophysiology and which proteases are involved, however, remains unknown.

14.9. CONCLUSION

In conclusion, once only a dream, numerous mouse models of HD have become available in the last 3 yr. The diversity of the mouse stains, constructs, and CAG repeat lengths complicates comparisons among these different mouse models. Furthermore, few studies have examined the same behavioral, cellular, or molecular effects across several mouse models. Nevertheless, several lines of consistent evidence are emerging from the available comparisons.

Most importantly, the mouse models have shed light on the respective importance of cell death versus cell dysfunction as the major cause of pathology in HD. The mutation seems to induces neuronal dysfunction long before it induces cell death and neuronal dysfunction appears sufficient to induce motor symptoms. Another emerging theme is the importance of protein aggregates, as opposed to nuclear inclusions, at early stages of the disease. In that respect, information obtained from the mouse models confirms data from cellular and postmortem studies that suggest a link between the basic pathophysiology of HD and that of other neurodegenerative diseases in which protein aggregation seems to be a key factor: not only other CAG repeat diseases, but also Alzheimer's, prion, and Parkinson's diseases. Finally, the mouse models are beginning to provide the most sought after information: a rational approach to the design of new therapies and a way to test them preclinically. In that respect, a critical contribution of the mouse models will be to identify the link between parameters that can be measured in humans (in accessible peripheral tissues or by brain imaging) and the progressive brain pathology. Once validated, these accessible measures will permit great improvement in the design of clinical trials, an essential step in bringing the benefit of bench science to the patients.

REFERENCES

Albin R. L., Young A. B., and Penney J. B. (1989) The functional anatomy of basal ganglia disorders.*Trends Neurosci.* **12,** 366–375.

Alexi, T., Hughes, P. E., Knüsel, B., and Tobin, A. J. (1998) Metabolic compromise with systemic 3-nitropropionic acid produces striatal apoptosis in Sprague-Dawley rats but not in BALB/c ByJ mice. *Exp. Neurol.* **153,** 74–93.

Becher, M. W., Kotzuk, J. A., Sharp, A. H., Davies, S. W., Bates, G. P., Price, D. L., et al. (1998) Intranuclear neuronal inclusions in Huntington's disease and dentatorubral and pallidoluysian atrophy, correlation between the density of inclusions and IT15 CAG triplet repeat length. *Neurobiol. Dis.* **4,** 387–397.

Bogdanov, M. B., Ferrante, R. J., Kuemmerle, S., Klivenyi, P., and Beal, M. F. (1998) Increased vulnerability to 3-nitropropionic acid in an animal model of Huntington's disease. *J. Neurochem.* **71,** 2642–2644.

Browne, S. E., Ferrante, R. J., and Beal, M. F. (1999) Oxidative stress in Huntington's disease. *Brain Pathol.* **9,** 147–163.

Carter, R. J., Lione, L. A., Humby, T., Mangiarini, L., Mahal, A., Bates, G. P., et al. (1999) Characterization of progressive motor deficits in mice transgenic for the human Huntington's disease mutation. *J. Neurosci.* **19,** 3248–3257.

Cha, J. H., Kosinski, C. M., Kerner, J. A., Alsdorf, S. A., Mangiarini, L., Davies, S. W., et al. (1998) Altered brain neurotransmitter receptors in transgenic mice expressing a portion of an abnormal human huntington disease gene. *Proc. Natl. Acad. Sci. USA* **95,** 6480–6485.

Cooper, J. K., Schilling, G., Peters, M. F., Herring, W. J., Sharp, A. H., Kaminsky, Z., et al. (1998) Truncated N-terminal fragments of huntingtin with expanded glutamine repeats form nuclear and cytoplasmic aggregates in cell culture. *Hum. Mol. Genet.* **7,** 783–790.

Davies, S. W., Turmaine, M., Cozens, B. A., DiFiglia, M., Sharp, A. H., Ross, C. A., et al. (1997) Formation of neuronal intranuclear inclusions underlies the neurological dysfunction in mice transgenic for the HD mutation. *Cell* **90,** 537–548.

Davies, S. W., Turmaine, M., Cozens, B. A., Raza, A. S., Mahal, A., Mangiarini, L., et al. (1999) From neuronal inclusions to neurodegeneration, neuropathological investigation of a transgenic mouse model of Huntington's disease. *Philos. Trans. Roy. Soc. London. Series B: Biol. Sci.* **354,** 981–989.

de la Monte, S. M., Vonsattel, J. P., and Richardson, E. P. J. (1988) Morphometric demonstration of atrophic changes in the cerebral cortex, white matter, and neostriatum in Huntington's disease. *J. Neuropathol. Exp. Neurol.* **47,** 516–525.

DiFiglia, M., Sapp, E., Chase, K., Schwarz, C., Meloni, A., Young, C., et al. (1995) Huntingtin is a cytoplasmic protein associated with vesicles in human and rat brain neurons. *Neuron* **14,** 1075–1081.

DiFiglia, M., Sapp, E., Chase, K. O., Davies, S. W., Bates, G. P., Vonsattel, J. P., et al. (1997) Aggregation of huntingtin in neuronal intranuclear inclusions and dystrophic neurites in brain. *Science* **277,** 1990–1993.

Ferrante, R. J., Kowall, N. W., Beal, M. F., Martin, J. B., Bird, E. D., and Richardson, E. P. J. (1987) Morphologic and histochemical characteristics of a spared subset of striatal neurons in Huntington's disease. *J. Neuropathol. Exp. Neurol.* **46,** 12–27.

File, S. E., Mahal, A., Mangiarini, L., and Bates, G. P. (1998) Striking changes in anxiety in Huntington's disease transgenic mice. *Brain Res.* **805,** 234–240.

Gourfinkel-An I, Cancel, G., Duyckaerts, C., Faucheux, B., Hauw, J. J., Trottier, Y., et al. (1998) Neural distribution of intranuclear inclusions in Huntington's Disease with adult onset. *NeuroReport* **9,** 1823–1826.

Graveland, G. A., Williams, R. S., and DiFiglia, M. (1985) Evidence for degenerative and regenerative changes in neostriatal spiny neurons in Huntington's disease. *Science* **227,** 770–773.

Gutekunst, C. A., Levey, A. I., Heilman, C. J., Whaley, W. L., Yi, H., Nash, N. R.,et al. (1995) Identification and localization of huntingtin in brain and human lymphoblastoid cell lines with anti–fusion protein antibodies. *Proc. Natl. Acad. Sci. USA* **92,** 8710–8714.

Gutekunst, C. A., Li, S. H., Yi, H., Mulroy, J. S., Kuemmerle, S., Jones, R., et al. (1999) Nuclear and neuropil aggregates in Huntington's disease, relationship to neuropathology. *J. Neurosci.* **19,** 2522–2534.

Hansson, O., Peters, N. A., Leist, M., Nicotera, P., Castilho, R. F., and Brundin, P. (1999) Transgenic mice expressing a Huntington's disease mutation are resistant to quinolinic acid–induced striatal excitotoxicity. *Proc. Natl. Acad. Sci. USA* **96,** 8727–8732.

Hedreen, J. C., Peyser, C. E., Folstein, S. E., and Ross, C. A. (1991) Neuronal loss in layers V and VI of cerebral cortex in Huntington's disease. *Neurosci. Lett.* **133,** 257–261.

Hedreen, J. C. and Folstein, S. E. (1995) Early loss of neostriatal striosome neurons in Huntington's disease. *J. Neuropathol. Exp. Neurol.* **54,** 105–120.

Hodgson, J. G., Agopyan, N., Gutekunst, C. A., Leavitt, B. R., LePiane, F., Singaraja, R., et al. (1999) A YAC mouse model for Huntington's disease with full-length mutant huntingtin, cytoplasmic toxicity, and selective striatal neurodegeneration. *Neuron* **23,** 181–192.

Huntington's Disease Collaborative Research Group (1993) A novel gene containing a trinucleotide repeat that is expanded and unstable in Huntington's disease chromosomes. *Cell* **72,** 971–983.

Hurlbert, M. S., Zhou, W., Wasmeier, C., Kaddis, F. G., Hutton, J. C., and Freed, C. R. (1999) Mice transgenic for an expanded CAG repeat in the Huntington's disease gene develop diabetes. *Diabetes* **48,** 649–651.

Kalchman, M. A., Koide, H. B., McCutcheon, K., Graham, R. K., Nichol, K., Nishiyama, K., et al. (1997) HIP1, a human homologue of S. cerevisiae Sla2p, interacts with membrane-associated huntingtin in the brain. *Nature Genet.* **16,** 44–53.

Kowall, N. W., Ferrante, R. J., and Martin, J. B. (1987) Patterns of cell loss in Huntington's Disease. *Trends Neurosci.* **10,** 24–29.

Kremer, H. P., Roos, R. A., Dingjan, G., Marani, E., and Bots, G. T. (1990) Atrophy of the hypothalamic lateral tuberal nucleus in Huntington's disease. *J. Neuropathol. Exp. Neurol.* **49,** 371–382.

Levine, M. S., Klapstein, G. J., Koppel, A., Gruen, E., Cepeda, C., Vargas, M. E., et al. (1999) Enhanced sensitivity to *N*–methyl–D–aspartate receptor activation in transgenic and knockin mouse models of Huntington's disease. *J. Neurosci. Res.* **58,** 515–532.

Li, H., Li, S. H., Cheng, A. L., Mangiarini, L., Bates, G. P., and Li, X. J. (1999) Ultrastructural localization and progressive formation of neuropil aggregates in Huntington's disease transgenic mice. *Hum. Mol. Genet.* **8,** 1227–1236.

Mangiarini, L., Sathasivam, K., Seller, M., Cozens, B., Harper, A., Hetherington, C., et al. (1996) Exon 1 of the HD gene with an expanded CAG repeat is sufficient to cause a progressive neurological phenotype in transgenic mice. *Cell* **87,** 493–506.

Menalled, L., Zanjani, H., MacKenzie, L., Koppel, A., Carpenter, E., Zeitlin, S., et al. (2000) Decrease in striatal enkephalin mRNA in mouse models of Huntington's disease. *Exp. Neurol.* **162,** 328–342.

Morris, M. and Scourfield, J. (1996) Psychiatric aspects of Huntington's Disease, in *Huntington's Disease* (Harper, P., ed.), The University Press. Cambridge.

Ona, V. O., Li, M., Vonsattel, J. P., Andrews, L. J., Khan, S. Q., Chung, W. M., et al. (1999) Inhibition of caspase-1 slows disease progression in a mouse model of Huntington's disease *Nature* **399,** 263–267.

Perutz, M. F. (1996) Glutamine repeats and inherited neurodegenerative diseases, molecular aspects. *Curr. Opin. Struct. Biol.* **6,** 848–858.

Petersén, A., Mani, K., and Brundin, P. (1999) Recent advances on the pathogenesis of Huntington's disease. *Exp. Neurol.* **157,** 1–18.

Portera-Cailliau, C., Hedreen, J. C., Price, D. L., and Koliatsos, V. E. (1995) Evidence for apoptotic cell death in Huntington disease and excitotoxic animal models. *J. Neurosci.* **15,** 3775–3787.

Quarell, O. and Harper, P. (1996) The clinical neurology of Huntington's disease, in *Huntington's Disease* (Harper, P., ed.), The University Press, Cambridge, pp. 31–72.

Reddy, P. H., Williams, M., Charles, V., Garrett, L., Pike-Buchanan, L., Whetsell, W. O. J., et al. (1998) Behavioural abnormalities and selective neuronal loss in HD transgenic mice expressing mutated full-length HD cDNA. *Nature Genet.* **20,** 198–202.

Reynolds, G. P., Dalton, C. F., Tillery, C. L., Mangiarini, L., Davies, S. W., and Bates, G. P. (1999) Brain neurotransmitter deficits in mice transgenic for the Huntington's disease mutation. *J. Neurochem.* **72,** 1773–1776.

Roizin, L., Stellar, S., and Liu, J. C. (1979) Neuronal nuclear-cytoplasmic changes in Huntington's chorea: electron microscope investigations, in *Huntington's Disease* Advances in Neurology Vol. 23, (Chase, T. S., Wexler, N. S., and Barbeau, A., eds.), Raven, New York, pp. 95–122.

Ross, C. A. (1995) When more is less, pathogenesis of glutamine repeat neurodegenerative diseases. *Neuron* **15,** 493–496.

Sathasivam, K., Hobbs, C., Turmaine, M., Mangiarini, L., Mahal, A., Bertaux, F., et al. (1999) Formation of polyglutamine inclusions in non-CNS tissue. *Hum. Mol. Genet.* **8,** 813–822.

Saudou, F., Finkbeiner, S., Devys, D., and Greenberg, M. E. (1998) Huntingtin acts in the nucleus to induce apoptosis but death does not correlate with the formation of intranuclear inclusions. *Cell* **95,** 55–66.

Sawa, A., Wiegand, G. W., Cooper, J., Margolis, R. L., Sharp, A. H., Lawler, J. F., Jr., et al. (1999) Increased apoptosis of Huntington disease lymphoblasts associated with repeat length–dependent mitochondrial depolarization. *Nat. Med.* **5,** 1194–1198.

Sánchez, I., Xu, C. J., Juo, P., Kakizaka, A., Blenis, J., and Yuan, J. (1999) Caspase-8 is required for cell death induced by expanded polyglutamine repeats. *Neuron* **22,** 623–633.

Schilling, G., Becher, M. W., Sharp, A. H., Jinnah, H. A., Duan, K., Kotzuk, J. A., et al. (1999) Intranuclear inclusions and neuritic aggregates in transgenic mice expressing a mutant N–terminal fragment of huntingtin. *Hum. Mol. Genet.* **8,** 397–407.

Shelbourne, P. F., Killeen, N., Hevner, R. F., Johnston, H. M., Tecott, L., Lewandoski, M., et al. (1999) A Huntington's disease CAG expansion at the murine Hdh locus is unstable and associated with behavioural abnormalities in mice. *Hum. Mol. Genet.* **8,** 763–774.

Thomas, L. B., Gates, D. J., Richfield, E. K., O'Brien, T. F., Schweitzer, J. B., and Steindler, D. A. (1995) DNA end labeling (TUNEL) in Huntington's disease and other neuropathological conditions. *Exp. Neurol.* **133,** 265–272.

Usdin, M. T., Shelbourne, P. F., Myers, R. M., and Madison, D. V. (1999) Impaired synaptic plasticity in mice carrying the Huntington's disease mutation. *Hum. Mol. Genet.* **8,** 839–846.

Velier, J., Kim, M., Schwarz, C., Kim, T. W., Sapp, E., Chase, K., et al. (1998) Wild-type and mutant huntingtins function in vesicle trafficking in the secretory and endocytic pathways. *Exp. Neurol.* **152,** 34–40.

Vonsattel, J. P., Myers, R. H., Stevens, T. J., Ferrante, R. J., Bird, E. D., and Richardson, E. P. J. (1985) Neuropathological classification of Huntington's disease. *J. Neuropathol. Exp. Neurol.* **44,** 559–577.

Weeks, R. A., Piccini, P., Harding, A. E., and Brooks, D. J. (1996) Striatal D1 and D2 dopamine receptor loss in asymptomatic mutation carriers of Huntington's disease. *Ann. Neurol.* **40,** 49–54.

Wellington, C. L., Ellerby, L. M., Hackam, A. S., Margolis, R. L., Trifiro, M. A., Singaraja, R., et al. (1998) Caspase cleavage of gene products associated with triplet expansion disorders generates truncated fragments containing the polyglutamine tract. *J. Biol. Chem.* **273,** 9158–9167.

Wexler, N. S., Young, A. B., Tanzi, R. E., Travers, H., Starosta-Rubinstein, S., Penney, J. B., et al. (1987) Homozygotes for Huntington's disease. *Nature* **326,** 194–197.

White, J. K., Auerbach, W., Duyao, M. P., Vonsattel, J. P., Gusella, J. F., Joyner, A. L., et al. (1997) Huntingtin is required for neurogenesis and is not impaired by the Huntington's disease CAG expansion. *Nature Genet.* **17,** 404–410.

Wu, Y., Zeitlin, S., Thomson, L., and Chesselet, M.–F. (1999) Processing of normal and mutated Huntingtin in human and mouse tissues. *Soc. Neurosci. Abst.* **25,** 588.

Zoghbi, H. Y. (1996) The expanding world of ataxins *Nature Genet.* **14,** 237–238.

15

Huntingtin-Associated Proteins

Marcy E. MacDonald, Lucius Passani,
and Paige Hilditch-Maguire

15.1. INTRODUCTION

Huntington's disease (HD), with its writhing dancelike movements (chorea) and cardinal loss of neurons in the striatum *(1)*, is the result of an unstable expanded CAG trinucleotide repeat that lengthens a variable glutamine tract in a novel protein called huntingtin *(HD) (2)*. HD shares elements of a common pathogenic mechanism with at least seven other inherited neurodegenerative diseases, including spinobulbar muscular atrophy (SBMA) *(3)*, dentato-rubral-pallidoluysian atrophy (DRPLA/Haw River syndrome) *(4–6)*, and several spinocerebellar ataxias (SCA1, SCA2, SCA3/MJD, SCA6 and SCA7) *(7–14)* (Fig. 1). Expanded glutamine segments in otherwise unrelated proteins cause specific neuronal cell loss in each case, suggesting unique protein context-dependent modulation of some intrinsic toxic property of polyglutamine *(15–17)*. In this view, some feature of huntingtin produces HD pathology by presenting the embedded toxic glutamine tract to cells in a manner that culminates in a graded loss of striatal neurons. One molecular possibility is a glutamine-induced conformational change that alters huntingtin's association with its normal or abnormal protein partners *(18,19)*.

Exploration of huntingtin-associated proteins offers a promising route to unraveling the HD pathogenic process, although the inherent activities of this large protein are not known and, *a priori*, it is difficult to predict the qualitative and/or quantitative features of a putative "toxic" interaction. However, these efforts can be guided by genetic correlates, from studies of HD patients and their families, which provide an essential framework for evaluating the role of huntingtin's potential partners in HD pathogenesis *(17)*.

Before cloning of the *HD* gene, biochemical models of HD pathology, such as *N*-methyl-D-asparate (NMDA) receptor-mediated glutamate excitotoxicity, free-radical damage, and deficits in mitochondrial energy

From: *Contemporary Clinical Neuroscience: Molecular Mechanisms of Neurodegenerative Diseases*
Edited by: M.-F. Chesselet © Humana Press Inc., Totowa, NJ

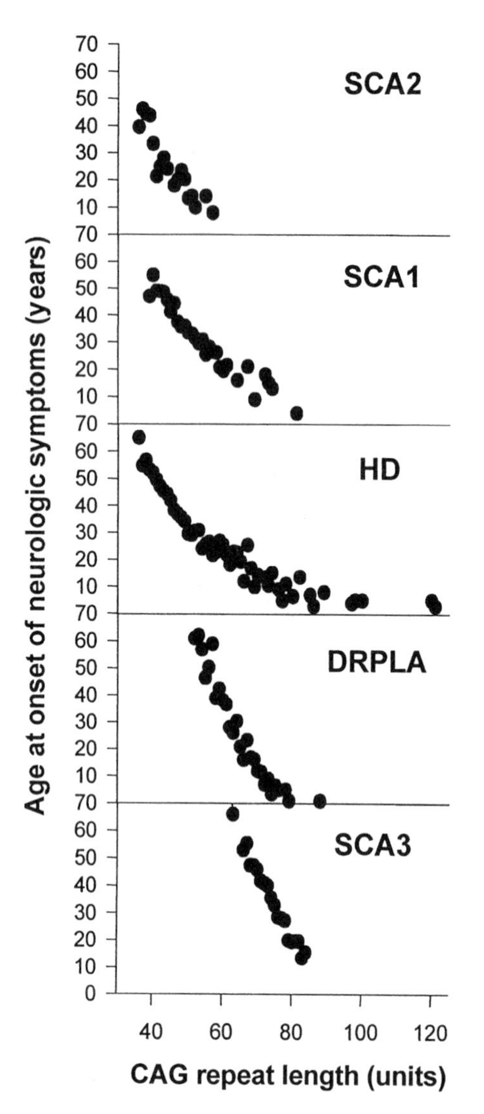

Fig. 1. Relationship of CAG repeat length to age at onset of neurologic symptoms in HD and other inherited polyglutamine diseases. The mean number of CAG repeats on the disease chromosome, taken from published data, is plotted against the age at onset of neurologic symptoms for each disease (filled circles) (SCA2, n = 92 cases; SCA1, n = 201; HD, n = 1226 cases; DRPLA, n = 149 cases; MJD/SCA3, n = 332 cases) *(17)*. HD occurs with more than approx 37 *HD* CAG repeats and above this threshold the age at onset of neurologic symptoms is inversely correlated with increasing CAG repeat size. The curves for SCA1 is remarkably similar, given that this disease affects Purkinje cells and is caused by lengthened glutamine segments in an unrelated protein (HD, huntingtin; SCA1, ataxin-1). When the expanded glutamine tract is embedded in any of the other unrelated protein contexts (DRPLA, atrophin; SCA2, ataxin-2; SCA3/MJD, ataxin-3), the threshold for disease symptoms and impact of each additional glutamine residue is distinct.

production, were derived from chemical lesion studies in experimental animals *(20–24)*. Identification of potential huntingtin interactors, however, has not provided direct links to any of these previously proposed models of neurodegeneration *(19)*. Instead, huntingtin's partners implicate it in a variety of cellular processes, rather than a single biochemical pathway, and suggest the involvement of novel cellular pathways in HD pathogenesis.

15.2. GENETIC CORRELATES OF HD

The *HD* CAG expansion lengthens a glutamine tract 17 residues from the amino-terminus of huntingtin, a novel approx 350-kDa protein of unknown function *(2)* that is required for embryonic development and neurogenesis *(25–28)*. The *HD* CAG repeat is normally polymorphic, varying from about 6 to 34 units, and is inherited in a Mendelian fashion *(2)*. By contrast, expanded *HD* CAG repeats, approx 37 to more than 100 units, cause HD with its progressively worsening symptoms; hallmark choreiform movements, psychiatric impairment, and cognitive decline *(2,29)*. The underlying pathology features a progressive loss of select neurons in the basal ganglia and cerebral cortex, with a characteristic gradient involving medium spiny neurons in the caudate nucleus that forms the neuropathologic grading of the HD postmortem brain *(30,31)*. The relatively rare *HD* CAG arrays that span these ranges, approx 36–39 CAG units, are found both in HD cases and in elderly unaffected individuals, revealing a reduced penetrance that implicates other factors in the precise timing of disease onset *(32,33)*.

Typically, expanded CAGs in the 40- to 50-unit range cause subtle signs of disease onset in mid-life (approx 40 yr) that worsens within approx 15 yr to incapacitating chorea and death *(29)*. Occasional large CAG expansions (>55 repeats), resulting from instability in spermatogenesis *(34)*, cause juvenile-onset HD, which features rigidity and a more severe clinical course that claims children and young adults. Indeed, genotype–phenotype studies demonstrate that *HD* CAG repeat size is inversely correlated with age at onset of clinical symptoms (Fig. 1) and age at death, but not with disease progression *(29)*. Moreover, the degree of striatal pathology in the postmortem brain is CAG length dependent with a correlation that implies that neuronal cell loss begins early in life, perhaps from birth *(35)*.

Genotype-phenotype studies also reveal that all confirmed cases of HD are the result of an *HD* CAG-repeat expansion, suggesting that HD's peculiar neuropathology involves a novel pathogenic property that can only be conferred by this mutational mechanism *(36)*. Elimination of one copy's worth of huntingtin, by translocation, is not associated with disease *(37)* and rare HD homozygotes with two mutant HD alleles are indistinguishable from typical HD individuals with their single *HD* CAG-repeat expansion *(38,39)*. The expanded *HD* CAG repeat, therefore, is likely to act by endowing a new property on the *HD* gene product, rather than by simple inactivation of its inherent activities *(28,37,40)*.

15.3. HUNTINGTIN

Huntingtin is an approx 350-kDa protein with a variable amino-terminal glutamine tract that is encoded by the polymorphic *HD* CAG repeat *(2,41,42)*. This large protein is highly conserved throughout evolution over its entire length, except for the amino-terminal glutamine–proline-rich segment *(43–47)* (Fig. 2), suggesting that the site of the HD mutation in man may participate in a species-specific manner or, alternatively, may be dispensable for huntingtin's activities. However, the striking identity of the initial 17 amino acids and residues immediately adjacent to the variable segment implies an important biological function for huntingtin's extreme amino-terminus (Fig. 2) *(40)*.

Despite its large size, huntingtin shares limited sequence similarity to reported proteins (Fig. 3) with the exception of 10 HEAT repeats, loosely conserved motifs found a number of otherwise unrelated proteins (Huntingtin, Elongation factor 3, A subunit of protein phosphatase 2A, TOR1) *(48)*. These motifs can form a flexible bipartite α-helical structure with intervening 'loops' that may mediate specific protein–protein interactions, including those involved in nuclear import *(48–50)*. Huntingtin also possesses a putative leucine zipper protein-association domain typically found in proteins that participate in transcription complexes *(2,51)*.

Huntingtin is broadly expressed in a variety of peripheral tissues and in the brain, throughout development and in the adult *(15,16)*. In most cells, including neurons, huntingtin is largely a soluble cytoplasmic protein *(52–57)*. However, a portion of the protein decorates microtubules and "vesicles" *(52–57)* and is loosely associated with membrane-fractions where it partially colocalizes with "markers" of endocytic and secretory vesicles *(54,55,58)*. In addition, a small but significant fraction of huntingtin (approx 5%) is found in the nucleus of diverse cell types, suggesting a function in this cellular compartment *(59,60)*. Indeed, huntingtin's unique subcellular distribution is consistent with a variety of cellular activities, although its potential role in microtubule-mediated vesicle transport, endocytosis, and/or secretory processes is emphasized *(54,55,58)*.

Elimination of huntingtin, by targeted disruption of the mouse's *HD* gene *(Hdh)*, causes embryonic lethality at gastrulation (25–27), resulting at least in part from defects in the extraembryonic tissue that provides nutritive

```
Human   MATLEKLMKAFESLKSFQQQQQQQQQQQQQQQQQQQQQPPPPPPPPPPP
Mouse   MATLEKLMKAFESLKSFQQQQQQQ                         PPPQPPPPPPPPPPP
Rat     MATLEKLMKAFESLKSFQQQQQQQQ                        PPPQPPPPPPPPPP
Fugu    MATMEKLMKAFESLKSFQQQQ

Human   QLPQPPPQAQPLLPQPQPPPPPPPPPPGPAVAEEPLHRP-KKELSATKKDRVN
Mouse   QPPQPPPQGQ.       PPPPPPPLPGPA   EEPLHRP-KKELSATKKDRVN
Rat     QPPQPPPQGQ.        PPPPPPLPGPA   EEPLHRP-KKELSATKKDRVN
Fugu                             GPPTAEEIVQRQ-KKEQATTKKDRVS
```

Fig. 2. Evolutionary conservation of huntingtin's amino-terminus. The amino-terminus of huntingtin and its homologs from mouse, rat, and pufferfish (fugu) are aligned to indicate regions of similarity and divergence. The first 88 amino acids of human huntingtin are encoded by *HD* exon 1 (before the dash) and include the variable glutamine tract, the site of the *HD* mutation in man, and the proline-rich segment that mediates binding of huntingtin interactors with SH3 and WW domains. Despite striking identity of the flanking residues, the glutamine/proline-rich segment is not conserved through evolution.

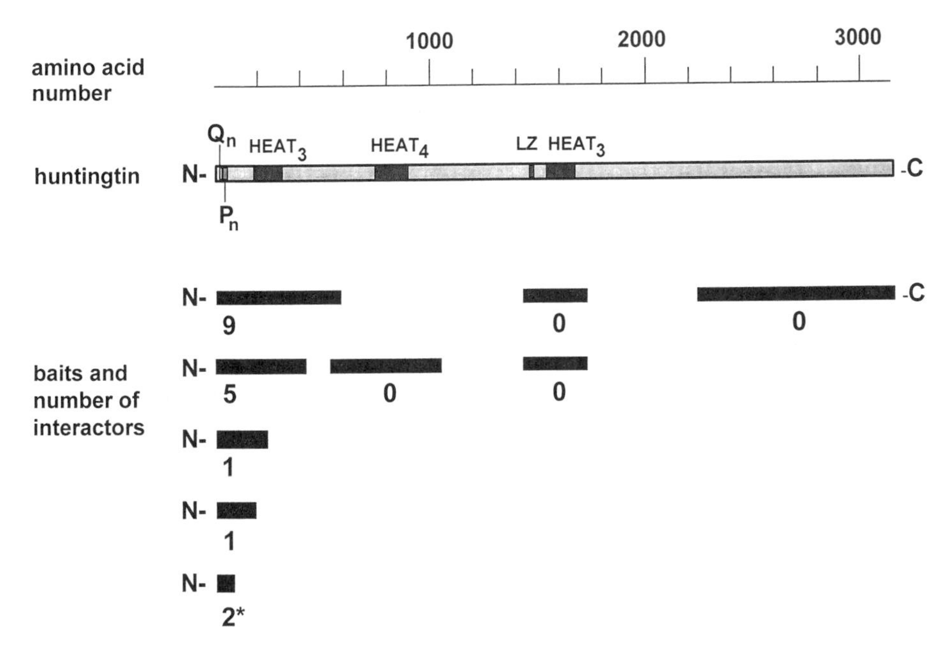

Fig. 3. Summary of huntingtin yeast two hybrid interaction screens. The full-length huntingtin protein (3,144 amino acids) is drawn to illustrate the locations of the variable amino-terminal glutamine tract (Q_n), adjacent proline-rich region (P_n), the single leucine zipper motif (LZ), and the clustered HEAT motifs (three or four repeats, as indicated). Huntingtin fragments used as baits in yeast two hybrid screens are depicted as filled bars with the number of independent interactors identified shown below. The asterisk denotes that an interactor was also isolated independently with a longer bait fragment.

support for the developing embryo *(61)*. Severely reduced huntingtin (between 0% and 50% of wild-type levels) is sufficient to overcome the gastrulation defect but is instead associated with perinatal lethality and abnormal brain development *(28)*. Its essential role in the developing organism also suggests that huntingtin may be needed in cells of the adult, including differentiated neurons. In cell culture, however, huntingtin is not required for the growth of totipotent embryonic stem (ES) cells or for their 'neuronal' differentiation *(62)*.

15.4. A PLETHORA OF HUNTINGTIN PARTNERS

A multiplicity of proteins (23 in total) can bind full-length huntingtin, or an amino-terminal fragment, providing possible clues to huntingtin's essential developmental activities and starting points for unraveling potentially novel cellular pathways (Table 1) *(18,19)*. One unidentified huntingtin-associated protein was deduced from purification experiments, whereas four candidate interactors derive from huntingtin's surmized cellular activities. The vast majority of putative huntingtin-associated proteins, however, were discovered because they bind an amino-terminal huntingtin fragment in yeast two hybrid–protein interaction trap assays (Fig. 3) *(19)*. This method, which does not presuppose function *(106)*, yielded genes encoding 18 different proteins, including 13 with previously reported activities that hint at huntingtin's participation in a variety of cellular processes, rather than a single biochemical pathway *(19)*.

15.5. PARTNERS FROM THE CANDIDATE APPROACH

Huntingtin–Calmodulin Binding Protein

In the presence of Ca^{2+}, highly purified native rat brain huntingtin is found in a large 700-kDa to 1-MDa calmodulin-containing complex that partially dissociates in the presence of chelating agents to yield an approx 500-kDa product in nondenaturing conditions *(63)*. Huntingtin indirectly binds calmodulin–Sepharose in a Ca^{2+}-dependent manner via an unidentified calmodulin-associated protein *(63)*, implicating huntingtin in a variety of Ca^{2+}-regulated signaling pathways in the developing embryo and adult nervous system, including those that regulate cell surface signaling cascades *(64)*.

GAPDH

Huntingtin is reported to bind to glyceraldehyde 3-phosphate dehydrogenase (GAPDH)-affinity columns, perhaps via its amino-terminal glutamine segment *(65)*. In addition to glycolysis and energy metabolism, this potential association implicates huntingtin in a large number of other cellular

Table 1
Summary of Huntingtin's Potential Partners

Huntingtin interactor	Method of identification	Identity	Suggested functions	Ref.
Calmodulin-binding protein	Purification	Unknown	Ca^{2+}-regulated processes?	*63,64*
GAPDH	Candidate	Glycolytic enzyme	Glycolysis, endocytosis, DNA repair, and more	*65,66*
Polymerized tubulin	Candidate	Microtubule protein	Cytoskeleton, transport, and more	*67*
Grb2	Candidate	Adaptor protein	Receptor-mediated signaling, endocytosis	*68,69*
RasGAP	Candidate	GTPase-activating protein	Receptor-mediated signaling cascades	*68*
HAP1	Yeast two hybrid	Novel; binds $P150^{Glued}$	Transport, cytoskeleton, signal transduction?	*70–76*
HIP1	Yeast two hybrid	Yeast Sla2p	Cytoskeleton, H^+-ATPase regulation, endocytosis	*77–82*
HIP2	Yeast two hybrid	E2-25kDa ubiquitin-conjugating enzyme	Protein catabolism	*83–85*
SH3GL3	Yeast two hybrid	Mouse Sh3d2c, rat SH3p13	Endocytosis, signaling	*86–88*
HYPA	Yeast two hybrid	Mouse FBP-11; WW domains; FF motif	Spliceosome	*89–94*
HYPB	Yeast two hybrid	*Drosophila* WW domain SET domain CG1716	Transcription	*89,93*
HYPC	Yeast two hybrid	FBP-11 related; WW domains; FF motif	Spliceosome	*89–94*
HYPD	Yeast two hybrid	MAGE-3	Tumor antigen	*89,95,96*

(continued)

Table 1 *(continued)*

Huntingtin interactor	Method of identification	Identity	Suggested functions	Ref.
HYPE	Yeast two hybrid	Novel	Unknown	*89*
HYPF	Yeast two hybrid	26S proteasome subunit	Protein catabolism	*89,97*
HYPJ	Yeast two hybrid	α-adaptin C	Endocytosis	*89,98–100*
HYPH	Yeast two hybrid	Yeast Akr1p	Endocytosis	*89,129*
HYPI	Yeast two hybrid	Symplekin	Tight junction plaque polyadenylation	*89,101,130*
HYPK	Yeast two hybrid	Novel	Unknown	*89*
HYPL	Yeast two hybrid	FIP-2	Apoptotic death-regulator	*89,102,103*
HYPM	Yeast two hybrid	Novel	Unknown	*89*
CBS	Yeast two hybrid	Cystathionine β-synthase	Catalyze formation of Cystathionine from homocysteine	*104,105*
N-COR	Yeast two hybrid	Nuclear receptor co-repressor	Transcription	*131*

processes that may involve GAPDH, including DNA repair and replication, and endocytosis *(66)*.

Polymerized Tubulin

Based on its hypothesized role in retrograde neuronal cell transport, huntingtin was found to bind to polymerized tubulin, but not tubulin-affinity columns, and to copurify with microtubules in successive polymerization–depolymerization cycles in in vitro experiments *(67)*. The interaction was not mediated by tested tubulin-binding proteins (GAPDH or MAP-2), although binding to polymerized tubulin in vivo may involve other microtubule-associated proteins, consistent with its proposed role in microtubule-mediated organelle transport.

SH3 Domain Proteins (Grb2 and RasGAP)

Huntingtin's amino-terminal proline segment is a likely target for Src homology 3 (SH3) domain-containing proteins, as this motif mediates binding to proline-rich ligands *(107,108)*. Two SH3 domain proteins involved in epidermal growth factor (EGF) receptor signaling, Grb2 and RasGAP, bind to huntingtin via their SH3 domains, suggesting huntingtin's participation in an EGF-regulated signaling cascade *(68)*. SH3-mediated interactions of Grb2 also drive clathrin-mediated endocytosis of the EGF receptor *(69)*, consistent with huntingtin's proposed role in endocytosis and membrane trafficking.

15.6. YEAST TWO HYBRID PARTNERS

HAP1

Rat HAP1 (huntingtin-associated protein 1), the first huntingtin yeast partner identified, is a novel protein with two isoforms, called HAP1-A and HAP1-B *(70)*, that binds huntingtin's amino-terminus, carboxy-terminal to the glutamine–proline-rich segment (residues 171–230) *(71)*. Human HAP1 (hHAP1) is a single approx 75-kDa protein found predominantly in the brain, particularly the subthalamic nucleus, amygdala, thalamus and substantia nigra, that in neurons is associated with membranous organelles in the cytoplasm *(72)*.

Rat HAP1 interacts with the $P150^{Glued}$ component of the cytoplasmic dynactin complex *(76,77)*, a protein that anchors this major microtubule-dependent transport motor to the membrane via its association with the intermediate chain of cytoplasmic dynein *(75,76)*. Rat HAP1 also binds Duo, a novel human Trio-like protein that possesses a (possibly) rac 1-specific guanine nucleotide exchange factor motif, a pleckstrin homology domain and spectrin-like repeats *(109)*. Potential HAP1-associated proteins may implicate HAP1, and indirectly huntingtin, in intracellular protein trafficking, cytoskeletal function, and signal transduction.

HIP1

HIP1 (huntingtin-interacting protein 1), the human homolog of *S. cerevisiae* Sla2p, binds truncated amino terminal huntingtin fragment, comprising the first 540 residues *(77)* or the short 88-amino-acid segment encoded by *HD* exon 1 *(78)*. Human HIP1 is enriched in the brain and is reported variously as an approx 100-kDa doublet *(77)* or a single approx 116-kDa protein *(78)* that, in neurons, is found in a punctate staining pattern at the periphery of the cytoplasm *(77)*. Yeast Sla2p is a transmembrane protein with a talin-like domain that is involved in cytoskeleton assembly, in the regulation of plasma membrane H+-ATPase abundance and in endocytosis *(79–82)*. The intracellular location of human HIP1 suggests similar activities in higher organisms, implicating huntingtin in plasma membrane-associated processes.

A chronic myelomonocytic leukemia (CML) translocation breakpoint t(5:7)(q33; q11.2) occurs with the *HIP1* gene (7q11.2), generating a *HIP1*/platelet-derived growth factor (PDGF)β receptor fusion protein capable of transforming cells in culture *(110)*. Binding of huntingtin to this approx 180-kDa CML fusion protein, although not tested, would suggest a role for huntingtin and HIP1 in hematopoietic malignancies *(110)*.

HIP2

Another huntingtin amino terminal (residues 1–540) interacting protein HIP2 (huntingtin-interacting protein 2), HIP2, is the E2-25kD ubiquitin-conjugating enzyme *(83)*. E2-25kD is a predominantly cytoplasmic 28-kDa protein that catalyzes the attachment of ubiquitin, a 76-amino-acid protein, to lysine residues of other proteins. Ubiquitination targets proteins for ATP-dependent degradation by the 26S proteasome, regulating levels of key cell cycle and signaling proteins and removing damaged molecules *(84)*. It also marks proteins for a variety of other fates, including altered intracellular compartmentalization *(85)*. E2-25kDa is expressed in peripheral tissues and, at higher levels, in the brain. Ubiquitinated huntingtin is detected in cell extracts *(83)*, consistent with its ability to interact with E2-25kDa and/or other ubiquitin-conjugating enzymes.

SH3GL3

The approx 41-kDa SH3GL3 protein associates with a short huntingtin amino-terminal fragment (residues 1–88) *(86)* and is a member of a family of SH3 domain-containing proteins found predominantly in the brain *(69,87)*. SH3GL3 binds huntingtin's proline-rich segment via its Grb2-like SH3 domain, consistent with the SH3-mediated interaction of Grb2 and huntingtin *(68)*. SH3GL3, and its mouse (Sh3d2c) and rat (SH3p13) counterparts, are enriched in the brain, testis, and thymus and are found in soluble synaptic fractions *(86,88)* but can be located in the nucleus of transfected cells *(86)*. SH3p13 binds both synaptojanin, a nerve-terminal inositol 5-phosphatase, and GTPase dynamin I. These proline-rich ligands in turn associate with amphiphysin, an essential component of endocytic reactions required for the recycling of clathrin-coated synaptic vesicles *(69,87)*. SH3GL3 binding, in brain extracts and in the yeast two hybrid system, therefore, provides indirect support for huntingtin's hypothesized role in endocytosis *(86)*.

WW Domain Proteins (HYPA, HYPB, HYPC)

The major class of huntingtin-interacting proteins comprises 3 WW domain proteins, HYPA, HYPB, and HYPC (huntingtin yeast partners) that bind huntingtin via its amino-terminal proline-rich segment *(89)*. HYPA is the human homolog of mouse formin-binding protein 11 (FBP11), an approx 130-kDa protein that associates with the SF1/mBBP component of the spliceosome complex *(90,91)*. HYPC shares extensive amino acid sequence identity with HYPA and, like this protein, is a mammalian homolog of *S. cerevisiae* splicing factor Prp40 *(89,92)*. HYPB is a SET domain transcription factor, homolog of *Drosophila* CG1716, with a WW domain that most closely resembles the tandem motifs found in HYPA and HYPC *(89)*. HYPA, HYPB, and HYPC are widely expressed in the brain and periphery and are detected in the nucleus and cytoplasm of neuronal cells *(89,93)*.

The WW domain, named for its critically spaced tryptophan residues, is a protein interaction motif that binds proline-rich recognition sites in specific protein ligands *(94)*. The tandem WW domains of HYPA and HYPC are accompanied by an FF motif (for two strictly conserved phenylalanine residues), forming a domain combination found in components of pre-mRNA splicing machinery and in the p190 family of GTPases important for organization of the cytoskeleton *(111)*. This set of related proteins suggest huntingtin's involvement in pre-mRNA splicing, transcription, and activities involving ribonucleoprotein particles in the cytoplasmic and nuclear compartments *(89–94,111)*.

HYPD

Huntingtin yeast partner HYPD is MAGE 3, a member of a large family of melanoma antigen proteins that are expressed in diverse tumor cells *(89)*. The physiologic function of this protein is unclear, although proteolytic degradation via the proteasome is required for its presentation at the surface of the tumor cells *(95,96)*. Association with MAGE 3, therefore, may indicate a role for huntingtin in antigen presentation at the level of the proteasome complex or the processes that mediate insertion of the peptide into the plasma membrane *(89)*.

HYPF

Another huntingtin yeast partner, HYPF, is the P31 subunit of the 26S proteasome complex *(89)* that mediates ATP-dependent degradation of ubiquitinated proteins *(97)*. Like HIP2/E2-25Ka and potentially HYPD/MAGE, HYPF suggests huntingtin's association with the cellular machinery that regulates the levels of either damaged or short-lived proteins required for cell-signaling cascades and cell–cell communication.

HYPH

Data base searches reveal that the novel amino terminal partner with ankyrin repeats, called HYPH *(89)*, is the human homolog of the yeast ankyrin repeat protein Akr1p, involved in endocytosis *(128)*. Loss of Akr1p blocks ubiquitination of the yeast alpha factor receptor, essential for rapid endocytosis and degradation of this pheromone mating factor, apparently due to abnormal lipid modification of other essential proteins, Yckp1 and Yckp2. HYPH/Akrp1, therefore, implies a role for huntingtin in regulating the endocytosis of specific cell surface proteins in signaling pathways that require rapid internalization and degradation.

HYPI

Symplekin (HYPI) binds huntingtin's amino-terminus (residues 1–540) *(89)*, and is a 150-kDa nuclear protein that, in cells forming stable cell–cell contacts, is recruited to the cytoplasmic side of the zonula occludens as a plasma membrane tight junction plaque protein *(101)*, recently shown to participate in nuclear polyadenylation complexes *(130)*. Other "dual" residence (plasma membrane/nucleus) junctional plaque proteins, including plakophilins, β-catenin, and plakoglobin, are involved in a myriad of cellular responses *(112)*, suggesting huntingtin's participation in junction structure formation, polyadenylation, and signaling complexes *(89)*.

HYPJ

Another huntingtin amino-terminal yeast partner (HYPJ) is α-adaptin C, a protein involved in endocytosis/membrane recycling *(89)*. This approx 104-kDa protein participates in an heterotetramer, dubbed assembly protein complex 2 (AP2) that is located at the plasma membrane on the cytoplasmic face of clathrin-coated pits *(98,100)*. It is widely expressed, and in neurons, it is found both at terminals and diffusely in cell bodies and dendrites *(98)*.

Components of the AP2 complex interact with amphiphysin, a molecule thought to target dynamin and SH3 proteins, such as synaptojanin *(69)*, to sites of synaptic vesicle endocytosis. α-Adaptin C implicates huntingtin in vesicle transport and endocytosis *(89)*.

HYPL

Huntingtin amino-terminal yeast partner HYPL *(89)* is human FIP-2, named for 14.7K-interacting protein *(102)*. This novel leucine zipper protein binds an adenovirus early region encoded protein, E3 14.7 kDa, to inhibit tumor necrosis factor-α (TNF-α)-induced cell death *(103)*. FIP-2 is in the cytoplasm, located near the nuclear membrane, where it prevents E3 14.7 kDa protein's reversal of TNF-α and Fas-mediated apopototic death in transfected cells *(102)*. Binding to FIP-2 provides a tantalizing unexplored link between huntingtin and cell death pathways.

CBS

Cystathionine β-synthase (CBS), the tetrameric enzyme that catalyzes the L-serine dependent formation of cystathionine from homocysteine, binds to a short huntingtin amino-terminal fragment (residues 1–171) *(104)*. CBS has multiple isoforms and is expressed throughout the brain and periphery *(105)*. This enzyme is deficient in homocystinuria patients who accumulate homocysteine and its potent excitotoxic amino acid metabolites. The association with CBS demonstrates an interaction with a protein possessing enzymatic activity and suggests huntingtin's involvement in homocystinuria while raising the possibility of CBS-mediated neuronal cell toxicity in HD *(104)*.

N-CoR

Another amino terminal partner is N-CoR, a factor known to repress transcription from ligand activated receptors, including the retinoid X-thyroid hormone receptor and Mad-Max receptor dimers *(130)*. The discovery of N-CoR's ability to bind huntingtin's amino terminus augments the category of interactors that point to a role for huntingtin in activities involving associations with large RNA-protein complexes in the cytoplasm and the nucleus.

Novel Proteins (HYPE, HYPK, HYPM)

Five genes encoding novel huntingtin amino-terminal interactor proteins (HYPE, HYPK, HYPM) also emerged from yeast two hybrid interaction trap experiments *(89)*. The amino acid sequences of each of these partners do not contain significant homologies to previously reported proteins or recognized structural motifs, precluding insights into their physiologic activities. Nevertheless, further investigation of these huntingtin interacting proteins may ultimately provide clues to huntingtin's normal or abnormal function(s).

15.7. HUNTINGTIN'S NORMAL ACTIVITIES?

The multitude of putative interacting proteins with reported cellular activities does not highlight any single pathway but rather suggests huntingtin's participation in a variety of processes. Some partners (and the partners of the partners) support huntingtin's hypothesized roles in endocytosis, intracellular trafficking, and membrane recycling (Grb2 and RasGAP SH3 proteins, tubulin, HAP1's partner P150^{Glued}, HIP1/Sla2p, SH3GL3's partner synaptojanin, HYPJ/α-adaptin C HYPH/Akrp1). Others implicate huntingtin in cytoskeletal function and signal transduction cascades (calmodulin–huntingtin-binding protein, HIP1/Sla2p, HYPD/MAGE, HYPH/Akrp1, HYPL/FIP2, Grb2 and RasGAP, HYPI/symplekin). Some partners participate in protein catabolism, providing links to processes that determine intracellular compartmentalization of critical signaling molecules (HIP2/E2-25k ubiquitin conjugating enzyme, HYPF/P31 subunit of the 26S proteasome HYPH/Akrp1). The WW domain proteins imply a role for huntingtin in spliceosome (HYPA/FBP-11, HYPC), transcription (HYPB, N-CoR) and polyadenylation (HYPI/symplekin) complexes in the cytoplasm and in the nucleus. Still others imply huntingtin's participation in cell–cell communication via proteins that may shuttle between the nucleus and the plasma membrane (HIP1/Sla2p, HYPI/symplekin, HYPL/FIP2), whereas some suggest a role in a specific enzymatic pathway (CBS, GAPDH).

Interference in any of these cellular processes could conceivably explain the embryonic lethality and abnormal brain development produced by mutant *Hdh* alleles that eliminate or severely reduce huntingtin's expression, respectively. Delineation of which of the interactions, and ensuing cellular processes, is relevant to huntingtin's biochemical activities will require a detailed comparison of huntingtin-deficient and normal cells. As huntingtin expression is not limited to neurons, its potentially diverse activities can be explored in a variety of cell types.

All 18 huntingtin yeast partners associate with an amino-terminal huntingtin fragment (ranging from 1–88 to 1–588 amino acids). Several interactors bind huntingtin's proline-rich stretch via their SH3 or WW domain protein–protein interaction motifs (Grb2, RasGAP, SH3GL3, HYPA/FBP-11, HYPB, HYPC), whereas sequences mediating binding to the majority of the partners have not been identified. Indeed, the large number of amino-terminal yeast partners strongly argues that this region of huntingtin participates in protein–protein interactions in vivo. Yeast two hybrid experiments did not, however, identify proteins that interact with segments comprising more than 80% of huntingtin's length, despite targeted analysis of the most conserved regions, including the HEAT repeats *(89)* (Fig. 3). Nevertheless, it is likely that other experimental strategies will identify proteins that bind more carboxy-terminal regions of huntingtin, perhaps revealing additional clues to the activities of this large protein.

15.8. HUNTINGTIN'S PATHOGENIC ACTIVITIES?

The lengthened amino-terminal glutamine tract in disease-causing versions of huntingtin, encoded by expanded *HD* CAG repeats, dramatically alters the protein's physical properties but does not impair its essential developmental function(s) (28). It unexpectedly decreases huntingtin's mobility on sodium dodecyl sulfate–polyacrylamide gel electrophoresis (SDS-PAGE) *(42)* and alters reactivity with specific monoclonal antibody reagents *(113–115)*. Lengthened glutamine segments also confer on truncated amino-terminal huntingtin fragments the capacity to aggregate in vitro, producing insoluble homotypic polymers *(115,116)*. In HD brain, the altered physical properties lead to the formation of abnormal morphologic amino-terminal–huntingtin deposits in the neuropil and neuronal nuclei (so-called intranuclear inclusions) *(117–121)*, as well as the production of insoluble amino terminal aggregate with the properties of amyloid *(115)*. In vitro experiments demonstrate that the property that compels the formation of insoluble huntingtin amino-terminal aggregate conforms to the criteria of threshold (approx 39 glutamines), glutamine length dependence, and doseage that are predicted for the HD pathogenic mechanism from genetic studies *(115)*. A glutamine-induced physical change in huntingtin, such as that which promotes aggregate formation, is, therefore, a likely pathogenic trigger.

It has been proposed that polyglutamine's ''toxic'' property acts via an amino terminal mutant huntingtin fragment *(16,116,122)*, with insoluble amino-terminal aggregate (amyloid) causing neuronal cell death *(116)*. However, the abnormal aggregate in HD brain may be either a cause or a consequence of the pathogenic process *(16)*, and other scenarios for the events that trigger HD toxicity are possible. Indeed, the inclusions and insoluble aggregate detected in HD brain are likely to be diverse, featuring different components depending on whether they are generated from an amino-terminal fragment or full-length protein *(114)*. In cell culture, the occasional complexes formed by mutant huntingtin exhibit soluble glutamine tracts, whereas the majority of the many amino-terminal-fragment-generated complexes possess insoluble glutamine segments *(113,123)*. Precise protein context, therefore, determines the conformation of the embedded expanded glutamine tract and impacts huntingtin's self-association properties as well its interactions with other proteins that are its normal or abnormal partners.

One alternative possibility, therefore, is that the same glutamine-induced conformational change that promotes self-aggregation, in the context of the entire huntingtin protein (rather than a truncated fragment), causes an aberrant interaction with another protein, triggering a cascade of events that culminates both in aggregate formation and specific neuronal cell death *(113)*. In this scenario, huntingtin's normal and abnormal protein partners are integral participants in the pathogenic trigger and are expected to mediate the peculiar sensitivity of medium-sized (spiny) striatal neurons in HD, although they may, but need not, be detected in amino-terminal aggregate.

No protein that interacts exclusively with mutant huntingtin has been identified, suggesting that if HD pathogenesis involves such a partner, it remains to be discovered *(89)*. However, a case can be made for any one of the "known" huntingtin-interacting proteins, as each is expressed in the brain and each possesses a physiologic activity that if perturbed could lead to neuronal cell death. Moreover, the vast majority of partners (Grb2 and RasGAP have not been assessed) also bind versions of huntingtin (or the amino terminus) with a lengthened disease-causing glutamine tract. In some cases, increased glutamine number subtly enhances binding (calmodulin–huntingtin-binding protein, GAPDH, HAP1, SH3GL3, HYPA, HYPB, HYPC) *(64,66,72,86,89,93,130)*, whereas, for other partners, the change mildly reduces the interaction (HIP1) *(77,78)* or fails to alter the association (polymerized tubulin, HIP2) *(67,83)*.

Many of huntingtin's putative binding proteins can be implicated in HD but are any likely to play a role in triggering pathogenesis? Genetic correlates specify a pathogenic process that increases in severity with increased glutamine number, above about 37 residues, but these criteria do not reveal whether a qualitative or a quantitative alteration (or the direction or magnitude of a quantitative change) is involved. Thus, a subtle alteration that increases or decreases the strength of an association with one (or more) of huntingtin's protein partners may explain the progressive nature and late onset of the disorder. On the other hand, none of huntingtin's potential partners has been rigorously tested as a candidate in HD. The threshold and glutamine-length dependence of the interaction with huntingtin has not been assessed to determine whether it conforms to the same criteria as the HD pathogenic mechanism and there is no direct evidence for the impact of any given gene product on the onset or progression of disease.

15.9. PROSPECTS

Deciphering the critical pathogenic events that are set in motion by the addition of a few more residues to a normally variable glutamine segment in a novel protein presents a formidable challenge. The HD pathogenic trigger is as subtle as it is specific, causing symptoms after decades and wreaking havoc on select striatal neurons. The pathogenic mechanism is shared by seven other inherited "polyglutamine neurodegenerative" diseases but is at the same time unique, causing symptoms in a glutamine-length-dependent fashion but only above a threshold of approx 39 residues in huntingtin *(17)*.

The availability of genetic HD mouse models, expressing the entire endogenous mutant huntingtin protein, provides an opportunity to determine whether any of huntingtin's putative partners is likely to be involved in abnormal phenotypes *(125–128)*. These animal models will also spur the identification of additional huntingtin-interacting proteins, perhaps revealing some that associate exclusively with mutant protein, and will lead to the discovery of genes whose regulation is affected by the mutant protein, regardless of whether they encode a protein or some other cellular constituent. Moreover, knowledge of the genes encoding many of huntingtin's potential partners can be used in genotype–phenotype studies in HD to determine whether any account for the variation in age at onset that is associated with a given *HD* CAG repeat expansion.

In the absence of a cellular component that is exclusive to mutant huntingtin, elucidation of huntingtin's inherent functions promises to shed light on its abnormal, ultimately toxic, activity. A multitude of potential huntingtin partners can now be explored to determine whether they account for huntingtin's essential developmental activities in gastrulation and neurogenesis. Knowledge of these physiological functions may augment the list of previously proposed neurodegenerative processes with novel pathways that are peculiar to HD, providing new opportunities for the development of rational therapeutic interventions for this tragic disease.

ACKNOWLEDGMENTS

The authors thank members of the laboratory for stimulating discussion and their contributions to HD research. L.P. receives fellowship support from the Hereditary Disease Foundation. The authors' research is supported by NIH grants NS16367 and NS32765, the Huntington's Disease Society of America, the Hereditary Disease Foundation, and the Foundation for the Care and Cure of Huntington's Disease.

REFERENCES

1. Folstein, S. (1989) *Huntington's Disease: A Disorder of Families.* The Johns Hopkins Press, Baltimore, pp. 13–64.
2. Huntington's Disease Collaborative Research Group (1993) A novel gene containing a trinucleotide repeat that is expanded and unstable on Huntington's disease chromosomes. *Cell* **72,** 971–983.
3. LaSpada, A. R., Wilson, E. M., Lubahn, D. B., Harding, A. E., and Fishbeck, H. (1991) Androgen receptor gene mutations in X–linked spinal and bulbar muscular atrophy. *Nature* **352,** 77–79.
4. Koide, R., Ikeuchi, T., Onodera, O., Tanaka, H., Igarashi, S., Endo, K., et al. (1994) Unstable expansion of CAG repeat in hereditary dentatorubral–pallidoluysian atrophy (DRPLA). *Nat. Genet.* **6,** 9–13.
5. Nagafuchi, S., Yanagisawa, H., Sato, K., Shirayama, T., Ohsaki, E., Bundo, M., et al. (1994) Dentatorubral and pallidoluysian atrophy expansion of an unstable CAG trinucleotide on chromosome 12p. *Nat. Genet.* **6,** 14–18.
6. Burke, J. R., Wingfield, M. S., Lewis, K. E., Roses, A. D., Lee, J. E., Hulette, C., et al. (1994) The Haw River Syndrome: dentatorubropallidoluysian atrophy (DRPLA) in an African-American family. *Nat. Genet.* **7,** 521–524.
7. Zoghbi, H. Y. and Orr, H. T. (1995) Spinocerebellar ataxia type 1. *Semin Cell Biol.* **6,** 29–35.
8. Imbert, G., Saudou, F., Yvert, G., Devys, D., Trottier, Y., Garnier, J. M., et al. (1996) Cloning of the gene for spinocerebellar ataxia 2 reveals a locus with high sensitivity to expanded CAG/glutamine repeats. *Nat. Genet.* **14,** 285–291.
9. Pulst, S. M., Nechiporuk, A., Nechiporuk, T., Gispert, S., Chen, X. N., Lopes-Cendes, I., et al. (1996) Moderate expansion of a normally biallelic trinucleotide repeat in spinocerebellar ataxia type 2. *Nat. Genet.* **14,** 269–276.
10. Sanpei, K., Takano, H., Igarashi, S., Sato, T., Oyake, M., Sasaki, H., et al. (1996) Identification of the spinocerebellar ataxia type 2 gene using a direct identification of repeat expansion and cloning technique, DIRECT. *Nat. Genet.* **14,** 277–284.
11. Kawaguchi, Y., Okamoto, T., Taniwaki, M., Aizawa, M., Inoue, M., Katayama, S., et al. (1994) CAG expansions in a novel gene for Machado-Joseph disease at chromosome 14q32. 1. *Nat. Genet.* **8,** 221–228.
12. Haberhausen, G., Damian, M. S., Leweke, F., and Muller, U. (1995) Spinocerebellar ataxia, type 3 (SCA3) is genetically identical to Machado-Joseph disease (MJD). *J. Neurol. Sci.* **132,** 71–75.
13. Zhuchenko, O., Bailey, J., Bonnen, P., Ashizawa, T., Stockton, D. W., Amos, C., et al. (1997) Autosomal dominant cerebellar ataxia (SCA6) associated with small polyglutamine expansions in the alpha1A-voltage-dependent calcium channel. *Nat. Genet.* **15,** 62–68.
14. Lindblad, K., Savontaus, M. L., Stevanin, G., Holmberg, M., Digre, K., Zander, C., et al. (1996) An expanded CAG repeat sequence in spinocerebellar ataxia type 7. *Genome Res.* **6,** 965–971.
15. Ross, C. A. (1995) When more is less, pathogenesis of glutamine repeat neurodegenerative diseases. *Neuron* **15,** 493–496.

16. Ross, C. A. (1997) Intranuclear neuronal inclusions, a common pathogenic mechanism for glutamine–repeat neurodegenerative diseases? *Neuron* **19,** 1147–1150.
17. Gusella, J. F., Persichetti, F., and MacDonald, M. E. (1997) The genetic defect causing Huntington's disease, repeated in other contexts? *Mol. Med.* **3,**238–246.
18. Ross, C. A., Becher, M. W., Colomer, V., Engelender, S., Wood, J. D., and Sharp, A. H. (1997) Huntington's disease and dentatorubral–pallidoluysian atrophy, proteins, pathogenesis and pathology. *Brain Pathol.* **7,** 1003–1016.
19. Gusella, J. F. and MacDonald, M. E. (1998) Huntingtin, A single bait hooks many species. *Curr. Opin. Neurobiol.* **8,** 425–430.
20. Choi, D. W. (1988) Glutamate neurotoxicity and diseases of the nervous system. *Neuron* **1,** 623–634.
21. Albin, R. L. and Greenamyre, J. T. (1992) Alternative excitotoxic hypotheses. *Neurology* **42,** 733–738.
22. Coyle, J. T. and Puttfarcken, P. (1993) Oxidative stress, glutamate, and neurodegenerative disorders. *Science* **262,** 689–695.
23. Beal, M. F. (1994) Huntington's disease, energy and excitotoxicity. *Neurobiol. Aging* **15,** 275–276.
24. Shulz, J. B., Matthews, R. T., Jenkins, B. G., Ferrante, R. J., Cipolloni, P. B., Kowall, N. W., et al. (1995) Involvement of free radicals in excitotoxicity in vivo. *J. Neurochem.* **64,** 2239–2247.
25. Nasir, J., Floresco, S. B., O'Kuskey, J. R., Diewert, V. M., Richman, J. M., Zeisler, J., et al. (1995) Targeted disruption of the Huntington's disease gene results in embryonic lethality and behavioral and morphological changes in heterozygotes. *Cell* **81,** 811–823.
26. Duyao, M. P., Auerbach, A. B., Ryan, A., Persichetti, F., Barnes, G. T., McNeil, S. M., et al. (1995) Inactivation of the mouse Huntington's disease gene homolog Hdh. *Science* **269,** 407–410.
27. Zeitlin, S., Liu, J. P., Chapman, D. L., Papaioannou, V. E., and Efstratiadis, A. (1995) Increased apoptosis and early embryonic lethality in mice nullizygous for the Huntington's disease gene homologue. *Nat. Genet.* **11,** 155–163.
28. White, J. K., Auerbach, W., Duyao, M. P., Vonsattel, J. P., Gusella, J. F., Joyner, A. L. et al. (1997) Huntingtin is required for neurogenesis and is not impaired by the Huntington's disease CAG expansion. *Nat. Genet.* **17,** 404–410.
29. Gusella, J. F., McNeil, S., Persichetti, F., Srinidhi, J., Novelletto, A., Bird, E., et al. (1996) Huntington's Disease. *Cold Spring Harbor Sympos. Quant. Biol.* **61,** 615–625.
30. Albin, R. L. (1995) Selective neurodegeneration in Huntington's disease. *Ann. Neurol.* **38,** 835–836.
31. Vonsattel, J. P. and DiFiglia, M. (1998) Huntington disease. *J. Neuropathol. Exp. Neurol.* **57,** 369–384.
32. McNeil, S. M., Novelletto, A., Srinidhi, J., Barnes, G., Kornbluth, I., Altherr, M. R., et al. (1997) Reduced penetrance of the Huntington's disease mutation. *Hum. Mol. Genet.* **6,** 775–779.
33. Rubinsztein, D. C., Leggo, J., Coles, R., Almqvist, E., Biancalana, V., Cassiman, J. J., et al. (1996) Phenotypic characterization of individuals with

30-40 CAG repeats in the Huntington disease (HD) gene reveals HD cases with 36 repeats and apparently normal elderly individuals with 36–39 repeats. *Am. J. Hum. Genet.* **9,** 16–22.
34. MacDonald, M. E., Barnes, G., Srinidhi, J., Duyao, M. P., Ambrose, C. M., Myers, R. H., et al. (1993) Gametic but not somatic instability of CAG repeat length in Huntington's disease. *J. Med. Genet.* **30,** 982–986.
35. Penney, J. B., Jr., Vonsattel, J. P., MacDonald, M. E., Gusella, J. F., and Myers, R. H. (1997) CAG repeat number governs the development rate of pathology in Huntington's disease. *Ann. Neurol.* **41,** 689–692.
36. Persichetti, F., Srinidhi, J., Kanaley, L., Ge, P., Myers, R. H., D'Arrigo, K., et al. (1994) Huntington's disease CAG trinucleotide repeats in pathologically confirmed post–mortem brains. *Neurobiol. Dis.* **1,** 159–166.
37. Ambrose, C. M., Duyao, M. P., Barnes, G., Bates, G. P., Lin, C. S., Srinidhi, J., et al. (1994) Structure and expression of the Huntington's disease gene, evidence against simple inactivation due to an expanded CAG repeat. *Somat. Cell. Mol. Genet.* **20,** 27–38.
38. Wexler, N. S., Young, A. B., Tanzi, R. E., Travers, H., Starosta-Rubenstein, S., Penney, J. B., et al. (1987) Homozygotes for Huntington's disease. *Nature* **326,** 194–197.
39. Myers, R. H., Leavitt, J., Farrer, L. A., Jagadeesh, J., McFarlane, H., Mark, R. J., et al. (1989) Homozygote for Huntington's disease. *Am. J. Hum. Genet.* **45,** 615–618.
40. MacDonald, M. E. and Gusella, J. F. (1996) Huntington's disease: translating a CAG repeat into a pathogenic mechanism. *Curr. Opin. Neurobiol.* **6,** 638–643.
41. Jou Y-S. and Myers, R. M. (1995) Evidence from antibody studies that the CAG repeat in the Huntington's disease gene is expressed in the protein. *Hum. Mol. Genet.* **4,** 465–469.
42. Ide, K., Nukina, N., Masuda, N., Goto, J., and Kanazawa, I. (1995) Abnormal gene product identified in Huntington's disease lymphocytes and brain. *Biochem. Biophys. Res. Commun.* **209,** 1119–1125.
43. Baxendale, S., Abdulla, S., Elgar, G., Buck, D., Berks, M., Micklem, G., et al. (1995) Comparative sequence analysis of the human and pufferfish Huntington's disease genes. *Nat. Genet,* **10,** 67–76.
44. Karlovich, C. A., John, R. M., Ramirez, L., Stainier, D. Y. R., and Myers, R. M. (1998) Characterization of the Huntington's disease (HD) gene homolog in the zebrafish *Danio rerio Gene* **217,** 117–125.
45. Lin, B., Nasir, J., MacDonald, H., Hutchinson, G., Graham, R. K., Rommens, J. M., et al. (1994) Sequence of the murine Huntington disease gene, evidence for conservation, alternate splicing and polymorphism in a triplet (CCG) repeat. *Hum. Mol. Genet.* **3,** 85–92; erratum: *Hum. Mol. Genet.* **3,** 530 (1994).
46. Barnes, G. T., Duyao, M. P., Ambrose, C. M., McNeil, S., Persichetti, F., Srinidhi, J., et al. (1994) Mouse Huntington's disease gene homolog (*Hdh*) *Somat .Cell. Mol. Genet.* **20,** 87–97.
47. Schmitt, I., Baechner, D., Megow, D., Henklein, P., Boulter, J., Hameister, H., et al. (1995) Expression of the Huntington disease gene in rodents: clon-

ing the rat homologue and evidence for down regulation in non–neuronal tissues during development. *Hum. Mol. Genet.* **4,** 1173–1182.
48. Andrade, M. A. and Bork, P. (1995) HEAT repeats in the Huntington's disease protein. *Nat. Genet.* **11,** 115–116.
49. Groves, M. R., Hanlon, N., Turowski, P., Hemmings, B. A., and Barford, D. (1999) The structure of the protein phosphatase 2A PR65/A subunit reveals the conformation of its 15 tandemly repeated HEAT motifs. *Cell* **96,** 99–110.
50. Cingolani, G., Petosa, C., Weis, K., and Muller, C. W. (1999) Structure of importin-beta bound to the IBB domain of importin-alpha. *Nature* **399,** 221–229.
51. Landschulz, W. H., Johnson, K. S., and McKnight, S. L. (1988) The leucine zipper. A hypothetical structure common to a new class of DNA binding proteins. *Science* **240,** 1759–1764.
52. Sharp, A. H. and Ross, C. A. (1996) Neurobiology of Huntington's disease. *Neurobiol. Dis.* **3,** 3–15.
53. Sharp, A. H., Loev, S. J., Schilling, G., Li, S–H., Li, X–J, Bao, J. et al. (1995) Widespread expression of the Huntington's disease gene (IT–15) protein product. *Neuron* **14,** 1065–1074.
54. DiFiglia, M., Sapp, E., Chase, K., Schwarz, C., Meloni, A., Young, C., et al. (1995) Huntingtin is a cytoplasmic protein associated with vesicles in human and rat brain neurons. *Neuron* **14,**1075–1081.
55. Gutekunst C.-A, Levey AI, Heilman CJ, Whaley WL, Hong Y, Nash NR, et al. (1995) Identification and localization of huntingtin in brain and human lymphoblastoid cell lines with anti-fusion protein antibodies. *Proc. Natl. Acad. Sci. USA* **92,** 8710–8714
56. Persichetti, F., Carlee, L., Faber, P. W., McNeil, S. M., Ambrose, C. M., Srinidhi, J., et al. (1996). Differential expression of normal and mutant Huntington's disease gene alleles. *Neurobiol. Dis.* **3,** 183–190.
57. Ferrante, R. J., Gutekunst, C. A., Persichetti, F., McNeil, S. M., Kowall, N. W., Gusella, J. F., et al. (1997) Heterogeneous topographic and cellular distribution of huntingtin expression in the normal human neostriatum. *J. Neurosci.* **17,** 3052–3063
58. Velier, J., Kim, M., Schwarz, C., Kim, T. W., Sapp, E., Chase, K., et al. (1998) Wild–type and mutant huntingtins function in vesicle trafficking in the secretory and endocytic pathways. *Exp. Neurol.* **152,** 34–40.
59. Hoogeveen, A. T., Willemsen, R., Meyer, N., de Rooij, K. E., Roos, R. A., van Ommen, G. J., et al. (1993) Characterization and localization of the Huntington disease gene product. *Hum. Mol. Genet.* **2,** 2069–2073.
60. De Rooij, K. E., Dorsman, J. C., Smoor, M. A., Den Dunnen, J. T., and Van Ommen, G. J. (1996) Subcellular localization of the Huntington's disease gene product in cell lines by immunofluorescence and biochemical subcellular fractionation. *Hum. Mol. Genet.* **5,**1093–1099.
61. Dragatsis, I., Efstratiadis, A., and Zeitlin, A. (1998). Mouse mutant embryos lacking huntingtin are rescued from lethality by wild-type extraembryonic tissues. *Development* **125,** 1529–1539.

62. Metzler, M., Chen, N., Helgason, C. D., Graham, R. K., Nichol, K., McCutcheon, K., et al. (1999) Life without huntingtin, normal differentiation into functional neurons. *J. Neurochem.* **72**, 1009–1018.
63. Bao, J., Sharp, A. H., Wagster, M. V., Becher, M., Schilling, G., Ross, C. A., et al. (1996) Expansion of polyglutamine repeat in huntingtin leads to abnormal protein interactions involving calmodulin. *Proc. Natl. Acad. Sci. USA* **93,** 5037–5042.
64. Beckingham, K., Lu, A. Q., and Andruss, B. F. (1998) Calcium-binding proteins and development. *Biometals* **11,** 359–73.
65. Burke, J. R., Enghild, J. J., Martin, M. E., Jou, Y. S., Myers, R. M., Roses, A. D., et al. (1996) Huntingtin and DRPLA proteins selectively interact with the enzyme GAPDH. *Nat. Med.* **2,** 347–350.
66. Sirover, M. A. (1996) Emerging new funcitons of the glycolytic protein, glyceraldehyde-3-phosphatate dehydrogenase, in mammalian cells. *Life Sci.* **58,** 2271–2277.
67. Tukamoto, T., Nukina, N., Ide, K., and Kanazawa, I. (1997) Huntington's disease gene product, huntingtin, associates with microtubules *in vitro*. *Brain Res. Mol. Brain Res.* **51,** 8–14.
68. Liu, Y. F., Deth, R. C., and Devys, D. (1997) SH3 domain–dependent association of huntingtin with epidermal growth factor receptor signaling complexes. *J. Biol. Chem.* **272,** 8121–8124.
69. Wang, Z. and Moran, M. F. (1996) Requirement for the adapter protein GRB2 in EGF receptor endocytosis. *Science* **272,** 1935–1939.
70. Li, X. J., Li, S. H., Sharp, A. H., Nucifora, F. C., Jr., Schilling, G., Lanahan, A., et al. (1995) A huntingtin–associated protein enriched in brain with implications for pathology. *Nature* **378,** 398–402.
71. Bertaux, F., Sharp, A., Ross, C. A., Lehrach, H., Bates, G. P., and Wanker, E. (1998) HAP1-huntingtin interactions do not contribute to the molecular pathology in Huntington's disease transgenic mice. *FEBS Letters* **426,** 229–232.
72. Li, S. H., Hossein, S. H., Gutekunst, C. A., Hersch, S. M., Ferrante, R. J., and Li, X. J. (1998) A human HAP1 homologue. *J. Biol. Chem.* **273,** 19,220–19,227.
73. Engelender, S., Sharp, A. H., Colomer, V., Tokito, M. K., Lanahan, A., Worley, P., et al. (1997) Huntingtin-associated protein 1 (HAP1) interacts with the p150Glued subunit of dynactin. *Hum. Mol. Genet.* **6,** 2205–2212.
74. Li, S. H., Gutekunst, C. A., Hersch, S. M., and Li, X. J. (1998) Interaction of huntingtin-associated protein with dynactin P150. *J. Neurosci.* **18,** 1261–1269.
75. Karki, S. and Holzbaur, E. L. (1995) Affinity chromatography demonstrates a direct binding between cytoplasmic dynein and the dynactin complex. *J. Biol. Chem.* **270,** 28,806–28,811.
76. Vaughan, K. T. and Vallee, R. B. (1995) Cytoplasmic dynein binds dynactin through a direct interaction between the intermediate chains and p150Glued. *J. Cell Biol.* **131,** 1507–1516.
77. Kalchman, M. A., Koide, H. B., McCutcheon, K., Graham, R. K., Nichol, K., Nishiyama, K., et al. (1997) HIP1, a human homologue of *S. cerevisiae* Sla2p, interacts with membrane–associated huntingtin in the brain. *Nat. Genet.* **16,** 44–53.

78. Wanker, E. E., Rovira, C., Scherzinger, E., Hasenbank, R., Walter, S., Tait, D., et al. (1997) HIP-I, a huntingtin interacting protein isolated by the yeast two- hybrid system. *Hum. Mol. Genet.* **6,** 487–495.
79. Holtzman, D. A., Yang, S., and Drubin, D. G. (1993) Synthetic–lethal interactions identify two novel genes, SLA1 and SLA2, that control membrane cytoskeleton assembly in *Saccharomyces cerevisiae. J. Cell Biol.* **122,** 635–644.
80. Li, R., Zheng, Y., and Drubin, D. G. (1995) Regulation of cortical actin cytoskeleton assembly during polarized cell growth in budding yeast. *J. Cell Biol.* **128,** 599–615.
81. Na, S., Hincapie, M., McCusker, J. H., and Haber, J. E. (1995) MOP2 (SLA2) affects the abundance of the plasma membrane H(+)–ATPase of *Saccharomyces cerevisiae. J. Biol. Chem.* **270,** 6815–6823.
82. Wesp, A., Hicke, L., Palecek, J., Lombardi, R., Aust, T., Munn, A. L., et al. (1997) End4p/Sla2p interacts with actin–associated proteins for endocytosis in *Saccharomyces cerevisiae. Mol. Biol. Cell.* **8,** 2291–2306.
83. Kalchman, M. A., Graham, R. K., Xia, G., Koide, H. B., Hodgson, J. G., Graham, K. C., et al. (1996) Huntingtin is ubiquitinated and interacts with a specific ubiquitin-conjugating enzyme. *J. Biol. Chem.* **271,** 19,385–19,394.
84. Hochstrasser, M. (1996) Ubiquitin-dependent protein degradation. *Annu. Rev. Genet.* **30,** 405–439.
85. Hicke, L. and Riezman, H. (1996) Ubiquitination of a yeast plasma membrane receptor signals its ligand-stimulated endocytosis. *Cell* **84,** 277–287.
86. Sittler, A., Walter, S., Wedemeyer, N., Hasenbank, R., Scherzinger, E., Eickhoff, H., et al. (1998) SH3GL3 associates with the huntingtin exon 1 protein and promotes the formation of polygln-containing protein aggegates. *Mol. Cell* **2,** 427–436.
87. Ringstad, N., Nemoto, Y., and De Camilli, P. (1997). the SH3p4/Sh3p8/SH3p13 protein family**,** binding partners for synaptojanin and dynamin via a Grb2–like Src homology 3 domain. *Proc. Natl. Acad. Sci. USA* **94,** 8569–8574.
88. Zechner, U., Scheel, S., Hemberger, M., Hopp, M., Haaf, T., Fundele, R, et al. (1998) Characterization of the mouse Src homology 3 domain gene Sh3d2c on Chr 7 demonstrates coexpression with huntingtin in the brain and identifies the processed pseudogene Sh3d2c-ps1 on Chr 2. *Genomics* **54,** 505–510.
89. Faber, P. W., Barnes, G. T., Srinidhi, J., Gusella, J. F., and MacDonald, M. E. (1998) Huntingtin interacts with a family of WW domain proteins. *Hum. Mol. Genet.* **7,** 1463–1474.
90. Bedford, M. T., Chan, D. C., and Leder, P. (1997) FBP WW domains and the Abl SH3 domain bind to a specific class of proline–rich ligands. *EMBO J.* **16,** 2376–2383.
91. Bedford, M. T., Reed, R., and Leder, P. (1998) WW domain-mediated interactions reveal a spliceosome-assocated protein that binds a third class of proline-rich motif**,** The proline glycine and methionine-rich motif. *Proc. Natl. Acad. Sci. USA* **95,** 10.602–10.607.
92. Kao, H. Y. and Siliciano, P. G. (1996) Identification of Prp40, a novel essential yeast splicing factor associated with the U1 small nuclear ribonucleoprotein particle. *Mol. Cell. Biol.* **16**, 960–967.

93. Passani, L. A., Bedford, M. T., Faber, P. W., McGinnis, K. M., Gusella, J. F., Vonsattel, J.-P., and MacDonald, M. E. (2000) Huntingtin's WW domain partners in HD post-mortem brain fulfill genetic criteria for direct involvement in HD pathogenesis. *Hum. Mol. Genet.*, in press.
94. Chan, D. C., Bedford, M. T., and Leder P. (1996) Formin binding proteins bear WWP/WW domains that bind proline-rich peptides and functionally resemble SH3 domains. *EMBO J.* **15,**1045–1054.
95. Gaugler, B., Van den Eynde, B., van der Bruggen, P., Romero, P., Gaforio, J. J., De Plaen, E., et al. (1994) Human MAGE-3 codes for an antigen recognized on a melanoma by autologous cytolytic T lymphocytes *J. Exp. Med.* **179**, 921–930.
96. Tahara, K., Mori, M., Sadanaga, N., Sakamoto, Y., Kitano, S., and Makuuchi, M. (1999) Expression of MAGE gene family in human hepatocellular carcinoma. *Cancer* **15,** 1234–1240.
97. Kominami, K., DeMartino, G. N., Moomaw, C. R., Slaughter, C. A., Shimbara, N., Fujimuro, M., et al. (1995) Nin1p, a regulatory subunit of the 26S proteasome, is necessary for activation of Cdc28p kinase of *Saccharomyces cerevisiae*. *EMBO J.* **14,** 3105–3115.
98. Robinson, M. S. (1989) Cloning of cDNAs encoding two related 100-kD coated vesicle proteins (alpha–adaptins). *J Cell Biol.* **108,** 833–842.
99. Ball, C. L., Hunt, S. P., and Robinson, M. S. (1995) Expression and localization of alpha-adaptin isoforms. *J. Cell Sci.* **108,** 2865–2875.
100. Dornan, S., Jackson, A. P., and Gay, N. J. (1997) Alpha-adaptin, a marker for endocytosis, is expressed in complex patterns during Drosophila development. *Mol. Biol.Cell* **8,** 1391–1403.
101. Keon, B. H., Schafer, S., Kuhn, C., Grund, C., and Franke, W. W. (1996) Symplekin, a novel type of tight junction plaque protein. *J. Cell Biol.* **134,** 1003–1018.
102. Yongan, L., Kang, J., and Horwitz, M. S. (1998) Interaction of an adenovirus E3 14. 7 kilodalton protein with a novel tumor necrosis factor alpha-inducible cellular protein containing leucine zipper domains. *Mol. Cell. Biol.* **18,** 1601–1610.
103. Wold, W. S. M., Hermiston, T. W., and Tollefson, A. E. (1994) Adenovirus proteins that subvert host defenses. *Trends Microbiol.* **2,** 437–443.
104. Boutell, J. M., Wood, J. D., Harper, P. S., and Jones, A. L. (1998) Huntingtin interacts with cystathionine beta-synthase. *Hum. Mol. Genet.* **7,** 371–378.
105. Bao, L., Vlcek, C., Paces, V., and Kraus, J. P. (1998) Identification and tissue distribution of human cystathionine beta- synthase mRNA isoforms. *Arch Biochem. Biophys.* **350,** 95–103.
106. Fields, S. and Sternglanz, R. (1994) The two–hybrid system**,** an assay for protein–protein interactions. *Trends Genet.* **10**, 286–292.
107. Pawson, T. (1995) Protein modules and signaling networks. *Nature* **373,** 573–578.
108. Ren, R., Mayer, B. J., Cicchetti, P., and Baltimore, D. (1993) Identification of a ten-amino acid proline rich SH3 binding site. *Science* **259,** 1157–1161.
109. Colomer, V., Engelender, S., Sharp, A. H., Duan, K., Cooper, J. K., Lanahan, A., et al. (1997) Huntingtin-associated protein 1 (HAP1) binds to a Trio-like

polypeptide, with a rac1 guanine nucleotide exchange factor domain. *Hum. Mol. Genet.* **6,**1519–1525.

110. Ross, T. S., Bernard, O. A., Berger, R., and Gilliland, D. G. (1998) Fusion of huntingtin interacting protein 1 to platelet-derived growth factor beta receptro (PDGFbetaR) in chronic myelomonocytic leukemia with t(5;7)(q33;q11. 2). *Blood* **91,** 4419–4426.
111. Bedford, M. and Leder, P. (1998) The FF domain, a novel motif that often accompanies WW domains. *Trends Biochem. Sci.* **24,** 264,265.
112. Barth, A. I., Nathke, I. S., and Nelson, J. (1997) Cadherins, catenins and APC protein, interplay between cytoskeletal complexes and signalling pathways. *Curr. Opin. Cell Biol.* **9,** 683–690.
113. Persichetti, F., Trettel, F., Huang, C. C., Fraefel, C., Timmers, H. T. M., Gusella, J. F., et al. (1999) Mutant huntingtin forms *in vivo* complexes with distinct context–dependent conformations of the polyglutamine segment. *Neurobiol. Dis.* **6,** 364–375.
114. Trottier, Y., Lutz, Y., Stevanin, G., Imbert, G., Devys, D., Cancel, G., et al. (1995). Polyglutamine expansion as a pathological epitope in Huntington's disease and four dominant cerebellar ataxias. *Nature* **378,** 403–406.
115. Huang, C. C., Faber, P. W., Persichetti, F., Mittal, V., Vonsattel, J-P., MacDonald, M. E., et al. (1998) Amyloid formation by mutant huntingtin, threshold, progressivity and recruitment of normal polyglutamine proteins. *Somat. Cell Mol. Genet.* in press.
116. Scherzinger, E., Lurz, R., Turmaine, M., Mangiarini, L., Hollenbach, B., Hasenbank, R., et al. (1997) Huntingtin–encoded polyglutamine expansions form amyloid–like protein aggregates *in vitro* and *in vivo*. *Cell* **90,** 549–558.
117. DiFiglia, M., Sapp, E., Chase, K. O., Davies, S. W., Bates, G. P., Vonsattel, J. P., et al. (1997). Aggregation of huntingtin in neuronal intranuclear inclusions and dystrophic neurites in brain. *Science* **277,** 1990–1993.
118. Sapp, E., Penney, B. A. J., Young, A., Aronin, N., Vonsattel, J-P., and DiFiglia, M. (1999) Axonal transport of N-terminal huntingtin suggests early pathology of corticostriatal projections in Huntington's disease. *J. Neuropathol. Exp. Neurol.* **58,** 165–173.
119. Sieradzan, K. A., Mechan, A. O., Jones, L., Wanker, E. E., Nukina, N., and Mann, D. M. (1999) Huntington's disease intranuclear inclusions contain truncated ubiquintated huntingtin protein. *Exp. Neurol.* **156,** 92–99.
120. Maat–Schieman, M. L., Dorsman, J. C., Smoor, M. A., Siesling, S., Van Duinen, S. G., Verschuuren, J. J., et al. (1999) Distribution of inclusions in neuronal nuclei and dystrophic neurites in Huntington's disease brain. *J. Neuropathol. Exp. Neurol.* **58,** 129–137.
121. Gutekunst, C.-A., Li, S-H., Yi, H., Mulroy, J. S., Kuemmerle, S., Jones, R., et al. (1999) Nuclear and neuropil aggregates in Huntington's disease, relationship to neuropathology. *J. Neurosci.* **19,** 2522–2534.
122. Goldberg, Y. P., Nicholson, D. W., Rasper, D. M., Kalchman, M. A., Koide, H. B., Graham, R. K., et al. (1996) Cleavage of huntingtin by apopain, a proapoptotic cysteine protease, is modulated by the polyglutamine tract. *Nat. Genet.* **13,** 442–449.

123. Lunkes, A. and Mandel, J. L. (1998) A cellular model that recapitulates major pathogenic steps of Huntington's disease. *Hum. Mol. Genet.* **7,** 1355–1361.
124. Reddy, P. H., Williams, M., Charles, V., Garrett, L., Pike–Buchanan, L., Whetsell, W. O. Jr., et al. (1998) Behavioral abnormalities and selective neuronal loss in HD transgenic mice expressing full-length *HD* cDNA. *Nature Genet.* **20,** 198–202.
125. Hodgson, J. G., Agopyan, N., Gutekunst, C–A., Leavitt, B. R., LePiane, F., Singaraja, R., et al. (1999) A YAC mouse model for Huntington's disease with full-length mutant huntingtin, cytoplasmic toxicity, and selective striatal neurodegeneration. *Neuron* **23,** 181–192.
126. Shelbourne, P. F., Killeen, N., Hevner, R. F., Johnston, H. M., Tecott, L., Lewandoski, M., et al. (1999) A Huntington's disease CAG expansion at the murine *Hdh* locus is unstable and associated with behavioural abnormalities in mice. *Hum. Mol. Genet.* **8,** 763–774.
127. Levine, M. S., Klaptein, G. J., Koppel, A., Gruen, E., Cepeda, C., Vargas, M. E., et al. (1999) Enhanced sensitivity to N-methyl-D-asparate receptor activation in transgenic and knockin mice models to Huntington's disease. *J. Neurosci. Res.* **58,** 515–532.
128. Wheeler, V. C., White, J. K., Gutekunst, C. A., Vrbanac, V., Weaver, M., Li, X. J., Li, S. H., Yi, H., Vonsattel, J. P., Gusella, J. F., Hersch, S., Auerbach, W., Joyner, A. L., and MacDonald, M. E. (2000) Long glutamine tracts cause nuclear localization of a novel form of huntingtin in medium spiny striatal neurons in Hdh^{Q92} and Hdh^{Q111} knock-in mice. *Hum. Mol. Genet.* **9,** 503–513.
129. Feng, Y. and Davis, N. G. (2000) Akr1p and the Type I casein kinases act prior to the ubiquitination step of yeast endocytosis: Akr1p is required for kinase localization to the plasma membrane. *Mol. Cell Biol.* **20,** 5350–5359.
130. Takagaki, Y. and Manley, J. L. (2000) Complex protein interactions within the human polyadenylation machinery identify a novel component. *Mol. Cell Biol.* **20,** 1515–1525.
131. Boutell, J. M., Thomas, P., Neal, J. W., Weston, V. J., Duce, J., Harper, P. S., and Jones, A. L. (1999) Aberrant interactions of transcriptional repressor proteins with the Huntington's disease gene product, huntingtin. *Hum. Mol. Genet.* **8,** 1647–1655.

16

Modeling Neurodegenerative Diseases in the Fruit Fly

George R. Jackson

16.1. INTRODUCTION

Genetic studies in *Drosophila* have the potential to provide valuable insight into fundamental pathophysiologic mechanisms of human disease. As compared to early-onset neurological diseases such as those associated with inborn errors of metabolism, the pathogenesis of late-onset neurodegenerative diseases such as amyotrophic lateral sclerosis (ALS) and Alzheimer's, Parkinson's, and Huntington's diseases is proving to be complex. Given that a fundamental understanding of disease mechanisms is crucial to the rational design of therapeutic strategies, the myriad genetic techniques available in *Drosophila* for the study of biological phenomena might prove a useful addition to our current armamentarium of transgenic mice, in vitro and unicellular techniques, and biochemical studies. Classical genetic screens in *Drosophila* have uncovered numerous mutations giving rise to both ectopic developmental cell death and late-onset neurodegeneration. Similarly, genetic study of fly homologs of human disease genes has provided important information about their function and pathological dysfunction. A more recent approach has utilized the targeted expression of mutant human disease genes in *Drosophila* to recapitulate certain aspects of human disease. Given that many aspects of cell death, signal transduction, cell cycle regulation, and pattern formation appear to be conserved in evolution from *Drosophila* to man, genetic screens for phenotypic modifiers of misexpressed mutant human genes may reveal similarly conserved modifier genes. Identification of such modifier genes might identify new therapeutic targets for otherwise incurable disorders.

As detailed elsewhere in this volume, Huntington's disease (HD) is an autosomal-dominant neurodegenerative disorder characterized by cognitive dysfunction, affective changes, and abnormal involuntary movements

From: *Contemporary Clinical Neuroscience: Molecular Mechanisms of Neurodegenerative Diseases*
Edited by: M.-F. Chesselet © Humana Press Inc., Totowa, NJ

(Martin and Gusella, 1986; Folstein, 1989). The disease is associated with expansion of an unstable trinucleotide repeat within exon 1 of the gene-encoding huntingtin (Huntington's Disease Collaborative Research Group, 1993). Although this gene is, to some extent, conserved in evolution from *Drosophila* to man (Baxendale et al., 1995; Djian et al., 1996; Karlovich et al., 1998; Li et al., 1999), apart from a role in embryogenesis and neurogenesis, its function remains unclear (Duyao et al., 1995; Nasir et al., 1995; White et al., 1997; Zeitlin et al., 1995). Current opinion holds that expansion of the unstable CAG repeats, resulting in expression of an elongated polyglutamine tract near the amino-terminus of huntingtin, imparts a toxic gain-of-function that does not compromise the normal developmental role of the protein. Although a number of proteins, both novel and previously identified, have been observed to interact physically with huntingtin using techniques such as the yeast two-hybrid screen (Boutell et al., 1998; Faber et al., 1998; Kalchman et al., 1996; Wanker et al., 1997), none of these has been implicated convincingly in HD pathogenesis. Thus, any technique that would add to our understanding of huntingtin-interacting genes might find a well-deserved place among more traditional experimental approaches to the disease.

Larger repeat expansions within the gene encoding huntingtin correlate with an earlier onset of disease. Nonetheless, for any given repeat length, age of onset may vary by more than a decade (Gusella and MacDonald, 1995). What factors govern the rate of phenoconversion or phenoemergence, the phenomenon in which a given genotype yields the HD phenotype? Although genetic factors have been suggested to modify age of onset, only one such factor, a polymorphism in the GluR6 kainate receptor gene, has been clearly established (MacDonald et al., 1999). If other such factors could be identified and their expression modified, substantial relief from disease burden could occur even in the absence of a "cure." Genetic modifier screens in *Drosophila* provide a means of searching for such modifiers in vivo.

Four published studies suggest that certain fundamental aspects of polyglutamine pathogenesis are widely conserved through evolution (Jackson et al., 1998; Warrick et al., 1998; Faber et al., 1999; Marsh et al., 2000). Undoubtedly, despite the similarities observed in fly models of polyglutamine diseases in man, it is naive to anticipate that such models will faithfully recapitulate human pathophysiology. In many respects, though, simple invertebrate systems supply an ideal system with which to study the pathophysiologic basis of neurodegenerative diseases. Although prokaryotes and yeast lack nervous tissue, the roundworm *Caenorhabditis elegans* and the fruit fly *Drosophila melanogaster* possess well-characterized nervous systems. In *C. elegans*, cell death pathways are well delineated. Identification of cell death genes in the worm has been instrumental in understanding their myriad homologs in vertebrates. Although cell death pathways to date have been less well characterized in *Drosophila*, a few crucial players have been identified that are homologs to those identified in man. In addition, a number of interesting proapoptotic genes have been identified in flies that appear to have no close mammalian homologs; nonetheless, despite the apparent absence of related genes, a number of these genes appear to be functional in mammalian systems. Perhaps the most important aspect of invertebrate approaches is the availability of a number of genetic manipulations that are impossible or impractical to carry out in mammals. Large numbers of flies and worms can be mutagenized and screened in a short period of time, thus permitting the identification of even rare mutations. Given the considerable success that fly genetic approaches have had in delineating processes such as cell cycle control, signal transduction, and pattern formation, it is reasonable to anticipate that similar approaches to the study of polyglutaminopathy may yield powerful insights into disease mechanisms.

This chapter will review briefly the current state of knowledge about mechanisms of cell death in *Drosophila*. I will then review mutations known to regulate developmental cell death in flies, as well as those associated with late-onset (here defined as post-eclosion) neurodegeneration. I will then discuss what is known about *Drosophila* homologs of human neurodegenerative disease-associated genes, as well as work done to date on loss-of-function mutations in such genes and how they may shed light on human pathology. In particular, recent study of *Drosophila* homologs of genes implicated in the pathogenesis of Alzheimer's disease has been insightful. Finally, I will discuss in some detail the established models of glutamine repeat diseases in flies, with attention toward practical aspects of further developing and interpreting such models.

16.2. CELL DEATH PATHWAYS IN *DROSOPHILA*

Programmed cell death is a phenomenon of vital importance in morphogenesis and development of the nervous and immune systems. An accumulating body of evidence suggests that inappropriate activation of intrinsic cell death programs may underlie neurodegeneration. Table 1 lists several identified *Drosophila* cell death genes.

The prototype for studies of cell death is *C. elegans*. In the nematode, a number of critical modulators of cell death have been described that have

Table 1
Selected Cell Death-Related Genes in *Drosophilia*

Gene	Function	Homologs
rpr	Proapoptotic	Unknown
hid	Proapoptotic	Unknown
grim	Proapoptotic	Unknown
ras	*hid* regulation	*ras*
dredd	Protease	Initiator caspases *(ced-3)*
Dcp-1	Protease	Effector caspases *(ced-3)*
drIce	Protease	Effector caspases *(ced-3)*
DRONC	Protease	Initiator caspases *(ced-3)*
DECAY	Protease	Effector caspases *(ced-3)*
Dapaf-1	Protease	Apaf-1 *(ced-9)*
DREP-1	Inhibits DNA fragmentation	DFF45
Dakt-1	Protein kinase	Akt/protein kinase B
DIAP-1/DIAP-2	Protease inhibitor	IAP/XIAP/survivin *(iap-1/iap-2)*

Where multiple designations exist, only one is listed; see text for alternate nomenclature. Mammalian homologs are listed where known; *C. elegans* homologs are given in parentheses.

human homologs: *ced-3* (caspases), *ced-4* (Apaf), and *ced-9* (*bcl*-2 and related mediators; Hengartner and Horvitz, 1994).

The deficiency line H99 provided one of the first insights into the regulation of cell death in *Drosophila* (White et al., 1994). In homozygous H99 embryos, acridine orange staining reveals an absence of developmental cell death. However, such homozygous embryos are still susceptible to apoptosis induced by ionizing radiation. Genetic analysis of this deficiency identified three pro-apoptotic genes: *reaper* (*rpr*), *head involution defective* (*hid*), and *grim* (White et al., 1994; Grether et al., 1995; Chen et al., 1996). Each of these genes contains a similar short motif near the amino-terminus, the RHG motif, but otherwise, they show no similarity (Wing et al., 1998). Although there are sequence similarities between *rpr* and the death domain of tumor necrosis factor receptor-1 family members, including the low-affinity neurotrophin receptor p75, mutational analysis does not support the functional similarity suggested by this sequence (Chen et al., 1996). Apart from a short, interrupted polyglutamine tract in *grim*, none of these gene products shows other motifs with known homology to vertebrate proteins. Nonetheless, cell death induced by expression of *rpr* and *hid* in mammalian cell culture systems suggests that certain aspects of cell death pathways utilized by these proteins are more widely conserved, despite the apparent absence of homologs (Claveria et al., 1998; McCarthy and Dixit, 1998;

Haining et al., 1999). Directed expression of each of these genes in the eye results in massive cell death with the appearance of reduced, rough eyes (Grether et al., 1995; White et al., 1996; Chen et al., 1996). The viral gene *P35*, which is required for apoptosis of insect cells by the baculovirus *Autographa californica* (Clem et al., 1991), serves as a strong suppressor of the cell death induced by eye-directed expression of *rpr*, *hid*, and *grim* in the eye and in vitro (Hay et al., 1994; Grether et al., 1995; Chen et al., 1996; Vucic et al., 1997). P35 functions as a broad-spectrum caspase-inhibitor substrate (Bump et al., 1995; Zhou et al., 1998), suggesting that caspase activity is a downstream mediator of *rpr*-, *hid*-, and *grim*-induced cell death. Mutagenesis screens for modifier of eye-specific *rpr* expression identified *Drosophila* inhibitor of apoptosis proteins (DIAP1 and DIAP2) (Hay et al., 1995). Both genes are homologous to apoptosis inhibitors found in a wide range of organisms, from viruses to *C. elegans* to mammals (Crook et al., 1993; Birnbaum et al., 1994; Duckett et al., 1996; Uren et al., 1996; Ambrosini et al., 1997; Farahani et al., 1997; Fraser et al., 1999). Inhibitors of apoptosis proteins (IAPs) may both inhibit caspase activity directly and by binding to the proapoptotic gene products of the H99 interval (Abrams, 1999); full details of the relationship among IAPs, caspases, and *rpr/hid/grim*, however, remain to be elucidated.

Further screening for suppressors of eye-specific expression of *hid* identified members of the *Ras/Raf*/MAP kinase signaling pathway as endogenous inhibitors of *hid* activity (Bergmann et al., 1998; Kurada and White, 1998). Loss-of-function mutations in this pathway enhance *hid*-induced cell death, whereas gain-of-functions mutations suppress it. Site-directed mutagenesis of MAP kinase phosphorylation sites in *hid* enhances its proapoptotic effects in the eye (Bergmann et al., 1998). Thus, intriguing links between regulation of widely conserved signaling pathways and cell death exist; others are likely to be identified.

The search for caspases in *Drosophila* homologous to those previously identified in vertebrates led to identification of the first *Drosophila* caspase, *Dcp-1* (Song et al., 1997). Homozygous loss-of-function alleles of *Dcp-1* show larval lethality and melanotic tumors. Impaired transfer of nurse cell cytoplasm to oocytes in *Dcp-1* mutants demonstrates a requirement for this caspase in oogenesis (McCall and Steller, 1998). Similar approaches were used to identify the caspase *dredd/Dcp-2* (Inohara et al., 1997; Chen et al., 1998). Loss of *dredd* function (using a deficiency line that deletes other genes, as well) serves as an enhancer of eye-specific *rpr* expression in vivo (Chen et al., 1998). *dredd* shows a large prodomain-containing death effector domain (DED) motifs, suggesting that may function as an apical caspase.

Further evidence implicating *dredd* as an apical caspase is provided by the observation that *rpr*, *hid*, and *grim* promote *dredd* activation even in the presence of caspase inhibitors. Interestingly, the sequence surrounding the catalytic cysteine of *dredd* differs from the canonical caspase sequence in that it contains a glutamate residue rather than a glycine, suggesting *dredd* may have novel substrate specificity. Another unique feature of *dredd* is that, unlike other fly caspases that show ubiquitous expression, embryonic *dredd* mRNA is found in cells destined for programmed cell death. This mRNA accumulation does not occur in homozygous H99 embryos but is rapidly induced on directed expression of *rpr*, *hid*, and *grim*.

A similar approach was used to identify the caspase *drICE* (Fraser and Evan, 1997; Fraser et al., 1997). Sequence information suggests that *drICE* is an effector caspase; however, genetic information is not available for *drICE*. Recently, two other fly caspases have been identified. *DECAY* (*Drosophila* executioner caspase related to apopain/YAMA) lacks a long prodomain and shows significant homology to caspase-3 and caspase-7, thus suggesting it acts as an effector caspase (Dorstyn et al., 1999). *DECAY* also shows substrate specificity in vitro similar to that of caspase-3 family members. Like *Dcp-1*, *DECAY* may play role in nurse cell apoptosis. *DECAY* mRNA is ubiquitously expressed in embryos but shows a more restricted pattern later in development. *DRONC* (*Drosophila* caspase similar to Nedd2) has a long prodomain containing a caspase recruitment domain (CARD; Dorstyn et al., 1999). Ecdysone treatment of larval tissues induces massive increases in *DRONC* mRNA, implicating this caspase in metamorphosis. Loss-of-function mutations have not been reported for *DRONC* or *DECAY*.

Recently, a *Drosophila* homolog of *C. elegans ced*-4/mammalian *Apaf*-1 has been identified (Kanuka, 1999; Rodriguez et al., 1999; Zhou, 1999). Designations for this homolog include *Dapaf-1* (Kanuka, 1999) and *DARK* (*Drosophila Apaf*-1-related killer; Rodriguez et al., 1999). Loss-of-function alleles show reduced embryonic PCD, abnormal wing and bristle morphology, melanotic tumors, and the persistence of supernumary cells in the nervous system, including photoreceptor neurons. Like *Apaf*-1, *DARK* contains an inhibitory WD domain that is thought to interact with cytochrome-*c*, in addition to a CARD. A splicing variant lacking the WD domain has also been identified; this variant, like *ced*-4, is not activated by cytochrome-*c*. The *Apaf*-1 and *ced*-4-like variants show different caspase specificities. Some, but not all, alleles of *DARK* suppress eye-directed expression of *rpr*, *hid*, *grim*, and *Dcp-1*. These gene products are thought to promote activation of apical caspases.

Akt or protein kinase B, a downstream target of phosphatidyl inositol-3 kinase, has been implicated in regulating the survival of neurons in response to extracellular signals (Hemmings, 1997). Genetic analysis of a loss-of-function allele of fly Akt, *Dakt*, shows ectopic apoptosis in embryos as assessed by acridine orange and terminal transferase-mediated biotinylated-UTP nick end-labeling (TUNEL) staining (Staveley et al., 1998). This cell death is rescued by an *Dakt* transgene. Given that Akt has been implicated in mediating the survival-promoting effects of growth factors such as IGF-1, which has been reported to enhance survival of striatal primary cultures transiently expressing polyglutamine-expanded huntingtin fragments (Saudou et al., 1998), the availability of loss-of-function alleles of *Dakt1*, as well as a means of ectopically expressing *Dakt1*, provide important tools for assessing the role of this kinase in polyglutaminopathy in vivo.

The means whereby caspase activation promotes cell death is poorly understood. One substrate of activated caspase-3 is the heterodimeric protein, DNA fragmentation factor (DFF; Liu et al., 1997). The 45-kDa subunit of DFF has a *Drosophila* homolog, *DREP*-1 (Inohara et al., 1998). In vitro, both *DREP*-1 and DFF45 inhibit apoptosis induced by related proteins, CIDE-A and CIDE-E (cell death-inducing DFF-like effectors; Inohara et al., 1998). The mouse homolog of DFF45 and *DREP*-1, ICAD (inhibitor of caspase-activated DNAse) is cleaved in response to the inducible expression of a polyglutamine-expanded huntingtin fragment in a neuroblastoma cell line (Wang et al., 1999). At least two other *DREP*-1-related genes are present in *Drosophila* (G. Nunez, personal communication). As information becomes available about the function of these genes, it will be interesting to determine if their products interact with polyglutamine-expanded proteins in vivo and, if so, how this affects neuronal dysfunction and degeneration in the fly.

16.3. INSIGHTS FROM DEVELOPMENTAL CELL DEATH

Ectopic cell death during *Drosophila* development may be triggered by a variety of mutations. Although such developmental cell death is not strictly comparable to late-onset neurodegenerative disease in humans, consideration of such mutants may provide valuable insight into mechanisms of neurodegenerative disease. Precise development of appropriate connections between neuronal processes and their targets is required for survival of neurons; such a target-dependent survival phenomenon also occurs in *Drosophila*. Photoreceptor neurons that fail to establish connections with the optic lobe degenerate. Conversely, optic ganglion neurons that fail to establish contact with photoreceptors in mutants of genes required for eye specification such as *sine oculis* differentiate normally but subsequently degenerate (Cheyette et al., 1994; Pignoni et al., 1997). In the *disconnected* mutant, there is a defect in the pathfinding of the larval optic nerve, resulting in the formation of photoreceptor neurons disconnected from the optic lobe (Steller et al., 1987; Campos et al., 1992). Such *disconnected* photoreceptors subsequently degenerate, as does the optic lobe target tissue, demonstrating a reciprocal requirement between neurons and their targets for survival.

As noted earlier, loss-of-function mutants in eye-specification genes such as *sine oculis* result in failure of eyes to develop, associated with increased cell death anterior to the morphogenetic furrow, prior to specification of imaginal disc epithelia as eye progenitors. Developmental cell death is also observed in mutants for genes required for progression of the morphogenetic furrow. For example, mutations in *hedgehog* interfere with furrow progression. This results in increased cell death anterior to the furrow, failure of differentiation, and the appearance of a scar in the adult eye (Heberlein et al., 1993; Ma et al., 1996). Mutants acting within the morphogenetic furrow may also show abnormal cell death during imaginal disc development, resulting in eye abnormalities in the adult. For example, mutants of *atonal*, a basic helix–loop–helix protein, show failure of furrow progression and cell specification, resulting in apoptosis posterior to the remains of the furrow and a small eye (Jarman et al., 1994; Jarman et al., 1995). One prominent period of developmental cell death occurs in pupal life. This wave of death eliminates supernumerary cells in each ommatidium to form the perfect crystalline lattice of the adult eye. Mutations in *roughest/irreC*, a cell adhesion molecule, show failure of this apoptosis, resulting in a rough eye (Wolff and Ready, 1991). Inhibition of pupal cell death by eye-specific expression of P35 also disrupts perfection of the lattice structure, resulting in a mildly rough eye (Hay et al., 1994).

How can an understanding of specific developmental mutants in *Drosophila* contribute to the treatment of human diseases? A dramatic example has come from study of Math1, the mouse *atonal* homolog (Bermingham et al., 1999). Hair cells of the inner ear fail to develop in Math1 null embryos, raising the intriguing possibility that ectopic expression of Math1 might provide a treatment for some cases of deafness. Without *Drosophila* genetic studies to lay the groundwork for analysis of Math1, such an exciting prospect might never have arrived.

16.4. CLASSICAL GENETIC APPROACHES TO NEURODEGENERATION

Adult-onset neurodegeneration mutations may be arbitrarily divided into those affecting the retina and those affecting the brain. A number of mutants have demonstrated that aberrant regulation of phototransduction results in death of photoreceptor neurons. Despite fundamental differences in the physiology of phototransduction between invertebrates and man, a number of fly retinal degeneration mutants have provided insight into mechanism of retinal degeneration in man. In invertebrates, rhodopsin activates transducin, a G-protein, which then activates phospholipase C. In contrast, transducin in vertebrate phototransduction activates cGMP phosphodiesterase. Although the relationship between the mutant alleles and photoreceptor degeneration is incompletely characterized for many mutants, an emerging theme in retinal degeneration is that either failure of phototransduction or its sustained activation can be deleterious to neurons. One illustrative mutant is *rdgA* (Hotta and Benzer, 1970; Harris and Stark, 1977). In most of these mutant alleles, photoreceptor neurons begin to degenerate the first week posteclosion, although the electroretinogram is small even at eclosion, when photoreceptors are morphologically normal. One allele has been described in which photoreceptor morphology is abnormal even at eclosion (Stark and Carlson, 1985). Degeneration of R7 photoreceptors is less severe than that of outer cells (R1–6; Harris and Stark, 1977). Structure of the lamina is relatively normal (Johnson, 1982; Stark and Carlson 1985), although degenerating photoreceptor axon terminals in the optic lobe undergo phagocytosis by glia (Stark and Carlson 1985). The locus encodes a diacylglyceryl kinase; the turnover of diacylgycerol is crucial in deactivating the light response (Inoue et al., 1989; Masai et al., 1993).

Another mutant of interest is *rdgB* (Harris and Stark, 1977). These mutants also have disruption in the turnover of diacylglycerol. In these mutants, photoreceptors undergo light-dependent degeneration. Retinal morphology is essentially normal on eclosion, but in the presence of light, the retina degenerates within 1 wk; ultrastructural abnormalities are apparent within 3 d (Stark and Carlson, 1982). These include electron-dense cytoplasm with liposomes, lysosomelike bodies, vacuoles, electron-dense endoplasmic reticulum, abnormal mitochondria, and electron opaque photoreceptor axons lacking synaptic vesicles containing seemingly no typical presynaptic structures. As is the case with *rdgA*, R7 and R8 are relatively spared as compared to outer R cells (Chang et al., 1997). Degeneration is rescued by rearing in the dark. The locus encodes a putative integral membrane protein with some sequence homology to a Ca^{2+}-ATPase, suggesting that it may function as a Ca^{2+} transporter (Vihtelic et al., 1991).

The *rdgC* locus encodes a serine–threonine phosphatase necessary for the deactivation of rhodopsin (Steele et al., 1992). Mutants undergo degeneration that is relatively delayed compared with other retinal degeneration mutants, but by 8 wk posteclosion, photoreceptor neurons show some ultrastructural features characteristic of apoptosis, including cytoplasmic condensation, and clumping of chromatin with relatively preserved mitochondria. Degeneration is light-dependent (Vinos et al., 1997). Either dark rearing or the presence of the *ninaE* mutation rescues degeneration, demonstrating that degeneration is a consequence of the light stimulation of rhodopsin (Kurada and O'Tousa, 1995). One *rdgC* mutant allele is rescued by eye-directed expression of P35 (Davidson and Steller, 1998).

Arrestin-2 normally serves to inactivate phosphorylated rhodopsin by blocking interaction with transducin (Dolph et al., 1993). *Arrestin-2* mutants undergo light-dependent photoreceptor degeneration by 10 d posteclosion (Dolph et al., 1993; Alloway and Dolph, 1999). The mutation is dominant. Thus, sustained, inappropriate activation of phototransduction can result in neurodegeneration.

The *rdgE* mutant shows retinal degeneration by 2 d posteclosion in constant light (Zars and Hyde, 1996). Electrophysiological abnormalities are present prior to degeneration. Degeneration is slowed but not eliminated in darkness. Ultrastructural analysis of mutants in trans to a deficiency deleting the locus show random loss and vesiculation of rhadomeres because of problems with stability and recycling of rhabdomere microvilli.

Another retinal degeneration mutant is encoded by the *ninaE* (neither inactivation nor afterpotential) locus (Kurada and O'Tousa, 1995). The *ninaE* protein encodes the opsin moiety of the Rh1 rhodopsin, which is localized to outer (R1–6) photoreceptors. Thus, degeneration occurs primarily in these outer R cells; in some alleles, rhadomeres degenerate but photoreceptor cell bodies are spared (Stark and Sapp, 1987). Both dominant and recessive alleles of *ninaE* have been described. The former are likely to act as dominant negative mutations by suppressing wild-type rhodopsin production. Consequently, such alleles suppress the rapid degeneration observed in *rdgC* mutants. Degeneration is present by 1 d posteclosion for some alleles; other alleles show a much more prolonged course (Stark and Sapp, 1987; O'Tousa et al., 1989). Both light-dependent and light-independent degeneration have been described for different alleles; most alleles show light-independent degeneration. The *ninaE* gene product is required for the degeneration observed in *rdgC* mutants (Kurada and O'Tousa, 1995). Degeneration of one allele of *ninaE* is suppressed by P35, as assessed by morphologic, behavioral, and electrophysiological criteria (Davidson and Steller, 1998).

The *ninaA* allele encodes a fly homolog of cyclophilin (Shieh et al., 1989; Stamnes et al., 1991). Mutants have impaired transport of opsin from the endoplasmic reticulom (ER) and show dramatic accumulations of ER cisternae in photoreceptors (Colley et al., 1991). Degeneration is gradual and light independent, suggesting that degeneration ensues from failure of phototransduction.

How can an understanding of retinal degeneration mutants in flies contribute to understanding of related diseases in man? In many cases, genes identified in fly retinal degenerations are found to be homologous to those in man, thus facilitating analysis of the relationship between mutations and their pathophysiologic effects (Huang and Honkanen, 1998; Aikawa et al., 1997; den Hollander et al., 1999). Human homologs of fly retinal degeneration proteins may serve to rescue the mutant fly phenotype, demonstrating functional homology (orthology) between the proteins (Chang et al., 1997). Even more intriguing is the observation that inhibition of apoptosis in vivo restores functional visual behavior in certain fly retinal degeneration mutants (Davidson and Steller, 1998). Clearly, further understanding of retinal degeneration in this relatively simple organism may further our ability to analyze and perhaps eventually treat related disorders in humans.

Many of the alleles for retinal degeneration mutants arose spontaneously. A more directed approach to the study of neurodegeneration has been required for analysis of brain degeneration, dating from early studies two decades ago using large-scale histological screens. The pioneering work of Seymour Benzer was helpful in establishing the utility of *Drosophila* as a model organism for the study of neurodegenerative diseases and in establishing the molecular basis of neurodegenerative mutants. In a screen for mutants showing reduced life span, Benzer and co-workers isolated a mutant that they dubbed *sponge cake*, in which the brains of mutant flies demonstrate normal appearance at eclosion (Min and Benzer, 1997). At 29°C; however, the brains of hemizygous males and heterozygous females develop vacuoles, initially in the optic lobes and later in the medulla and lobula. Glia and neuronal cell bodies are unaffected, but axon terminals in the optic lobe become swollen and, at times, coalesce to form vacuoles. The molecular basis of the mutant phenotype remains to be clarified. The same approach was also used to identify *eggroll* (Min and Benzer, 1997). In this mutant, hemizygous males and heterozygous females are normal at eclosion but show degeneration by 4–5 d posteclosion. By 12 d, retinal degeneration is also seen. Both neurons and glia develop cytoplasmic lamellated inclusions that resemble those seen in storage diseases such as Tay-Sachs. Inclusions are present even in third-instar larvae, well before the onset of obvious degeneration. As is the case with *sponge cake*, the molecular basis of the mutation has not been clarified.

Benzer and co-workers clarified the molecular basis of *swiss cheese*, a mutant originally isolated using mass screening for morphologic defects of the brain (Kretzschmar et al., 1997). Beginning in late pupal development, neurons begin to show multilayered glial sheaths enveloping neurons. By 5 d posteclosion, vacuolization has begun throughout the brain. This is progressive and by 20 d results in cortical atrophy. A subset of neurons is intensely stained with toluidine blue; this staining also increases with age. TUNEL staining suggests that apoptotic cell death occurs for both glia and neurons. The phenotype is temperature dependent such that the changes described at 25°C are accelerated at 29°C. Ultrastructural analysis also reveals apoptotic features such as cell shrinkage and nuclear pyknosis. Homozygous flies show reduced life span. The *swiss cheese* gene product is expressed in cortical neurons and shows 41% amino acid identity with the human protein neuropathy target esterase (NTE). Intoxication with organophosphate esters results in covalent modification of NTE and a subacute distal axonal neuropathy. Both NTE and *swiss cheese* have a region of homology with the regulatory subunit of protein kinase A (Lush et al., 1998). *swiss cheese* is believed to regulate neuronal–glial interactions.

The gene *piroutte* was isolated during a screen for defective auditory responses (Eberl et al., 1997). Within days after eclosion, mutants begin circling, performing in progressively smaller circles as they age. Optic lobes coalesce into amorphous masses. Connections from retina to optic lobe degenerate and photoreceptors separate from one another. The molecular basis of the mutation remains to be clarified.

The mutant *vacuolar medulla* exhibits very rapid neurodegeneration posteclosion (Coombe and Heisenberg, 1986). Vacuoles begin to appear in the distal medulla within the first hour posteclosion; all mutants have vacuoles by 1 h. At later stages, vacuoles appear in the lobula and medulla and occasionally in the central brain. Degeneration is semidominant in that heterozygotes show some vacuoles beginning the sixth day posteclosion and less degeneration in the lamina. The molecular basis of the mutation is unknown.

The *drop dead* mutant shows normal development but begins to show uncoordinated motor activity within 11 d, followed rapidly by death (Buchanan and Benzer, 1993). Glia have shortened cytoplasmic processes that fail to envelop surrounding neurons properly. Neuronal morphology is relatively normal, although some neurons show abnormal lamellated cytoplasmic inclusions. As they age, mutants begin to show accelerated expression of age-dependent markers (Rogina et al., 1997). The wild-type gene product has been suggested to function in normal glial maintenance of neurons. Following death of laminar neurons, retinal cells undergo retrograde degeneration.

Another glial mutant that results in neuronal degeneration is *reversed polarity* (*repo*). *repo* is a glial homeodomain protein expressed within the optic lobes (Campbell et al., 1994; Xiong et al., 1994). Loss of laminar neurons in the optic lobe occurs secondary to the mutation impairing normal factor delivery from glia to neurons (Xiong and Montell, 1995). Photoreceptor neurons in the *repo* mutant undergo retrograde degeneration. Glia also degenerate, suggesting either that *repo* suppresses an intrinsic glial cell death program or that neurons are required to provide trophic support to glia.

The *bubblegum* mutant was isolated during a screen for P element insertion lines showing a reduced life span at 29°C (Min and Benzer, 1999). Histological analysis of homozygous mutants showed normal morphology at eclosion; however, by 10 d, a bubbly the lamina assumes a bubbly appearance, with accompanying degeneration of photoreceptor cell bodies. Ultrastructural analysis of these mutants showed dilated photoreceptor axons. The responsible mutation was identified as a gene with homology to a very-long-chain fatty acid (VLCFA) CoA synthetase. In adrenoleukodystrophy, an early-onset disorder of myelination, the activity of VLCFA CoA synthetase is reduced as a result of a mutation in a transporter protein thought to be necessary for transport or stabilization of the enzyme. Analysis of male flies shows that VLCFAs are elevated in *bubblegum* mutants, although, for unclear reasons, similar alterations are not identified in females. Most surprisingly, feeding of flies with glyceryl trioleate, a component of Lorenzo's oil, from the larval stages onward largely prevents neurodegeneration and restores visual behavior. This finding is astounding given that myelination is a much later phylogenetic innovation and the fact that ALD is primarily a disorder of myelin formation. The *bubblegum* mutant highlights the potential utility of *Drosophila* as a means of studying neurogenetic disorders and potential therapeutics, despite fundamental differences in the processes by which such mutations result in pathology.

16.5. *DROSOPHILA* HOMOLOGS OF HUMAN NEURODEGENERATION-ASSOCIATED GENES

Huntington's Disease

One use of *Drosophila* akin to the classical screens used to identify neurodegenerative mutants is the generation of loss-of-function alleles in fly homologs of genes mutated in human neurodegenerative diseases. Although targeted inactivation of identified *Drosophila* genes remains technically problematic, brute force approaches using chemical mutagenesis and taking advantage of the wealth of available P transposable element lines and deficiency libraries make such efforts feasible. Targeted inactivation of the mouse homolog of huntingtin, *Hdh*, has demonstrated a requirement for this gene in embryogenesis and morphogenesis (Duyao et al., 1995; Nasir et al., 1995; White et al., 1997; Zeitlin et al., 1995). Moreover, the demonstration that mutant huntingtin alleles with expanded repeats rescue the null phenotype has provided compelling in vivo evidence for the toxic gain-of-function hypothesis (White et al., 1997). These insights notwithstanding, mouse models have failed to elucidate the developmental function of huntingtin. Recently, a *Drosophila* homolog of huntingtin has been identified (Li et al., 1999). The exon–intron structure of the fly homolog differs significantly form that of human huntingtin. The fly homolog lacks the polyglutamine and polyproline tracts. Northern blot analysis indicates that the fly huntingtin, like human and mouse, is ubiquitously expressed. Now that the fly homolog has been identified, *Drosophila* genetics may be applied to identification of the native protein's function, in addition to the mechanism of the gain-of-function conferred by the unstable CAG tract.

ALS

A minority of cases of ALS are familial; of these, less than 20% are caused by mutations in Cu^{2+}/Zn^{2+} superoxide dismutase (SOD) (Rosen et al., 1993). Although neuronal cell death in diseases such as stroke, trauma, and Parkinson's disease has been thought for many years to depend in part on susceptibility to oxidative stress (e.g., [Jackson et al., 1990]), prior to discovery of such familial ALS mutations, evidence supporting this hypothesis in vivo has been scant. Even so, current opinion holds that the dominant ALS SOD mutations act as novel gain-of-function mutations (Cleveland, 1999). *Drosophila* SOD mutants act recessively to cause reduced life span and increased susceptibility to oxidative stress (Phillips, 1988; Staveley et al., 1990; Orr and Sohal, 1993). SOD transgenes act to rescue these phenotypes, even when expressed only in adult motoneurons (Orr and Sohal, 1993; Orr and Sohal, 1994; Parkes et al., 1998). Age-dependent photoreceptor degeneration has been described in SOD mutants, but such effects are not dominant (Phillips et al., 1995). Both the wild-type and a familial ALS-associated transgene associated with rapid degeneration function equally well to rescue early death, suggesting that the mechanism underlying the putative gain-of-function seen in FALS SOD mutants is not conserved in *Drosophila* (Elia et al., 1999). Nor does misexpression in the eye under *glass* control yield any obvious phenotype for either wild-type or FALS SOD transgenes (unpublished observations). It is conceivable that other potent FALS mutations might yield a misexpression phenotype or that provocative measures such as paraquat feeding might provoke a neurodegenerative phenotype; nonetheless, despite the intriguing findings observed in *Drosophila* studies of SOD, their relevance to the neurodegeneration observed in FALS remains unclear.

Alzheimer's Disease

Dominant mutations in the β-amyloid precursor protein (β-APP) and presenilins 1 and 2 have been described in early-onset familial Alzheimer's disease (Alzheimer's Disease Collaborative Research Group, 1995; Tanzi et al., 1992). Homologs of β-APP and presenilin are present in *Drosophila*. The fly β-APP homolog, *Appl*, lacks the segment of β-APP that is cleaved to generate pathogenic amyloid peptides; nonetheless, classical genetic approaches have been informative regarding the function of *Appl* in flies. Despite an absence of morphological defects, flies homozygous for large deletions in *Appl* show defective locomotor behavior; a human β-APP transgene rescues this behavior, demonstrating functional similarity between fly and human homologs (Luo et al., 1992). A role for *Appl* in synaptogenesis in flies is suggested by the loss of synaptic boutons in *Appl* null larval neuromuscular junctions, as well as the finding of increased synaptic bouton density observed when *Appl* or human β-APP is overexpressed in motoneurons (Torroja et al., 1999).

The *Drosophila* presenilin homolog, *Psn*, has also been informative regarding certain aspects of AD pathogenesis. The fly mutations are recessive larval lethal, but result in *Notch*-like phenotypes, including embryonic neuroblast hyperplasia (Struhl and Greenwald, 1999; Ye et al., 1999). Mosaic analysis of homozygous *Psn* clones in the wing also demonstrate *Notch*-like phenotypes, including scalloping and wing vein thickening (Ye et al., 1999). *Psn* is required for normal proteolytic processing of *Notch*, and mutations prevent proteolytic cleavage and nuclear access of *Notch* necessary for its signaling.

Misexpression approaches have also been applied to study of *Psn* in *Drosophila* (Ye and Fortini, 1999). Eye-specific overexpression of wild-type *Psn* using the GAL4/UAS transactivation system yields a rough eye, a defect suppressed by both P35 and DIAP-1. This phenotype affects pigment cells and yields missing bristles and fused ommatidia; although ommatidial structure is disrupted, photoreceptor neurons are intact. The rough eye phenotype is dominantly enhanced by strong *Notch* alleles. Wild-type *Psn* misexpression in larval eye discs results in increased TUNEL staining. These data suggest that aberrant *Psn* expression activates apoptosis the result of impaired *Notch* signaling, perhaps resulting from a dominant negativelike effect impairing *Notch* activity, resulting in abnormal developmental programming of cells that are then eliminated by apoptosis. One aspect of these transgenic *Psn* experiments that is difficult to reconcile with the dominant nature of human presenilin mutations is that misexpression of such mutants in the eye is generally less potent at generating rough eyes than the wild-type protein. Thus, the consequences of presenilin mutations may be complex, involving both dominant effects, as in humans, and loss-of-function effects, as in *C. elegans* (Levitan et al., 1996; Baumeister et al., 1997).

16.6. MODELS OF POLYGLUTAMINOPATHY IN *DROSOPHILA*

In 1998, two *Drosophila* models of polyglutaminopathy were published (Warrick et al., 1998; Jackson et al., 1998). The first of these (Warrick et al., 1998) expressed a truncated form of ataxin-3/MJD, the mutant protein in spinocerebellar ataxia-3/Machado-Joseph disease (Kawaguchi et al., 1994), using the binary GAL4/UAS transactivation system (Brand and Perrimon, 1993). In the MJD protein, the unstable polyglutamine repeat lies toward the carboxyl-terminus. A hemagglutinin epitope tag was engineered at the amino-terminus, followed by 12 amino acids of MJD. Forty-three amino acids of MJD lie on the carboxy terminal side of the repeats. Two repeat lengths were used: 27, corresponding to a "wild-type" repeat length, and 78. These cDNA constructs were engineered into the UAS vector. Expression of these constructs was driven by crossing with existing lines expressing the yeast transcription factor GAL4 under control of various tissue-specific enhancers.

These include the pan-neuronal *elav (embryonic lethal abnormal vision)*-GAL4, which is expressed in all neurons from the embryonic period into adulthood (Robinow and White, 1988). Non-neural drivers included 24B-GAL4, which expresses in presumptive mesoderm and muscle (Brand and Perrimon, 1993), as well as *dpp* (*decapentaplegic*)-GAL4 (Wilder and Perrimon, 1995), which is expressed in all imaginal discs at the anterior–posterior boundary (Blackman et al., 1991; Raftery et al., 1991). The authors also used the pGMR-GAL4 driver (*glass* multimer reporter; Hay et al., 1994; Freeman, 1996). *glass* is a zinc-finger transcription factor that is expressed in the eye disc beginning in third-instar larvae (Ellis et al., 1993). During this stage, the homogeneous eye epithelium differentiates into cell types that eventually form the specialized cell types comprising the adult compound eye, such as photoreceptor and mechanosensory neurons, pigment cells, and cone cells. Differentiation occurs in an orderly process as the morphogenetic furrow progresses, with successive differentiation of cells into ommatidial preclusters within and posterior to the furrow.

Warrick and co-workers reported on the effects of one UAS-MJDtrQ27 line and three UAS-MJDtrQ78 lines, which were classified as strong, medium, and weak (Warrick et al., 1998). Q27 constructs expressed in presumptive mesoderm and muscle cells had no effect; however, weak and medium-strength Q78 lines were larval lethal with this 24B driver, whereas the strong Q78 insertion was embryonic/early larval lethal. Expression using the *dpp*-GAL4 driver yielded no phenotypic effects for either short or expanded repeat constructs. Directed expression to all cells of the peripheral and central nervous system using the *elav*-GAL driver had no phenotypic effects for Q27. The strongest insertion for Q78 was lethal with the *elav*-GAL4 driver, whereas other Q78 insertions were viable but resulted in early adult death. Using the medium-strength Q78 insertion line, plastic sections revealed abnormal retinal morphology with the *elav*-GAL4 driver even at eclosion; this progressed to widespread degeneration by 4 d posteclosion. The Q27 line showed normal morphology of the retina at eclosion with the *elav*-GAL4 driver.

The most dramatic results were obtained using the pGMR-GAL4 driver. The Q27 line showed no effects on external morphology of the compound eye. The weak Q78 line showed a mildly rough external appearance at eclosion but progressive depigmentation thereafter, presumably the result of pigment cell degeneration. The medium and strong insertions showed rough, depigmented eye at eclosion; low-power plastic sections showed cell loss in the retinal cell layer that was progressively more severe with the stronger Q78 insertion lines examined.

Using the epitope tag, localization of MJD transgene products was examined in third-instar larval eye discs by immunohistochemistry. Q27 showed a cytoplasmic pattern. The strong Q78 transgene product was cytoplasmic just posterior to the morphogenetic furrow but showed progressive nuclear localization in older cells farther from the furrow. Using the *dpp*-GAL4 driver, Q78 was shown to be nuclear in larval leg discs and salivary glands, whereas the Q27 product showed diffuse cytoplasmic localization.

The second article to appear used amino-terminal fragments of huntingtin cDNA encoding 2, 75, or 120 glutamine residues (Jackson et al., 1998). This fragment is slightly larger than the exon-1 constructs used by Bates and co-workers to create their transgenic mice, including exons 2, 3, and a portion of 4. Using the numbering system of the published IT15 sequence encoding 23 repeats, these constructs encode the first 171 amino acids of human huntingtin. These cDNAs were subcloned directly into pGMR (Hay et al., 1994), which uses a minimal hsp70 promoter and the *glass* enhancer to permit expression in all cells of the eye from third-instar larvae through adulthood without the need for crossing with a GAL4 driver line. Q2 lines showed no phenotypic effects. Q75 lines showed some very late-onset degeneration of photoreceptor neurons; however, demonstration of such effects required examination of flies bearing four copies of the Q75 transgene. Q120 lines bearing single insertions of the transgene showed normal external morphology of the eye and normal development of the retina at eclosion as assessed by light and transmission electron microscopy; however, a subset of Q120 lines showed massive photoreceptor neuron degeneration beginning at 4 d and progressing thereafter. This progressive degeneration can be assessed in living flies by retrodromic illumination using a technique called the deep pseudopupil (Franceschini, 1972; *see also* Fig. 1). Toluidine blue-stained plastic sections at 10 d showed severe disruption of retinal morphology with intense staining in degenerating photoreceptor cell bodies and loss of rhabdomeres (*see* Fig. 2). Ultrastructural analysis of degenerating photoreceptors bearing Q120 transgenes showed some features of apoptosis, including nuclear and cytoplasmic condensation and chromatin clumping, with relative preservation of subcellular organelles such as mitochondria. Degeneration occurs independently of phototransduction, as it is observed in dark-reared flies (unpublished observations). Immunohistochemical analysis using an antibody recognizing the amino-terminus of human huntingtin showed cytoplasmic localization in third-instar larval eye discs, irrespective of repeat length. Huntingtin immunoreactivity was cytoplasmic in Q2 lines at both eclosion and 10 d. The Q75 transgenes product was cytoplasmic at eclosion and 10 d, but it showed some nuclear staining, as well, at 10 days. The Q120 product showed both cytoplasmic and diffuse nuclear staining at eclosion, but by ten days showed some punctate immunoreactivity suggestive of inclusion formation. More severe degeneration is observed in lines bearing multiple copies of the Q120 transgene, with degeneration apparent even at eclosion with two or more copies as assessed by pseudopupil or plastic sections; however, a normal external appearance is observed even in flies bearing six copies of the transgene *(unpublished observations).*

More recently, similar constructs encoding amino acids 1–171 containing 2 or 120 repeats in the UAS expression vector have been generated *(unpublished observations).* UAS-Q2 expressed using a relatively strong pGMR-GAL4 driver line (derived by transposition from Hu and Zipursky [unpublished]) showed no phenotypic effects for any insertion line examined; however, the strongest Q120 insertion obtained shows a mildly rough external eye by light microscopy. Little, if any, depigmentation is observed. Scanning electron microscopy shows missing bristles and mild irregularities in bristle placement with some fusion of ommatidia. This phenotype is comparable to, but slightly milder than that obtained for the weak UAS-MJDQ78 insertion. Other drivers obtain a similarly subtle phenotypic effect; using a strong *elav*-GAL4 driver line (Lin and Goodman, 1994), UAS-Q2 and UAS–Q120 show normal retinal morphology at eclosion; by contrast, the strong UAS-MJDQ78 line is lethal with this driver. Even at 10 d

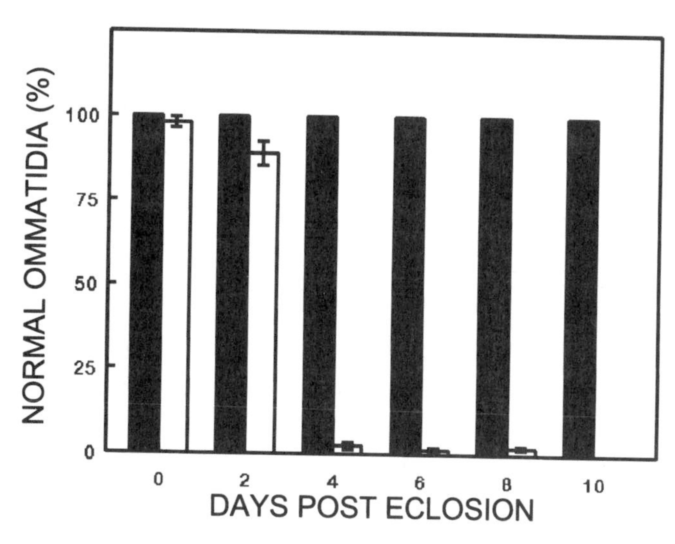

Fig. 1. Time-course for degeneration of rhabdomeres in wild-type (closed bars) and pGMR-Q120 flies (open bars) as determined by the pseudopupil technique. Data shown are the mean + SEM for five eyes. (Modified from Jackson, et al., 1998).

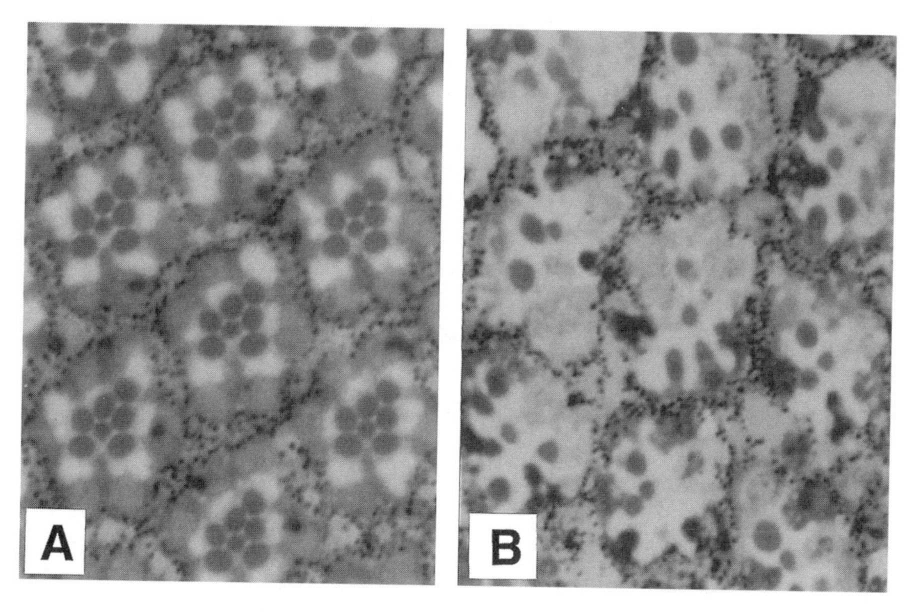

Fig. 2. Toluidine blue-stained plastic sections of wild-type (**A**) and pGMR-Q120 (**B**) transgenic flies 10 d posteclosion. Massive degeneration is observed in transgenic eyes at this time-point, with features including cytoplasmic condensation and rhabdomere loss apparent at the light level.

posteclosion, little, if any, loss of rhabdomeres or photoreceptor cell bodies is seen when UAS-Q120 is driven by *elav*-GAL4 (*unpublished observations*); these changes are much more subtle than those observed using direct Q120 expression with the pGMR construct.

Thus, the protein context in which expanded polyglutamine tracts appear is critical to phenotypic effects observed in *Drosophila* models, with even longer repeat lengths (e.g., UAS-Q120) showing much milder phenotypic effects when expressed within larger protein fragments (compare 120 repeats in 250 amino acids for UAS-Q120 to 78 repeats in 142 amino acids for UASMJD-Q78). Other investigators have directly addressed the importance of protein context by expressing essentially pure polyglutamine peptides in transgenic flies. Marsh and co-workers (Marsh et al., 2000) expressed CAG tracts encoding 22 or 108 repeats flanked by six additional amino- and four carboxy-terminal amino acids using the GAL4/UAS system. Additional constructs were generated containing a carboxy-terminal myc-FLAG epitope tag. Phenotypic effects were assessed using a range of driver lines, including pGMR-GAL4, *sev*-GAL4, and *dpp*-GAL4. An *elav*-GAL4 driver distinct from that employed by Warrick and co-workers (Lin and Goodman, 1994) was used by Marsh and co-workers (Ellis et al., 1993); this insertion on the second chromosome appears to be weaker than the X-chromosome insertion used to drive UAS-MJD constructs, as it gives viable adult flies when used with the strong MJDQ78 insertion, as compared to lethality with the X-chromosome insertion *(unpublished observations)*. Phenotypic effects of the Q108 transgene obtained were generally much more severe than those obtained with the truncated MJD construct; indeed, in many cases, lethality was complete, so that morphogenetic effects in adults could not be assessed. Using the *elav*-GAL4 driver to express UAS-Q108 in all neurons, incomplete lethality was observed; survivors showed premature adult death. This lethality was almost completely rescued by addition of the epitope tag. UAS-Q22 constructs showed no effect with the *elav*-GAL4 or indeed any driver, irrespective of the presence of the epitope tag; however, despite polymerase chain reaction (PCR) evidence for presence of genomic Q22 insertions, expression of these constructs could not be demonstrated by immunohistochemical techniques. Using the *sev*-GAL4 driver (Sun and Artavanis-Tsakonas, 1997), which drives expression beginning in third-instar larval eye discs in a subset of photoreceptor precursors (R3, R4, and R7), as well as cone cell precursors (Tomlinson et al., 1987), the UAS-Q108 transgene was completely lethal; this lethality was incompletely rescued by the epitope tag. Sensory bristles, which are of neural origin, were absent in surviving UAS-Q108 eyes. Using the same weak pGMR-GAL4 driver employed for

UAS-MJD constructs, significant lethality was obtained. Surviving eyes showed a very severe external phenotype, including complete absence of pigmentation, reduced size, and necrotic patches. The epitope tag ameliorated both lethality and the rough eye phenotype. In contrast to the completely benign effects of expanded repeats expressed within the MJD fragment, *dpp*-GAL4-driven UAS-Q108 resulted in pupal lethality. Dissected pupae showed abnormal morphology of the head and mesothorax, as well as missing third legs. Again, the addition of the epitope tag rescued lethality. A minority of these UASQ-108-myc-FLAG flies showed aberrant development of the mesothorax.

Marsh and co-workers also examined the effects of polyglutamine repeats of varying length within the context of *disheveled* (*dsh*), a ubiquitously expressed gene required for *wingless* signal transduction and containing a stretch of 28 glutamine residues near its amino-terminus (Klingensmith et al., 1994; Theisen et al., 1994). Constructs were engineered with the *dsh* cDNA containing 22 and 108 polyglutamine repeats in the UAS expression vector. (The 108 repeat construct actually contains an arginine interruption resulting from a PCR error generated during manipulation of the huntingtin exon-1 construct from which it was derived.) Constructs were also engineered that deleted the native polyglutamine-encoding tract within *dsh*, hereafter referred to as deleted Q. All *dsh* constructs resulted in some lethality with the *sev*- and *elav*-GAL4 drivers, presumably the result of ectopic activation of *wingless* signaling. Survivors of crosses with the *elav*-GAL4 driver line showed no external phenotype. Using the *sev*-GAL4 driver, survivors had reduced eye size and extra wing veins. Using the *dpp*-GAL4 driver, the wild-type and Q-deleted *dsh* transgenes showed complete lethality; animals dissected out of pupal cases showed shortened, thick legs and abnormal bristle morphology in wing blades. In contrast, *dpp*-GAL4-driven expression of *dsh*-Q108 resulted in the rescue of lethality and normal or minimally disordered wing morphology. Experiments also used the pCaSpER expression vector (Pirrotta, 1988) to drive expression of wild-type, Q-deleted, or Q108 *dsh* transgenes in a loss-of-function *dsh* background. This construct contains the *dsh* minimal promoter, allowing expression in the pattern of the wild-type *dsh* gene. The wild-type *dsh* transgene rescues the *dsh* phenotype, as does the deleted Q transgene; the latter showed only mild wing and bristle phenotypes, suggesting a mild loss-of-function effect due to deletion of the native *dsh* polyglutamine tract. By contrast, a much more dramatic loss of the transgene's ability to rescue the mutant *dsh* phenotype was caused by the expanded repeats, including notched wings, disordered bristle morphology, and bifurcated legs. Thus, the expanded repeats within *dsh* appear to impart a loss-of-function phenotype, impairing the ability of the transgene to rescue the *dsh* phenotype and yielding effects suggestive of impaired *dsh* function rather than polyglutaminopathy.

Using a *dsh* antibody, both the *dsh*-Q27 and *dsh*-Q108 transgenes showed cytoplasmic immunoreactivity in wing discs using a *ptc* (*patched*)-GAL4 driver (Speicher et al., 1994), which expresses in a pattern similar to that of *dpp*. Neither the Q108 nor Q22 transgene products could be detected without epitope tags, although genomic analysis identified the transgenes for Q22. The epitope tag was used to identify the Q108 peptide, which was cytoplasmic and possibly nuclear and perinuclear in wing discs; in salivary glands, staining was exclusively nuclear.

16.7. PERSPECTIVES ON *DROSOPHILA* MODELS OF MJD, HD, AND PURE POLYGLUTAMINOPATHY

Now that a variety of approaches have been used to study polyglutaminopathy in *Drosophila*, the relative merits and potential pitfalls of each approach may be examined. The most severe phenotypes obtained are those of lethality for UAS-Q108 with the majority of drivers (Marsh et al., 2000) and for UAS-MJDQ78 with the mesoderm/muscle (24B) and the pan-neuronal drivers (*elav*; Warrick et al., 1998). Screens for rescue from lethality provide an obvious advantage over the more laborious screens for modifiers of a visible phenotype. However, there are practical limitations to the design of screens for rescue from lethality, given the difficulty of raising enough healthy, fertile flies to develop a high-throughput screen.

The findings of lethality with a variety of non-neuronal drivers are surprising, given the relatively specific neuronal degeneration observed in various glutamine repeat diseases despite ubiquitous expression of the relevant genes. Some, but by no means all, of the phenotypes obtained using "pure" polyglutamine peptides or those within the context of a MJD fragment show neuronal selectivity. For example, a primarily neuronal phenotype, complete loss of photoreceptor cell bodies, is observed when pGMR-GAL4 is used to drive UAS-MJDQ78 (Warrick et al., 1998) or UAS-Q108 (Marsh et al., 2000). However, deleterious effects of polyglutamine on other cell types such as cone and pigment cells are indicated by the rough external lattice and depigmentation observed in each model. Even the UAS-huntingtin-Q120 construct, by far the weakest described UAS line, yields some effects on cone cells when expression is driven by pGMR-GAL4, as evidenced by ommatidial fusion visualized by scanning electron microscopy. Direct fusion of huntingtin constructs to the *glass*-binding region show highly specific effects on photoreceptor neurons; even with six copies of the pGMR-Q120 transgene, a normal appearance of the external lattice is obtained, despite severe abnormalities of the underlying retina even at eclosion *(unpublished observations)*. Although expanded MJD constructs show some neuronal specificity, their lethality with drivers restricting expression to embryonic mesoderm and muscle cells would seem to question the purportedly benign effects of these transgene products in non-neural cells. Marsh and co-workers attempted to rationalize findings of lethality obtained using a variety of putatively tissue-specific drivers by examining the effects of each driver on a UAS-GFP construct (Marsh et al., 2000). Using this technique, "leaky" expression with the pGMR-GAL4 driver was indeed observed around the esophagus and mouth hooks. Similarly, the *sev*-GAL4 driver was found to drive GFP expression in non-neuronal tissues, including the gut and trachea. The authors concluded that lethality observed in many cases with the UAS-Q108 construct was due to this leaky expression in non-neuronal tissues.

16.8. UTILITY OF THE FLY EYE IN STUDY OF POLYGLUTAMINOPATHY: FUTURE PROSPECTS AND POTENTIAL PITFALLS

Although in the aggregate, the effects of polyglutamine tracts on in the fly appear to be relatively more deleterious to neuronal tissues, this selectivity appears to be incomplete. The search for neuronal specificity is by no means a purely academic exercise, as interpretation of the effects of potential modifiers requires careful analysis of their effects on both neural and non-neural tissues. As an example, effects of polyglutamine tracts on pigment cells of the eye may be used to infer ameliorative effects of coexpressed transgenes. Careful analysis of such suppression requires an understanding of the variables affecting eye color in transgenic flies. The technique used to generate transgenic flies relies on the selection of transformants bearing the *w*+ transgene in a mutant *w*– background (i.e., identifying a few red-eyed flies among thousands of white-eyed ones). However, the mini-*w* gene used as a selectable marker in vectors such as pCaSpER, pGMR, and pUAS usually does not completely restore pigmentation to the wild-type level. Moreover, because of varying genomic insertion sites, different lines will have different eye colors. In many cases, crosses containing several different transgenes will show an additive effect on pigmentation. Variations in pigmentation also occur depending on the expression vector used; as an example, pGMR-huntingtin lines show light to dark red pigmentation, whereas UAS-huntingtin lines have eye colors varying from pale yellow to dark red.

Interpreting effects of transgenes on pigmentation may be simplified by expressing polyglutamine-encoding transgenes in a wild-type background (i.e., lacking the *w* mutation). By this technique, partial rescue of the pGMR-GAL4/UASMJDQ78 (strong) depigmentation phenotype is observed *(unpublished observations)* but not that of pGMR-GAL4/UASQ108 (Marsh et al., 2000). In the latter case, this was interpreted as evidence for complete degeneration of pigment cells independent of genetic background. Given the complexity of interpreting the effects of putative modifier genes on pigmentation, it seems prudent to use other criteria in addition to depigmentation in interpreting effects of modifier genes, such as external or internal morphology.

Another potential pitfall in the use of rough eye phenotypes for screens is the pleiotropic effects of GAL4 when expressed at high levels in the eye under *glass* regulation. The majority of pGMR-GAL4 lines generated (Freeman, 1996) show a rough phenotype, emphasizing the potential neurotoxic effects of GAL4 apart from those of other UAS transgenes. Even the weakest pGMR-GAL4 line (Freeman, 1996) is rough and depigmented with two copies. The external appearance of this line is essentially wild type with one insertion, but mild abnormalities of rhabdomeres are obvious using the deep pseudopupil; more complex abnormalities of ommatidial morphology are observed in plastic sections *(unpublished observations)*. Moreover, the pGMR-GAL4-driven eye phenotypes are highly subject to variation because of temperature effects (rescued at 18°C) and changes in genetic background, such as the presence of various balancer chromosomes *(unpublished observations)*. Nonetheless, pGMR-GAL4-driven rough eye phenotypes provide a highly sensitive means of screening for modifier genes. Ideally, putative modifiers identified by such screens can then be evaluated further using other drivers such as *elav*-GAL4 or by direct fusion constructs.

16.9. USING REVERSE GENETICS TO EVALUATE POTENTIAL MODIFIERS

A second approach to the dominant modifier screens, which requires no prior assumptions about the identity of potential modifiers, is to investigate the effects of candidate modifiers on the available phenotypes using misexpression approaches or *Drosophila* mutants in modifier homologs. As an example, the literature suggests that polyglutamine effects might be mediated by the action of caspases, chaperonins, or the ubiquitin–proteasome pathway (Cummings et al., 1998; Wellington et al., 1998; Chai et al., 1999; Chai et al., 1999; Ona et al., 1999). Indeed, such a reverse genetic approach to candidate modifiers has already proved fruitful:

overexpression of human hsp70 suppresses the phenotype of UAS-MJDQ78 using both *elav*- and pGMR-GAL4 drivers (Warrick et al., 1999). Conversely, misexpression of a dominant negative fly hsc4 containing a mutation in the ATP-binding domain (Elefant and Palter, 1999) enhances the rough phenotype. These findings confirm the potential utility of reverse genetic approaches to evaluate candidate modifiers in vivo. A complementary approach would be to use known mutants in endogenous flies genes to assess modifier effects; as an example, a corresponding dominant negative mutation in the fly hsc4 has been identified as a modifier of *Notch* signaling (Hing et al., 1999). With the imminent completion of the *Drosophila* genome project, evaluation of such loss-of-function alleles of candidate modifiers will further increase the utility of flies as model organisms for the study of polyglutaminopathy.

ACKNOWLEDGMENTS

This work was supported in part by the Cure HD Initiative of the Hereditary Disease Foundation and the NIH (NS02116). Thanks to Larry Zipursky for his longstanding interest in the use of *Drosophila* to study HD, and to Iris Salecker for helpful comments.

REFERENCES

Abrams, J. M. (1999) An emerging blueprint for apoptosis in *Drosophila*. *Trends Cell Biol.* **9,** 435–440.

Aikawa, Y., Hara, H., and Watanabe, T. (1997) Molecular cloning and characterization of mammalian homologs of the *Drosophila* retinal degeneration B gene. *Biochem. Biophys. Res. Commun.* **236,** 559–564.

Alloway, P. G. and Dolph, P. J. (1999) A role for the light-dependent phosphorylation of visual arrestin. *Proc. Natl. Acad. Sci. USA* **96,** 6072–6077.

Alzheimer's Disease Collaborative Group (1995) The structure of the presenilin 1 (S182) gene and identification of six novel mutations in early onset AD families. *Nat. Genet.* **11,** 219–222.

Ambrosini, G., Adida, C., and Altieri, D. C. (1997) A novel anti-apoptosis gene, survivin, expressed in cancer and lymphoma. *Nat. Med.* **3,** 917–921.

Baumeister, R., Leimer, U., Zweckbronner, I., Jakubek, C., Grunberg, J., and Haass, C. (1997) Human presenilin-1, but not familial Alzheimer's disease (FAD) mutants, facilitate *Caenorhabditis elegans* Notch signalling independently of proteolytic processing. *Genes Funct.* **1,** 149–159.

Baxendale, S., Abdulla, S., Elgar, G., Buck, D., Berks, M., Micklem, G., et al. (1995) Comparative sequence analysis of the human and pufferfish Huntington's disease genes. *Nat. Genet.* **10,** 67–76.

Bergmann, A., Agapite, J., McCall, K., and Steller, H. (1998) The *Drosophila* gene hid is a direct molecular target of Ras-dependent survival signaling. *Cell* **95,** 331–341.

Bermingham, N. A., Hassan, B. A., Price, S. D., Vollrath, M. A., Ben-Arie, N., Eatock, R. A., et al. (1999) Math1: an essential gene for the generation of inner ear hair cells. *Science* **284**, 1837–1841.

Birnbaum, M. J., Clem, R. J., and Miller, L. K. (1994) An apoptosis-inhibiting gene from a nuclear polyhedrosis virus encoding a polypeptide with Cys/His sequence motifs. *J. Virol.* **68,** 2521–2528.

Blackman, R. K., Sanicola, M., Raftery, L. A., Gillevet, T., and Gelbart, W. M. (1991) An extensive 3' cis-regulatory region directs the imaginal disk expression of decapentaplegic, a member of the TGF–beta family in *Drosophila. Development* **111,** 657–666.

Boutell, J. M., Wood, J. D., Harper, P. S., and Jones, A. L. (1998) Huntingtin interacts with cystathionine beta-synthase. *Hum. Mol. Genet.* **7,** 371–378.

Brand, A. H. and Perrimon, N. (1993) Targeted gene expression as a means of altering cell fates and generating dominant phenotypes. *Development* **118,** 401–415.

Buchanan, R. L. and Benzer, S. (1993) Defective glia in the Drosophila brain degeneration mutant drop-dead. *Neuron* **10,** 839–850.

Bump, N. J., Hackett, M., Hugunin, M., Seshagiri, S., Brady, K., Chen, P., et al. (1995) Inhibition of ICE family proteases by baculovirus antiapoptotic protein p35. *Science* **269**, 1885–1888.

Campbell, G., Goring, H., Lin, T., Spana, E., Andersson, S., Doe, C. Q., et al. (1994) RK2, a glial-specific homeodomain protein required for embryonic nerve cord condensation and viability in *Drosophila. Development* **120,** 2957–2966.

Campos, A. R., Fischbach, K. F., and Steller, H. (1992) Survival of photoreceptor neurons in the compound eye of *Drosophila* depends on connections with the optic ganglia. *Development* **114,** 355–366.

Chai, Y., Koppenhafer, S. L., Bonini, N. M., and Paulson, H. L. (1999) Analysis of the role of heat shock protein (Hsp) molecular chaperones in polyglutamine disease. *J. Neurosci.* **19,** 10,338–10,347.

Chai, Y., Koppenhafer, S. L., Shoesmith, S. J., Perez, M. K., and Paulson, H. L. (1999) Evidence for proteasome involvement in polyglutamine disease: localization to nuclear inclusions in SCA3/MJD and suppression of polyglutamine aggregation in vitro. *Hum. Mol. Genet.* **8,** 673–682.

Chang, J. T., Milligan, S., Li, Y., Chew, C. E., Wiggs, J., Copeland, N. G., et al. (1997) Mammalian homolog of Drosophila retinal degeneration B rescues the mutant fly phenotype. *J. Neurosci.* **17,** 5881–5890.

Chen, P., Lee, P., Otto, L., and Abrams, J. (1996) Apoptotic activity of REAPER is distinct from signaling by the tumor necrosis factor receptor 1 death domain. *J. Biol. Chem.* **271,** 25,735–25,737.

Chen, P., Nordstrom, W., Gish, B., and Abrams, J. M. (1996) grim, a novel cell death gene in *Drosophila. Genes Dev.* **10,** 1773–1782.

Chen, P., Rodriguez, A., Erskine, R., Thach, T., and Abrams, J. M. (1998) Dredd, a novel effector of the apoptosis activators reaper, grim, and hid in *Drosophila. Dev. Biol.* **201,** 202–216.

Cheyette, B. N., Green, P. J., Martin, K., Garren, H., Hartenstein, V., and Zipursky, S. L. (1994) The *Drosophila* sine oculis locus encodes a homeodomain-containing protein required for the development of the entire visual system. *Neuron* **12,** 977–996.

Claveria, C., Albar, J. P., Serrano, A., Buesa, J. M., Barbero, J. L., Martinez, A. C., et al. (1998) Drosophila grim induces apoptosis in mammalian cells. *EMBO J.* **17,** 7199–7208.

Clem, R. J., Fechheimer, M., and Miller, L. K. (1991) Prevention of apoptosis by a baculovirus gene during infection of insect cells. *Science* **254,** 1388–1390.

Cleveland, D. W. (1999) From Charcot to SOD1: mechanisms of selective motor neuron dath in ALS. *Neuron* **23,** 515–520.

Colley, N. J., Baker, E. K., Stamnes, M. A., and Zuker, C. S. (1991) The cyclophilin homolog ninaA is required in the secretory pathway. *Cell* **67,** 255–263.

Coombe, P. E. and Heisenberg, M. (1986) The structural brain mutant Vacuolar medulla of *Drosophila melanogaster* with specific behavioral defects and cell degeneration in the adult. *J. Neurogenet.* **3,** 135–158.

Crook, N. E., Clem, R. J., and Miller, L. K. (1993) An apoptosis-inhibiting baculovirus gene with a zinc finger-like motif. *J. Virol.* **67,** 2168–2174.

Cummings, C. J., Mancini, M. A., Antalffy, B., DeFranco, D. B., Orr, H. T., and Zoghbi, H. Y. (1998) Chaperone suppression of aggregation and altered subcellular proteasome localization imply protein misfolding in SCA1. *Nat. Genet.* **19,** 148–154.

Davidson, F. F. and Steller, H. (1998) Blocking apoptosis prevents blindness in Drosophila retinal degeneration mutants. *Nature* **391,** 587–591.

den Hollander, A. I., ten Brink, J. B., de Kok, Y. J., van Soest, S., van den Born, L. I., van Driel, M. A., et al. (1999) Mutations in a human homolog of Drosophila crumbs cause retinitis pigmentosa (RP12) *Nat. Genet.* **23,** 217–221.

Djian, P., Hancock, J. M., and Chana, H. S. (1996) Codon repeats in genes associated with human diseases: fewer repeats in the genes of nonhuman primates and nucleotide substitutions concentrated at the sites of reiteration. *Proc. Natl. Acad. Sci. USA* **93,** 417–421.

Dolph, P. J., Ranganathan, R., Colley, N. J., Hardy, R. W., Socolich, M., and Zuker, C. S. (1993) Arrestin function in inactivation of G protein-coupled receptor rhodopsin in vivo. *Science* **260,** 1910–1916.

Dorstyn, L., Colussi, P. A., Quinn, L. M., Richardson, H., and Kumar, S. (1999) DRONC, an ecdysone-inducible *Drosophila* caspase. *Proc. Natl. Acad. Sci. USA* **96,** 4307–4312.

Dorstyn, L., Read, S. H., Quinn, L. M., Richardson, H., and Kumar, S. (1999) DECAY, a novel *Drosophila* caspase related to mammalian caspase-3 and caspase-7. *J. Biol. Chem.* **274,** 30,778–30,783.

Duckett, C. S., Nava, V. E., Gedrich, R. W., Clem, R. J., Van Dongen, J. L., Gilfillan, M. C. et al. (1996) A conserved family of cellular genes related to the baculovirus iap gene and encoding apoptosis inhibitors. *EMBO J.* **15,** 2685–2694.

Duyao, M. P., Auerbach, A. B., Ryan, A., Persichetti, F., Barnes, G. T., McNeil, S. M., et al. (1995) Inactivation of the mouse Huntington's disease gene homolog Hdh. *Science* **269,** 407–410.

Eberl, D. F., Duyk, G. M., and Perrimon, N. (1997) A genetic screen for mutations that disrupt an auditory response in *Drosophila melanogaster. Proc. Natl. Acad. Sci. USA* **94,** 14,837–14,842.

Elefant, F. and Palter, K. B. (1999) Tissue-specific expression of dominant negative mutant *Drosophila* HSC70 causes developmental defects and lethality. *Mol. Biol. Cell* **10,** 2101–2117.

Elia, A. J., Parkes, T. L., Kirby, K., St. George-Hyslop, P., Boulianne, G. L., Phillips, J. P., et al. (1999) Expression of human FALS SOD in motorneurons of *Drosophila. Free Radical Biol. Med.* **26,** 1332–1338.

Ellis, M. C., O'Neill, E. M., and Rubin, G. M. (1993) Expression of *Drosophila* glass protein and evidence for negative regulation of its activity in non-neuronal cells by another DNA-binding protein. *Development* **119,** 855–865.

Faber, P. W., Alter, J. R., MacDonald, M. E., and Hart, A. C. (1999) Polyglutamine-mediated dysfunction and apoptotic death of a *Caenorhabditis elegans* sensory neuron. *Proc. Natl. Acad. Sci. USA* **96,** 179–184.

Faber, P. W., Barnes, G. T., Srinidhi, J., Chen, J., Gusella, J. F., and MacDonald, M. E. (1998) Huntingtin interacts with a family of WW domain proteins. *Hum. Mol. Genet.* **7,** 1463–1474.

Farahani, R., Fong, W. G., Korneluk, R. G., and MacKenzie, A. E. (1997) Genomic organization and primary characterization of miap-3: the murine homolog of human X-linked IAP. *Genomics* **42,** 514–518.

Folstein, S. E. (1989) *Huntington's Disease: A Disorders of Families.* Johns Hopkins University Press, Baltimore.

Franceschini, N. (1972) Pupil and pseudopupil in the compound eye of *Drosophila*, in *Information Processing in the Visual System of Drosophila*, (Wehner, R., ed.), Springer-Verlag, Berlin, pp. 75–82.

Fraser, A. G. and Evan, G. I. (1997) Identification of a *Drosophila melanogaster* ICE/CED-3-related protease, drICE. *EMBO J.* **16,** 2805–2813.

Fraser, A. G., James, C., Evan, G. I., and Hengartner, M. O. (1999) *Caenorhabditis elegans* inhibitor of apoptosis protein (IAP) homolog BIR-1 plays a conserved role in cytokinesis. *Curr. Biol.* **9,** 292–301.

Fraser, A. G., McCarthy, N. J., and Evan, G. I. (1997) drICE is an essential caspase required for apoptotic activity in *Drosophila* cells. *EMBO J.* **16,** 6192–6199.

Freeman, M. (1996) Reiterative use of the EGF receptor triggers differentiation of all cell types in the *Drosophila* eye. *Cell* **87,** 651–660.

Grether, M. E., Abrams, J. M., Agapite, J., White, K., and Steller, H. (1995) The head involution defective gene of *Drosophila melanogaster* functions in programmed cell death. *Genes Dev.* **9,** 1694–1708.

Gusella, J. F. and MacDonald, M. E. (1995) Huntington's disease: CAG genetics expands neurobiology. *Curr. Opin. Neurobiol.* **5,** 656–662.

Haining, W. N., Carboy-Newcomb, C., Wei, C. L., and Steller, H. (1999) The proapoptotic function of *Drosophila* Hid is conserved in mammalian cells. *Proc. Natl. Acad. Sci. USA* **96,** 4936–4941.

Harris, W. A. and Stark, W. S. (1977) Hereditary retinal degeneration in *Drosophila melanogaster.* A mutant defect associated with the phototransduction process. *J. Gen. Physiol.* **69,** 261–291.

Hay, B. A., Wassarman, D. A., and Rubin, G. M. (1995) *Drosophila* homologs of baculovirus inhibitor of apoptosis proteins function to block cell death. *Cell* **83,** 1253–1262.

Hay, B. A., Wolff, T., and Rubin, G. M. (1994) Expression of baculovirus P35 prevents cell death in *Drosophila. Development* **120,** 2121–2129.

Heberlein, U., Wolff, T., and Rubin, G. M. (1993) The TGF beta homolog dpp and the segment polarity gene hedgehog are required for propagation of a morphogenetic wave in the *Drosophila* retina. *Cell* **75,** 913–926.

Hemmings, B. A. (1997) Akt signaling: linking membrane events to life and death decisions. *Science* **275,** 628–630.

Hengartner, M. O., and Horvitz, H. R. (1994) Programmed cell death in *Caenorhabditis elegans. Curr. Opin. Genet. Dev.* **4,** 581–586.

Hing, H. K., Bangalore, L., Sun, X., and Artavanis-Tsakonas, S. (1999) Mutations in the heatshock cognate 70 protein (hsc4) modulate Notch signaling. *Eur. J. Cell Biol.* **78,** 690–697.

Hotta, Y. and Benzer, S. (1970) Genetic dissection of the *Drosophila* nervous system by means of mosaics. *Proc. Natl. Acad. Sci. USA* **67,** 1156–1163.

Huang, X. and Honkanen, R. E. (1998) Molecular cloning, expression, and characterization of a novel human serine/threonine protein phosphatase, PP7, that is homologous to *Drosophila* retinal degeneration C gene product (rdgC). *J. Biol. Chem.* **273,** 1462–1468.

Huntington's Disease Collaborative Research Group (1993) A novel gene containing a trinucleotide repeat that is expanded and unstable on Huntington's disease chromosomes. The Huntington's Disease Collaborative Research Group. *Cell* **72,** 971–983.

Inohara, N., Koseki, T., Chen, S., Wu, X., and Nunez, G. (1998) CIDE, a novel family of cell death activators with homology to the 45 kDa subunit of the DNA fragmentation factor. EMBO J. **17,** 2526–2533.

Inohara, N., Koseki, T., Hu, Y., Chen, S., and Nunez, G. (1997) CLARP, a death effector domain-containing protein interacts with caspase-8 and regulates apoptosis. P*Proc. Natl. Acad. Sci. USA* **94,** 10,717–10,722.

Inoue, H., Yoshioka, T., and Hotta, Y. (1989) Diacylglycerol kinase defect in a *Drosophila* retinal degeneration mutant rdgA. *J. Biol. Chem.* **264,** 5996–6000.

Jackson, G. R., Apffel, L., Werrbach-Perez, K., and Perez-Polo, J. R. (1990) Role of nerve growth factor in oxidant-antioxidant balance and neuronal injury. I. Stimulation of hydrogen peroxide resistance. *J. Neurosci. Res.* **25,** 360–368.

Jackson, G. R., Salecker, I., Dong, X., Yao, X., Arnheim, N., Faber, P. W., et al. (1998) Polyglutamine-expanded human huntingtin transgenes induce degeneration of *Drosophila* photoreceptor neurons. *Neuron* **21,** 633–642.

Jarman, A. P., Grell, E. H., Ackerman, L., Jan, L. Y., and Jan, Y. N. (1994) Atonal is the proneural gene for *Drosophila* photoreceptors. *Nature* **369,** 398–400.

Jarman, A. P., Sun, Y., Jan, L. Y., and Jan, Y. N. (1995) Role of the proneural gene, atonal, in formation of *Drosophila* chordotonal organs and photoreceptors. *Development* **121,** 2019–2030.

Johnson, M. A., Frayer, K.L., Stark, W.S. (1982) Characterization of rdgA: mutants with retinal degeneration in *Drosophila. J. Insect Physiol.* **28,** 233–242.

Kalchman, M. A., Graham, R. K., Xia, G., Koide, H. B., Hodgson, J. G., Graham, K. C., et al. (1996) Huntingtin is ubiquitinated and interacts with a specific ubiquitin-conjugating enzyme. *J. Biol. Chem.* **271,** 19,385–19,394.

Kanuka, H., Sawamoto, K., Inohara, N., Matsuno, K., Okano, H., and Miura, M. (1999) Control of the cell death pathway by Dapaf-1, a *Drosophila* Apaf-1 CED-4-related caspase activator. *Mol. Cell* **4,** 757–769.

Kawaguchi, Y., Okamoto, T., Taniwaki, M., Aizawa, M., Inoue, M., Katayama, S., et al. (1994) CAG expansions in a novel gene for Machado-Joseph disease at chromosome 14q32.1. *Nat. Genet.* **8,** 221–228.

Klingensmith, J., Nusse, R., and Perrimon, N. (1994) The *Drosophila* segment polarity gene dishevelled encodes a novel protein required for response to the wingless signal. *Genes Dev.* **8,** 118–130.

Kretzschmar, D., Hasan, G., Sharma, S., Heisenberg, M., and Benzer, S. (1997) The swiss cheese mutant causes glial hyperwrapping and brain degeneration in *Drosophila. J. Neurosci.* **17,** 7425–7432.

Kurada, P. and O'Tousa, J. E. (1995) Retinal degeneration caused by dominant rhodopsin mutations in *Drosophila. Neuron* **14,** 571–579.

Kurada, P. and White, K. (1998) Ras promotes cell survival in *Drosophila* by downregulating hid expression. *Cell* **95,** 319–329.

Levitan, D., Doyle, T. G., Brousseau, D., Lee, M. K., Thinakaran, G., Slunt, H. H., et al. (1996) Assessment of normal and mutant human presenilin function in *Caenorhabditis elegans. Proc. Natl. Acad. Sci. USA* **93,** 14,940–14,944.

Li, Z., Karlovich, C. A., Fish, M. P., Scott, M. P., and Myers, R. M. (1999) A putative *Drosophila* homolog of the Huntington's disease gene. *Hum. Mol. Genet.* **8,** 1807–1815.

Lin, D. M. and Goodman, C. S. (1994) Ectopic and increased expression of Fasciclin II alters motoneuron growth cone guidance. *Neuron* **13,** 507–523.

Liu, X., Zou, H., Slaughter, C., and Wang, X. (1997) DFF, a heterodimeric protein that functions downstream of caspase-3 to trigger DNA fragmentation during apoptosis. *Cell* **89,** 175–184.

Luo, L., Tully, T., and White, K. (1992) Human amyloid precursor protein ameliorates behavioral deficit of flies deleted for Appl gene. *Neuron* **9,** 595–605.

Lush, M. J., Li, Y., Read, D. J., Willis, A. C., and Glynn, P. (1998) Neuropathy target esterase and a homologous *Drosophila* neurodegeneration-associated mutant protein contain a novel domain conserved from bacteria to man. *Biochem. J.* **332,** 1–4.

Ma, C., Liu, H., Zhou, Y., and Moses, K. (1996) Identification and characterization of autosomal genes that interact with glass in the developing *Drosophila* eye. *Genetics* **142,** 1199–1213.

MacDonald, M. E., Vonsattel, J. P., Shrinidhi, J., Couropmitree, N. N., Cupples, L. A., Bird, E. D., et al. (1999) Evidence for the GluR6 gene associated with younger onset age of Huntington's disease. *Neurology* **53,** 1330–1332.

Marsh, J. L., Walker, H., Theisen, H., Zhu, Y. Z., Fielder, T., Purcell, J., et al. (2000) Expanded polyglutamine peptides alone are intrinsically cytotoxic and cause neurodegeneration in *Drosophila. Hum. Mol. Genet.* **9,** 13–25.

Martin, J. B. and Gusella, J. F. (1986) Huntington's disease. Pathogenesis and management. *N. Engl. J. Med.* **315,** 1267–1276.

Masai, I., Okazaki, A., Hosoya, T., and Hotta, Y. (1993) *Drosophila* retinal degeneration A gene encodes an eye-specific diacylglycerol kinase with cysteine-rich zinc-finger motifs and ankyrin repeats. *Proc. Natl. Acad. Sci. USA* **90,** 11,157–11,161.

McCall, K. and Steller, H. (1998) Requirement for DCP–1 caspase during *Drosophila* oogenesis. *Science* **279,** 230–234.

McCarthy, J. V. and Dixit, V. M. (1998) Apoptosis induced by *Drosophila* reaper and grim in a human system. Attenuation by inhibitor of apoptosis proteins (cIAPs) *J. Biol. Chem.* **273,** 24,009–24,015.

Min, K. T. and Benzer, S. (1999) Preventing neurodegeneration in the *Drosophila* mutant bubblegum. *Science* **284,** 1985–1988.

Min, K. T. and Benzer, S. (1997) Spongecake and eggroll: two hereditary diseases in *Drosophila* resemble patterns of human brain degeneration. *Curr. Biol.* **7,** 885–888.

Nasir, J., Floresco, S. B., O'Kusky, J. R., Diewert, V. M., Richman, J. M., Zeisler, J., et al. (1995) Targeted disruption of the Huntington's disease gene results in embryonic lethality and behavioral and morphological changes in heterozygotes. *Cell* **81,** 811–823.

Ona, V. O., Li, M., Vonsattel, J. P., Andrews, L. J., Khan, S. Q., Chung, W. M., et al. (1999) Inhibition of caspase-1 slows disease progression in a mouse model of Huntington's disease. *Nature* **399,** 263–267.

Orr, W. C. and Sohal, R. S. (1993) Effects of Cu-Zn superoxide dismutase overexpression of life span and resistance to oxidative stress in transgenic *Drosophila melanogaster. Arch. Biochem. Biophys.* **301,** 34–40.

Orr, W. C. and Sohal, R. S. (1994) Extension of life-span by overexpression of superoxide dismutase and catalase in *Drosophila melanogaster. Science* **263,** 1128–1130.

O'Tousa, J. E., Leonard, D. S., and Pak, W. L. (1989) Morphological defects in photoreceptors caused by mutation in R1-6 opsin gene of *Drosophila. J. Neurogenet.* **6,** 41–52.

Parkes, T. L., Elia, A. J., Dickinson, D., Hilliker, A. J., Phillips, J. P., and Boulianne, G. L. (1998) Extension of *Drosophila* lifespan by overexpression of human SOD1 in motorneurons. *Nat. Genet.* **19,** 171–174.

Phillips, J. P. (1988) A cSOD-null mutant of *Drosophila* confers sensitivity to paraquat and reduced longevity. *Basic Life Sci.* **49,** 689–693.

Phillips, J. P., Tainer, J. A., Getzoff, E. D., Boulianne, G. L., Kirby, K., and Hilliker, A. J. (1995) Subunit-destabilizing mutations in *Drosophila* copper/zinc superoxide dismutase: neuropathology and a model of dimer dysequilibrium. *Proc. Natl. Acad. Sci. USA* **92,** 8574–8578.

Pignoni, F., Hu, B., Zavitz, K. H., Xiao, J., Garrity, P. A., and Zipursky, S. L. (1997) The eye-specification proteins So and Eya form a complex and regulate multiple steps in *Drosophila* eye development. *Cell* **91,** 881–891.

Pirrotta, V. (1988) Vectors for P-mediated transformation in *Drosophila. Biotechnology* **10,** 437–456.

Raftery, L. A., Sanicola, M., Blackman, R. K., and Gelbart, W. M. (1991) The relationship of decapentaplegic and engrailed expression in *Drosophila* imaginal disks: do these genes mark the anterior-posterior compartment boundary? *Development* **113,** 27–33.

Robinow, S. and White, K. (1988) The locus elav of *Drosophila melanogaster* is expressed in neurons at all developmental stages. *Dev. Biol.* **126,** 294–303.

Rodriguez, A., Oliver, H., Zou, H., Chen, P., Wang, X., and Abrams, J. M. (1999) Dark is a *Drosophil*a homolog of Apaf-1/CED-4 and functions in an evolutionarily conserved death pathway. *Nat. Cell Biol.* **1**, 272–279.

Rogina, B., Benzer, S., and Helfand, S. L. (1997) *Drosophila* drop-dead mutations accelerate the time course of age- related markers. *Proc. Natl. Acad. Sci. USA* **94,** 6303–6306.

Rosen, D. R., Siddique, T., Patterson, D., Figlewicz, D. A., Sapp, P., Hentati, A., et al. (1993) Mutations in Cu/Zn superoxide dismutase gene are associated with familial amyotrophic lateral sclerosis. *Nature* **362,** 59–62.

Rubin, G. M. (1998) The *Drosophila* genome project: a progress report. *Trends Genet.* **14,** 340–343.

Saudou, F., Finkbeiner, S., Devys, D., and Greenberg, M. E. (1998) Huntingtin acts in the nucleus to induce apoptosis but death does not correlate with the formation of intranuclear inclusions. *Cell* **95,** 55–66.

Shieh, B. H., Stamnes, M. A., Seavello, S., Harris, G. L., and Zuker, C. S. (1989) The ninaA gene required for visual transduction in Drosophila encodes a homolog of cyclosporin A-binding protein. *Nature* **338,** 67–70.

Song, Z., McCall, K., and Steller, H. (1997) DCP-1, a *Drosophila* cell death protease essential for development. *Science* **275,** 536–540.

Speicher, S. A., Thomas, U., Hinz, U., and Knust, E. (1994) The Serrate locus of *Drosophila* and its role in morphogenesis of the wing imaginal discs: control of cell proliferation. *Development* **120,** 535–544.

Stamnes, M. A., Shieh, B. H., Chuman, L., Harris, G. L., and Zuker, C. S. (1991) The cyclophilin homolog ninaA is a tissue-specific integral membrane protein required for the proper synthesis of a subset of *Drosophila* rhodopsins. *Cell* **65,** 219–227.

Stark, W. S. and Carlson, S. D. (1985) Retinal degeneration in rdgA mutants of *Drosophila melanogaster. Int. J. Insect Morph. Embryol.* **14,** 243–254.

Stark, W. S. and Carlson, S. D. (1982) Ultrastructural pathology of the compound eye and optic neuropiles of the retinal degeneration mutant (w rdg BKS222) D*rosophila melanogaster. Cell Tissue Res.* **225,** 11–22.

Stark, W. S. and Sapp, R. (1987) Ultrastructure of the retina of *Drosophila melanogaster*: the mutant ora (outer rhabdomeres absent) and its inhibition of degeneration in rdgB (retinal degeneration-B) *J. Neurogenet.* **4,** 227–240.

Staveley, B. E., Phillips, J. P., and Hilliker, A. J. (1990) Phenotypic consequences of copper-zinc superoxide dismutase overexpression in *Drosophila melanogaster. Genome* **33,** 867–872.

Staveley, B. E., Ruel, L., Jin, J., Stambolic, V., Mastronardi, F. G., Heitzler, P., et al. (1998) Genetic analysis of protein kinase B (AKT) in *Drosophila. Curr. Biol.* **8,** 599–602.

Steele, F. R., Washburn, T., Rieger, R., and O'Tousa, J. E. (1992) *Drosophila* retinal degeneration C (rdgC) encodes a novel serine/threonine protein phosphatase. *Cell* **69,** 669–676.

Steller, H., Fischbach, K. F., and Rubin, G. M. (1987) Disconnected: a locus required for neuronal pathway formation in the visual system of *Drosophila*. *Cell* **50,** 1139–1153.

Struhl, G. and Greenwald, I. (1999) Presenilin is required for activity and nuclear access of Notch in *Drosophila*. *Nature* **398,** 522–525.

Sun, X. and Artavanis-Tsakonas, S. (1997) Secreted forms of DELTA and SERRATE define antagonists of Notch signaling in *Drosophila*. *Development* **124,** 3439–3448.

Tanzi, R. E., Vaula, G., Romano, D. M., Mortilla, M., Huang, T. L., Tupler, R. G., et al. (1992) Assessment of amyloid beta–protein precursor gene mutations in a large set of familial and sporadic Alzheimer disease cases. *Am. J. Hum. Genet.* **51,** 273–282.

Theisen, H., Purcell, J., Bennett, M., Kansagara, D., Syed, A., and Marsh, J. L. (1994) dishevelled is required during wingless signaling to establish both cell polarity and cell identity. *Development* **120,** 347–360.

Tomlinson, A., Bowtell, D. D., Hafen, E., and Rubin, G. M. (1987) Localization of the sevenless protein, a putative receptor for positional information, in the eye imaginal disc of *Drosophila*. *Cell* **51,** 143–150.

Torroja, L., Packard, M., Gorczyca, M., White, K., and Budnik, V. (1999) The *Drosophila* beta-amyloid precursor protein homolog promotes synapse differentiation at the neuromuscular junction. *J. Neurosci.* **19,** 7793–7803.

Uren, A. G., Pakusch, M., Hawkins, C. J., Puls, K. L., and Vaux, D. L. (1996) Cloning and expression of apoptosis inhibitory protein homologs that function to inhibit apoptosis and/or bind tumor necrosis factor receptor-associated factors. *Proc. Natl. Acad. Sci. USA* **93,** 4974–4978.

Vihtelic, T. S., Hyde, D. R., and O'Tousa, J. E. (1991) Isolation and characterization of the *Drosophila* retinal degeneration B (rdgB) gene. *Genetics* **127,** 761–768.

Vinos, J., Jalink, K., Hardy, R. W., Britt, S. G., and Zuker, C. S. (1997) A G protein-coupled receptor phosphatase required for rhodopsin function. *Science* **277,** 687–690.

Vucic, D., Kaiser, W. J., Harvey, A. J., and Miller, L. K. (1997) Inhibition of reaper-induced apoptosis by interaction with inhibitor of apoptosis proteins (IAPs) *Proc. Natl. Acad. Sci. USA* **94,** 10,183–10,188.

Wang, G. H., Mitsui, K., Kotliarova, S., Yamashita, A., Nagao, Y., Tokuhiro, S., et al. (1999) Caspase activation during apoptotic cell death induced by expanded polyglutamine in N2a cells. *NeuroReport* **10,** 2435–2438.

Wanker, E. E., Rovira, C., Scherzinger, E., Hasenbank, R., Walter, S., Tait, D., et al. (1997) HIP–I: a huntingtin interacting protein isolated by the yeast two-hybrid system. *Hum. Mol. Genet.* **6,** 487–495.

Warrick, J. M., Chan, H. Y., Gray-Board, G. L., Chai, Y., Paulson, H. L., and Bonini, N. M. (1999) Suppression of polyglutamine-mediated neurodegeneration in *Drosophila* by the molecular chaperone HSP70. *Nat. Genet.* **23,** 425–428.

Warrick, J. M., Chan, H. Y., Gray-Board, G. L., Chai, Y., Paulson, H. L., and Bonini, N. M. (1999) Suppression of polyglutamine-mediated neurodegeneration in *Drosophila* by the molecular chaperone HSP70. *Nat. Genet.* **23,** 425–428.

Warrick, J. M., Paulson, H. L., Gray-Board, G. L., Bui, Q. T., Fischbeck, K. H., Pittman, R. N., et al. (1998) Expanded polyglutamine protein forms nuclear inclusions and causes neural degeneration in *Drosophila*. *Cell* **93,** 939–949.

Wellington, C. L., Ellerby, L. M., Hackam, A. S., Margolis, R. L., Trifiro, M. A., Singaraja, R., et al. (1998) Caspase cleavage of gene products associated with triplet expansion disorders generates truncated fragments containing the polyglutamine tract. *J. Biol. Chem.* **273,** 9158–9167.

White, J. K., Auerbach, W., Duyao, M. P., Vonsattel, J. P., Gusella, J. F., Joyner, A. L., et al. (1997) Huntingtin is required for neurogenesis and is not impaired by the Huntington's disease CAG expansion. *Nat. Genet.* **17,** 404–410.

White, K., Grether, M. E., Abrams, J. M., Young, L., Farrell, K., and Steller, H. (1994) Genetic control of programmed cell death in *Drosophila*. *Science* **264,** 677–683.

White, K., Tahaoglu, E., and Steller, H. (1996) Cell killing by the *Drosophila* gene reaper. *Science* **271,** 805–807.

Wilder, E. L. and Perrimon, N. (1995) Dual functions of wingless in the *Drosophila* leg imaginal disc. *Development* **121,** 477–488.

Wing, J. P., Zhou, L., Schwartz, L. M., and Nambu, J. R. (1998) Distinct cell killing properties of the *Drosophila* reaper, head involution defective, and grim genes. *Cell Death Differ.* **5,** 930–939.

Wolff, T. and Ready, D. F. (1991) Cell death in normal and rough eye mutants of *Drosophila*. *Development* **113,** 825–839.

Xiong, W. C. and Montell, C. (1995) Defective glia induce neuronal apoptosis in the repo visual system of *Drosophila*. *Neuron* **14,** 581–590.

Xiong, W. C., Okano, H., Patel, N. H., Blendy, J. A., and Montell, C. (1994) repo encodes a glial-specific homeo domain protein required in the *Drosophila* nervous system. *Genes Dev.* **8,** 981–994.

Ye, Y. and Fortini, M. E. (1999) Apoptotic activities of wild-type and Alzheimer's disease-related mutant presenilins in *Drosophila melanogaster*. *J. Cell Biol.* **146,** 1351–1364.

Ye, Y., Lukinova, N., and Fortini, M. E. (1999) Neurogenic phenotypes and altered Notch processing in *Drosophila* Presenilin mutants. *Nature* **398,** 525–529.

Zars, T. and Hyde, D. R. (1996) rdgE: a novel retinal degeneration mutation in *Drosophila melanogaster*. *Genetics* **144,** 127–138.

Zeitlin, S., Liu, J. P., Chapman, D. L., Papaioannou, V. E., and Efstratiadis, A. (1995) Increased apoptosis and early embryonic lethality in mice nullizygous for the Huntington's disease gene homologue. *Nat. Genet.* **11,** 155–163.

Zhou, L., Song, Z., Tittel, J., and Steller, H. (1999) HAC-1, a *Drosophila* Homolog of APAF-1 and CED-4, functions in developmental and radiation-induced apoptosis. *Mol. Cell* **4,** 745–755.

Zhou, Q., Krebs, J. F., Snipas, S. J., Price, A., Alnemri, E. S., Tomaselli, K. J., et al. (1998) Interaction of the baculovirus anti-apoptotic protein p35 with caspases. Specificity, kinetics, and characterization of the caspase/p35 complex. *Biochemistry* **37,** 10,757–10,765.

Index